Cellular Therapy for Neurological Diseases

Cellular Therapy for Neurological Diseases

Editor: Katelyn Galbraith

FOSTER ACADEMICS
www.fosteracademics.com

www.fosteracademics.com

FOSTER ACADEMICS

Cataloging-in-Publication Data

Cellular therapy for neurological diseases / edited by Katelyn Galbraith.
p. cm.
Includes bibliographical references and index.
ISBN 978-1-64646-576-7
1. Nervous system--Diseases. 2. Nervous system--Diseases--Treatment.
3. Cellular therapy. I. Galbraith, Katelyn.
RC349.8 .C45 2023
616.804 6--dc23

Foster Academics,
118-35 Queens Blvd., Suite 400,
Forest Hills, NY 11375, USA

ISBN 978-1-64646-576-7 (Hardback)

Contents

Preface

Neurological diseases refer to a group of diseases which occur in the peripheral and central nervous system. These diseases include neuroinfections, multiple sclerosis, brain tumors, epilepsy, Parkinson's disease, etc. The causes of neurological diseases include a wide variety of factors such as viral or bacterial infections, genes, and accidents or traumatic injuries. Brain scans, laboratory screening tests and genetic testing are some of the most common screening tests used to diagnose these diseases. The prevention and treatment of neurological diseases differ depending on their severity. Several types of stem cells are used in the treatment of neurological diseases. The application of cell replacement therapy in neurologic diseases is beneficial for treating various diseases like demyelinating diseases, Parkinson's disease and Huntington disease. This book is compiled in such a manner, that it will provide in-depth knowledge on the importance of cellular therapy for the treatment of neurological diseases. Its extensive content provides the readers with a thorough understanding of the subject.

After months of intensive research and writing, this book is the end result of all who devoted their time and efforts in the initiation and progress of this book. It will surely be a source of reference in enhancing the required knowledge of the new developments in the area. During the course of developing this book, certain measures such as accuracy, authenticity and research focused analytical studies were given preference in order to produce a comprehensive book in the area of study.

This book would not have been possible without the efforts of the authors and the publisher. I extend my sincere thanks to them. Secondly, I express my gratitude to my family and well-wishers. And most importantly, I thank my students for constantly expressing their willingness and curiosity in enhancing their knowledge in the field, which encourages me to take up further research projects for the advancement of the area.

Editor

1

Differentiation of Retinal Glial Cells from Human Embryonic Stem Cells by Promoting the Notch Signaling Pathway

Sook Hyun Chung[1†], Weiyong Shen[1], Kathryn C. Davidson[2†], Alice Pébay[2,3,4], Raymond C. B. Wong[2,4,5], Belinda Yau[1] and Mark Gillies[1]*

[1] Save Sight Institute, Department of Clinical Ophthalmology and Eye Health, The University of Sydney, Sydney, NSW, Australia, [2] Centre for Eye Research Australia, Royal Victorian Eye and Ear Hospital, East Melbourne, VIC, Australia, [3] Department of Anatomy and Neuroscience, The University of Melbourne, Parkville, VIC, Australia, [4] Department of Surgery, The University of Melbourne, Parkville, VIC, Australia, [5] Shenzhen Eye Hospital, Shenzhen, China

***Correspondence:**
Mark Gillies
mark.gillies@sydney.edu.au

†Present address:
Sook Hyun Chung, Department of Ophthalmology, UC Davis Eye Centre, University of California Davis, Davis, CA, United States
Kathryn C. Davidson, Australian Regenerative Medicine Institute, Monash University, Clayton, VIC, Australia

Dysfunction of retinal glial cells, particularly Müller cells, has been implicated in several retinal diseases. Despite their important contribution to retinal homeostasis, a specific way to differentiate retinal glial cells from human pluripotent stem cells has not yet been described. Here, we report a method to differentiate retinal glial cells from human embryonic stem cells (hESCs) through promoting the Notch signaling pathway. We first generated retinal progenitor cells (RPCs) from hESCs then promoted the Notch signaling pathway using Notch ligands, including Delta-like ligand 4 and Jagged-1. We validated glial cell differentiation with qRT-PCR, immunocytochemistry, western blots and fluorescence-activated cell sorting as we promoted Notch signaling in RPCs. We found that promoting Notch signaling in RPCs for 2 weeks led to upregulation of glial cell markers, including glial fibrillary acidic protein (GFAP), glutamine synthetase, vimentin and cellular retinaldehyde-binding protein (CRALBP). Of these markers, we found the greatest increase in expression of the pan glial cell marker, GFAP. Conversely, we also found that inhibition of Notch signaling in RPCs led to upregulation of retinal neuronal markers including cone-rod homeobox (CRX) and orthodenticle *homeobox* 2 (OTX2) but with little expression of GFAP. This retinal glial differentiation method will help advance the generation of stem cell disease models to study the pathogenesis of retinal diseases associated with glial dysfunction such as macular telangiectasia type 2. This method may also be useful for the development of future therapeutics such as drug screening and gene editing using patient-derived retinal glial cells.

Keywords: human pluripotent cells, retinal glial cells, notch signaling pathway, differentiation, Müller cells in retina

INTRODUCTION

Retinal glial cells consist of astrocytes, Müller cells and microglia. Their functions include providing anatomical, metabolic and functional support for neurons and surrounding compartments, phagocytosing of cell debris, taking up neurotransmitters and ions and maintaining the blood retinal barrier (Vecino et al., 2016). Activation of retinal glia (gliosis) and dysfunction have been found in animal models of retinal diseases as well as studies on post mortem human tissues

(Rungger-Brandle et al., 2000; Fletcher et al., 2005; Powner et al., 2010, 2013; Arroba et al., 2011; Ly et al., 2011; Ma et al., 2012).

Müller cell deficiency is a striking and consistent pathological feature observed in *post mortem* specimens of eyes with macular telangiectasia type 2 (Mactel2), a bilateral macular disease that damages the central vision by causing characteristic alterations in retinal photoreceptors and blood vessels (Charbel Issa et al., 2013; Wu et al., 2013). Previous studies found absence of Müller cell markers, including vimentin, glutamine synthetase (GS) and cellular retinaldehyde binding protein (CRALBP), in the affected macular region of MacTel2 donor eyes (Powner et al., 2010, 2013). Selective disruption of Müller cells in transgenic mice leads to photoreceptor degeneration, retinal vascular leak, and, later, the development of subretinal neovascularization, all of which are important features of MacTel2 in humans (Shen et al., 2012, 2014). These observations indicate that Müller cell dysfunction may play an important role in the pathogenesis of MacTel2.

The rapid progress in stem cell research has made it possible to generate several retinal cell types, including retinal pigment epithelial (RPE) cells, photoreceptors and ganglion cells, from human pluripotent stem cells (Lamba et al., 2010; Gill et al., 2014, 2016; Lidgerwood et al., 2016, 2018). Retinal glial cell differentiation has previously been described in a method for differentiation of retinal organoids (Sasai et al., 2012; Zhong et al., 2014). However, these organoids consist of heterogenous cell types in suspension culture, which limits the downstream analysis assays that can be performed. To date, there is no report of a differentiation method in adherent culture that specifically produces retinal glial cells. Here, we report a method to differentiate retinal glial cells from human embryonic stem cells (hESCs) by promoting the Notch signaling pathway.

The Notch signaling pathway, which is highly conserved in embryogenesis, regulates cell-fate decisions, such as self-renewal and survival, and cellular differentiation in various organs, including the central and peripheral nervous systems (Gaiano and Fishell, 2002; Taylor et al., 2007). It also promotes glial cell differentiation during retinogenesis (Jadhav et al., 2006a,b) and several animal studies have reported its critical role in driving RPCs to differentiate into retinal glial cells in rodents (Furukawa et al., 2000; Bernardos et al., 2005). When the Notch signaling pathway is activated in retinal progenitor cells (RPCs), the downstream effector genes, including Hairy and Enhancer of Split (Hes) 1 and Hes 5, suppress transcription of pro-neural genes and activate glial-specific genes such as GFAP (Vetter and Moore, 2001). Here, we tested the hypothesis that activation of the Notch signaling pathway in human RPCs will generate human retinal glial cells.

MATERIALS AND METHODS

hESCs Culture, Retinal Progenitor Cell Differentiation and the Notch Signaling Pathway Promotion

Undifferentiated hESCs (WA-09 alias H9, WiCell) were maintained in mTeSR1^{TM} media (Stem Cell Technologies, 85850) without feeders and passaged weekly. hESCs were differentiated into RPCs using a published method with minor modifications (Lamba et al., 2006, 2010). Briefly, hESCs were seeded into Matrigel-coated 6 well plates (around 15–20 /well) and cultured in neural induction media (NIM; DMEM/F12 with 10% Knockout Serum Replacement, B27 and N2) containing a cocktail consisting of recombinant human proteins including insulin-like growth factor-1 (IGF-1, 1 ng/ml, PeproTech, cat# 100-11), Dickkopf Wnt signaling pathway inhibitor (DKK-1, 1 ng/ml, PeproTech, cat# 120–30) and the bone morphogenetic protein antagonist Noggin (R&D systems, 3344-NG-050) for 4 days. We increased the concentrations of all recombinant proteins to 10 ng/ml from the 5th day on wards.

We promoted the Notch signaling pathway in the resultant RPCs by adding recombinant human Notch ligands. RPCs were passaged at 3 weeks of differentiation and re-seeded on matrigel-coated 6 well plates (20–25 clumps/well) and cultured in NIM without N2 supplement but including Notch ligands Delta-like ligand 4 (DLL-4, R&D systems, cat# 15106-D4-050) and Jagged-1 (R&D systems, cat# 1277-JD-050). Media were changed every 2–3 days. All recombinant proteins including IGF-1, Noggin, Dkk-1 for RPC generation and Notch ligands including DLL4 and Jagged-1 for Notch stimulation, were freshly added into culture media every time when the media were changed. The Notch ligand treatment lasted for as long as 6 weeks. Cell were collected for analysis 2, 4, and 6 weeks after promoting the Notch signaling pathway as described below.

qRT-PCR

Total RNA were extracted from cell pallets using a RNeasy mini kit (Qiagen, 74104) according to the manufacturer's instructions. The quality and quantity of extracted RNA were assessed with a Bioanalyzer (Agilent). An equal amount of RNA were reverse transcribed into cDNA with SuperScript Vilo cDNA synthesis kit (Invitrogen, 11704050). A SYBR GreenER qPCR Supermix (Invitrogen, 11784-200) was used for qRT-PCR reaction. The PCR cycling temperatures were 95°C for 5 min, 95°C for 10 s, 60°C for 15 s, and 72°C for 20 s followed by a melting curve analysis. A total of 40 cycles were conducted for qRT-PCR. Quantitative analyses were performed by Relative Expression Software Tool 2009 (REST2009) and a built-in analysis software in the PCR machine (Bio-Rad CFX96). The primers used in this study are listed in **Table 1**.

Immunocytochemistry

Cells growing on plastic coverslips placed in 24 well plates were fixed with 4% paraformaldehyde for 20 min on ice and washed with PBS for times times, with 5 min/time. The fixed cells were permeabilized with 0.1% Triton X100 in PBS followed by blocking with 10% normal donkey serum in PBS for 1 h at room temperature. Primary antibodies were applied over night at 4°C, followed by an incubation with corresponding secondary antibodies-conjugated with Alexa Fluor 488 or 594 at room temperature for 2 h on the next day. The primary antibodies we used in the study are GFAP (DAKO, cat# Z3044), PAX6 (DSHB, cat# pax6), OTX2 (R&D Systems, AF1979), and CRX (Santa Cruz, cat# sc-30150). The stained coverslips were mounted

TABLE 1 | Primers used in this study.

Primer	Forward (5 - >3)	Reverse (5 - >3)
OCT4	GTG GAG GAA GCT GAC AAC AA	ATT CTC CAG GTT GCC TCT CA
NANOG	CAA AGG CAA ACA ACC CAC TT	TCT GCT GGA GGC TGA GGT AT
CHX10	GGC GAC ACA GGA CAA TCT TTA	TTC CGG CAG CTC CGT TTT C
PAX6	AACGATAACATACCAAGCGTGT	GGTCTGCCCGTTCAACATC
HES1	CCTGTCATCCCCGTCTACAC	CACATGGAGTCCGCCGTAA
HES5	CTCAGCCCCAAAGAGAAAAA	GACAGCCATCTCCAGGATGT
GS	AAGAGTTGCCTGAGTGGAATTTC	AGCTTGTTAGGGTCCTTACGG
CRALBP	TGCAGGCATATTGCTTCATCC	GCTTGACCACATTGTAGGTCG
VIMENTIN	TGCCGTTGAAGCTGCTAACTA	CCAGAGGGAGTGAATCCAGATTA
GFAP	CTGCGGCTCGATCAACTCA	TCCAGCGACTCAATCTTCCTC
CRX	TAT TCT GTC AAC GCC TTG GCC CTA	TGC ATT TAG CCC TCC GGT TCT TGA
RECOVERIN	CCAGAGCATCTACGCCAAGTT	CCGTCGAGGTTGGAATCGAAG

for confocal laser scanning microscopy (Zeiss) as described previously (Shen et al., 2012; Chung et al., 2016).

Western Blots

Western blots were performed as previously described (Chung et al., 2016). Briefly, proteins were extracted from cell pallets with RIPA buffer (Sigma, cat# R0278) containing protease inhibitor (Roche, cat# 04693159001). The concentrations of extracted cellular proteins were measured using a QuantiPro BCA assay kit (Sigma-Aldrich, cat# QPBCA). Equal amounts of protein were loaded into NuPage Bis-Tris gels (Life Technologies, cat# NP3023BOX) and transferred to a polyvinylidene difluoride membrane with iBlot semi-dry transfer system (Invitrogen, cat# IB21001). The membranes were blocked with 5% BSA in TBST and primary antibody was incubated overnight at 4°C. After incubation with secondary antibodies conjugated with horseradish peroxidase, protein bands were visualized using the G:Box BioImaging systems and quantified using the GeneTools image scanning and analysis package. Protein expression was normalized to α/β-tubulin (rabbit polyclonal, 1:2000; Cell Signaling #2148), which serves as a loading control.

Fluorescence Activated Cell Sorting (FACS) Analysis

Cells were collected with Accutase (Sigma, cat# A6964) to produce single cell suspension. After washing with PBS for 2 times, we labeled the cells with Fixable Viability Dye eFluor 780 (eBioscience, cat# 65-0865) minutes at 4°C for 30 min in order to label dead cells. The cells were washed 2 times with flow staining buffer (eBioscience, cat# 00-42226) and fixed with IC fixation buffer (eBioscience, cat# 822249) at room temperature for 20 min. Cells were then washed 2 times with ×1 permeabilization buffer (eBioscience, cat# 833356), blocked with 2% normal donkey serum at room temperature for 15 min and labeled with GFAP antibody (Rabbit Polyclonal, 1:500, Dako, cat# Z0334) at room temperature for 1 h. After washing cells 2 times with ×1 permeabilization buffer, GFAP-labeled cells were incubated with a secondary antibody-conjugated with Alexa Fluor 488 (1:1000, Invitrogen, cat# A21206) at room temperature for 40 min. Cells were washed 2 times with flow staining buffer and FACS was performed with LXR Fortessa X-20 (BD). Ten thousand cells were counted per sample and a rabbit IgG was used as a control for FACS analysis in this study. Data was analyzed with FlouJo software.

Statistical Analysis

Statistical analysis was performed by Student *t*-test with *p* values <0.05 deemed significant. In each characterization study, 3–6 biological replicates were used.

RESULTS

Differentiation of H9 hESCs Into RPCs

We first differentiated hESCs into human RPCs using a previously reported method (Lamba et al., 2006, 2010) by culturing them in neural induction media containing recombinant human IGF1, DKK1, and Noggin for 3 weeks (**Figure 1A**). We performed RT-PCR to study the expression profiles of a panel of stem cell and RPC markers from D0 to D12 of RPC differentiation. We found that pluripotent stem cell markers including *OCT4, NANOG*, and *SOX 2* were suppressed during RPC differentiation, whereas RPC markers including *CHX10, LHX, RX, SIX6*, and *PAX6* were increased as early as 3 days after differentiation (**Supplementary Figure S1**). hESCs formed rosette-like structures from 10 days after treatment with the cocktail of recombinant human proteins (**Figure 1B** and **Supplementary Figure S2A**), consistent with previous reports that the formation of rosette-like structures was one of the characteristic stages of RPC differentiation (Lamba et al., 2006, 2010). By 3 weeks after treatment with Notch ligands, stem cell pluripotency markers including *OCT4* and *NANOG* were markedly suppressed, whereas *PAX6* and *CHX10* were significantly upregulated for more than 50 times than that of undifferentiated ESCs (**Figure 1C**).

Long-Term Activation of Notch Signaling in RPCs Leads to Upregulation of Retinal Glial Cell-Related Genes

Once RPC differentiation had been confirmed, we promoted the Notch signaling pathway in RPCs by culturing them with

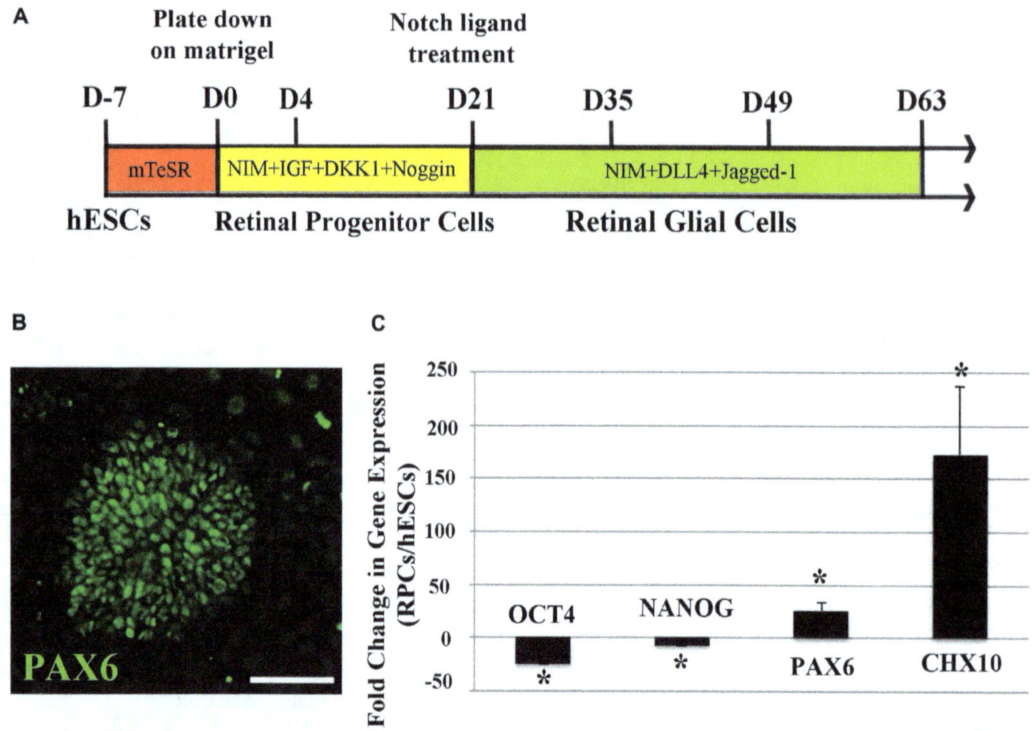

FIGURE 1 | Differentiation of H9 human embryonic stem cells (hESCs) into retinal progenitor cells (RPCs). **(A)** A schematic diagram illustrating the method of differentiating hESCs to RPCs and glial cells. **(B)** Immunocytochemistry results showing formation of PAX6 positive retinal progenitors with rosette-like morphology 15 days after culture in the neural induction medium. **(C)** qRT-PCR analyses 3 weeks after directing hESCs to differentiate into RPCs. Our results indicated significant upregulation of the eye field markers *PAX6* and *CHX10* along with decreased expression of stem cell markers of *OCT4* and *NANOG* after culturing H9 hESCs in the neural induction medium for 3 weeks $^*p < 0.05$ vs. undifferentiated human ESCs, error bars represent SEM, $n = 3$/group. Scale bar: 100 μm.

Notch ligands including DLL4 and Jagged-1. qRT-PCR analyses indicated that promoting the Notch signaling pathway with DLL4 and Jagged-1 (both 50 ng/ml) for 9 days significantly increased the expression of Notch downstream genes, *HES1* and *HES5*, while the increases in retinal glial cell-related genes including *CRALBP*, *GFAP*, and *GS* were not significant (**Figure 2A**), suggesting that the differentiation of human RPCs into retinal glial cells may require longer term Notch signaling activation.

We next treated RPCs for more extended periods ranging from 2 to 6 weeks with a cocktail of DLL4 and Jagged-1 (50 ng/ml). Transcription of glial genes including *GFAP, GS, CRALBP* and *VIMENTIN* started to increase after 2 weeks of Notch promotion when compared with untreated RPCs (**Figure 2B**). qRT-PCR analysis indicated promoting the Notch signaling pathway significantly increased the expression of *GFAP* from 2 weeks after treatment. By 6 weeks, the level of *GFAP* expression in Notch-treated group was more than 200 times higher than that in un-treated RPCs (**Figure 2C**). We also noted that there was a decrease in up-regulation of GFAP at 6 weeks compared with 2 weeks after Notch treatment (**Figure 2C**).

We also studied morphological changes of RPCs after activating the Notch signaling pathway. While bright field images from an Epi-fluorescence microscope revealed relatively consistent morphology throughout the culture (**Supplementary Figure S2B**), immunohistochemical analysis revealed that $GFAP^+$ cells appeared from 2 weeks after treatment and this glia-phenotypic change became more obvious after 4 to 6 weeks of promoting Notch signaling (**Figures 2D–F**). Interestingly, confocal microscopy showed that most $GFAP^+$ cells had long processes 4 and 6 weeks after Notch treatment, which is consistent with the typical morphology of Müller glial cells in the retina (**Figures 2E,F**). Double labeling with the glial cell marker GFAP and the photoreceptor precursor marker cone-rod homeobox protein (CRX) indicated that most treated RPCs lost this precursor marker of CRX and became positive for the glial cell marker of GFAP 4 weeks after Notch promotion (**Figures 2G–I**). Consistent with the results of qRT-PCR and immunohistochemistry, Western blots indicated that Notch ligand treatment for 6 weeks significantly increased GFAP expression compared with RPCs cultured in media without Notch ligands (**Figures 2J,K**). qRT-PCR analyses of Notch receptors and Notch downstream effector genes indicate that the upregulated *HES1* and *HES5* were decreased at 6 weeks (**Supplementary Figure S3**) compared with 9 days after Notch treatment (**Figure 2A**). We performed FACS to analyze the number of $GFAP^+$ cells 4 and 6 weeks after Notch treatment (**Figures 2L,M**). Treatment of RPCs with Notch ligands for 4 weeks significantly increased the number of $GFAP^+$ cells compared with RPCs cultured in a medium without Notch ligands for the same period (**Figure 2L**

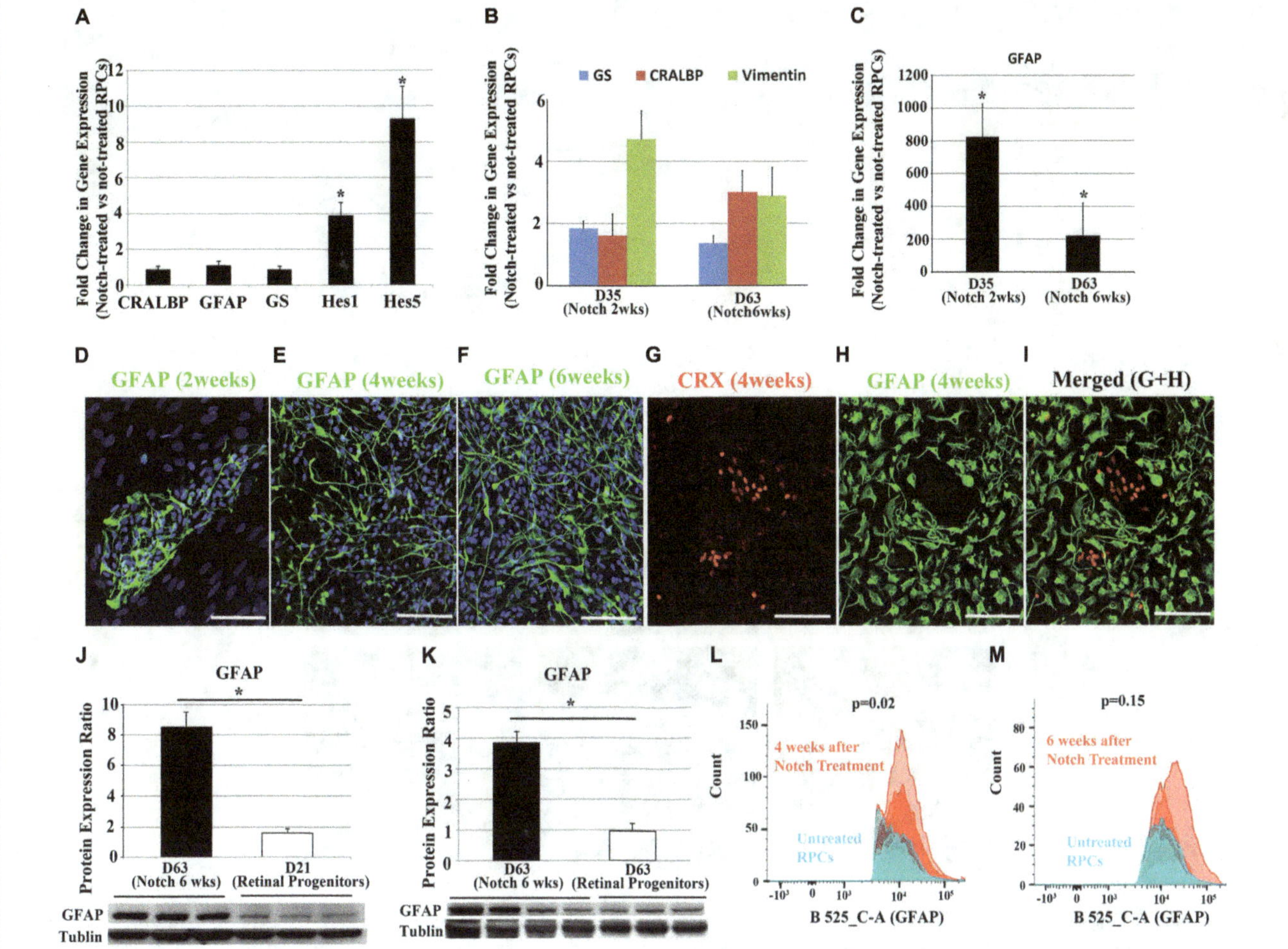

FIGURE 2 | Changes in Notch target genes and glia cell-associated markers after Notch ligand treatment in RPCs. **(A)** qRT-PCR analyses indicate Notch ligand treatment of RPCs for 9 days significantly upregulated the expression of Notch downstream effector genes including *HES1* and *HES5*, but with much less effect on glia cell-associated makers *CRALBP, GFAP* and *GS*. **(B,C)** Gene expression levels of *CRALBP, VIMENTIN, GS* and *GFAP* over the course of Notch ligand treatment. **(D–F)** Immunostaining for the glial cell marker GFAP 2, 4, and 6 weeks after Notch ligand treatment in RPCs. **(G–I)** Double label immunostaining for the photoreceptor precursor marker CRX and glial cell marker GFAP 4 weeks after Notch ligand treatment. Scale bars: 100 μm. **(J,K)** Western blots for GFAP using proteins extracted from un-treated RPCs and those treated with Notch ligands for 6 weeks. $n = 3–4$. **(L,M)** FACS analyses showed increased numbers of GFAP positive cells after treating RPCs with Notch ligands for 4 and 6 weeks. $*p < 0.05$, error bars represent SEM, $n \geq 3$.

and **Supplementary Figure S4**). Consistent with our qRT-PCR results revealed in **Figure 2C**, we also found that the number of $GFAP^+$ cells decreased during Notch ligand treatment (**Figure 2M** and **Supplementary Figure S4**). Interestingly, 22% of RPCs became $GFAP^+$ 4 weeks after culturing in the medium without Notch ligands and the number of $GFAP^+$ cells also decreased over time (**Figures 2L,M** and **Supplementary Figure S4**), indicating that some un-treated RPCs may undergo spontaneous differentiation without the promotion of the Notch signaling pathway.

Notch Promotion and Inhibition Have Opposite Effects on Retinal Precursor Cell Differentiation

We also studied the effects of promotion or inhibition of Notch signaling on neuro-gliogenesis further by treating RPCs with Notch ligands (DLL4 and Jagged-1) or a Notch inhibitor DAPT (5 μM in neural induction media) for 2 weeks. Consistent with what we observed earlier, Notch promotion led to upregulation of *GFAP* and *CRALBP* (**Figure 3A**), whereas Notch inhibition with DAPT in RPCs upregulated the expression of photoreceptor-associated genes, including *CRX* and *RECOVERIN* (**Figure 3B**). Immunocytochemical studies confirmed that the majority of DAPT-treated RPCs were positive for neuronal markers CRX and OTX2 but mostly negative for GFAP, whereas most Notch ligand-treated RPCs became positive for GFAP with only a few cells positive for CRX and OTX2 (**Figure 3C**).

DISCUSSION

In this study we developed a method to differentiate hESCs into human retinal glial cells. Our two-step differentiation

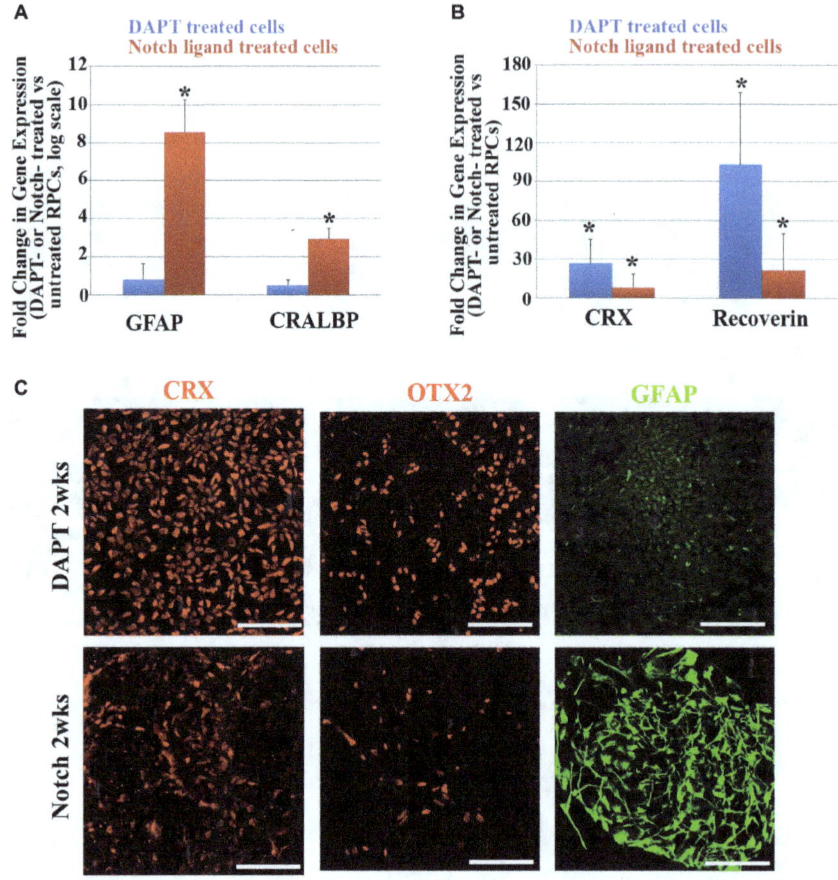

FIGURE 3 | Changes in glial cell- and retinal neuron-associated markers after Notch promotion or inhibition in RPCs. **(A,B)** qRT-PCR analyses of changes in glial cell-associated markers including *GFAP* and *CRALBP* **(A)** and retinal neuron-associated markers including *CRX* and recoverin **(B)** in RPCs 2 weeks after treatment with Notch ligands or Notch inhibitor DAPT. $^{*}p < 0.05$, RPCs treated with Notch ligands or Notch inhibitor DAPT vs. untreated RPCs, n = 3–4/group. **(C)** Immunostaining showing that most RPCs treated with the Notch inhibitor DAPT for 2 weeks were positive for the neuronal markers CRX and OTX2 but they expressed little of the glial cell marker GFAP **(upper panel)**. In contrast, 2 weeks after Notch ligand treatment only a few cells were positive for CRX and OTX2 and the majority of cells became positive for the glial cell marker GFAP **(lower panel)**. Scale bar: 100 μm. Error bars represent SEM.

method consists of directing hESCs to differentiate first into RPCs, followed by promotion of Notch signaling using DLL4 and Jagged-1 to facilitate retinal glial cell differentiation. We performed qRT-PCR, immnunocytochemistry, Western blots and FACS analyses to study changes in both gene and protein expression of retinal glial cell markers over the course of retinal glial cell differentiation. We found that promoting Notch signaling in RPCs led to upregulation of GFAP and cells displayed Müller-glia like morphology. Our findings suggest that promotion of the Notch signaling pathway in human RPCs leads to differentiation of retinal glial cells.

Müller cells, which account for 90% of macroglial cells in the retina, are the only retinal glial cell type that originates from RPCs (Arroba et al., 2011). Previous cell lineage tracing studies indicate that retinal astrocytes are derived from glial progenitor cells and migrate into the retina through the optic nerve (Stone and Dreher, 1987; Liu et al., 2002, 2004). Microglia cells originate from circulating monocytes as a part of immune defense mechanism (Djukic et al., 2006; Ginhoux et al., 2013). We used a well established method to differentiate RPCs from hESCs (Lamba et al., 2006, 2010) and then treated the resultant RPCs with Notch ligands to drive them to differentiate into glial cells. As there is no sole Müller cell marker available and GFAP is expressed by both Müller cells and astrocytes in the neural retina, we are unable to identify the exact subtype of retinal glial cells we have differentiated in this study. However, since the $GFAP^{+}$ cells were derived from RPCs, we believe most $GFAP^{+}$ cells that we produced represent a Müller cell population rather than astrocytes. Future studies to identify Müller cell-specific markers and conduct Müller cell functional assays, such as measurement of aquaporin 4-mediated osmotic water permeability and electrophysiological recording of Kir4.1 current-to-voltage relationship (Reichenbach and Bringmann, 2013), will help determine the composition and function of differentiated glial cells resulting from Notch ligand treatment.

Among the markers we used to characterize glial cell differentiation, GFAP, a pan glial cell marker, showed the most profound upregulation over the course of promoting the Notch signaling pathway. We also noticed that the level of other glial cell markers, including GS, CRALBP and vimentin, remained relatively low in RPCs after Notch ligand treatment. Perhaps RPCs require a longer-term of manipulation of the Notch signaling pathway to express the full panel of Müller cell markers. Previous studies indicate that, among different populations of retinal cells during development, Müller cells are the last retinal cell type to be formed (Cepko, 1993; Vecino and Acera, 2015). Consistent with these findings, it has been reported that it takes approximately 4–5 months to form Müller cells during the differentiation of 3 dimensional retinal organoids from hESCs (Zhong et al., 2014).

While Notch treatment substantially increased GFAP gene expression in RPCs, we also observed a decrease in GFAP expression 6 weeks after Notch treatment compared with earlier treatment (**Figure 2C**). Interestingly, the Notch downstream effector genes including *Hes1* and *Hes5*, also decreased over time (**Supplementary Figure S3**). As mature Müller glia in the normal retina express little GFAP (Bringmann et al., 2006; Beach et al., 2017), the reduced GFAP expression over time suggest that RPCs-derived retinal glia may become more mature at 6 weeks than 2 weeks after Notch treatment.

We observed that $GFAP^+$ cells showed different morphologies after RPCs were treated with Notch ligands for the same duration (**Figures 2E,H**). Previous studies reported that Müller cells isolated from human retinas can have diverse morphologies during culture, including long spindle shape with processes and trapezoidal shape (Lupien et al., 2004; Giannelli et al., 2011). It has been reported that Müller cells in the macula display different morphology and function from peripheral Müller cells (Yamada, 1969; Gass, 1999; Bringmann et al., 2006; Zhang et al., 2019). We also recently found that Müller cells isolated from the macula are small spindle to stellate shaped cells with lower cytoplasm/nucleus ratios and shorter processes, while the Müller cells from peripheral retinas have larger cell bodies, multiple cytoplasmic processes, higher cytoplasm/nucleus ratios (Zhang et al., 2019).

We observed that inhibiting Notch signaling in RPCs led to upregulation of retinal neuronal markers including CRX and OTX2, with glial cell markers hardly detectable. This is consistent with a previous observation that inhibiting Notch signaling in postnatal murine eyes suppresses glial cell maturation (Vecino et al., 2016). These results indicate that both humans and rodents may share similar mechanisms in differentiating RPCs into retinal glial cells.

There are a number of limitations in this study. We used one glial cell surface marker (GFAP) for FACS analyses to study the number and proportion of glial cells differentiated from RPCs. This can be improved by using a panel of glial cell surface markers such as CD44 and CD29 in future experiments (Shinoe et al., 2010; Eastlake et al., 2019). Our study indicates that the Notch signaling pathway plays a pivotal role in driving hESCs-derived RPCs to differentiate into glial cells. Further studies are warranted to validate our differentiation method in multiple stem cell lines including induced pluripotent stem cells (iPSCs).

In summary, we have developed a method to differentiate human ESC-derived RPCs into retinal glial cells by promoting the Notch signaling pathway. Application of this differentiation method to patient-derived iPSCs will help advance the generation of stem cell disease models to study the cellular and molecular mechanisms of retinal diseases associated with glial dysfunction such as MacTel2. Our method may also be useful for developing therapeutic strategies such as drug screening and gene editing using patient-derived retinal glial cells.

AUTHOR CONTRIBUTIONS

SC and MG conceived the project. SC and BY conducted the experiments. SC, WS, and MG analyzed the results and wrote the manuscript. SC, WS, KD, AP, RW, and MG critically revised the manuscript.

SUPPLEMENTARY MATERIAL

FIGURE S1 | RT-PCR analyses of changes in expression of stem cell and retinal progenitor genes during RPC differentiation. **(A,B)** Stem cells markers were gradually decreased from day 0 to day 12, whereas increased expression of a panel of RPC markers were observed as early as at day 3 of RPC differentiation.

FIGURE S2 | Bright field images showing cell morphologies over the course of RPC differentiation. **(A)** Morphology of RPCs differentiated from H9 ESCs during culture without Notch ligand treatment. **(B)** Morphology of RPCs treated with Notch ligands for 1–6 weeks.

FIGURE S3 | qPCR analysis of Notch downstream effector genes 6 weeks after Notch stimulation in RPCs. We found decreased expression of Hes1 and Hes5 6 weeks after Notch promotion despite adding fresh Notch ligands every 2–3 days while changing media. Boxes represents the interquartile range, or the middle 50% of observations. The dotted line represents the median gene expression. Whiskers represent the minimum and maximum observations $n = 3$, $p > 0.05$.

FIGURE S4 | FACS analyses of $GFAP^+$ cells 4 and 6 weeks after Notch ligand treatment in H9-derived RPCs. The number of $GFAP^+$ cell was counted from 10,000 events. **(A,C)** Notch treatment for 4 weeks increased the number of $GFAP^+$ cells compared with un-treated RPCs cultured for 7 weeks. **(B,D)** There was a decrease in the number of $GFAP^+$ cells in both groups after incubation in test media for further 2 weeks. Error bars represent SEM, $n = 4$/group.

REFERENCES

Arroba, A. I., Alvarez-Lindo, N., van Rooijen, N., and de la Rosa, E. J. (2011). Microglia-mediated IGF-I neuroprotection in the rd10 mouse model of retinitis pigmentosa. *Invest. Ophthalmol. Vis. Sci.* 52, 9124–9130. doi: 10.1167/iovs.11-7736

Beach, K. M., Wang, J., and Otteson, D. C. (2017). Regulation of stem cell properties of müller glia by JAK/STAT and MAPK signaling in the mammalian retina. *Stem. Cells Int.* 2017:1610691. doi: 10.1155/2017/1610691

Bernardos, R. L., Lentz, S. I., Wolfe, M. S., and Raymond, P. A. (2005). Notch-delta signaling is required for spatial patterning and müller glia differentiation in the zebrafish retina. *Dev. Biol.* 278, 381–395. doi: 10.1016/j.ydbio.2004.11.018

Bringmann, A., Pannicke, T., Grosche, J., Francke, M., Wiedemann, P., Skatchkov, S. N., et al. (2006). Müller cells in the healthy and diseased retina. *Prog. Retin. Eye Res.* 25, 397–424. doi: 10.1016/j.preteyeres.2006.05.003

Cepko, C. (1993). Lineage versus environment in the embryonic retina. *Trends Neurosci.* 16, 96–97.

Charbel Issa, P., Gillies, M. C., Chew, E. Y., Bird, A. C., Heeren, T. F., Peto, T., et al. (2013). Macular telangiectasia type 2. *Prog. Retin. Eye Res.* 34, 49–77. doi: 10.1016/j.preteyeres.2012.11.002

Chung, S. H., Gillies, M., Yam, M., Wang, Y., and Shen, W. (2016). Differential expression of microRNAs in retinal vasculopathy caused by selective müller cell disruption. *Sci. Rep.* 6:28993. doi: 10.1038/srep28993

Djukic, M., Mildner, A., Schmidt, H., Czesnik, D., Bruck, W., Priller, J., et al. (2006). Circulating monocytes engraft in the brain, differentiate into microglia and contribute to the pathology following meningitis in mice. *Brain* 129, 2394–2403. doi: 10.1093/brain/awl206

Eastlake, K., Wang, W., Jayaram, H., Murray-Dunning, C., Carr, A. J. F., Ramsden, C. M., et al. (2019). Phenotypic and functional characterization of müller glia isolated from induced pluripotent stem cell-derived retinal organoids: improvement of retinal ganglion cell function upon transplantation. *Stem. Cells Transl. Med.* 8, 775–784. doi: 10.1002/sctm.18-0263

Fletcher, E. L., Phipps, J. A., and Wilkinson-Berka, J. L. (2005). Dysfunction of retinal neurons and glia during diabetes. *Clin. Exp. Optom.* 88, 132–145. doi: 10.1111/j.1444-0938.2005.tb06686.x

Furukawa, T., Mukherjee, S., Bao, Z. Z., Morrow, E. M., and Cepko, C. L. (2000). rax, Hes1, and notch1 promote the formation of müller glia by postnatal retinal progenitor cells. *Neuron* 26, 383–394. doi: 10.1016/s0896-6273(00)81171-x

Gaiano, N., and Fishell, G. (2002). The role of notch in promoting glial and neural stem cell fates. *Annu. Rev. Neurosci.* 25, 471–490. doi: 10.1146/annurev.neuro.25.030702.130823

Gass, J. D. (1999). Müller cell cone, an overlooked part of the anatomy of the fovea centralis: hypotheses concerning its role in the pathogenesis of macular hole and foveomacualr retinoschisis. *Arch. Ophthalmol.* 117, 821–823. doi: 10.1001/archopht.117.6.821

Giannelli, S. G., Demontis, G. C., Pertile, G., Rama, P., and Broccoli, V. (2011). Adult human müller glia cells are a highly efficient source of rod photoreceptors. *Stem. Cells* 29, 344–356. doi: 10.1002/stem.579

Gill, K. P., Hewitt, A. W., Davidson, K. C., Pebay, A., and Wong, R. C. (2014). Methods of retinal ganglion cell differentiation from pluripotent stem cells. *Transl. Vis. Sci. Technol.* 3:7. doi: 10.1167/tvst.3.3.7

Gill, K. P., Hung, S. S., Sharov, A., Lo, C. Y., Needham, K., Lidgerwood, G. E., et al. (2016). Enriched retinal ganglion cells derived from human embryonic stem cells. *Sci. Rep.* 6:30552. doi: 10.1038/srep30552

Ginhoux, F., Lim, S., Hoeffel, G., Low, D., and Huber, T. (2013). Origin and differentiation of microglia. *Front. Cell Neurosci.* 7:45. doi: 10.3389/fncel.2013.00045

Jadhav, A. P., Cho, S. H., and Cepko, C. L. (2006a). Notch activity permits retinal cells to progress through multiple progenitor states and acquire a stem cell property. *Proc. Natl. Acad. Sci. U.S.A.* 103, 18998–19003. doi: 10.1073/pnas.0608155103

Jadhav, A. P., Mason, H. A., and Cepko, C. L. (2006b). Notch 1 inhibits photoreceptor production in the developing mammalian retina. *Development* 133, 913–923. doi: 10.1242/dev.02245

Lamba, D. A., Karl, M. O., Ware, C. B., and Reh, T. A. (2006). Efficient generation of retinal progenitor cells from human embryonic stem cells. *Proc. Natl. Acad. Sci. U.S.A.* 103, 12769–12774. doi: 10.1073/pnas.0601990103

Lamba, D. A., McUsic, A., Hirata, R. K., Wang, P. R., Russell, D., and Reh, T. A. (2010). Generation, purification and transplantation of photoreceptors derived from human induced pluripotent stem cells. *PLoS One* 5:e8763. doi: 10.1371/journal.pone.0008763

Lidgerwood, G. E., Lim, S. Y., Crombie, D. E., Ali, R., Gill, K. P., Hernandez, D., et al. (2016). Defined medium conditions for the induction and expansion of human pluripotent stem cell-derived retinal pigment epithelium. *Stem. Cell Rev.* 12, 179–188. doi: 10.1007/s12015-015-9636-2

Lidgerwood, G. E., Morris, A. J., Conquest, A., Daniszewski, M., Rooney, L. A., Lim, S. Y., et al. (2018). Role of lysophosphatidic acid in the retinal pigment epithelium and photoreceptors. *Biochim. Biophys. Acta Mol. Cell Biol. Lipids* 1863, 750–761. doi: 10.1016/j.bbalip.2018.04.007

Liu, Y., Han, S. S., Wu, Y., Tuohy, T. M., Xue, H., Cai, J., et al. (2004). CD44 expression identifies astrocyte-restricted precursor cells. *Dev. Biol.* 276, 31–46. doi: 10.1016/j.ydbio.2004.08.018

Liu, Y., Wu, Y., Lee, J. C., Xue, H., Pevny, L. H., Kaprielian, Z., et al. (2002). Oligodendrocyte and astrocyte development in rodents: an in situ and immunohistological analysis during embryonic development. *Glia* 40, 25–43. doi: 10.1002/glia.10111

Lupien, C., Brenner, M., Guerin, S. L., and Salesse, C. (2004). Expression of glial fibrillary acidic protein in primary cultures of human müller cells. *Exp. Eye Res.* 79, 423–429. doi: 10.1016/j.exer.2004.05.008

Ly, A., Yee, P., Vessey, K. A., Phipps, J. A., Jobling, A. I., and Fletcher, E. L. (2011). Early inner retinal astrocyte dysfunction during diabetes and development of hypoxia, retinal stress, and neuronal functional loss. *Invest. Ophthalmol. Vis. Sci.* 52, 9316–9326. doi: 10.1167/iovs.11-7879

Ma, W., Zhao, L., and Wong, W. T. (2012). Microglia in the outer retina and their relevance to pathogenesis of age-related macular degeneration. *Adv. Exp. Med. Biol.* 723, 37–42. doi: 10.1007/978-1-4614-0631-0_6

Powner, M. B., Gillies, M. C., Tretiach, M., Scott, A., Guymer, R. H., Hageman, G. S., et al. (2010). Perifoveal müller cell depletion in a case of macular telangiectasia type 2. *Ophthalmology* 117, 2407–2416. doi: 10.1016/j.ophtha.2010.04.001

Powner, M. B., Gillies, M. C., Zhu, M., Vevis, K., Hunyor, A. P., and Fruttiger, M. (2013). Loss of müller's cells and photoreceptors in macular telangiectasia type 2. *Ophthalmology* 120, 2344–2352. doi: 10.1016/j.ophtha.2013.04.013

Reichenbach, A., and Bringmann, A. (2013). New functions of müller cells. *Glia* 61, 651–678. doi: 10.1002/glia.22477

Rungger-Brandle, E., Dosso, A. A., and Leuenberger, P. M. (2000). Glial reactivity, an early feature of diabetic retinopathy. *Invest. Ophthalmol. Vis. Sci.* 41, 1971–1980.

Sasai, Y., Eiraku, M., and Suga, H. (2012). In vitro organogenesis in three dimensions: self-organising stem cells. *Development* 139, 4111–4121. doi: 10.1242/dev.079590

Shen, W., Fruttiger, M., Zhu, L., Chung, S. H., Barnett, N. L., Kirk, J. K., et al. (2012). Conditional müller cell ablation causes independent neuronal and vascular pathologies in a novel transgenic model. *J. Neurosci.* 32, 15715–15727. doi: 10.1523/JNEUROSCI.2841-12.2012

Shen, W., Lee, S. R., Araujo, J., Chung, S. H., Zhu, L., and Gillies, M. C. (2014). Effect of glucocorticoids on neuronal and vascular pathology in a transgenic model of selective müller cell ablation. *Glia* 62, 1110–1124. doi: 10.1002/glia.22666

Shinoe, T., Kuribayashi, H., Saya, H., Seiki, M., Aburatani, H., and Watanabe, S. (2010). Identification of CD44 as a cell surface marker for müller glia precursor cells. *J. Neurochem.* 115, 1633–1642. doi: 10.1111/j.1471-4159.2010.07072.x

Stone, J., and Dreher, Z. (1987). Relationship between astrocytes, ganglion cells and vasculature of the retina. *J. Comp. Neurol.* 255, 35–49. doi: 10.1002/cne.902550104

Taylor, M. K., Yeager, K., and Morrison, S. J. (2007). Physiological Notch signaling promotes gliogenesis in the developing peripheral and central nervous systems. *Development* 134, 2435–2447. doi: 10.1242/dev.005520

Vecino, E., and Acera, A. (2015). Development and programed cell death in the mammalian eye. *Int. J. Dev. Biol.* 59, 63–71. doi: 10.1387/ijdb.150070ev

Vecino, E., Rodriguez, F. D., Ruzafa, N., Pereiro, X., and Sharma, S. C. (2016). Glia-neuron interactions in the mammalian retina. *Prog. Retin. Eye Res.* 51, 1–40. doi: 10.1016/j.preteyeres.2015.06.003

Vetter, M. L., and Moore, K. B. (2001). Becoming glial in the neural retina. *Dev. Dyn.* 221, 146–153. doi: 10.1002/dvdy.1145

Wu, L., Evans, T., and Arevalo, J. F. (2013). Idiopathic macular telangiectasia type 2 (idiopathic juxtafoveolar retinal telangiectasis type 2A. Mac Tel 2). *Surv. Ophthalmol.* 58, 536–559. doi: 10.1016/j.survophthal.2012.11.007

Yamada, E. (1969). Some structural features of the fovea centralis in the human retina. *Arch. Ophthalmol.* 82, 151–159. doi: 10.1001/archopht.1969.00990020153002

Zhang, T., Zhu, L., Madigan, M. C., Liu, W., Shen, W., Cherepanoff, S., et al. (2019). Human macular Müller cells rely more on serine biosynthesis to combat oxidative stress than those from the periphery. *eLife* 8:e43598. doi: 10.7554/eLife.43598

Zhong, X., Gutierrez, C., Xue, T., Hampton, C., Vergara, M. N., Cao, L. H., et al. (2014). Generation of three-dimensional retinal tissue with functional photoreceptors from human iPSCs. *Nat. Commun.* 5:4047. doi: 10.1038/ncomms5047

2

Exosomes Secreted from Bone Marrow Mesenchymal Stem Cells Attenuate Oxygen-Glucose Deprivation/Reoxygenation-Induced Pyroptosis in PC12 Cells by Promoting AMPK-Dependent Autophagic Flux

Qing Zeng[1,2†], *Yuqing Zhou*[1,2†], *Donghui Liang*[3], *He He*[1,2], *Xiaoli Liu*[1,2], *Rui Zhu*[1,2], *Meimei Zhang*[1,2], *Xun Luo*[4,5], *Yao Wang*[6] *and Guozhi Huang*[1,2*]

[1]*Department of Rehabilitation Medicine, Zhujiang Hospital, Southern Medical University, Guangzhou, China,* [2]*Rehabilitation Medical School, Southern Medical University, Guangzhou, China,* [3]*Department of Traditional Chinese Medicine, Zhujiang Hospital, Southern Medical University, Guangzhou, China,* [4]*Kerry Rehabilitation Medicine Research Institute, Shenzhen, China,* [5]*Shenzhen Sanming Project Group, Spaulding Rehabilitation Hospital, Harvard Medical School, Charlestown, MA, United States,* [6]*Department of Rehabilitation Medicine, Shenzhen Dapeng New District Nan'ao People's Hospital, Shenzhen, China*

Correspondence:
Guozhi Huang
drhuang66@163.com

†These authors have contributed equally to this work

Background: Cerebral ischemia–reperfusion (I/R) injury can lead to severe dysfunction, and its treatment is difficult. It is reported that nucleotide-binding domain and leucine-rich repeat family protein 3 (NLRP3) inflammasome-mediated cell pyroptosis is an important part of cerebral I/R injury and the activation of autophagy can inhibit pyroptosis in some tissue injury. Our previous study found that the protective effects of bone marrow mesenchymal stem cells (BMSCs) in cerebral I/R injury may be associated with the regulation of autophagy. Recent studies have demonstrated that exosomes secreted from BMSCs (BMSC-Exos) may play an essential role in the effective biological performance of BMSCs and the protective mechanism of BMSC-Exos is associated with the activation of autophagy and the remission of inflammation, but it has not been reported in studies of cerebral I/R injury. We aimed to investigate the effects of BMSC-Exos on cerebral I/R injury and determine if the mechanism is associated with the regulation of pyroptosis and autophagic flux.

Method: PC12 cells were subjected to oxygen-glucose deprivation/reoxygenation (OGD/R) to induce cerebral I/R *in vitro* and were cocultured with BMSC-Exos. Cell viability was determined with CCK-8 and lactate dehydrogenase (LDH) detection kits. Scanning electron microscopy (SEM), Hoechst 33342/propidium iodide (PI) double staining, 2′,7′-dichlorodihydrofluorescein diacetate assay, immunofluorescence, Western blot, and Enzyme-linked immunosorbent assay (ELISA) were used to detect cell pyroptosis. Furthermore, transmission electron microscopy (TEM), GFP-RFP-LC3 adenovirus transfection, and Western blot were used to detect autophagic flux and its influence on pyroptosis. Finally, coimmunoprecipitation was used to detect the binding interaction between NLRP3 and LC3.

Results: BMSC-Exos increased cell viability in OGD/R. The inhibitory effect of BMSC-Exos on pyroptosis was comparable to the NLRP3 inhibitor MCC950 and was reversed by NLRP3 overexpression. Furthermore, BMSC-Exos promoted autophagic flux through the AMP-activated kinase (AMPK)/mammalian target of the rapamycin pathway, whereas chloroquine, AMPK silencing, and compound C blocked the inhibitory effect on pyroptosis.

Conclusions: BMSC-Exos can protect PC12 cells against OGD/R injury *via* attenuation of NLRP3 inflammasome-mediated pyroptosis by promoting AMPK-dependent autophagic flux.

Keywords: autophagy, exosomes secreted from bone marrow mesenchymal stem cells, cerebral ischemia/reperfusion, pyroptosis, nucleotide-binding domain leucine-rich repeats family protein 3

INTRODUCTION

Ischemic stroke accounts for the majority of stroke cases and is the second leading cause of death worldwide. Ischemic stroke is a common cerebrovascular disease with high morbidity, mortality, and disability (Stonesifer et al., 2017; Wang et al., 2017; Campbell et al., 2019). In the clinical setting, thrombolytic therapy to restore the blood supply is a common treatment for ischemic stroke. However, blood reperfusion after cerebral ischemia often causes oxidative stress, inflammation, and other adverse effects, which can aggravate the cerebral injury and lead to further dysfunction (Jayaraj et al., 2019; Lambertsen et al., 2019). Therefore, it is necessary to explore more effective treatments to reduce the cerebral ischemia/reperfusion (I/R) injury and promote the repair of nerve function. Cerebral I/R injury involves complex pathophysiological processes, including autophagy (Zhang D. M. et al., 2019), apoptosis (Chen et al., 2017), oxidative stress (Rana and Singh, 2018), pyroptosis (Zhu et al., 2019), ion homeostasis imbalance (Gu et al., 2019), and acidosis (Fan et al., 2014). Among them, pyroptosis has recently been discovered as a pro-inflammatory programmed cell death process that plays an important role in cerebral I/R injury (Xia et al., 2018; Zhu et al., 2019). Recent studies have suggested that activation of autophagy has a potential therapeutic effect on cerebral I/R injury (Yao et al., 2019), and the mechanism of autophagy may be related to the inhibition of pyroptosis in some tissue injury (Li et al., 2019), which still remains unclear in cerebral I/R injury.

Bone marrow mesenchymal stem cell (BMSC) transplantation has been shown to promote the recovery of nerve function after cerebral ischemia (Stonesifer et al., 2017; Zhang Q. et al., 2019). Our previous study found that the protective effects of BMSCs in cerebral I/R injury may be associated with the regulation of autophagy *via* the PI3K/Akt/mTOR signaling pathway (He et al., 2019). Recent studies have demonstrated that exosomes secreted from BMSCs (BMSC-Exos) may play important roles in the effective biological performance of BMSCs (McBride et al., 2017; Lazar et al., 2018). Moreover, BMSC-Exos affect the biological characteristics of target cells through their interaction with specific ligand receptors, the transfer of receptors between cells, and the transfer of proteins and RNAs to target cells (Hou et al., 2020). Moreover, without any cytological characteristics such as proliferation and differentiation of BMSCs, BMSC-Exos have relatively stable biological characteristics, which reduce the risk of BMSC transplantation and make it possible to replace BMSCs for more effective treatment of cerebral I/R injury. In addition, BMSC-Exos have been shown to protect against myocardial I/R injury and inhibit myocardial infarction pathogenesis by regulating autophagy (Zou et al., 2019). Therefore, we hypothesized that BMSC-Exos may have a similar effect in cerebral I/R injury and the mechanism may be related to autophagy and pyroptosis.

In the present study, PC12 cells were subjected to OGD/R to stimulate cerebral I/R injury *in vitro* to investigate the effect of BMSC-Exos in cerebral I/R injury as well as the role of the AMP-activated kinase (AMPK)-dependent autophagic flux in the protection of BMSC-Exos against nucleotide-binding domain and leucine-rich repeat family protein 3 (NLRP3) inflammasome-mediated pyroptosis.

MATERIALS AND METHODS

Cell Culture

Rat pheochromocytoma (PC12) cells were obtained from Jennio Biotech (Guangzhou, China) and were maintained in RPMI-1640 (Gibco, Gaithersburg, MD, USA) medium supplemented with 10% fetal bovine serum (FBS), 100 U/ml penicillin, and 100 mg/ml streptomycin in a 37°C incubator with 5% CO_2. When the cell density reached approximately 90%, the cells were detached with 0.02% EDTA/0.25% trypsin. Cells in the logarithmic phase and those that demonstrated good growth were used for subsequent experiments.

Oxygen-Glucose Deprivation/Reoxygenation (OGD/R) for *in vitro* Cerebral I/R

OGD/R has been recognized as an *in vitro* model for simulating I/R injury (Chen et al., 2020). PC12 cells have been widely used in neurophysiological and pathological research (Koubi et al., 2005). To mimic cerebral I/R *in vitro*, PC12 cells were subjected to OGD/R according to prior demonstrations (Ren et al., 2019). Briefly, cells were washed three times with phosphate-buffered saline (PBS; Gibco, Gaithersburg, MD, USA) and incubated for 12 h in glucose-free Dulbecco's modified Eagle's medium (DMEM; Gibco, Gaithersburg, MD, USA) medium under hypoxic conditions (1% O_2, 94% N_2, and 5% CO_2). The cells were then incubated under normoxic conditions (95% air and 5% CO_2) in RPMI-1640 medium for 1 h for reoxygenation. The cells were cocultured with BMSC-Exos (10 μg/ml), chloroquine (Cq; Sigma, Georgetown, SC, USA; 5 μM) or compound C (Sigma, Georgetown, SC, USA; 5 μM) during OGD/R.

BMSC Isolation and Characterization

Primary BMSCs were isolated from male rats (80–100 g) as described previously (He et al., 2019). Cell pellets were cultured in DMEM/F-12 (1:1; Gibco, Gaithersburg, MD, USA) supplemented with 10% FBS, 100 U/ml penicillin, and 100 mg/ml streptomycin in a 37°C incubator with 5% CO_2. When the cells reached 90% confluence, the BMSCs were detached with 0.02% EDTA/0.25% trypsin. For phenotypic analysis, the expression of CD29, CD90, CD44, and CD45 were evaluated. An immunoglobulin G (IgG)-matched isotype was used as the internal control for each antibody. BMSCs at passages 3 (P3)–P8 were used for subsequent experiments.

BMSC-Exos Isolation and Characterization

BMSC-Exos were isolated from the cell culture supernatants of BMSCs. Before collecting the culture medium, the BMSCs were washed twice with PBS and the medium was changed to serum-free medium. The medium was then collected and centrifuged at 2,000× *g* for 30 min, 10,000× *g* for 30 min, and 100,000× *g* for 4 h at 4°C using an ultracentrifuge. The isolated exosomes were washed once with PBS and resuspended for further characterization by transmission electron microscopy (TEM), Western blot, and NanoSight NTA technology according to international standards (Théry et al., 2018).

Cell Viability Assays by CCK-8

Cell viability was detected using the CCK-8 kit, according to the manufacturer's instructions (TransGen Biotech, China). Briefly, the culture medium was removed and CCK-8 (10%, 100 μl/well) was added to each well and incubated for 1 h. A microplate reader was used to measure the absorbance OD value at 450 nm.

Evaluation of Lactate Dehydrogenase (LDH) Release

LDH released into the cell culture supernatants was determined with the LDH detection kit, according to the manufacturer's instructions (KeyGENBioTECH, China). A microplate reader was used to measure the absorbance OD value at 440 nm.

Assessment of Reactive Oxygen Species (ROS) Levels

A 2′,7′-dichlorodihydrofluorescein diacetate (DCFH-DA) kit (Sigma, Georgetown, SC, USA) was used to detect ROS levels. After exposure to OGD/R, the cells were washed with PBS and treated with DCFH-DA (20 μM) for 30 min at 37°C in the darkness. Images were collected using a confocal microscope (Leica Microsystems, Wetzlar, Germany).

Scanning Electron Microscopy (SEM)

For SEM, cells were seeded on glass slides. After treatment, three glass slides were selected for each group and the 2.5% glutaraldehyde solution was added to the slides for fixation. For inspection, the specimens were commissioned to the Central Laboratory of Southern Medical University for post-processing.

Hoechst 33342/Propidium Iodide (PI) Double Staining

The Hoechst 33342 stain can penetrate the complete cell membranes, whereas the PI stain cannot penetrate normal or apoptotic cells with intact cell membranes but can enter cells through pyroptosis-related pores in the membranes. PC12 cells in each group were stained with Hoechst 33342 (Sigma, Georgetown, SC, USA; 10 μg/ml), followed by PI (Sigma, Georgetown, SC, USA; 1 μg/ml) for 15 min at 37°C. An inversion fluorescence microscope (Leica Microsystems, Wetzlar, Germany) was used to collect the images and the percentage of PI-positive cells was calculated.

Enzyme-Linked Immunosorbent Assay (ELISA)

The levels of interleukin-1β (IL-1β) in the cell culture supernatants were detected by ELISA kits, according to the manufacturer's instructions (Cloud-Clone Crop, China).

Immunofluorescence Staining

After being treated with 4% buffered paraformaldehyde for 15 min, the cells were blocked with 0.2% Triton-X and 1% BSA for 1 h. The cells were then incubated overnight with the primary antibody against the N-terminal of gasdermin D (GSDMD-N; CST, USA) at 4°C. The cells were subsequently incubated for 2 h with the appropriate secondary antibody at room temperature. DAPI (Sigma, Georgetown, SC, USA) was used to stain the nuclei, and images were captured using a fluorescence microscope (Leica Microsystems, Wetzlar, Germany).

TEM

TEM was used to observe the autophagic flux of the cells. The medium was removed, and the cells were scraped off, followed by centrifugation at 2,000 rpm for 5 min. The cells were then fixed in 2.5% glutaraldehyde for 2 h at room temperature and kept at 4°C until further analysis. The specimens were subsequently commissioned to the Central Laboratory of Southern Medical University for post-processing.

Transfection of GFP-RFP-LC3 Adenoviruses

Cells were cultured in confocal dishes with 8×10^4 cells per well and then transiently transfected with GFP-RFP-LC3 adenoviruses, according to the manufacturer's instructions (GeneChem, China). After treatment, the cells were washed with PBS, fixed with 4% buffered paraformaldehyde, stained with Hoechst 33342, and observed under a confocal microscope (Leica Microsystems, Wetzlar, Germany). The numbers of yellow puncta (GFP+RFP+) and red puncta (GFP-RFP+) were counted for each cell, representing the autophagosomes and autolysosomes, respectively.

Protein Extraction and Western Blot Analysis

After being washed with cold PBS, the cells were lysed with RIPA buffer containing protease-inhibitor cocktail and PMSF. SDS-PAGE was then used to separate equal amounts of protein, and the gels were transferred onto PVDF membranes. The membranes were then blocked with 5% nonfat milk for 1 h at room temperature, followed by an overnight incubation with the primary antibody solution at 4°C. After three washes with TBST, the membranes were incubated with the horseradish peroxidase-linked secondary antibodies for 2 h at room temperature. Immunoreactive bands were visualized by enhanced chemiluminescence and quantified by ImageJ software. β-Actin was used as the internal standard. The antibodies used in this study included anti-NLRP3 (1:1,000, Abcam, Eugene, OR, USA), anti-caspase-1 (1:1,000, Proteintech, China), anti-GSDMD (1:1,000, CST, USA), anti-AMPK (1:1,000, Bioss, China), anti-p-AMPK (1:1,000, CST, USA), anti-mTOR (1:1,000, Bioss, China), anti-p-mTOR (1:1,000, Abcam, Eugene, OR, USA), anti-LC3B (1:1,000, CST, USA), and anti-P62 (1:1,000, CST, USA).

Construction of Plasmids and Transfection

To overexpress NLRP3, pGV102-NLRP3 plasmids were constructed by GeneChem (Shanghai, China) and transfected into PC12 cells. The pCMV-MCS-SV40-Neomycin plasmids acted as the control group. Transient transfection was performed by the Lipofectamine 3000 reagent (Invitrogen, USA) according to the manufacturer's instructions.

Real-Time Quantitative Polymerase Chain Reaction (RT-qPCR)

The effect of pNLRP3 transfection was determined by RT-qPCR. Total RNA was extracted using the TRIzol regent kit (TaKaRa Bio, Japan), and cDNA was prepared by reverse transcription. PCR was performed on a 7500 FAST Real-Time PCR System (Applied Biosystems, Foster City, CA, USA) with specific primers and the RT-qPCR Assay Kit (TaKaRa Bio, Japan). Primer sequences of the targeted genes used in this study were as follows: NLRP3 (5′-ATTACCCACCCGAGAAAGG-3′, forward; 5′-CATGAGTGTGGCTAGATCCAAG-3′, reverse) and β-actin (5′-CACCCGCGAGTACAACCTTC-3′, forward; 5′-CCCATACCCACCATCACACC-3′, reverse). NLRP3 expression levels were normalized for expression of β-actin and expressed as the fold ratio compared with the control group.

Short Interfering RNA (siRNA) Transfection

PC12 cells were transfected with siRNA directed against AMPK, with scrambled siRNA used as control. Both specific and control siRNAs were obtained from Obio Technology (Shanghai, China). After siRNA transfection for 48 h, the cells were subjected to OGD/R and then collected for subsequent analyses. Transient transfection was performed using Lipofectamine 3000 reagent (Invitrogen, USA) according to the manufacturer's instructions.

Coimmunoprecipitation (Co-IP) Assays

For Co-IP assays, the PC12 cells were lysed in RIPA buffer containing protease inhibitors. Approximately 100 μg of total protein was mixed with a suitable amount (2 μg) of the anti-LC3B antibody (Abcam, Eugene, OR, USA) or the anti-normal IgG antibody. Protein A+G agarose was added to the lysate mixture, and the solution was incubated with gentle rocking at 4°C overnight. The samples were then pelleted by centrifugation, retrieved and suspended in 50 μl of SDS-PAGE buffer for immunoblotting. The obtained samples were then subjected to Western blot with the anti-LC3B (Abcam, Eugene, OR, USA) and anti-NLRP3 antibodies (Abcam, Eugene, OR, USA).

Statistical Analyses

All experiments in the "Materials and Methods" section were conducted in triplicate. GraphPad Prism 5.0 was used for statistical analysis. A p-value < 0.05 was considered statistically significant. Homogeneity testing and one-way analysis of variance were used to evaluate the differences among the groups. The data are expressed as the mean ± SEM.

RESULTS

Characterization of BMSCs and BMSC-Exos

Our results indicated that the expansion potential of P4 was positive for CD29, CD44, and CD90 but negative for CD45 (**Figure 1A**). The purity of the BMSCs in the culture was determined to be up to 99%, confirming that they met the standard for defining BMSCs and their use in further experiments.

TEM, Western blot, and NanoSight NTA were performed to comprehensively characterize the purified nanoparticles derived from BMSCs. As shown in **Figure 1B**, TEM demonstrated that the BMSC-Exos exhibited a round-shaped morphology, indicating the presence of exosomes. In addition, Western blot showed that the BMSC-Exos were positive for the specific exosome surface markers TSG101 and CD9 (**Figure 1C**). NanoSight NTA indicated that the diameter of the BMSC-Exos was around 100 nm (**Figure 1D**). Collectively, these analyses confirmed that the BMSC-Exos were successfully isolated and identified.

BMSC-Exos Increase Cell Viability in OGD/R

CCK-8 assay results showed that the cell viability of the OGD/R group decreased significantly in a time-dependent manner. Compared with the control group, the cell viability was reduced significantly in the OGD 6-h, 9-h, 12-h, and 24-h groups (**Figure 1E**). The cell viability of the OGD 12-h group decreased to about 50%; thus, this group was selected as the OGD/R group. These results indicated that treatment with BMSC-Exos effectively enhanced the viability of PC12 cells in a dose-dependent manner (**Figure 1F**). There was no significant difference between BMSC-Exos at 10 and 20 μg/ml; thus, the former was selected as the intervention

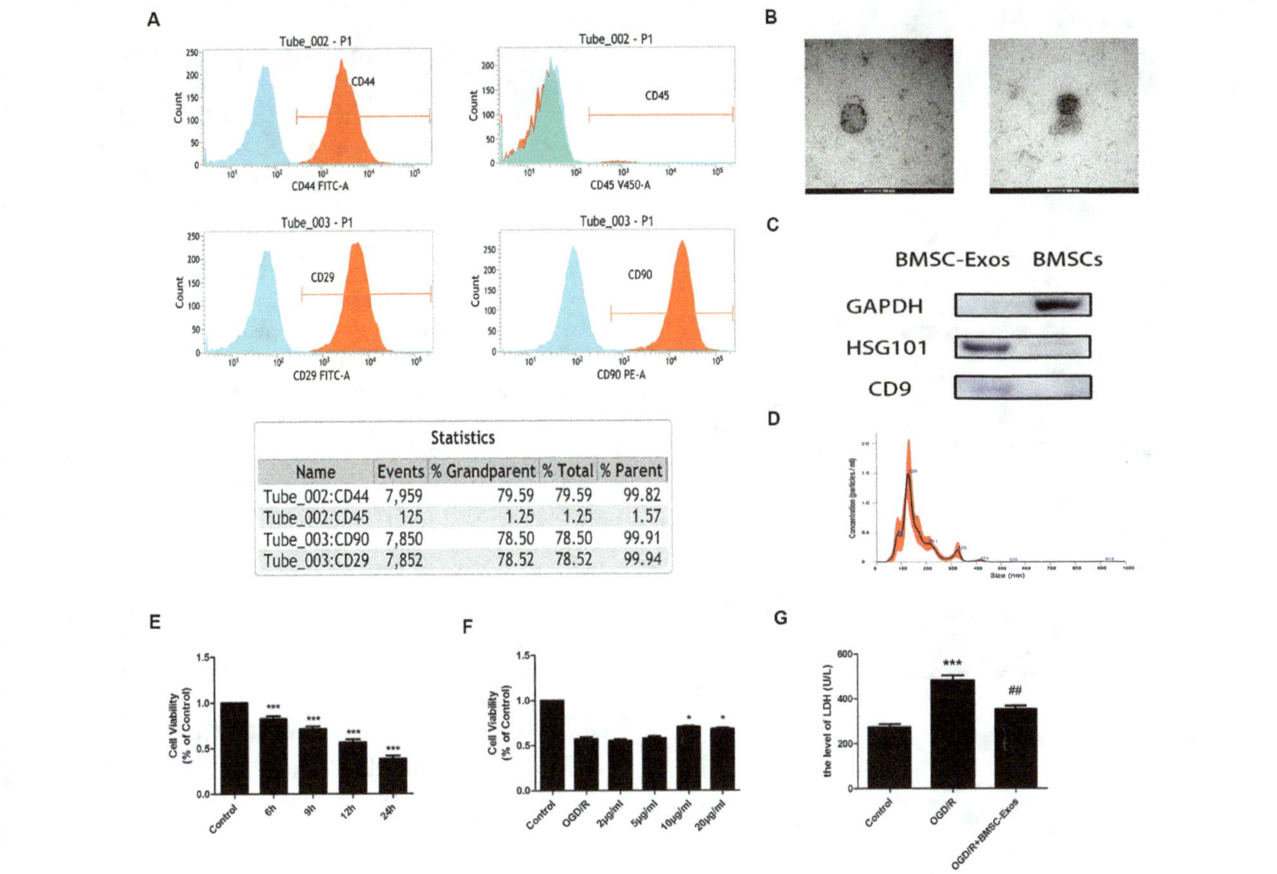

FIGURE 1 | Characterization of bone marrow mesenchymal stem cells (BMSCs) and exosomes from BMSCs (BMSC-Exos); BMSC-Exos increase cell viability following oxygen-glucose deprivation/reoxygenation (OGD/R). **(A)** Flow cytometry analyses indicated that BMSCs were positive for CD29, CD44, and CD90 but were negative for CD45. **(B)** Round morphologies of BMSC-Exos were demonstrated by tandem electron microscopy (TEM); scale bar = 100 nm. **(C)** Western blot analyses showed that the BMSC-Exos were positive for the specific exosome surface markers TSG101 and CD9. **(D)** NanoSight NTA analysis indicated that the diameters of BMSC-Exos were around 100 nm. **(E)** A significant reduction in cell viability was observed in OGD 6-h, 9-h, 12-h, and 24-h groups ($n = 3$). **(F)** BMSC-Exos enhanced cell viability in a dose-dependent manner ($n = 3$). **(G)** BMSC-Exos reduced lactate dehydrogenase (LDH) release following OGD/R ($n = 3$). $^{*}p < 0.05$, $^{***}p < 0.001$ vs. control group; $^{\#\#}p < 0.01$ vs. OGD/R group.

dose. In addition, the LDH value was shown to increase after OGD/R treatment. Furthermore, the LDH release of the cell supernatant was reduced after BMSC-Exos treatment when compared with the OGD/R group (**Figure 1G**). These results showed that BMSC-Exos could significantly improve the viability of PC12 cells after OGD/R, suggesting that BMSC-Exos had a protective effect against OGD/R injury in PC12 cells.

BMSC-Exos Reduce the NLRP3 Inflammasome-Mediated Pyroptosis Induced by OGD/R

Morphological Changes Observed by SEM

As shown in **Figure 2A**, SEM showed that the PC12 cells of the control group had clear outlines, long protuberances, and tight cell connections. However, in the OGD/R group, the cell membranes had lost their integrity. In addition, the cells had swollen and became flat and the surface villi were significantly reduced, which is consistent with the previously reported characteristics of late pyroptosis (Herr et al., 2020). In the BMSC-Exos treatment group, the cell contours became clearer; the surface villi increased and became denser, with longer protuberances, indicating that the characteristics of pyroptosis were significantly reduced.

Hoechst 33342/PI Double Staining

In the process of pyroptosis, pores can be formed in the cell membranes and result in the release of cellular contents and positive staining of dead cells, which can be determined by Hoechst 33342/PI double staining (Wu et al., 2018). Our results indicated that the PI staining proportion was increased significantly in the OGD/R group when compared with the control group. In addition, the PI staining proportion was decreased significantly in the BMSC-Exos group when compared with the OGD/R group, indicating that pyroptosis was significantly inhibited (**Figures 2B,C**).

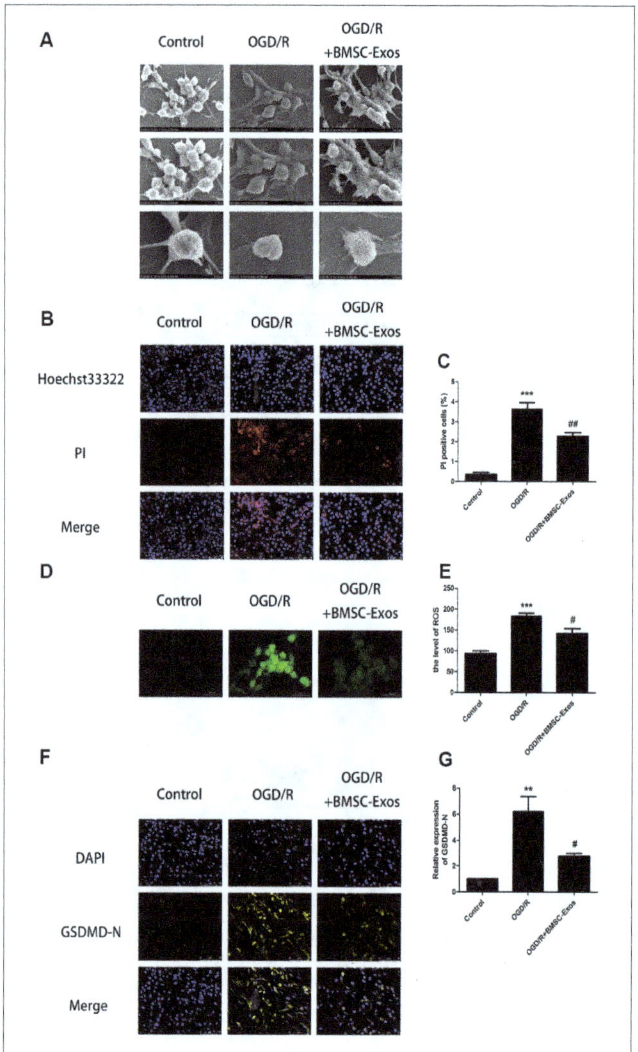

FIGURE 2 | BMSC-Exos reduce pyroptosis induced by OGD/R. **(A)** Morphological changes of PC12 cells were observed using scanning electron microscopy (SEM); scale bar = 20/10/5 μm. **(B)** Representative images of Hoechst 33342/propidium iodide (PI) staining; scale bar = 50 μm. **(C)** PI-stained cells (red puncta) were quantified as the percentage of red puncta/total puncta signals in merged images ($n = 3$). PI-stained areas were lower in the BMSC-Exos group than in the OGD/R group ($n = 3$). **(D)** Representative images of reactive oxygen species (ROS); scale bar = 10 μm. **(E)** ROS levels are presented relative to control levels ($n = 3$). ROS levels in the BMSC-Exos group were lower than in the OGD/R group. **(F)** Representative immunofluorescence images; scale bar = 75 μm. **(G)** Relative expression levels of N-terminal of gasdermin D (GSDMD-N) were lower in the BMSC-Exos group than in the OGD/R group ($n = 3$). $^{**}p < 0.01$, $^{***}p < 0.001$ vs. control group, $^{\#}p < 0.05$; $^{\#\#}p < 0.01$ vs. OGD/R group.

ROS Levels

ROS is essential for NLRP3 inflammasome activation and ROS activation-induced NLRP3 inflammasome triggering caspase-1-dependent pyroptosis plays an important role in I/R injury (Liao et al., 2019). Our results showed that ROS levels were significantly increased in the OGD/R group. In addition, ROS levels were significantly decreased in the BMSC-Exos group compared with the OGD/R group, suggesting that BMSC-Exos significantly inhibited the activation of NLRP3 (**Figures 2D,E**).

Activities of NLRP3, Cleaved Caspase-1, GSDMD-N, and IL-1β

Western blot results showed that the expression levels of NLRP3, cleaved caspase-1, and IL-1β were increased significantly in the OGD/R group and these increased levels were alleviated with BMSC-Exos treatment. We also used immunofluorescence and Western blot to detect the expression levels of GSDMD-N, a key enzyme in the process of pyroptosis. Our results suggested that BMSC-Exos inhibited the expression of GSDMD-N when compared with the OGD/R group (**Figures 2F,G, 3A,G**).

To identify whether NLRP3 inflammasome-mediated pyroptosis was involved in the neuroprotection of BMSC-Exos, the PC12 cells were treated with and without the NLRP3 inhibitor MCC950 under OGD/R conditions. Our results showed that suppressing the activation of NLRP3 with BMSC-Exos was similar to the effect of the inhibitor MCC950 (**Figures 3A–G**). However, NLRP3 overexpression gave rise to a significant increase in pyroptosis-related proteins, but the levels of cleaved caspase-1, GSDMD-N, and IL-1β were decreased in the OGD/R+BMSC-Exos+pNLRP3 group when compared with the OGD/R+pNLRP3 group (**Figures 3H–N**). These results suggested that BMSC-Exos play protective roles in the NLRP3 inflammasome-mediated pyroptosis induced by OGD/R.

BMSC-Exos Promote Autophagic Flux in OGD/R Through the AMPK/mTOR Pathway

While TEM indicated the presence of a few autophagosomes in the control group, there were more autophagosomes in the OGD/R group. In contrast, a larger number of autolysosomes were detected in the BMSC-Exos group (**Figures 4A,B**).

Compared with the control group, the numbers of yellow puncta (GFP+RFP+) and red puncta (GFP-RFP+) increased in the OGD/R group. The numbers of red puncta further increased with BMSC-Exos treatment when compared with the OGD/R group, suggesting that BMSC-Exos treatment may promote autophagic flux. Furthermore, the autophagy inhibitor Cq reversed the results of GFP-RFP-LC3 staining and CCK-8 and LDH assays (**Figures 4C–F**).

The expression levels of LC3 II/I and p-AMPK/AMPK increased, whereas those of P62 and p-mTOR/mTOR decreased in the OGD/R group when compared with the control group. LC3 II/I expression did not change significantly after BMSC-Exos treatment, whereas p-AMPK/AMPK expression increased further and p-mTOR/mTOR and P62 expression levels decreased (**Figures 4G–K**). Moreover, AMPK knockdown by compound C and siRNA not only negated the effects of BMSC-Exos on AMPK, autophagic flux-associated proteins (**Figures 4G–K, 5A–H**), and GFP-RFP-LC3 staining (**Figures 5K,L**) but also abrogated their ability to rescue PC12 cells from OGD/R-induced cell death and LDH release (**Figures 4L,M, 5I,J**).

Blocked Autophagic Flux Reversed the Protective Effect of BMSC-Exos Against OGD/R-Induced Pyroptosis

To identify whether AMPK-dependent autophagic flux was involved in the effect of BMSC-Exos on the inhibition of cell

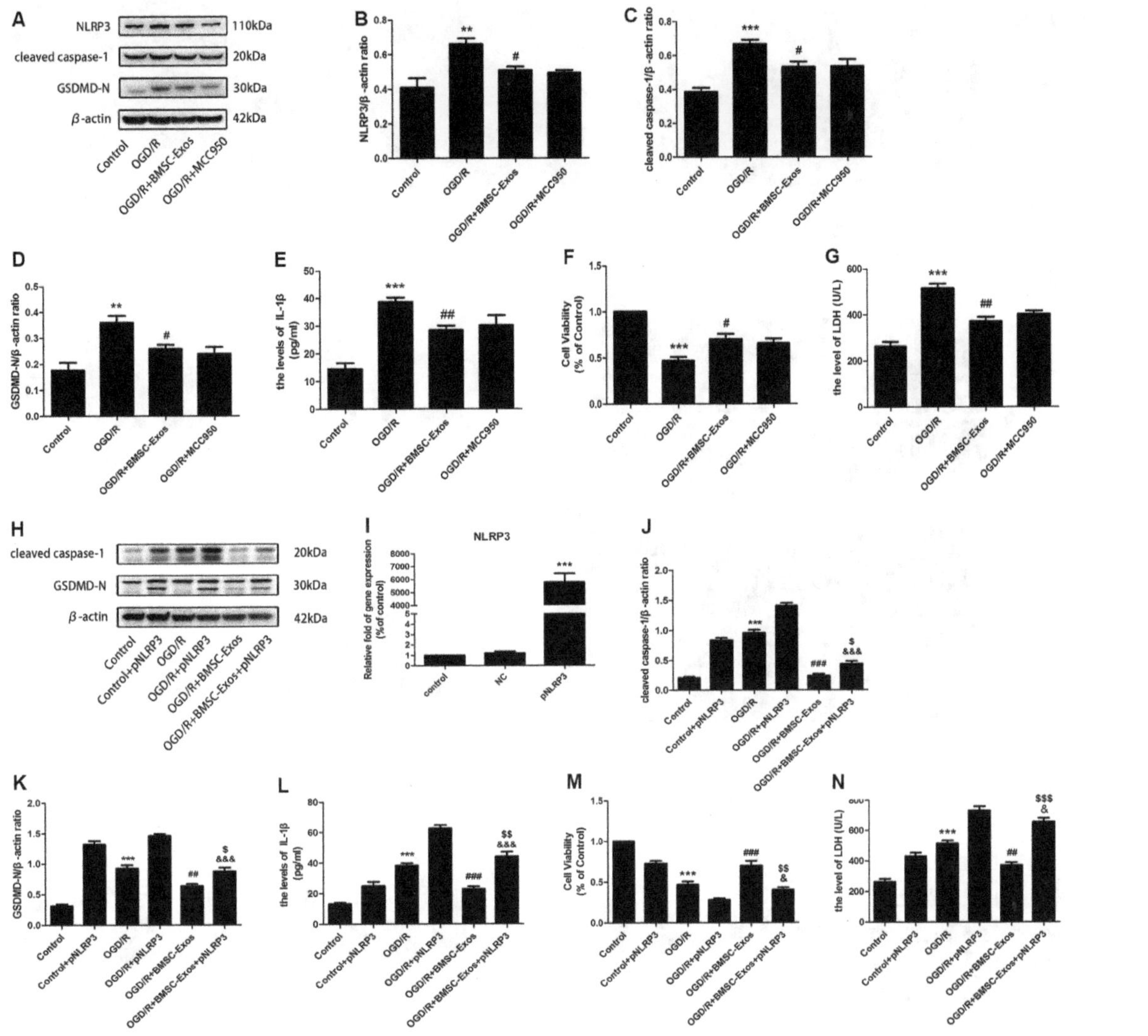

FIGURE 3 | BMSC-Exos reduce nucleotide-binding domain and leucine-riçh repeat family protein 3 (NLRP3) inflammasome-mediated pyroptosis following OGD/R. **(A)** Representative Western blots of NLRP3, cleaved caspase-1, and GSDMD-N. **(B–D)** Expression levels of NLRP3, cleaved caspase-1, and GSDMD-N in the BMSC-Exos group were higher than in the OGD/R group, whereas no significant difference was observed between BMSC-Exos and MCC950 groups ($n = 3$). **(E)** Interleukin-1β (IL-1β) levels in the BMSC-Exos group were higher than in the OGD/R group, whereas no significant difference was observed between BMSC-Exos and MCC950 groups ($n = 3$). **(F)** No significant difference in cell viability was observed between BMSC-Exos and MCC950 groups ($n = 3$). **(G)** No significant difference in LDH release was observed between BMSC-Exos and MCC950 groups ($n = 3$). **(H)** Representative Western blots of cleaved caspase-1 and GSDMD-N. **(I)** Real-time quantitative polymerase chain reaction (RT-qPCR) showing the effects of pNLRP3 transfection ($n = 3$). **(J,K)** Expression levels of cleaved caspase-1 and GSDMD-N were higher in the BMSC-Exos+pNLRP3 group than in the BMSC-Exos group ($n = 3$). **(L)** IL-1β levels were higher in the BMSC-Exos+pNLRP3 group than in the BMSC-Exos group ($n = 3$). **(M)** Cell viability in the BMSC-Exos+pNLRP3 group was lower than in the BMSC-Exos group ($n = 3$). **(N)** LDH release in the BMSC-Exos+pNLRP3 group was higher than in the BMSĆ-Exos group ($n = 3$). $^{**}p < 0.01$, $^{***}p < 0.001$ vs. control group; $^{\#}p < 0.05$, $^{\#\#}p < 0.01$, $^{\#\#\#}p < 0.001$ vs. OGD/R group; $^{\$}p < 0.05$, $^{\$\$}p < 0.01$, $^{\$\$\$}p < 0.001$ vs. OGD/R+BMSC-Exos group; $^{\&}p < 0.05$, $^{\&\&\&}p < 0.001$ vs. OGD/R+pNLRP3 group.

pyroptosis, PC12 cells were pretreated with Cq, compound C, or transfected with AMPK-specific siRNA. Our results showed that Cq, compound C, and AMPK siRNA increased the expression of NLRP3, cleaved caspase-1, GSDMD-N, and IL-1β and reversed the results of the Hoechst 33342/PI double staining when compared with the OGD/R+BMSC-Exos group (**Figures 6, 7**).

We next detected the specific binding interaction between NLRP3 and LC3 in Co-IP assays. As shown in **Figure 8**, there was a direct interaction between NLRP3 and LC3. The direct interaction was significantly decreased under OGD/R, and BMSC-Exos could increase it.

DISCUSSION AND CONCLUSION

The pathological mechanism of cerebral I/R injury is complex and not well understood (Kalogeris et al., 2016). Among all the pathological factors, NLRP3 inflammasome-mediated pyroptosis

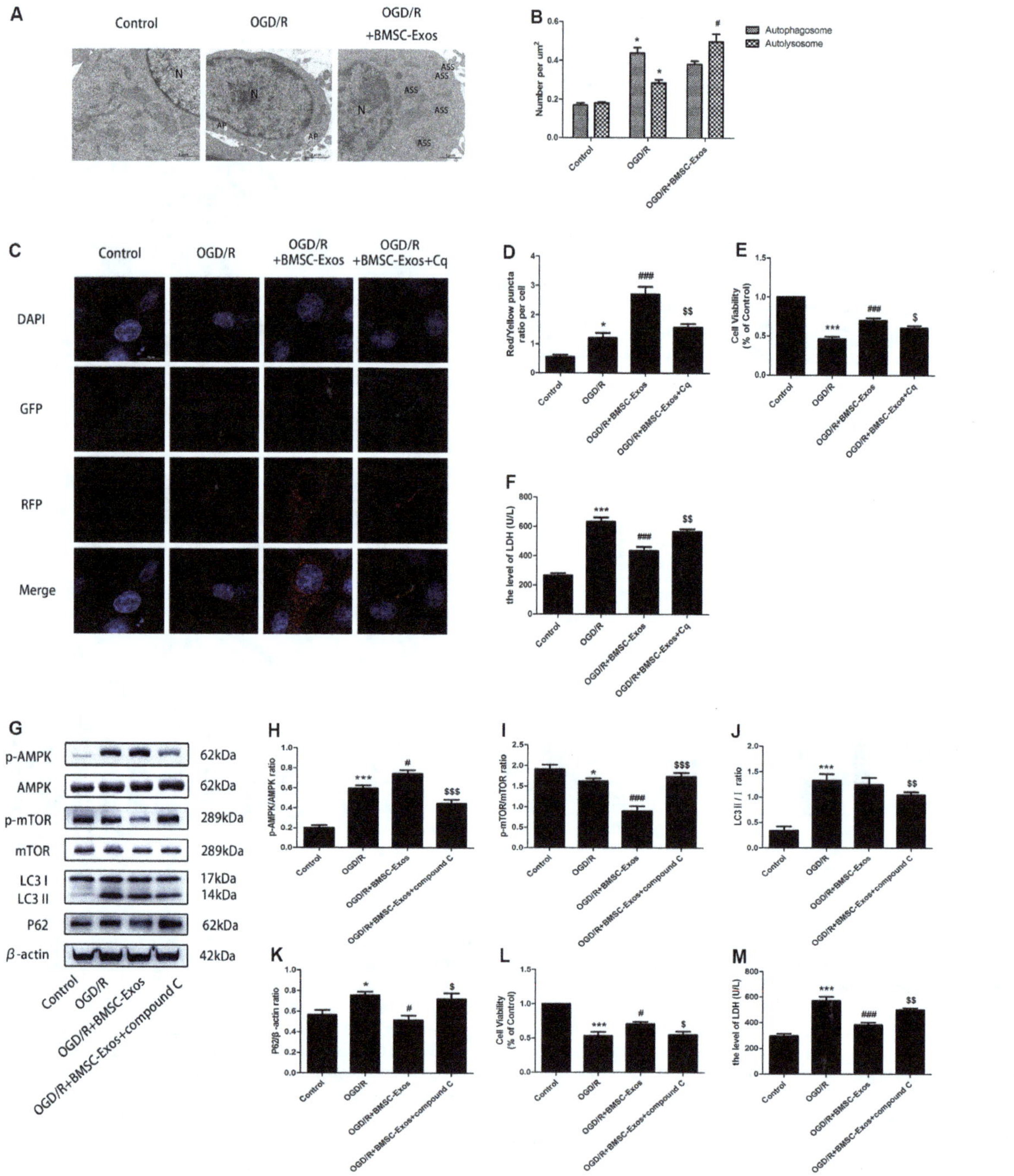

FIGURE 4 | BMSC-Exos promote autophagic flux through the AMP-activated kinase (AMPK)/mammalian target of rapamycin (mTOR) pathway following OGD/R. **(A)** Autophagic flux was detected using TEM. Typical cytoplasms and nuclei (N) in the control group; double-membrane autophagosomes (AP) were observed in the OGD/R group. Autolysosomes were darkly stained, indicating that autolysosomes (ASS) are activated in the OGD/R+BMSC-Exos group; scale bar = 1 μm. **(B)** Quantitative analysis of numbers of autophagosomes and autolysosomes in each treatment group (*n* = 3). Autophagosomes and autolysosomes were more numerous in the OGD/R group than in the control group, whereas larger numbers of autolysosomes were detected in the BMSC-Exos group. **(C)** Representative images of GFP-RFP-LC3 staining; scale bar = 10 μm. **(D)** Autophagy was quantified as the ratio of red puncta (GFP-RFP+) to yellow puncta (GFP+RFP+) in each cell. This ratio was higher in the BMSC-Exos group than in control and OGD/R groups (*n* = 3). **(E,F)** The autophagy inhibitor Cq reversed cell viability and LDH release (*n* = 3). **(G)** Representative Western blots of p-AMPK, AMPK, p-mTOR, mTOR, LC3 II/I and P62. **(H–K)** LC3 II/I and p-AMPK/AMPK expression increased, although P62 and p-mTOR/mTOR expression levels decreased in the OGD/R group when compared with the control group (*n* = 3). LC3 II/I expression did not change significantly after BMSC-Exos treatment, whereas p-AMPK/AMPK expression increased further and p-mTOR/mTOR and P62 expression levels decreased (*n* = 3). **(L,M)** The AMPK inhibitor compound C reversed the effects of OGD/R on cell viability and LDH release (*n* = 3). $^{*}p < 0.05$, $^{***}p < 0.001$ vs. control group; $^{\#}p < 0.05$, $^{\#\#\#}p < 0.001$ vs. OGD/R group; $^{\$}p < 0.05$, $^{\$\$}p < 0.01$, $^{\$\$\$}p < 0.001$ vs. OGD/R+BMSC-Exos group.

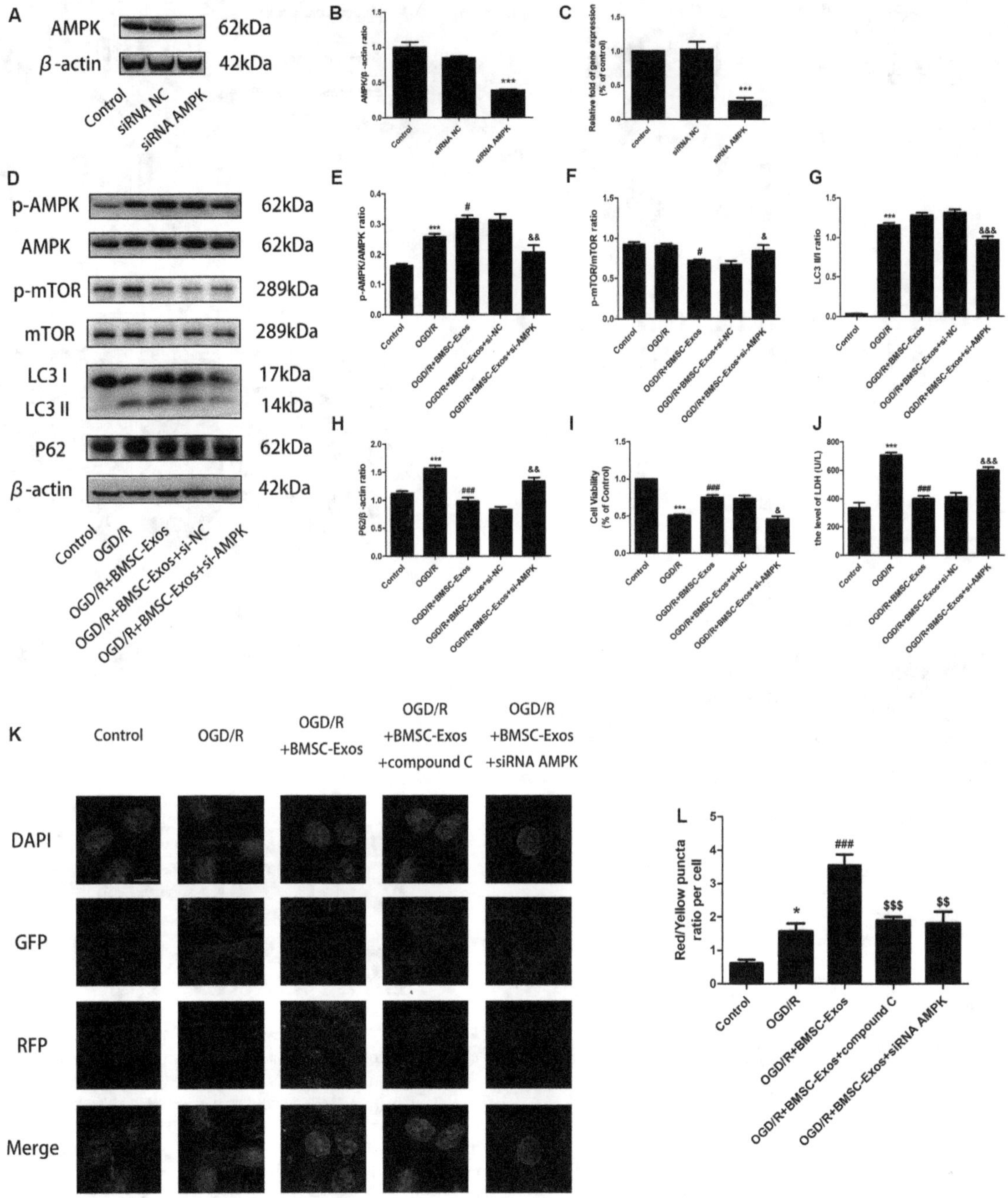

FIGURE 5 | AMPK knockdown negated the effects of BMSC-Exos on autophagic flux. **(A–C)** AMPK protein levels were reduced following gene silencing. AMPK knockdown reversed the effects of BMSC-Exos on **(D–H)** activation of p-AMPK/AMPK and LC3 II/I, inhibition of p-TOR/mTOR and P62, **(I)** increase in cell viability, and **(J)** reduction in LDH release (n = 3). **(K,L)** AMPK knockdown reversed the effects of BMSC-Exos on the ratio of red puncta to yellow puncta (n = 3). $^{*}p < 0.05$, $^{***}p < 0.001$ vs. control group; $^{\#}p < 0.05$, $^{\#\#\#}p < 0.001$ vs. OGD/R group; $^{\$\$}p < 0.01$, $^{\$\$\$}p < 0.001$ vs. OGD/R+BMSC-Exos group; $^{\&}p < 0.05$, $^{\&\&}p < 0.01$, $^{\&\&\&}p < 0.001$ vs. OGD/R+BMSC-Exos+si-NC group.

has recently been considered as an important process in cerebral I/R injury (Zhou et al., 2019). In this study, we demonstrated that the protective effect of BMSC-Exos on OGD/R injury is related to the inhibition of NLRP3 inflammasome-mediated pyroptosis by promoting AMPK-dependent autophagic flux (**Figure 9**).

As an important component of paracrine signaling in stem cells, exosomes appear to be promising candidates for repairing tissue injury in place of stem cells (Lazar et al., 2018; Williams et al., 2020). Previous studies have found that BMSC-Exos can promote nerve function and improve the nervous system in ischemic stroke (Deng et al.,

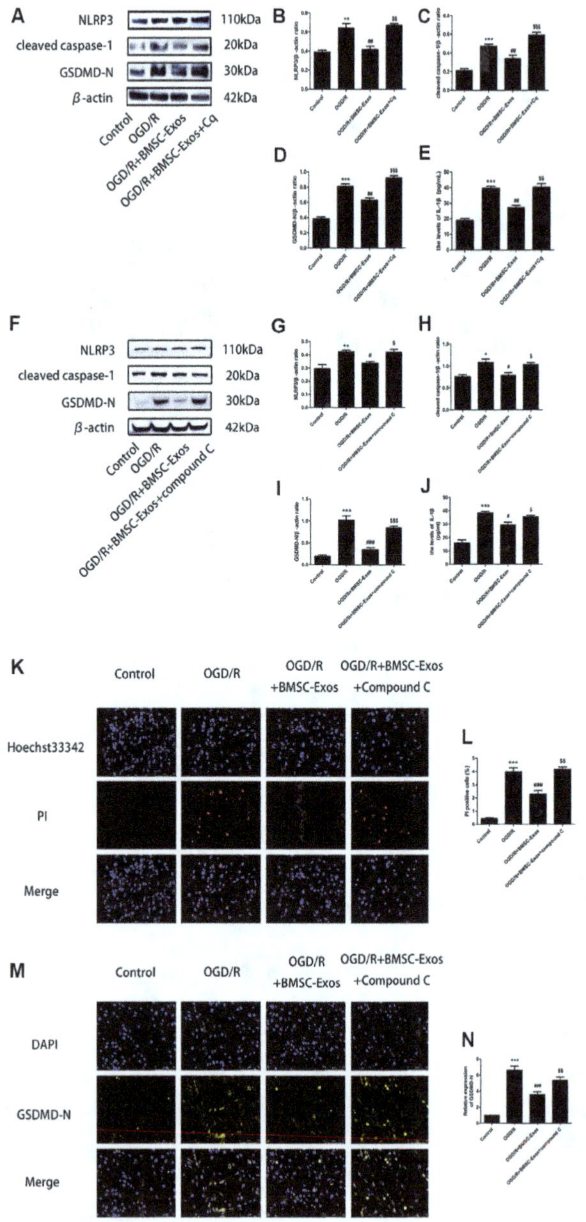

FIGURE 6 | Blocked autophagic flux reverses the protective effect of BMSC-Exos against OGD/R-induced pyroptosis. **(A)** Representative Western blots of NLRP3, cleaved caspase-1, and GSDMD-N. **(B–D)** NLRP3, cleaved caspase-1, and GSDMD-N expression levels in the BMSC-Exos+Cq group were higher than in the BMSC-Exos group ($n = 3$). **(E)** IL-1β expression was higher in the BMSC-Exos+Cq group than in the BMSC-Exos group ($n = 3$). **(F)** Representative Western blots of NLRP3, cleaved caspase-1, and GSDMD-N. **(G–I)** NLRP3, cleaved caspase-1, and GSDMD-N expression levels in the BMSC-Exos+compound C group were higher than in the BMSC-Exos group ($n = 3$). **(J)** IL-1β expression was higher in the BMSC-Exos+compound C group than in the BMSC-Exos group ($n = 3$). **(K)** Representative images of Hoechst 33342/PI staining; scale bar = 75 μm. **(L)** Proportions of PI stained areas were higher in the BMSC-Exos+compound C group than in the BMSC-Exos group ($n = 3$). **(M)** Representative immunofluorescence images; scale bar = 75 μm. **(N)** Relative expression levels of GSDMD-N were higher in the BMSC-Exos+compound C group than in the BMSC-Exos group ($n = 3$). $^{*}p < 0.05$, $^{**}p < 0.01$, $^{***}p < 0.001$ vs. control group; $^{\#}p < 0.05$, $^{\#\#}p < 0.01$, $^{\#\#\#}p < 0.001$ vs. OGD/R group; $^{\$}p < 0.05$, $^{\$\$}p < 0.01$, $^{\$\$\$}p < 0.001$ vs. OGD/R+BMSC-Exos group.

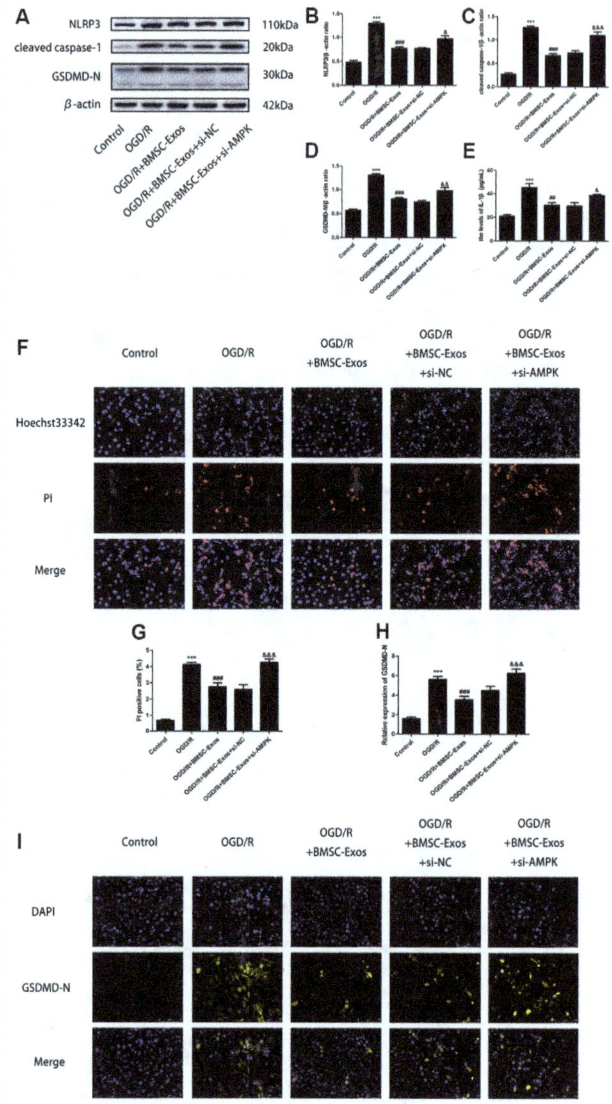

FIGURE 7 | AMPK knockdown negated the protective effect of BMSC-Exos against pyroptosis. **(A–E)** AMPK knockdown reversed the effect of BMSC-Exos on inhibition of NLRP3, cleaved caspase-1, GDDMD-N, and IL-1β ($n = 3$). **(F,G)** AMPK knockdown reversed the effects of BMSC-Exos on the proportions of PI stained areas ($n = 3$). **(H,I)** AMPK knockdown reversed the effects of BMSC-Exos on the relative expression levels of GSDMD-N ($n = 3$). $^{***}p < 0.001$ vs. control group; $^{\#\#}p < 0.01$, $^{\#\#\#}p < 0.001$ vs. OGD/R group; $^{\&}p < 0.05$, $^{\&\&}p < 0.01$; $^{\&\&\&}p < 0.001$ vs. OGD/R+BMSC-Exos+si-NC group.

2019; Xiao et al., 2019; Hou et al., 2020; Safakheil and Safakheil, 2020), but the specific mechanism remains to be explored. In the current study, TEM, NanoSight NTA, and the surface marker proteins confirmed that the BMSC-Exos were successfully isolated and met international standards (Théry et al., 2018). CCK-8 and LDH assays further showed that BMSC-Exos at a dose of 10 μg/ml improved the viability of cells subjected to OGD/R, indicating that BMSC-Exos had a protective effect.

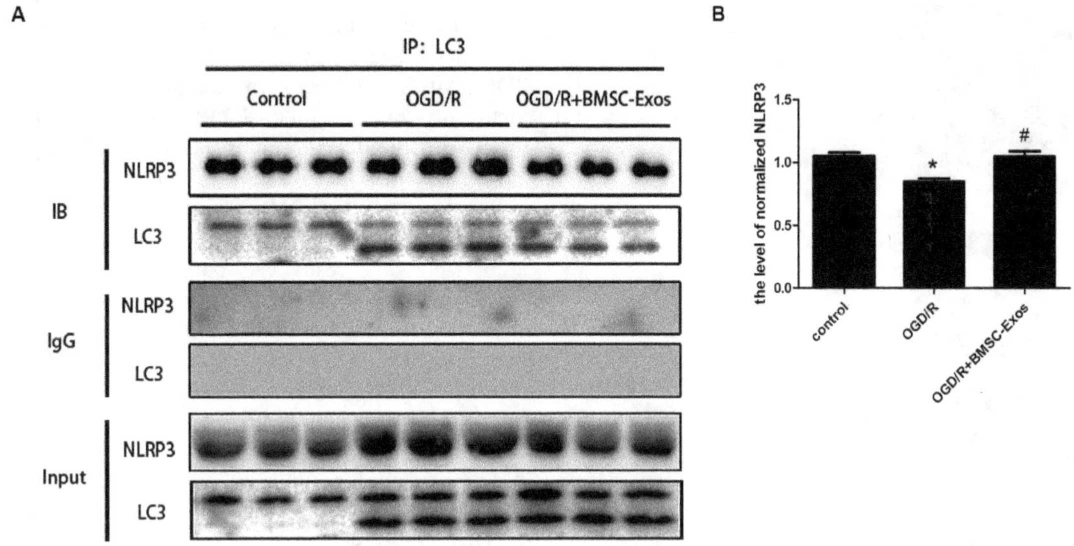

FIGURE 8 | Coimmunoprecipitation of NLRP3 and LC3. **(A)** Cell lysates from PC12 cells under normal condition or OGD/R with or without BMSC-Exos were incubated with anti-LC3B antibody. NLRP3 and LC3 in precipitates were detected by Western blot. IgG was used as a negative control. **(B)** The NLRP3 was precipitated by anti-LC3B antibody, and the relative amount of NLRP3 was compared (n = 3 biological replicates per group). The direct interaction between NLRP3 and LC3 was significantly decreased under OGD/R, and BMSC-Exos could increase the interaction. $^{*}p < 0.05$ vs. control group, $^{\#}p < 0.05$ vs. OGD/R group.

Pyroptosis, also known as inflammatory necrosis, is a new type of pro-inflammatory programmed cell death that plays an important role in cerebral I/R injury (Xia et al., 2018; Zhou et al., 2019; Zhu et al., 2019). This mode of cell death is mediated by caspase-1 and GSDMD, which induces the formation of pores in the cell membranes, resulting in the release of a large number of inflammatory cytokines (Coll et al., 2011; Lacey et al., 2018). Recent study has indicated that BMSC-Exos could protect against ischemic stroke through anti-inflammation and anti-apoptosis effects (Hou et al., 2020; Safakheil and Safakheil, 2020). In this study, we found that the protective effect of BMSC-Exos was associated with a reduced pyroptosis. Western blot and ELISA results further showed that BMSC-Exos inhibited the high expression of the key proteins associated with pyroptosis such as cleaved caspase-1, GSDMD-N, and IL-1β induced by OGD/R. Moreover, the PI staining proportion was decreased in the BMSC-Exos treatment group and LDH release was reduced significantly, indicating that the cell membranes of the cells that received the BMSC-Exos intervention were more complete than those of the OGD/R group. Importantly, SEM indicated that the cells treated with BMSC-Exos had clearer outlines, more dense surface villi, and longer protrusions, which were in contrast to the characteristics of pyroptosis in the OGD/R group.

As a component of pyroptosis, NLRP3 inflammasome is of concern because it can be activated by various PAMPs and DAMPs and it is closely related to various human diseases (Gong et al., 2018; Liu et al., 2018; Eren and Özõren, 2019). When NLRP3 is assembled and activated, caspase-1 can be activated and turn pro-IL-1β and GSDMD into the bioactive cytokines IL-1β and GSDMD-N, respectively (Hou et al., 2018). In this study, we found that the use of BMSC-Exos inhibited the high expression of NLRP3 and other key proteins associated with pyroptosis in OGD/R injury. This effect was similar to the NLRP3 inhibitor MCC950 and was reversed by NLRP3 overexpression. Various studies have found that ROS represents a key signal that regulates the activation of the NLRP3 inflammasome (Tschopp and Schroder, 2010; Eren and Özõren, 2019). We found that the ROS level in the cells subjected to OGD/R was increased significantly and that BMSC-Exos could weaken this response, indicating that BMSC-Exos reduced pyroptosis mediated by the NLRP3 inflammasome.

Autophagy is a dynamic process of degradation (Cardinal et al., 2009). The protective effect of BMSC-Exos on myocardial I/R injury appears to be closely related to the activation of autophagy (Zou et al., 2019), but their mechanism of autophagy in cerebral I/R injury has never been reported. In the current study, both the results of TEM and GFP-RFP-LC3 indicated that there was a large increase in number of autolysosomes in the OGD/R+BMSC-Exos group, further suggesting that BMSC-Exos was involved in the promotion of autophagic flux. In addition, the autophagy inhibitor Cq was shown to reverse the effect of activation and protection, indicating that the role of BMSC-Exos in reducing OGD/R injury was related primarily to their promotion of autophagic flux.

AMPK is a heterologous silk/threonine kinase distributed in many tissues where it regulates cellular energy, whose phosphorylation can also activate inhibitors of mTOR, thus activating autophagy (Qi and Young, 2015). Therefore, the AMPK/mTOR signaling pathway is an important regulatory pathway for autophagy (Puissant and Auberger, 2010; Jiang et al., 2018). In this study, BMSC-Exos treatment was found to activate

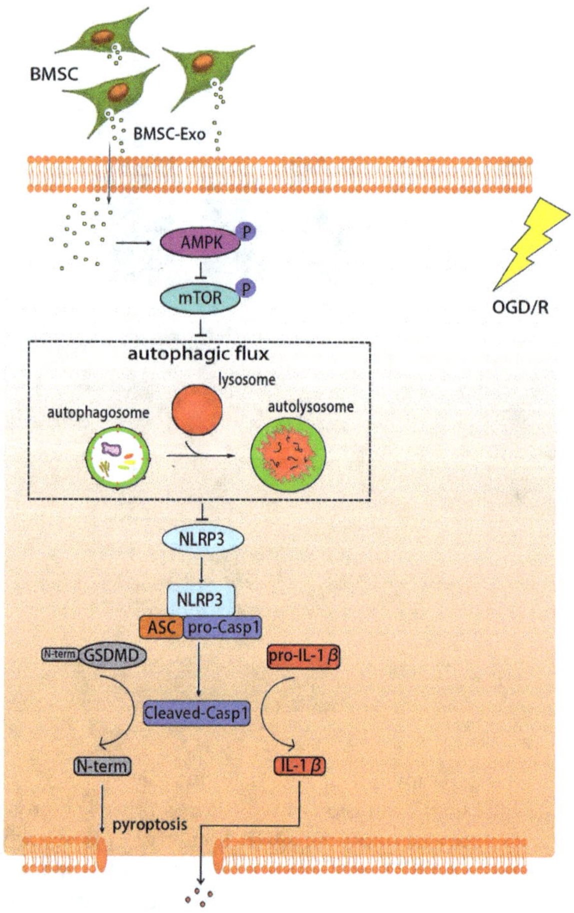

FIGURE 9 | Graphical abstract demonstrating how BMSC-Exos inhibit NLRP3 inflammasome-mediated pyroptosis by promoting AMPK-dependent autophagic flux in OGD/R.

AMPK phosphorylation and suppress mTOR phosphorylation during OGD/R. Both compound C and AMPK silencing reversed the promoting effect of BMSC-Exos on autophagic flux, which suggested that autophagic flux promoted by BMSC-Exos was AMPK-dependent.

Some recent studies have revealed that the protective effect of AMPK-dependent autophagy is partially related to the inhibition of pyroptosis. Li et al. (2019) found that adrenomedullin may protect the steroidogenic functions of Leydig cells against pyroptosis by activating autophagy *via* the ROS/AMPK/mTOR axis. In addition, Yang et al. verified that metformin could inhibit the NLRP3 inflammasome by activating the AMPK/mTOR pathway in diabetic cardiomyopathy (Yang et al., 2019). Our results showed that Cq reversed the inhibitory effect of BMSC-Exos on pyroptosis, suggesting that BMSC-Exos may alleviate cell pyroptosis by promoting autophagic flux. Moreover, both compound C and AMPK silencing reversed the effect described above, which suggested that autophagic flux promoted by BMSC-Exos for alleviating cell pyroptosis was AMPK-dependent. Co-IP experiments further showed a direct interaction between NLRP3 and LC3. The interaction was significantly decreased under OGD/R, and BMSC-Exos could increase it, which confirmed the impact of BMSC-Exos on the interaction between NLRP3 inflammasome and autophagy.

This study demonstrated that BMSC-Exos can protect PC12 cells against OGD/R injury through the attenuation of NLRP3 inflammasome-mediated pyroptosis by promoting AMPK-dependent autophagic flux. Further *in vivo* studies are warranted to verify the effect of BMSC-Exos on cerebral I/R injury.

AUTHOR CONTRIBUTIONS

QZ, YZ, and GH planned and conducted all the experiments. QZ, YZ, DL, HH, XL, RZ, and MZ contributed to performing the experiments. QZ and YZ contributed to performing the data analysis and drafting the manuscript. XL and YW revised the manuscript. All authors approved the final manuscript.

ACKNOWLEDGMENTS

We would like to give the sincere appreciation to Qing Mei Wang, MD, PhD, Director of Stroke Biological Recovery Laboratory, Physiatrist of Spaulding Rehabilitation Hospital, the teaching affiliate of Harvard Medical School, for her helpful comments on this article.

REFERENCES

Campbell, B. C. V., De Silva, D. A., Macleod, M. R., Coutts, S. B., Schwamm, L. H., Davis, S. M., et al. (2019). Ischaemic stroke. *Nat. Rev. Dis. Primers* 5:70. doi: 10.1038/s41572-019-0118-8

Cardinal, J., Pan, P., and Tsung, A. (2009). Protective role of cisplatin in ischemic liver injury through induction of autophagy. *Autophagy* 5, 1211–1212. doi: 10.4161/auto.5.8.9972

Chen, X., Lu, M., He, X., Ma, L., Birnbaumer, L., and Liao, Y. (2017). TRPC3/6/7 knockdown protects the brain from cerebral ischemia injury *via* astrocyte apoptosis inhibition and effects on NF-small ka, CyrillicB translocation. *Mol. Neurobiol.* 54, 7555–7566. doi: 10.1007/s12035-016-0227-2

Chen, Z., Yang, J., Zhong, J., Luo, Y., Du, W., Hu, C., et al. (2020). MicroRNA-193b-3p alleviates focal cerebral ischemia and reperfusion-induced injury in rats by inhibiting 5-lipoxygenase expression. *Exp. Neurol.* 327:113223. doi: 10.1016/j.expneurol.2020.113223

Coll, N. S., Epple, P., and Dangl, J. L. (2011). Programmed cell death in the plant immune system. *Cell Death Differ.* 18, 1247–1256. doi: 10.1038/cdd.2011.37

Deng, Y., Chen, D., Gao, F., Lv, H., Zhang, G., Sun, X., et al. (2019). Exosomes derived from microRNA-138-5p-overexpressing bone marrow-derived mesenchymal stem cells confer neuroprotection to astrocytes following ischemic stroke *via* inhibition of LCN2. *J. Biol. Eng.* 13:71. doi: 10.1186/s13036-019-0193-0

Eren, E., and Özören, N. (2019). The NLRP3 inflammasome: a new player in neurological diseases. *Turk. J. Biol.* 43, 349–359. doi: 10.3906/biy-1909-31

Fan, Y. Y., Shen, Z., He, P., Jiang, L., Hou, W. W., Shen, Y., et al. (2014). A novel neuroprotective strategy for ischemic stroke: transient mild acidosis treatment by CO2 inhalation at reperfusion. *J. Cereb. Blood Flow Metab.* 34, 275–283. doi: 10.1038/jcbfm.2013.193

Gong, Z., Pan, J., Shen, Q., Li, M., and Peng, Y. (2018). Mitochondrial dysfunction induces NLRP3 inflammasome activation during cerebral ischemia/reperfusion injury. *J. Neuroinflammation* 15, 242–242. doi: 10.1186/s12974-018-1282-6

Gu, H., Hou, Q., Liu, Y., Cai, Y., Guo, Y., Xiang, H., et al. (2019). On-line regeneration of electrochemical biosensor for *in vivo* repetitive measurements of striatum Cu^{2+} under global cerebral ischemia/reperfusion events. *Biosens. Bioelectron.* 135, 111–119. doi: 10.1016/j.bios.2019.03.014

He, H., Zeng, Q., Huang, G., Lin, Y., Lin, H., Liu, W., et al. (2019). Bone marrow mesenchymal stem cell transplantation exerts neuroprotective effects following cerebral ischemia/reperfusion injury by inhibiting autophagy *via* the PI3K/Akt pathway. *Brain Res.* 1707, 124–132. doi: 10.1016/j.brainres.2018.11.018

Herr, D. R., Yam, T. Y. A., Tan, W. S. D., Koh, S. S., Wong, W. S. F., Ong, W. Y., et al. (2020). Ultrastructural characteristics of DHA-induced pyroptosis. *Neuromolecular Med.* 22, 293–303. doi: 10.1007/s12017-019-08586-y

Hou, K., Li, G., Zhao, J., Xu, B., Zhang, Y., Yu, J., et al. (2020). Bone mesenchymal stem cell-derived exosomal microRNA-29b-3p prevents hypoxic-ischemic injury in rat brain by activating the PTEN-mediated Akt signaling pathway. *J. Neuroinflammation* 17:46. doi: 10.1186/s12974-020-1725-8

Hou, Y., Wang, Y., He, Q., Li, L., Xie, H., Zhao, Y., et al. (2018). Nrf2 inhibits NLRP3 inflammasome activation through regulating Trx1/TXNIP complex in cerebral ischemia reperfusion injury. *Behav. Brain Res.* 336, 32–39. doi: 10.1016/j.bbr.2017.06.027

Jayaraj, R. L., Azimullah, S., Beiram, R., Jalal, F. Y., and Rosenberg, G. A. (2019). Neuroinflammation: friend and foe for ischemic stroke. *J. Neuroinflammation* 16:142. doi: 10.1186/s12974-019-1516-2

Jiang, S., Li, T., Ji, T., Yi, W., Yang, Z., Wang, S., et al. (2018). AMPK: potential therapeutic target for ischemic stroke. *Theranostics* 8, 4535–4551. doi: 10.7150/thno.25674

Kalogeris, T., Baines, C. P., Krenz, M., and Korthuis, R. J. (2016). Ischemia/Reperfusion. *Compr. Physiol.* 7, 113–170. doi: 10.1002/cphy.c160006

Koubi, D., Jiang, H., Zhang, L., Tang, W., Kuo, J., Rodriguez, A. I., et al. (2005). Role of Bcl-2 family of proteins in mediating apoptotic death of PC12 cells exposed to oxygen and glucose deprivation. *Neurochem. Int.* 46, 73–81. doi: 10.1016/j.neuint.2004.06.006

Lacey, C. A., Mitchell, W. J., Dadelahi, A. S., and Skyberg, J. A. (2018). Caspase-1 and caspase-11 mediate pyroptosis, inflammation and control of brucella joint infection. *Infect. Immun.* 86:e00361-18. doi: 10.1128/iai.00361-18

Lambertsen, K. L., Finsen, B., and Clausen, B. H. (2019). Post-stroke inflammation-target or tool for therapy? *Acta Neuropathol.* 137, 693–714. doi: 10.1007/s00401-018-1930-z

Lazar, E., Benedek, T., Korodi, S., Rat, N., Lo, J., and Benedek, I. (2018). Stem cell-derived exosomes—an emerging tool for myocardial regeneration. *World J. Stem Cells* 10, 106–115. doi: 10.4252/wjsc.v10.i8.106

Li, M. Y., Zhu, X. L., Zhao, B. X., Shi, L., Wang, W., Hu, W., et al. (2019). Adrenomedullin alleviates the pyroptosis of Leydig cells by promoting autophagy *via* the ROS-AMPK-mTOR axis. *Cell Death Dis.* 10:489. doi: 10.1038/s41419-019-1728-5

Liao, Y., Hussain, T., Liu, C., Cui, Y., Wang, J., Yao, J., et al. (2019). Endoplasmic reticulum stress induces macrophages to produce IL-1β during *Mycobacterium bovis* infection *via* a positive feedback loop between mitochondrial damage and inflammasome activation. *Front. Immunol.* 10, 268–268. doi: 10.3389/fimmu.2019.00268

Liu, D., Zeng, X., Li, X., Mehta, J. L., and Wang, X. (2018). Role of NLRP3 inflammasome in the pathogenesis of cardiovascular diseases. *Basic Res. Cardiol.* 113:5. doi: 10.1007/s00395-017-0663-9

McBride, J. D., Rodriguez-Menocal, L., Guzman, W., Candanedo, A., Garcia-Contreras, M., and Badiavas, E. V. (2017). Bone marrow mesenchymal stem cell-derived $CD63^+$ exosomes transport Wnt3a exteriorly and enhance dermal fibroblast proliferation, migration, and angiogenesis *in vitro*. *Stem Cells Dev.* 26, 1384–1398. doi: 10.1089/scd.2017.0087

Puissant, A., and Auberger, P. (2010). AMPK- and p62/SQSTM1-dependent autophagy mediate resveratrol-induced cell death in chronic myelogenous leukemia. *Autophagy* 6, 655–657. doi: 10.4161/auto.6.5.12126

Qi, D., and Young, L. H. (2015). AMPK: energy sensor and survival mechanism in the ischemic heart. *Trends Endocrinol. Metab.* 26, 422–429. doi: 10.1016/j.tem.2015.05.010

Rana, A. K., and Singh, D. (2018). Targeting glycogen synthase kinase-3 for oxidative stress and neuroinflammation: Opportunities, challenges and future directions for cerebral stroke management. *Neuropharmacology* 139, 124–136. doi: 10.1016/j.neuropharm.2018.07.006

Ren, Q., Hu, Z., Jiang, Y., Tan, X., Botchway, B. O. A., Amin, N., et al. (2019). SIRT1 protects against apoptosis by promoting autophagy in the oxygen glucose deprivation/reperfusion-induced injury. *Front. Neurol.* 10:1289. doi: 10.3389/fneur.2019.01289

Safakheil, M., and Safakheil, H. (2020). The effect of exosomes derived from bone marrow stem cells in combination with rosuvastatin on functional recovery and neuroprotection in rats after ischemic stroke. *J. Mol. Neurosci.* 70, 724–737. doi: 10.1007/s12031-020-01483-1

Stonesifer, C., Corey, S., Ghanekar, S., Diamandis, Z., Acosta, S. A., and Borlongan, C. V. (2017). Stem cell therapy for abrogating stroke-induced neuroinflammation and relevant secondary cell death mechanisms. *Prog. Neurobiol.* 158, 94–131. doi: 10.1016/j.pneurobio.2017.07.004

Théry, C., Witwer, K. W., Aikawa, E., Alcaraz, M. J., Anderson, J. D., Andriantsitohaina, R., et al. (2018). Minimal information for studies of extracellular vesicles 2018 (MISEV2018): a position statement of the International Society for Extracellular Vesicles and update of the MISEV2014 guidelines. *J. Extracell. Vesicles* 7:1535750. doi: 10.1080/20013078.2018.1535750

Tschopp, J., and Schroder, K. (2010). NLRP3 inflammasome activation: the convergence of multiple signalling pathways on ROS production? *Nat. Rev. Immunol.* 10, 210–215. doi: 10.1038/nri2725

Wang, W., Jiang, B., Sun, H., Ru, X., Sun, D., Wang, L., et al. (2017). Prevalence, incidence, and mortality of stroke in china: results from a nationwide population-based survey of 480 687 adults. *Circulation* 135, 759–771. doi: 10.1161/circulationaha.116.025250

Williams, A. M., Bhatti, U. F., Brown, J. F., Biesterveld, B. E., Kathawate, R. G., Graham, N. J., et al. (2020). Early single-dose treatment with exosomes provides neuroprotection and improves blood-brain barrier integrity in swine model of traumatic brain injury and hemorrhagic shock. *J. Trauma Acute Care Surg.* 88, 207–218. doi: 10.1097/ta.0000000000002563

Wu, X., Zhang, H., Qi, W., Zhang, Y., Li, J., Li, Z., et al. (2018). Nicotine promotes atherosclerosis *via* ROS-NLRP3-mediated endothelial cell pyroptosis. *Cell Death Dis.* 9:171. doi: 10.1038/s41419-017-0257-3

Xia, P., Pan, Y., Zhang, F., Wang, N., Wang, E., Guo, Q., et al. (2018). Pioglitazone confers neuroprotection against ischemia-induced pyroptosis due to its inhibitory effects on HMGB-1/RAGE and Rac1/ROS pathway by activating PPAR. *Cell. Physiol. Biochem.* 45, 2351–2368. doi: 10.1159/000488183

Xiao, Y., Geng, F., Wang, G., Li, X., Zhu, J., and Zhu, W. (2019). Bone marrow-derived mesenchymal stem cells-derived exosomes prevent oligodendrocyte apoptosis through exosomal miR-134 by targeting caspase-8. *J. Cell. Biochem.* 120, 2109–2118. doi: 10.1002/jcb.27519

Yang, F., Qin, Y., Wang, Y., Meng, S., Xian, H., Che, H., et al. (2019). Metformin inhibits the NLRP3 inflammasome *via* AMPK/mTOR-dependent effects in diabetic cardiomyopathy. *Int. J. Biol. Sci.* 15, 1010–1019. doi: 10.7150/ijbs.29680

Yao, X., Yao, R., Huang, F., and Yi, J. (2019). LncRNA SNHG12 as a potent autophagy inducer exerts neuroprotective effects against cerebral ischemia/reperfusion injury. *Biochem. Biophys. Res. Commun.* 514, 490–496. doi: 10.1016/j.bbrc.2019.04.158

Zhang, D. M., Zhang, T., Wang, M. M., Wang, X. X., Qin, Y. Y., Wu, J., et al. (2019). TIGAR alleviates ischemia/reperfusion-induced autophagy and ischemic brain injury. *Free Radic. Biol. Med.* 137, 13–23. doi: 10.1016/j.freeradbiomed.2019.04.002

Zhang, Q., Zhou, M., Wu, X., Li, Z., Liu, B., Gao, W., et al. (2019). Promoting therapeutic angiogenesis of focal cerebral ischemia using thrombospondin-4 (TSP4) gene-modified bone marrow stromal cells (BMSCs) in a rat model. *J. Trans. Med.* 17:111. doi: 10.1186/s12967-019-1845-z

Zhou, Y., Gu, Y., and Liu, J. (2019). BRD4 suppression alleviates cerebral ischemia-induced brain injury by blocking glial activation *via* the inhibition of inflammatory response and pyroptosis. *Biochem. Biophys. Res. Commun.* 519, 481–488. doi: 10.1016/j.bbrc.2019.07.097

3

Mesenchymal Stromal Cells' Therapy for Polyglutamine Disorders: Where do we Stand and where Should we Go?

Inês Barros[1,2,3], *Adriana Marcelo*[1,2†], *Teresa P. Silva*[1,2†], *João Barata*[1,2†], *David Rufino-Ramos*[1,2,4], *Luís Pereira de Almeida*[1,2,4,5*] *and Catarina O. Miranda*[1,2,3*]

[1]CNC—Center for Neuroscience and Cell Biology, University of Coimbra, Coimbra, Portugal, [2]CIBB—Center for Innovative Biomedicine and Biotechnology, University of Coimbra, Coimbra, Portugal, [3]III—Institute for Interdisciplinary Research, University of Coimbra, Coimbra, Portugal, [4]Faculty of Pharmacy, University of Coimbra, Coimbra, Portugal, [5]Viravector—Viral Vector for Gene Transfer Core Facility, University of Coimbra, Coimbra, Portugal

***Correspondence:**
Catarina O. Miranda
csmiranda@cnc.uc.pt
Luís Pereira de Almeida
luispa@cnc.uc.pt

† Present address:
Adriana Marcelo,
Algarve University, Faro, Portugal
Teresa P. Silva,
IBB-Institute for Bioengineering and Biosciences and Department of Bioengineering, Instituto Superior Técnico, Universidade de Lisboa, Lisbon, Portugal
João Barata,
Institute for Genome Stability in Ageing and Disease, Cologne Cluster of Excellence in Cellular Stress Response in Aging-Associated Diseases (CECAD), University Hospital Cologne, Cologne, Germany

Polyglutamine (polyQ) diseases are a group of inherited neurodegenerative disorders caused by the expansion of the cytosine-adenine-guanine (CAG) repeat. This mutation encodes extended glutamine (Q) tract in the disease protein, resulting in the alteration of its conformation/physiological role and in the formation of toxic fragments/aggregates of the protein. This group of heterogeneous disorders shares common molecular mechanisms, which opens the possibility to develop a pan therapeutic approach. Vast efforts have been made to develop strategies to alleviate disease symptoms. Nonetheless, there is still no therapy that can cure or effectively delay disease progression of any of these disorders. Mesenchymal stromal cells (MSC) are promising tools for the treatment of polyQ disorders, promoting protection, tissue regeneration, and/or modulation of the immune system in animal models. Accordingly, data collected from clinical trials have so far demonstrated that transplantation of MSC is safe and delays the progression of some polyQ disorders for some time. However, to achieve sustained phenotypic amelioration in clinics, several treatments may be necessary. Therefore, efforts to develop new strategies to improve MSC's therapeutic outcomes have been emerging. In this review article, we discuss the current treatments and strategies used to reduce polyQ symptoms and major pre-clinical and clinical achievements obtained with MSC transplantation as well as remaining flaws that need to be overcome. The requirement to cross the blood-brain-barrier (BBB), together with a short rate of cell engraftment in the lesioned area and low survival of MSC in a pathophysiological context upon transplantation may contribute to the transient therapeutic effects. We also review methods like pre-conditioning or genetic engineering of MSC that can be used to increase MSC survival *in vivo*, cellular-free approaches—i.e., MSC-conditioned medium (CM) or MSC-derived extracellular vesicles (EVs) as a way of possibly replacing the use of MSC and methods required to standardize the potential of

MSC/MSC-derived products. These are fundamental questions that need to be addressed to obtain maximum MSC performance in polyQ diseases and therefore increase clinical benefits.

Keywords: mesenchymal stromal cells, polyglutamine disorders, neurodegenerative disorders, cell therapy, extracellular vesicles, secretome

INTRODUCTION

Polyglutamine (polyQ) diseases are a group of nine inherited neurodegenerative disorders including dentatorubral pallidoluysian atrophy, spinal bulbar muscular atrophy (SBMA), Huntington's disease (HD), and spinocerebellar ataxias (SCAs) type 1, 2, 3, 6, 7 and 17. All of them are autosomal dominant, apart from SBMA, which is X-linked (Orr and Zoghbi, 2007).

PolyQ disorders are associated with the unstable expansion of the cytosine-adenine-guanine (CAG) trinucleotide repeat in the respective causative gene. Such mutation encodes for extended glutamine (Q) tract, which occurs in different proteins according to the disorder, thus having unrelated functions and distinct cellular and subcellular locations. Nonetheless, all polyQ diseases share common features suggesting that the polyQ stretch directly contributes to the toxic properties of these proteins through a "toxic gain of function."

Although these disorders express distinct symptoms, all of them greatly impact a patient's quality of life by leading to both physical and psychological complications. Some of the shared common features include: onset typically during midlife; the inverse correlation between the number of repeats and the age of onset; degeneration of specific neuronal subpopulations that slowly progress over 10–20 years after the onset, culminating in premature death (Maciel et al., 1995; Ranum et al., 1995; Dürr et al., 1996; Zoghbi and Orr, 2000). The cerebellum, basal ganglia, brainstem nuclei, and spinal motor nuclei are some of the regions of the nervous system that are transversely affected in polyQ diseases (Zoghbi and Orr, 2000; Ross et al., 2003). The neuropathological alterations are the result of cellular changes such as accumulation of polyQ proteins in the cytoplasm and nucleus, the formation of inclusions, disturbance of the quality control systems of the cell, mitochondrial dysfunction, axonal transport disruption, synaptic activity decline, and neuroinflammation (Shao and Diamond, 2007; Havel et al., 2009). Additionally, the generation of toxic fragments resulting from the proteolytic cleavage has also been described in a few polyQs. The existence of common pathological mechanisms opens windows for the design of possible common therapies. In that sense, cell therapies and cellular-derived treatments are good candidates to be used for that purpose.

Currently, besides palliative care, treatments for polyQ disorders are only able to suppress or reduce specific symptoms. Levodopa or a dopamine agonist have been successful in cases displaying parkinsonian features, with dystonia and bradykinesia also being reduced through this approach (Tuite et al., 1995; Paulson, 2007). Furthermore, gait symptoms plus dysarthria can be improved by occupational and physical therapy, and dysphagia might be ameliorated by implementing strategies on what and how patients eat to avoid complications (Paulson, 2007). On the other hand, since pre-symptomatic testing already exists, medical and ethical guidance regarding treatment and parenthood counseling can be given, not only to post-symptomatic patients but also to pre-symptomatic individuals (Sequeiros et al., 1998; Drüsedau et al., 2004; Bettencourt and Lima, 2011; Schuler-Faccini et al., 2014).

Despite the efforts that have been made to develop effective therapies, there is still no strategies that can prevent, cure, or delay the progression of polyQ disorders.

Promising molecular and cellular therapies have been investigated in the last decades. Regarding gene therapy, the most frequently used method consists of silencing the defective protein resulting from the CAG expansion in a given gene (depending on the polyQ), mostly using RNA interference. However, there are challenges regarding delivery to the central nervous system and the risk of off-target effects. Significant progress has been made regarding this approach and such strategies will probably be implemented soon. Still, since the genetic defects differ from disorder to disorder, the strategy will require adaptations for each disease. As this subject is not the main focus of this review, we will not discuss it in detail, but this subject is reviewed in Matos et al. (2018).

Cellular therapies, on the other hand, may exert a pan effect on the common defective mechanisms that polyQ disorders share, as described above. For that, neural precursor and neural stem cells, embryonic stem cells, induced pluripotent stem cells, and mesenchymal stem cells (MSC) can be envisioned. Naturally, each source of the referred stem cells has its advantages, drawbacks, and success upon usage in polyQ diseases, as summarized elsewhere (Mendonça et al., 2018). Presently, MSC is not the first choice for cellular therapies in neurodegenerative conditions. Nonetheless, they are undoubtedly the safest kind among the different stem cells that can be used for therapeutic purposes since they are adult cells, and thus, not prone to develop tumors in a stable organism. Therefore, they have been widely used in studies of neurodegenerative disorders, including polyQs. However, despite some promising results, some aspects need to be improved to amplify MSCs healing effects and achieve translational significance. These concerns will be discussed in the present review article.

MESENCHYMAL STROMAL CELLS (MSC): PROMISING TOOLS FOR THE TREATMENT OF PolyQ DISORDERS?

Mesenchymal stromal cells (MSC) are adult multipotent progenitor cells capable of giving rise to tissues from the mesenchymal lineage. They were described in the 70's decade by Friedenstein et al. (1970, 1974) and since the discovery of their multipotency in 1997 (Prockop, 1997), their use in cellular therapies for neurodegenerative disorders suffered a steadfast evolution, due to the high potential/low-risk balance (**Figure 1**).

In neuronal context, their role was initially thought to be associated with the replacement of aged and damaged cells (Wakitani et al., 1995; Pittenger et al., 1999) but, currently, there is consensus that this is not their principal mode of action. Cell fusion was also described to occur by some authors. Of note, this mechanism was described in a SCA1 mouse model, where Purkinje cells (PCs) of the host animals fused with MSC (Huda et al., 2016). However, this mechanism is rare in the literature and does not appear to account significantly for MSC's overall outcomes. So MSC's main effects are probably mainly mediated by influencing neighboring cells, through secretion of bioactive factors, or by inducing secretion of these factors in host cells, thus being able to modulate the immune system and to promote tissue repair (Gao et al., 2001; Tremain et al., 2001). The establishment of strong communication through chemical signals between MSC and surrounding cells, which can be either other MSC, neurons, astrocytes, or glia, is thought to be the anchor for MSC's therapeutic capabilities. Indeed, MSC can respond to stress "signals" of the nearby cell by modulating their molecular pathways towards the production of specific factors, which can support these cells under stress (Millard and Fisk, 2013). As an example, a mechanism called "bio bridge" in which MSC seem to form a stream of extracellular matrix (bridge) between the lesion site and neurogenic niches was described for some neurological contexts such as traumatic brain injury and stroke, and could occur in other neurological disorders (Tajiri et al., 2013; Sullivan et al., 2015). This bridge is concomitant with higher expression levels of metalloproteinases (MMP) such as MMP9, leading to better communication between MSC and endogenous neuronal precursors and increment in neurogenesis post-lesion. Precursors cells from neurogenic niches then reach the brain-lesioned tissue, replacing injured cells. For all these mechanisms, there is evidence of bidirectional communication exerted through bioactive soluble molecules [neurotrophins, cytokines, chemokines, microRNAs (miRNAs), etc.], extracellular vesicles (EVs; that can contain proteins, RNAs, or even DNA) or by direct contact (e.g., *via* tunneling nanotubes or through mechanotransduction). Therefore, depending on the defects in the host damaged tissue, MSC may: (1) modulate inflammatory processes; (2) reduce oxidative stress, either by inducing survival pathways or by the direct transfer of healthy mitochondria to the host cells (*via* nanotubes); (3) favor neurogenesis by the secretion of neurotrophins and by the formation of "bio bridges"; (4) induce gliogenesis and remyelination; and (5) increase axonal survival and plasticity, thus inducing synaptogenesis (Paul and Anisimov, 2013; **Figure 2**). These exquisite cross-talks lead to a wide evaluation of MSC for the therapy of neurological diseases in preclinical and clinical models.

From Alzheimer's (AD) to Parkinson's (PD) or HD, the encouraging effects of MSC in a few pre-clinical studies prompted clinicians to perform preliminary clinical trials to evaluate their safety and/or effectiveness. However, this process started before fundamental issues were properly addressed at the pre-clinical level, which led to some disappointing results relative to the ones expected. Due to the initial lack of information, strategies did not contemplate solutions for problems such as the challenge of surpassing the blood-brain barrier (BBB), the low rate of cells engraftment in the lesioned tissue, the low survival of MSC, or the unidentified mechanisms involved in MSC's positive effects. Finally, the standardization of MSC source of cells or even of methods capable of evaluating their potential, are imperative to make translation possible.

The investigation in this field is therefore currently aiming at resolving these difficulties and giving answers to the urgent need of efficacious therapies for neurodegenerative disorders for which therapeutic tools are presently scarce.

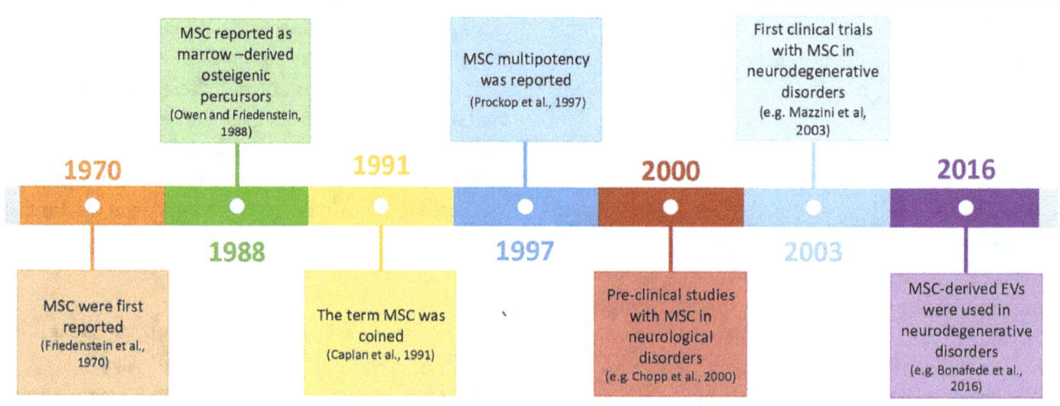

FIGURE 1 | Timeline of mesenchymal stromal cells (MSC)'s history and their use in neurodegenerative disorders (Friedenstein et al., 1970; Owen and Friedenstein, 1988; Caplan, 1991; Prockop, 1997; Chopp et al., 2000; Mazzini et al., 2003; Bonafede et al., 2016).

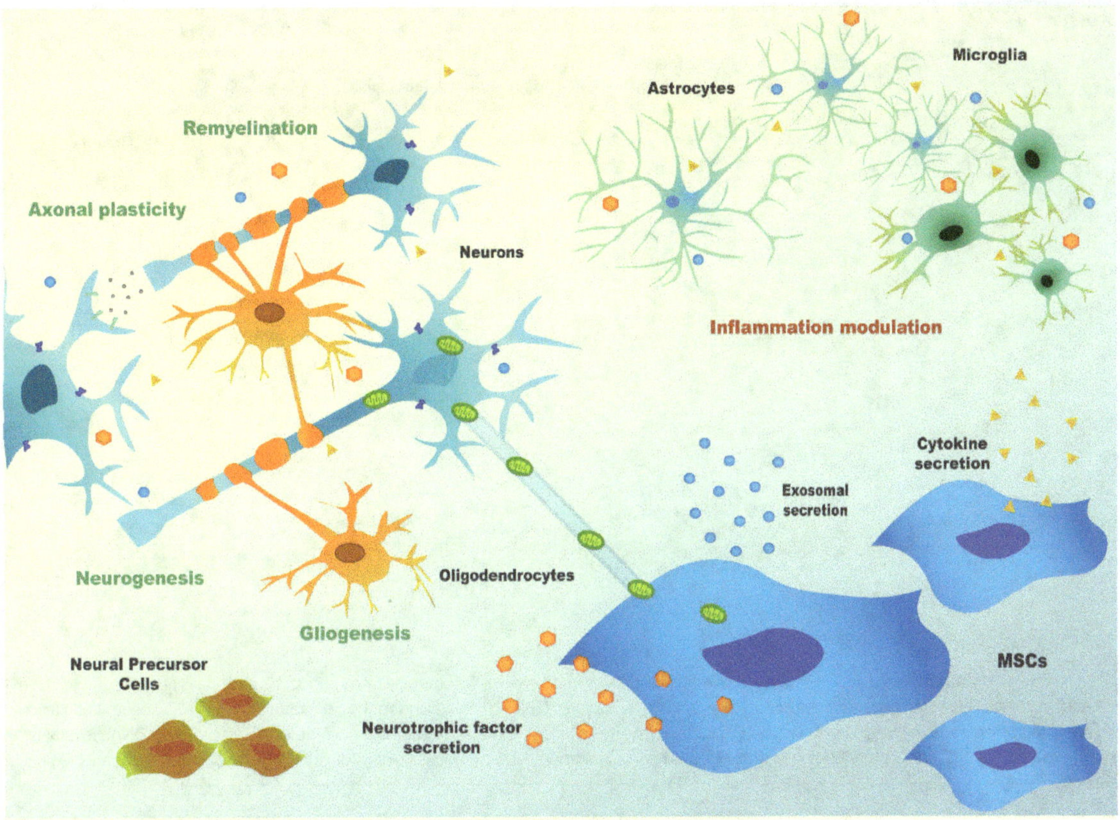

FIGURE 2 | MSC's paracrine mechanism(s) in neuronal cells.

This review gives an overview of this subject with a particular focus on polyQ disorders, which besides HD, are scarcely referred to in the literature.

Pre-clinical Studies Assessing MSC's Therapeutic Potential in PolyQs

Several pre-clinical studies have investigated the therapeutic efficiency of MSC isolated from different sources, including bone marrow (BM-MSC), adipose tissue (AD-MSC), and umbilical cord (UC-MSC), in rodent models of HD. HD is the polyQ disease with the highest prevalence worldwide affecting about 1 in 7,500 individuals (Evans et al., 2013; Fisher and Hayden, 2014). HD causes brain atrophy in several regions such as the striatum, thalamus, cerebellum, brain stem, and cortex (Harper et al., 2005; Hassel et al., 2008; Labbadia and Morimoto, 2013; Chao et al., 2017) leading to progressive motor dysfunction and incoordination, cognitive impairment and psychiatric symptoms.

Over the last decade, it has been demonstrated that MSC can relieve phenotype and neuropathology of HD in both transgenic (Lee et al., 2009; Im et al., 2010; Snyder et al., 2010; Lin et al., 2011; Yu-Taeger et al., 2019) and chemically-induced models (Lee et al., 2009; Edalatmanesh et al., 2011; Rossignol et al., 2011, 2015; Hosseini et al., 2015; Ebrahimi et al., 2018). These studies show that animals treated with MSC displayed improved behavioral performance, cognitive functions, and, in the excitotoxic Quinolinic Acid (QA)-induced HD rats, reduction of apomorphine-induced rotation. Hosseini et al. (2015) also showed that MSC was able to reduce anxiety levels in treated QA-induced HD rats. Also, the administration of MSC was able to increase the survival of the R6/2 mouse model (Lee et al., 2009; Lin et al., 2011; Yu-Taeger et al., 2019). Importantly, one of these studies pointed out the importance of well-dosing the number of MSC administered. The authors compared two different doses of MSC (2×10^5 and 4×10^5) and surprisingly only the group treated with the lowest dose presented motor improvements (Rossignol et al., 2011). The authors concluded that a high number of MSC may cause tissue damage to striatal architecture.

Regarding neuropathological improvement, MSC preserved the volume of the striatum, induced neurogenesis, and differentiation of the host endogenous neural stem cells, and reduced neuronal death (Snyder et al., 2010). Lin et al. (2011) saw a decrease in the levels of Bax and caspase-3 as well as an up-regulation in the expression of Erk1/2, which suggests that MSC can inhibit apoptosis and promote neuronal survival. Furthermore, this study points out the strong capability of MSC for recruiting microglia, neuroblasts, and bone marrow-derived cells to the lesion site which helps tissue regeneration (Lin et al., 2011). MSC administration has also

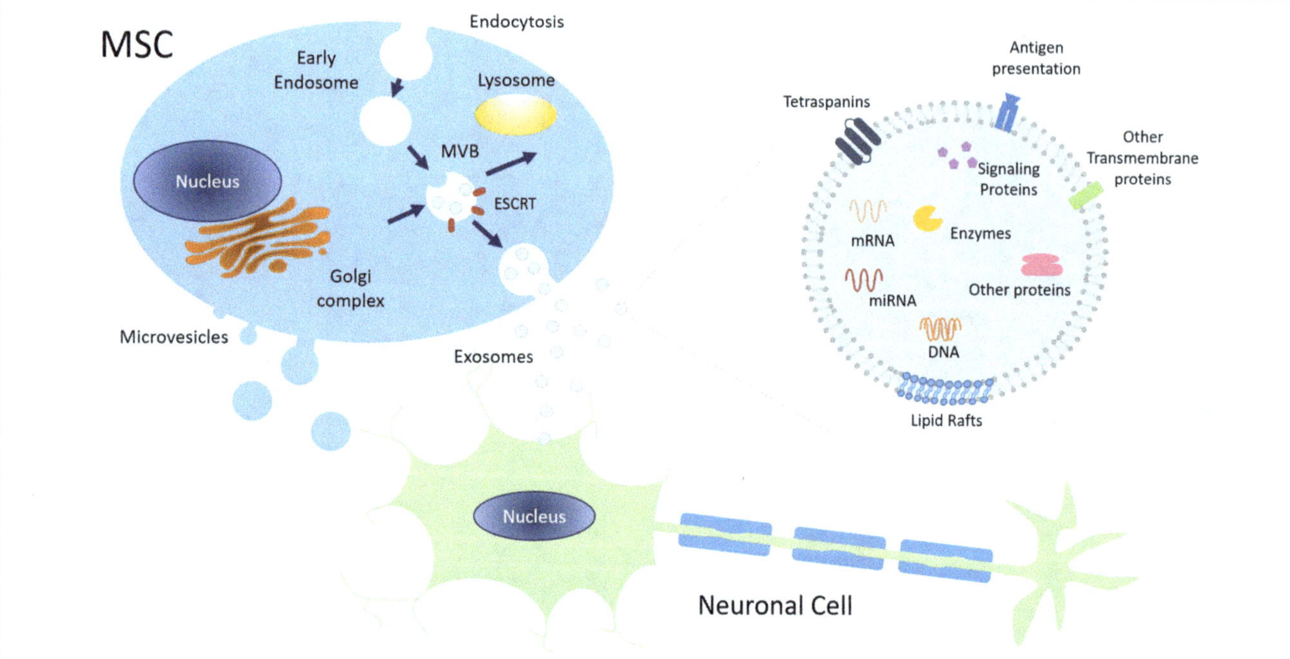

FIGURE 3 | Schematic representation of exosome biogenesis and molecular composition. Exosomes have an endocytic origin and are released to the extracellular milieu upon the fusion of MVB with the plasma membrane. Other MVB may fuse with lysosomes leading to their degradation. Overall, exosomes surface markers, cytoskeleton, and cellular metabolism proteins are common to those of the cell of origin. Plus, exosomes also include signaling proteins, DNA, several types of RNA, and lipids. Abbreviations: ESCRT, Endosomal Sorting Complex Required for Transport; MVB, Multivesicular bodies; mRNA, messenger RNA, miRNA, microRNA.

led to the reduction of the number of huntingtin aggregates (Lee et al., 2009).

Notably, Yu-Taeger et al. (2019) showed that MSC could ameliorate disruptions in the dopaminergic signaling cascade, increasing the expression of DARPP-32 and tyrosine hydroxylase in the striatum, known to be reduced in R6/2 HD mice (Bibb et al., 2000). This study also evidenced the immunomodulatory potential of MSC as treated mice restored the expression of Iba1, and microglia morphology in the striatum, plus inducing the downregulation of several inflammatory modulators (Yu-Taeger et al., 2019).

The feasibility of performing autologous transplants in HD has also been assessed. Researchers compared the efficacy of AD-MSC isolated from the subcutaneous adipose tissue of an HD patient and a healthy donor in the treatment of YAC128 transgenic mice, a long-lived HD model (Im et al., 2010). Even though both kinds of cells expressed the same cell surface markers and showed a similar secretory potential in terms of neurotrophic factors, normal AD-MSC could reduce striatal atrophy when they were transplanted into 8 month-old YAC128 mice, whereas HD AD-MSC failed to promote such benefits. Nonetheless, the injection of cells in this time-window did not produce improvements in any of the treated groups regarding rotarod performance or body weight. On the contrary, when transplanting normal AD-MSC at 12 months of age, there were significant improvements in motor performance. These results show that AD-MSC could modify the course of HD in long-lived mice, though their effect is dependent on the donor condition and time window of delivery.

More recently, two studies assessing how the number of passages of MSC in *ex vivo* cultures can affect their therapeutic effects were performed in R6/2 transgenic mice (Fink et al., 2013; Rossignol et al., 2015). In the first study, BM-MSC isolated from mice were grown for a low (3–8) or high (40–50) number of passages, and no neuronal or glial differentiation was observed. The striatum of R6/2 mice transplanted with BM-MSC at higher passage showed increased metabolic activity compared with non-treated R6/2 mice and significantly-improved behavioral sparing compared to either untreated R6/2 mice or R6/2 mice receiving low-passage BM-MSC. Thus, the passage number of BM-MSC expansion could alter their efficacy.

In the second study, they compared the therapeutic potential of low- (3–8) and high-passage (40–50) UC-MSC in the R6/2 transgenic HD mouse model (Fink et al., 2013). Interestingly, high-passage UC-MSC showed a higher survival rate *in vivo*, but contrary to the previous study, both transplanted groups showed increased metabolic activity and reduced brain damage in the striatal area. Moreover, no major differences were observed in the motor and non-motor behavioral tests compared with non-treated HD mice. In contrast, Ebrahimi and collaborators showed that the administration of UC-MSC at a low passage in 3-NP lesioned rat models led to significant improvement of motor performance and muscle activity. Plus, they also observed neuropathological ameliorations as UC-MSC were able to decrease gliosis and increase both the striatal volume and dendritic length of striatal cells (Ebrahimi et al., 2018). So, as a conclusion, the time of MSC expansion and concomitant number of passages is an important parameter and should be further and carefully addressed in future assessments.

The relief of phenotypical and neuropathological defects in HD models has been associated with MSC trophic effect. Several studies reported that MSC transplantation increased the expression of both messenger RNA (mRNA) and protein levels of brain-derived neurotrophic factor (BDNF) in the striatum (Snyder et al., 2010). Additionally, Snyder et al. (2010) reported that BM-MSC transplantation led to increased expression of the neurotrophic factors fibroblast growth factor (FGF), ciliary neurotrophic factor (CNTF), vascular endothelial growth factor (VEGF), and nerve growth factor (NGF). Over-expression of collagen type-I and fibronectin in the striatum of MSC-treated animals has also been observed which contributes to the stabilization of the extracellular matrix, helping tissue regeneration (Rossignol et al., 2011).

Regarding other polyQ disorders, few preclinical studies have been performed in mouse models of SCA1, SCA2, and SCA3. Despite the causative mutation being located in different genes, an expansion of the CAGs in *ATXN1*, *ATXN2*, and *ATXN3* (also known as *MJD1*), respectively, many symptoms and major mechanisms involved in disease progression are similar among SCAs (Paulson, 2009; Klockgether et al., 2019; Sullivan et al., 2019). Over time, patients develop motor incoordination as the main progressive clinical feature, along with tremors, muscle rigidity, dysarthria, dysphagia, peripheral neuropathy, oculomotor problems (such as double vision, nystagmus or ophthalmoplegia), and parkinsonism in some cases (Paulson, 2009). Cognitive impairments have also been pointed out to occur.

MSC's effect has shown to be equally promising in these groups of polyQs, demonstrating that they can alleviate motor function deterioration in SCAs. A pre-symptomatic study was performed in a transgenic mouse model of SCA-1 (SCA1-Tg mice). Intrathecal (IT) injection of an MSC line (KUM10) in SCA1-Tg mice led to neuropathological improvements. MSC was shown to migrate towards the cerebellum and mitigate neuronal disorganization. The cerebellum of these SCA1 transgenic mice had ectopically-located PCs bodies that resulted in a multilayer of PC, and mice showed atrophy of the cerebellar molecular layer (ML). In the referred study, MSC-treated mice displayed a single layer of PC and the ML was thicker than that of the untreated mice. Motor function assessed by the rotarod test was also shown to be improved in treated animals (Matsuura et al., 2014).

Additionally, MSC showed to mitigate motor functions deterioration in a SCA2 transgenic mouse model (Chang et al., 2011). The study compared the efficacy of repeated intravenous (IV) or intracranial (IC) injections of human BM-MSC and concluded that IV infusions effectively delayed diseases' onset and improved motor performance by rescuing cerebellar PC, whereas IC injections failed to improve motor skills. In this study, more grafted BM-MSC were found in the IV transplantation group, located in the cerebellum and cerebral cortex (CC). The authors explained the higher success of IV delivered BM-MSC due to the extravasation capacity as cells migrated from intracranial vessels through white matter (WM) into several regions of the cerebellum.

Our group has recently demonstrated that mouse BM-MSC transplantation can ameliorate both phenotypical and neuropathological MJD symptoms in Machado–Joseph disease (MJD)/SCA3 (Oliveira Miranda et al., 2018). In agreement with the previous study, our investigation showed that the administration of several consecutive IV injections of BM-MSC sustainably ameliorated motor impairments, when delivered post-symptomatically. Using intracerebroventricular administration (ICV), a single injection could only alleviate MJD symptoms transiently in the same Tg-ATXN3-69Q transgenic mouse model of MJD (Torashima et al., 2008). This suggests that repeated treatment through the IV route led to better outcomes. Notably, our results were in line with Li et al. (2018) findings, which showed that repeated systemic treatment of pre-symptomatic transgenic homozygous MJD mice [SCA3-YAC-84Q; (Cemal et al., 2002)] with human UC-MSC was able to alleviate motor impairments and neuropathology of MJD.

These findings were corroborated with low survival and poor engraftment in host tissues, which represent a common problem in these therapies. In our study, we demonstrated that after ICV transplantation, despite cells surviving up to 8 weeks after transplantation, the SPION-labeled MSC signal was drastically reduced from 1 to 8 weeks *in vivo*. Moreover, BM-MSC delivered IV was detected in the brain of mice 30 min after transplantation, but completely disappeared from circulation before completion of 7 days (Oliveira Miranda et al., 2018). Similarly, other studies in HD animal models also observed that MSC remained only for a short time in the lesion site (Snyder et al., 2010), or in other cases, although the authors could detect cells in the brain, only a small percentage of the implanted cells could survive (Lee et al., 2009; Im et al., 2010; Snyder et al., 2010; Lin et al., 2011; Yu-Taeger et al., 2019). A concise review of these studies is given in (**Table 1**).

Interestingly, these studies in SCAs also relate the action of MSC to neuronal survival/activity pathways, which is following the previous studies performed in HD rodent models. In the study of Li et al. (2018), the neuroprotective action of MSC was linked with the upregulation of the insulin-like growth factor 1/heat shock protein 70 (IGF-1/HSP70) pathway. In our study, we demonstrated that MSC protected GABAergic and Glutamatergic neurons, as by proton magnetic resonance spectroscopy (1H-MRS) we detected an increased expression of Gaba and the complex Glutamine-Glutamate upon a treatment, along with increased levels of mRNA for the receptors Gabrb2 and Grml in the cerebellum, denoting a higher synaptic activity (Oliveira Miranda et al., 2018).

In conclusion, despite the general agreement on the fact that MSC promotes beneficial outcomes in polyQ diseases, the different severity grades of the rodent models used as well as the ability to simulate human diseases make the interpretation of the potency scale of the treatments difficult. The development of models that better recapitulate polyQ diseases neuropathology in humans would be essential. Thus, knock-in models of larger mammals like the HD CRISPR-Cas9 knock-in pig model that shows strong neurodegeneration patterns (Yan et al., 2018) could

TABLE 1 | Sum up of the survival, grafting, and integration of mesenchymal stromal cells (MSC) in pre-clinical studies in animal models of polyglutamine (polyQ) disorders.

PolyQ	Study	Animal model	MSC's source	N° of cells injected	Site of injection	Time of injection	Tracking time	Grafting sites	Integration and neuronal differentiation
HD	Bantubungi et al. (2008)	QA-induced rat striatal model	rat BM-MSC	1×10^5	Unilateral injection in the striatum	1-week after the QA lesion	8 weeks after transplantation	$1.65 \pm 0.18 \times 10^5$ BM-MSC found at the lesion site in the striatum at 3 and 8 weeks	BM-MSC remained undifferentiated
	Lee et al. (2009)	QA-induced rat striatal model	human AD-MSC	1×10^6	Unilateral injection in the striatum	immediately after the QA injection	4 weeks after transplantation	AD-MSC found near the initial transplantation foci, forming a lump	Co-labeling with BDNF, calbindin, GABA, and GAD
		R6/2 transgenic mice*	human AD-MSC	0.5×10^6	Bilateral injection in the striatum	8.5-weeks-old mice (post-symptomatic treatment)	4 weeks after transplantation	Few AD-MSC migrated toward the dorsolateral subventricular area, or into the striatum; mean survival rate: 13.5 ± 1.4% (*n* = 3)	low rate of *in vivo* proliferation of AD-MSC; expression of nestin (~70%), doublecortin (~80%), Tuj-1 (~25%), and GAD (~80%)
	Im et al. (2010)	YAC128 transgenic mice	human AD-MSC isolated from an HD patient and a healthy donor	5×10^5	Bilateral injection in the striatum	8-months-old (pre-symptomatic) and 12-months-old (symptomatic)	4 weeks after transplantation	No cells detected extremely low % of cell survival AD-MSC found in the striatum and in the periventricular area	ND
	Snyder et al. (2010)	N171–82Q transgenic mice	human BM-MSC isolated from healthy donors	0.1×10^5	Unilateral injection in the striatum	8-week-old mice (pre-symptomatic)	2–3 weeks after transplantation	Grafted BM-MSC do not survive long in the striatum (disappeared over 3–15 days); Only 15.1% survived in the first 24 h (15.063 ± 3.776)	BM-MSC did not proliferate, but induced neurogenesis and neural differentiation of the host endogenous neural stem cells
	Lin et al. (2011)	QA-induced mouse striatal model	Immortalized human BM-MSC	2×10^5	Unilateral injections in the striatum	7 days after the QA lesion	8 weeks after transplantation	BM-MSC found in the striatum; a small amount of BM-MSC survived for 16 weeks	Some BM-MSC co-localized with GFAP, NeuN, and DARPP-32

(Continued)

TABLE 1 | Continued

PolyQ	Study	Animal model	MSC's source	N° of cells injected	Site of injection	Time of injection	Tracking time	Grafting sites	Integration and neuronal differentiation
		R6/2 transgenic mice	Immortalized human BM-MSC	2×10^5	Unilateral injections in the striatum	12-week-old mice (post-symptomatic)	12 weeks after transplantation	BM-MSC found in the striatum 12 weeks after transplantation	Some BM-MSC differentiated: co-localization with DARPP-32, GFAP, and F4/80.
	Rossignol et al. (2011)	3NP-induced rat model	Rat BM-MSC	0.2×10^6 or 0.4×10^6	Bilateral injection in the striatum	28 days after the first administration of 3NTP	~10 weeks after transplantation	BM-MSC were visible in the implantation site	No neuronal differentiation of BM-MSC was observed
	Fink et al. (2013)	R6/2 transgenic mice	UC-MSC (from WT C57/BL6 mice pups)	0.4×10^5 per hemisphere	Bilateral injection in the striatum	5-week-old mice (post-symptomatic)	6 weeks after transplantation	UC-MSC found in the striatum 6 weeks post-transplantation; Mice transplanted with high-passage UC-MSC showed more surviving cells than mice treated with low-passage UC-MSC	Barely any co-localization of UC-MSC with NeuN was found
	Hosseini et al. (2015)	QA-induced rat striatal model	Human AD-MSC purified from healthy male donors	2×10^5	Unilateral injection in the striatum	7 days after the QA lesion	7 weeks	AD-MSC found in the brain	AD-MSCs far migrated into other brain regions (even into the cortex)
	Fink et al. (2013)	QA-induced rat cerebellar model	rat BM-MSC	2.5×10^5	Injection into the cerebellum (folia VI)	48 h after the QA lesion	6 weeks after transplantation	BM-MSC found in the cerebellum; a few cells could migrate to deep layers	BM-MSC seemed to be integrating within host tissue
	Rossignol et al. (2015)	R6/2 transgenic mice	mouse BM-MSC	2 injections of 0.2×10^5 (total of 0.4×10^5)	Bilateral injection in the striatum	5-week-old mice (post-symptomatic)	6 weeks after transplantation	Both low-passage and high-passage BM-MSC were capable of surviving in the brain for up to the 6 weeks of following	No neuronal or glial differentiation was observed, as BM-MSC could not co-localize with the markers NeuN or GFAP.

(Continued)

TABLE 1 | Continued

PolyQ	Study	Animal model	MSC's source	N° of cells injected	Site of injection	Time of injection	Tracking time	Grafting sites	Integration and neuronal differentiation
	Ebrahimi et al. (2018)	3NP rat model	UC-MSC	2.5×10^5	Bilateral injection in the striatum	7 days after 3-NP administration	1 month after injection	UC-MSC found in the striatum 30 days post-transplantation;	ND
	Yu-Taeger et al. (2019)	R6/2 transgenic Model	mouse BM-MSC	1×10^6	Intranasal transplantation	4-week-old mice (post-symptomatic)	5 days and 7.5 weeks post-delivery of MSC	BM-MSC were found in the midbrain, striatum, and olfactory bulb 5 days post-transplantation At 7.5 weeks no signal was detected	ND
SCA1	Matsuura et al. (2014)	SCA1 transgenic mice (B05 line)	mouse BM-MSC	6×10^5	Injection into the Meninges Covering the Cerebellum	5-week-old mice (early symptomatic)	1 h and 3 days after administration	1 h—BM-MSC were in between the meninges 3d— BM-MSC were essentially confined to the cerebellum in lobules 3, 4, 5, and 6 and in the spaces between the folia	ND
SCA2	Chang et al. (2011)	SCA2 transgenic mice	Human clonally derived BM-MSC	IV: 4.2×10^7/kg	IV	IV: 12, 23, 33 and 42-week-old mice	IV: 8 weeks after the last injection	IV: BM-MSC found in the cerebellar WM, ML, and lumens of blood vessels in the WM. Large clusters of grafted hMSC were detected in the CC.	ND
		SCA2 transgenic mice	Human clonally derived BM-MSC	IC: 8.4×10^6/kg	IC (through the foramen magnum, into the position of the cerebellum)	IC: 12, 23, and 33-week-old mice	IC: 17 weeks after the last injection	IC: BM-MSC not detected over cerebellar WM, ML, or PC layer, but limited to a few lumens of blood vessels and a few scattered cells in the CC.	ND

(Continued)

TABLE 1 | Continued

PolyQ	Study	Animal model	MSC's source	N° of cells injected	Site of injection	Time of injection	Tracking time	Grafting sites	Integration and neuronal differentiation
SCA3/MJD	Oliveira Miranda et al. (2018)	SCA3/MJD transgenic mice (Tg-ATXN3-69Q model)	mouse BM-MSC	ICV: 3×10^5	ICV (unilateral injection in the lateral ventricle of the brain)	ICV: Unilateral injection in the lateral ventricle of the brain (4–7.5-week-old mice)	ICV: 7 weeks after administration	ICV: BM-MSC found in the lateral ventricles, but the volume of MSC graft drastically reduced from 1 to 4 weeks after injection (55% volume reduction)	ND
		SCA3/MJD transgenic mice (Tg-ATXN3-69Q model)	mouse BM-MSC	IV: $4.5–8 \times 10^7$/Kg	IV	IV: every 2–3 weeks, four consecutive times; first treatment at 4–6.5-weeks-old	Iv: 7 days	IV: BM-MSC engrafted mainly in the lungs but could also reach the brain parenchyma however remained there just for the first hours.	ND
	Li et al. (2018)	(YAC transgenic MJD model, homozygous-84Q)	Human UC-MSC	2×10^6	IV	Treatment started at 3-months-old mice (pre-symptomatic) and consisted of 2×10^6/week, within 12 weeks	3 months	UC-MSC found in the pons, PCL, and ML of the cerebellum 3 months after transplantation	UC-MSC did not co-localize with NeuN, GFAP, or calbindin

List of abbreviations: HD, Huntington's disease; QA, quinolinic acid; BM-MSC, Bone marrow-derived mesenchymal stromal cells; AD-MSC, adipose tissue-derived mesenchymal stromal cells; BDNF, Brain-derived neurotrophic factor; GABA, gamma-aminobutyric acid; GAD, glutamic acid decarboxylase; YAC, yeast artificial chromosome; ND, not determined; GFAP, glial fibrillary acidic protein; NeuN, neuronal nuclei; DARP-32, dopamine- and cAMP-regulated neuronal phosphoprotein; F4/80, major macrophage marker; UC-MSC, umbilical cord-derived mesenchymal stromal cells; 3-NP, 3-nitropropionic acid; SCA1, Spinocerebellar ataxia type-1; SCA2, Spinocerebellar ataxia type-2; IV, Intravenous injections; IC, Intracerebral injection; WM, white matter; ML, molecular layer; PCL, Purkinje cell layer; CC, cerebral cortex; SCA3/MJD, Spinocerebellar ataxia type-3 or Machado-Joseph disease; ICV, Intracerebroventricular injection.

be an important contribution to the field to validate findings achieved with rodent models and test a possible replacement of degenerated neurons promoted by MSC.

Furthermore, different cell sources, heterogeneous methods of isolation, discrepancies in MSC origins (i.e., from healthy/disease tissues) and characteristics (cell passage number, conditions used during cell growth) have been used throughout the studies, which makes it difficult to reach a correct comprehension and make assertive conclusions with enough confidence to implement these therapies in the clinics.

Clinical Studies in PolyQ Patients: Achievements and Flaws

Despite the extensive number of unanswered issues from the pre-clinical assessments herein referred, the promising results obtained from research on the use of MSC in polyQ soon led to the investigation of their therapeutic application in clinical studies. The clinical trials, mostly phase I/II, aiming at testing the safety and effectiveness of MSC in the treatment of polyQ disorders (completed and ongoing clinical trials—ClinicalTrials.gov Identifiers: NCT02728115, NCT03252535, NCT04219241, NCT03378414 for HD and NCT01360164, NCT01489267, NCT01649687, NCT02540655 for other polyQs/SCAs) are summarized in **Table 2**.

Data collected from four of these clinical trials have demonstrated that the therapeutic application of MSC is safe, does not produce severe side effects and might delay the progression of disease symptoms (Dongmei et al., 2011; Jin et al., 2013; Miao et al., 2015; Tsai et al., 2017). In one of these studies, fourteen patients diagnosed with SCA and ten with multiple system atrophy-cerebellar type C (MSA-C) were treated weekly with IT injections of UC-MSC at a dose of 1×10^6/kg during four weeks (except for three patients who received two courses of treatment, all the other patients received only one course). Furthermore, the neuronal function and quality of daily life were assessed through the International Cooperative Ataxia Rating Scale (ICARS) and Activity of Daily Living Scale (ADL). Improvements in ICARS and ADL scores were observed after 1 month of treatment (Dongmei et al., 2011). In another study, Jin et al. (2013) demonstrated that MSC treatment could alleviate SCA symptoms for at least half a year. In this study, 16 patients diagnosed with SCA1, 2 or 3, received IV or IT injections of UC-MSC (first treatment: 4×10^7 UC-MSC were infused iv.; following three treatments: 2×10^7 UC-MSC were infused IV and simultaneously 2×10^7 UC-MSC were infused by IT administration; all patients received only one-course treatment). Results showed that most patients obtained improved scores on the Berg Balance Scale (BBS), which evaluates patients' balance, and ICARS at 3 and 6 months after treatment (Jin et al., 2013). A different clinical trial using UC-MSC was performed to evaluate the technical difficulties underlying IT UC-MSC injection *via* lumbar puncture and assess the effects of the cell in different neurodegenerative disorders, including eight SCA patients (Miao et al., 2015). The UC-MSC was injected 4–6 times between the L4 and L5 interspace within 5–7 days (single course of treatment). All patients were followed-up for more than 1 year for clinical status. Three of the eight patients showed motor function improvement after treatment. Recently, Tsai et al. (2017) intravenously administered ($5–7 \times 10^7$) allogeneic MSC derived from the adipose tissue (AD-MSC, single injection) into six patients with MJD and one with MSA, who showed a slight trend of improvement and stabilization of the disease according to the Scale for Assessment and Rating of Ataxia (SARA) and the sensory organization testing (SOT).

Importantly, no major side effects were reported, besides dizziness, headache, back pain in case of IT injections, or fever (Dongmei et al., 2011; Jin et al., 2013; Miao et al., 2015; Tsai et al., 2017). Consistently with these results, no adverse effects were reported so far in any of the pre-clinical studies. Moreover, in our study, we assessed for the levels of hepatic aminotransferases, which were not increased when compared to non-treated or WT mice, denoting the absence of liver toxicity in mice treated repeatedly with MSC (Oliveira Miranda et al., 2018).

Altogether, the studies reinforce the safety and tolerability of MSC. Furthermore, they demonstrate that treatment with allogenic UC-MSC may delay disease progression, improving motor performance, and patients' quality of life without severe adverse reactions. However, the regression of some patients to the status before the treatment was reported (Dongmei et al., 2011; Jin et al., 2013; Miao et al., 2015; Tsai et al., 2017; summarized in **Table 2**). This was also following the need for performing several treatments suggested by some of the pre-clinical studies (Chang et al., 2011; Li et al., 2018; Oliveira Miranda et al., 2018). Nonetheless, in the clinical context, periodic MSC injections may not be feasible as they could lead to unwanted side effects. Indeed, some studies highlighted the long-term risk that could emerge from the repeated therapeutic transplantation of these cells, namely maldifferentiation, immunosuppression (increasing the risk of opportunistic infections), liver and lung accumulation, and malignant tumor growth promotion in patients with pre-existing malignancy (Sundin et al., 2006; Dierks et al., 2007; Kunter et al., 2007; Ning et al., 2008). On the other hand, as previously discussed, the local administration of MSC may cause brain damage if administered above a certain number (Rossignol et al., 2011). Therefore, it would be important to further clarify the mechanism involved in MSC therapeutic effects, uncovering new potential treatments that do not require the infusion of cells.

On the other hand, these clinical trials often mix patients with different conditions/stages of the disease and frequently include all disorders when reporting results, making reports not as informative as desired. Future clinical studies should focus not only on the feasibility of repeated treatments for sustained benefits but also on the appropriate timing, dosage and best method of injection to be used, while also performing longer follow-ups of all patients to evaluate treatment efficacy and safety, and with placebo-controls being used. For that purpose, established cohorts, well-defined and characterized must be performed so that clinical studies can be well-designed and as instructive as possible. An example of such a cohort has been implemented by the European Consortium ESMI ("European Spinocerebellar Ataxia Type 3/Machado-Joseph

TABLE 2 | Concise information on the outcomes obtained from clinical trials performed with MSCs in patients with PolyQ diseases.

Trial identification	Disease	Type of MSC	Route of administration	Outcomes
Dongmei et al. (2011)	SCA and MSA-C	UC-MSC	IT	ICARS and ADL (Katona et al. (2008)) scores significantly decreased 1 month after treatment. Unstable walking and standing, slow movement, fine motor disorders of the upper limbs and writing difficulties, and dysarthria were greatly improved except for 1 patient, who had no response. Ten cases (42%) remained stable for half a year or longer, while 14 cases (58%) had regressed to the status before the treatment within 1–14 months (an average of 3 months).
NCT01360164 Jin et al. (2013)	SCA1, SCA2, and SCA3	UC-MSC	IV and IT	After 1 year of treatment: 44% of the patients exhibited improved Berg Balance Scale (BBS) over baseline, and only 31% of the patients suffered from disease aggravation; 63% of the patients exhibited improved ICARS over baseline, and only 25% of the patients suffered from disease exacerbation.
Miao et al. (2015)	Several neurodegenerative disorders, including SCAs	UC-MSC	IT	Patients were followed-up for more than 1 year. No significant side effects were reported. Three of the 8 patients with SCAs (37.5%) showed motor function improvement 1 year after the treatments.
NCT01489267	SCA1	UC-MSC	IT	Unknown
NCT01649687 Tsai et al. (2017)	SCA3 or MSA-C	AD-MSC	IV	Patients were monitored for 1 year with no reported adverse events. They showed an initial trend toward brief improvement, followed by a slight stabilization and a progression of the disease in later stages. At 6 months follow-up, SARA scores were improved in 2 (33.3%), unaltered in 2 (33.3%), and worsened in 2 (33.3%) patients as compared with the baseline. At 1-year follow-up, SARA scores were improved in 1 (16.6%), unaltered in 3 (50%), and worsened in 2 (33.3%) patients as compared with the baseline.
NCT02540655	Cerebellar Ataxias	AD-MSC (Stemchymal®)	IV	Unknown.
NCT02728115	HD	Cellavita*	IV	Active, not recruiting
NCT03252535	HD	Cellavita	IV	Active, not recruiting
NCT04219241	HD	Cellavita	IV	This study is not yet open for participant recruitment.
NCT03378414	SCA1, SCA2, SCA3 and SCA6	UC-MSC	IV	Unknown

*List of abbreviations: SCA, Spinocerebellar ataxia; MSA-C, and multiple system atrophy-cerebellar types; SCA1, Spinocerebellar ataxia type-1; SCA2, Spinocerebellar ataxia type-2; SCA3, Spinocerebellar ataxia type-3 or Machado-Joseph disease; SCA6, Spinocerebellar ataxia type-6; UC-MSC, umbilical cord-derived mesenchymal stromal cells; AD-MSC, adipose tissue-derived mesenchymal stromal cells; IT, Intrathecal injection; IV, Intravenous injections; ICARS, International Cooperative Ataxia Rating Scale; ADL, Activity of Daily Living; BBS, Berg Balance Scale. *Stem cells isolated from the dental pulp.*

Disease Initiative," Joint Programme on Neurodegenerative disorders), which put together eight European cohorts that over the last 4 years integrated all the existing data in a common database to allow for standardized and quality-controlled protocols in the forthcoming studies.

Major Questions to be Addressed to Increase Clinical Benefits

From the clinical studies referred so far, we can conclude that the major weakness of the cellular therapies using MSC (or another adult/progenitor stem cell source) is that their effect is not prolonged throughout time. However, the cause of these transitory effects of MSC *in vivo* remains elusive. In our study, we could determine that MSC survival was limited in time after administration, which is consistent with other pre-clinical studies (as shown in **Table 1**). This may result from the inhospitable environment they face—the high oxidative stress of the damaged tissue, together with the low glucose availability—that leads MSC to choose an anerobic metabolic pathway, forcing cells to rapidly undergo apoptosis (Moya et al., 2018). Methods for increasing MSC fitness are described in the following section.

Another important question is whether MSC can effectively pass through the BBB upon intravenous injection. Several pre-clinical studies reported the finding of MSC in the central nervous system, including ours, as referred in **Table 1**. However, whether MSC crosses this barrier by mechanisms that are similar to leukocytes such as transmigration (Chamberlain et al., 2007), or whether this barrier needs to be impaired to let MSC pass-through remains undetermined. BBB leakage was already described for HD (Drouin-Ouellet et al., 2015). Due to the strong presence of inflammatory mechanisms in polyQ disorders, it can be speculated that this is also the case for the remaining polyQs. In accordance, we recently reported evidence that the BBB is damaged in MJD (Lobo et al., 2020). Still, there are no indications that the BBB opening is mandatory for MSC to reach the brain parenchyma. Studies accessing MSC biodistribution inside the nervous system in non-neurodegenerative rodents (wild type animals), should be performed to answer this intriguing question. In case MSC cannot cross the BBB when it is fully preserved, then other strategies should be explored such as the expression of specific receptors on the MSC surface that would allow their passage, or consider the use of MSC products (either naïve or engineered) that can effectively reach brain cells.

Finally, the prophylactic administration of this cellular treatment is also an important issue to be discussed. Since these disorders are monogenic and can be passed to the offspring, the patients can choose to know whether they carry the disease mutation before the manifestation of disease symptoms. Therefore, successful treatment could be implemented in pre-symptomatic patients. Accordingly, there is evidence that this would favor a later appearance of symptoms (Snyder et al., 2010; Li et al., 2018), supporting such early intervention as a feasible clinical approach for polyQs patients in the future. Nonetheless, Im and collaborators showed that MSC only modified the course of the disease when administered after its phenotypic expression in long-lived HD transgenic mice (Im et al., 2010). Therefore, the time window of delivery should be further investigated.

Overall, both pre-clinical and clinical studies show the therapeutic relevance of MSC for polyQ but also highlight the need to design better clinical approaches and more adjusted to the real biology and behavior of cells upon transplantation. Possible solutions are discussed in the following sections.

METHODS TO INCREASE MSC's EFFICACY *IN VIVO*

Strategies to overcome some of these shortcomings and improve MSC's efficiency have been explored and will be described next, such as pre-conditioning of cells, the use of adjuvant factors, or the genetic enhancement of MSC. A concise review of that is given in **Table 3**.

Pre-conditioning of MSC

Knowing that the secretion of soluble and insoluble factors is an adaptive mechanism, which allows MSC to regulate intracellular stress and influence their surroundings, researchers have investigated whether the pre-conditioning of MSC could improve their therapeutic abilities. Pre-conditioning is nothing more than manipulating cells, so they perform a specific function or undergo a desirable differentiation pathway. Some studies showed that pre-conditioning of MSC using hypoxia, different culture conditions, biomolecules (cytokines and growth factors), or pharmacological compounds, may lead to greater MSC survival *in vivo* and enhanced regenerative and immunomodulatory effects (Ferreira et al., 2018; Noronha et al., 2019).

Pre-conditioning: Hypoxia

Oxygen tension under standard cell culture conditions (21%) is much higher than physiological oxygen tension in tissues, which can vary from 1% to 12% depending on vascularization and metabolic activity (Das et al., 2010). Notably, when we speak of hypoxia in the context of cell culture, usually it means that oxygen availability ranges from 0 to 10% to better mimic physiological conditions. MSC is usually found in tissues with low oxygen tension (1% to 7%), is naturally able to endure in *in vivo* hypoxic environments (Deschepper et al., 2011).

When cultured under hypoxic conditions, MSC increases their fitness to adapt to adverse circumstances which will make them better prepared to face inhospitable environments. Indeed, under hypoxia MSC changes their metabolism, increasing their glucose consumption, decreasing the production of oxygen species, and shortening telomeres (Das et al., 2010). In these conditions, MSC presents a greater proliferation rate and increased secretion of soluble factors without changing their multipotency, which improves MSC capacity to survive in damaged tissues upon transplantation (Das et al., 2010; Bader et al., 2015; Lee et al., 2017). Accordingly, Lee et al. (2017) showed that culturing MSC in hypoxic conditions leads to the up-regulation of Hypoxia-inducible factor (HIF)-α, which induces survival and proliferation rate.

TABLE 3 | Summary of results using modified-MSCs in PolyQ diseases in comparison to naive cells—evidence from pre-clinical studies.

PolyQ	Study	Animal model	MSC's source	Enhancement strategy	N° of cells injected	Site of injection	Time of injection	Outcomes compared with naïve cells
HD	Bantubungi et al. (2008); Edalatmanesh et al. (2011); Sadan et al. (2012)	QA-induced rat striatal model	Human BM-MSC isolated from healthy donors and HD patients differentiated into NTF⁺ secreting cells*	Pre-conditioning with supplemented medium	4.2×10^5	Bilateral delivered at the lesion site	On the day of QA induction	NTF⁺ MSC led to less striatal volume loss and decreased T2 levels, as well as behavior improvements (reduced rotational behavior)
	Linares et al. (2016)	N171–82Q transgenic mice	Mouse BM-MSC	Pre-conditioning with Lithium and VPA	3×10^5	Intranasal administration	8-week old mice (early symptomatic)	Pre-conditioned cells survived longer *in vivo* (60% more at 9 weeks after transplants), led to behavioral improvements, and reduced neuronal loss and aggregate formation
	Elbaz et al. (2019)	3NP-induced rat model	Rat BM-MSC	Combined administration of MSC and LER	1×10^6	IV	1 h before 3-NP injections	Combined treatment decreased weight loss, and mortality rate to a greater extent; improved rotarod performance; decreased neuronal loss, inflammation, astroglia activation, and apoptosis
	Dey et al. (2010)	YAC 128 transgenic mice	BM-MSC	genetic modification to overexpress BDNF (MSC-BDNF) and NGF (MSC-NGF)	3×10^5	Bilateral injections into the striata	4 Month old mice (post-symptomatic)	Both MSC-BDNF or MSC-NGF treated mice showed improved rotarod performances with MSC-BDNF having a greater effect and reduced neuronal loss
	Pollock et al. (2016)	YAC 128 transgenic Mice	human BM-MSC	genetic modification to overexpress BDNF (MSC-BDNF)	5×10^5 per hemisphere	Bilateral injections into the striata	8.5-Month-old mice (post-symptomatic)	Only MSC-BDNF reduced mice anxiety levels; MSC-BDNF had a greater effect in decreasing striatal atrophy
		R6/2 transgenic mice	human BM-MSC	genetic modification to overexpress BDNF (MSC-BDNF)	5×10^5 per hemisphere	Bilateral injections into the striata	7 Month old mice (symptomatic)	MSC-BDNF increased mice lifespan to a greater extent

*List of abbreviations: HD, Huntington's disease; QA, quinolinic acid; VPA, valproic acid; AD-MSC, adipose tissue-derived mesenchymal stromal cells; BM-MSC, Bone marrow-derived mesenchymal stromal; LER, lercanidipine; BDNF, Brain-derived neurotrophic factor; NTF, neurotrophic factors; NGF, Nerve growth factor IV, Intravenous injections; *NTF secreting cells produce and secrete large amounts of neurotrophic factors like BDNF and GDNF.*

Hence, UC-MSC pre-conditioned with hypoxia (MSC-h) showed to have neuroprotective, anti-apoptotic, and anti-inflammatory actions in a rat model of spinal cord injury. When compared with MSC cultured under standard conditions, MSC-h exhibited increased expression of neuroprotective trophic factors, such as hepatocyte growth factor (HGF), BDNF, and VEGF. Rats transplanted with MSC-h into the spinal cord immediately after lesion induction showed increased axonal preservation, even higher as compared to MSC not submitted to hypoxia. MSC-h also reduced the number of caspase-3 (apoptotic marker) positive cells, microglia, and macrophage infiltration (Zhilai et al., 2016).

Furthermore, Wang J. W. et al. (2017) demonstrated that hypoxia pre-conditioning enhanced the migration and integration of MSC upon transplantation into a rat model of cerebral ischemia, induced by cardiac arrest. These actions were linked with the activation of PI3K/AKT pathways and increased expression of HIF-1α and C-X-C chemokine receptor type 4. Plus, the IV injection of MSC-h was able to diminish neuronal death and inflammation in the cortex.

Even though these studies suggest hypoxia pre-conditioning may improve MSC neuroprotective effect and increase MSC survival after transplant, we found no studies that investigated the potential use of this approach in the treatment of neurodegenerative/polyQ disorders. Therefore, in the future efforts should be made to better explore this strategy.

Pre-conditioning: Biomolecules/Chemicals

Pre-conditioning of MSC with biomolecules such as cytokines, growth factors, or hormones can improve MSC survival and efficiency. Indeed, Sadan and collaborators showed that MSC that were pre-conditioned in media supplemented with epidermal growth factor, human basic FGF, cyclic adenosine monophosphate, human neuregulin1-β1, and platelet-derived growth factor (Sadan et al., 2008, 2009) differentiated into NTF secreting cells (NTF+) cells, which could secrete high levels of BDNF and glial cell line-derived neurotrophic factor (GDNF). NTF+ MSC could improve both the neuropathology and behavior patterns in the QA rat model of HD (Sadan et al., 2012). QA-injected rats transplanted with NTF+ MSC lost less striatal volume than rats transplanted with naïve MSC. Moreover, NTF+ MSC decreased T2 values when compared with PBS-treated QA rats, as detected by MRI acquisitions. Phenotypically, NTF+ cells were more efficient in reducing rotational behavior than naïve MSC in QA-injected rats, thus suggesting these cells to be more efficient.

In a recent study, the authors assessed the impact of culturing MSC with different sera. MSC was cultured in medium with 10% of the following sera: fetal bovine serum, serum from healthy controls (NS-MSC), or serum from stroke patients (SS-MSC). Interestingly, SS-MSC had a higher proliferation rate and lower senescence as well as increased expression of VEGF, GDNF, and FGF. Additionally, SS-MSC promoted neurogenesis and angiogenesis in stroke rat models, leading to improved behavioral performance (Moon et al., 2018).

Alternatively, MSC can be primed using pharmacological or chemical agents. Linares et al. (2016) investigated whether pre-conditioning MSC with lithium and valproic acid (VPA) could enhance their therapeutic effect. Lithium and VPA are mood stabilizers known to exert neuroprotective effects (Chiu et al., 2013) and pre-conditioning of MSC with these factors led to an increase in the expression of genes involved in trophic effects, as well as in pro-survival and immunomodulatory pathways. Accordingly, N171–82Q transgenic HD mice were treated with an intranasal administration of MSC or MSC pre-conditioned with Lithium and VPA. The transplanted cells migrated into the brain, with pre-conditioned MSC surviving for a longer period. Moreover, mice treated with pre-conditioned MSC showed a greater amelioration of motor and behavior performance, decreased neuronal death, and reduction of huntingtin aggregates in the striatum (Linares et al., 2016). This strategy could be used in other polyQ diseases where Lithium and VPA showed promising results (Saute et al., 2014; Esteves et al., 2015; Lei et al., 2016; Lopes-Ramos et al., 2016).

Therefore, though pre-conditioning needs to be further explored, the results obtained so far seem promising, which opens good prospects for the near future.

Combined Treatment With Pharmacological Agents

A less common strategy recently evaluated involved the use of pharmacological agents in combination with MSC to treat polyQ disorders. Elbaz et al. (2019) showed that the combined administration of Lercanidipine (LER), an antihypertensive drug, and MSC boosted their therapeutic efficiency in a rat model for HD. LER is a vasoselective dihydropyridine calcium channel blocker that can modulate calcium levels and, therefore may be able to influence the calcineurin (CaN)/nuclear factor of activated T cells c4 (NFATc4) pathway, which is deregulated in neurodegenerative disorders (Sompol and Norris, 2018). Treatment with LER, BM-MSC, or a combination of LER and BM-MSC was given to 3-NP rats. The combined treatment with LER and BM-MSC had the most promising results by leading to greater amelioration of motor and behavioral performance. Moreover, the levels of BDNF, forkhead box P3, Wnt, and β-catenin increased in the striatum, along with a decrease of CaN, tumor necrosis factor-α, and NFATc4 protein expression and the Bax/B-cell lymphoma 2. Their results suggest that this combined therapy can promote neuroprotective effects by, at least in part, suppressing Ca^{2}+/CaN/NFATc4 and Wnt/β-catenin signaling pathways activation, which is dysregulated in the 3-NP rats as well as in other neurodegenerative disorders (Sompol and Norris, 2018).

Hence, this study shows how a combination of factors can produce a synergistic effect, so it would be interesting to further investigate the neuroprotective potential of therapies combining different drugs with MSC.

Genetically Engineered MSC

Another strategy that has been studied to enhance MSC's therapeutic efficiency is the production of genetically engineered MSC that can be used as delivery vehicles (Park et al., 2015). For the treatment of neurodegenerative diseases, the most common approaches include the use of MSC that were either virally or non-virally modified to overexpress trophic factors known to have neuroprotective actions (Huang et al., 2012; Deng et al., 2016).

Concerning polyQ disorders, only two studies are available in the literature. The safety and efficacy of genetically engineered MSC have been tested in HD animal models. In 2010, BM-MSC were genetically engineered to over-express BDNF (MSC-BDNF) and/or NGF (MSC-NGF), thus allowing these cells to deliver these factors in higher amounts than normal MSC (Dey et al., 2010). YAC128 transgenic mice were treated with MSC-BDNF, MSC-NGF, or with both MSC-BDNF and MSC-NGF. MSC was administered through bilateral injections into the striatum. From all the tested groups, YAC128 transgenic mice transplanted with MSC-BDNF exhibited the best outcomes, showing the least neuronal loss and the highest latencies to fall values in rotarod.

Another study demonstrated that human MSC modified by lentiviral transduction to overexpress BDNF (MSC-BDNF) could ameliorate HD symptoms in two different animal models, YAC128 and R6/2 transgenic mice in a higher extent (Pollock et al., 2016). The intrastriatal transplantation of MSC-BDNF decreased striatal atrophy and reduced anxiety levels in YAC128 mice. Additionally, MSC-BDNF administration promoted neurogenesis-like activity and led to an increase in mice lifespan. Notably, mice treated with MSC-BDNF had overall better outcomes than those treated with non-modified MSC.

Despite the few studies using genetically engineered MSC to treat polyQ disorders, this type of approach has been further explored in other neurodegenerative diseases, namely in AD and PD, also with promising results (Li et al., 2008; Moloney et al., 2010; Ren et al., 2013; Hoban et al., 2015).

Importantly, since neurodegenerative disorders have common pathogenic mechanisms, current evidence obtained from these studies indicate that genetically modified MSC to overexpress various trophic factors is strong therapeutic candidates for several neurodegenerative diseases, including polyQ disorders.

CELLULAR-FREE APPROACHES

The Success of Studies Using MSC's Secretome in PolyQ/Other Neurodegenerative Disorders

Several studies detected reduced numbers of MSCs in the targeting tissues suggesting that MSCs can also act through paracrine mechanisms. Conditioned medium (CM) of MSC (CM-MSC), i.e., MSC's secretome, can reproduce the therapeutic effects obtained with MSC treatment in several neurological diseases.

Bai et al. (2012) demonstrated that CM-MSC mitigates functional abnormalities in a mouse model of multiple sclerosis by promoting oligodendrocyte and neuron development. Suto et al. (2016) showed that injections of MSC released factors, such as HGF, improved motor coordination in an SCA-1-Knock-in mouse model. Indeed, the administration of CM-MSC was able to protect axons and myelin of spinal motor neurons (Suto et al., 2016). Importantly, such improvements were similar to those obtained in previous work using the same model of disease in which animals were treated with MSC (Mieda et al., 2016).

An antibody-based protein array analysis and ELISA showed that, in addition to HGF, IGF-1, VEGF, and transforming growth factor (TGF)-β-1 are present in the CM. However, the combination of these factors is not as effective as treatment with CM in promoting neuronal survival and neurite outgrowth, which suggests that CM contains other complementary key factors (Nakano et al., 2010). The secretion of BDNF and NGF by MSC has also been demonstrated to promote neuronal/glial survival and neurogenesis (Crigler et al., 2006). Likewise, the beneficial effect of MSC may also be mediated by secretion of GDNF, basic FGF-2, CNTF, angiopoietin-1, and neurotrophin-3 (Chen et al., 2005; Onda et al., 2008). Besides, the extracellular matrix proteins, including collagen, fibronectin, and laminin (Hidalgo-Bastida and Cartmell, 2010) provide physical guidance, support neurogenesis, and may exert trophic activity themselves (Maltman et al., 2011).

Recently, Ebrahimi et al. (2018) demonstrated that CM collected from UC-MSC protect PC12 cells against superoxide-induced oxidative stress, increasing cell viability, and neurite outgrowth. GDNF and VEGF were detected in CM of UC-MSC and were linked to the neuroprotective effects observed.

Moreover, it has been demonstrated that both human adipose-derived MSC and their CM could reduce neuronal cell damage in a model of glutamate excitotoxicity, a mechanism that plays a role in the pathogenesis of neurodegenerative diseases. Indeed, co-cultures with MSC using a transwell system or addition of CM similarly promoted axonal regeneration, energy metabolism improvements, and prevented apoptosis of cortical neurons (Hao et al., 2014). In another study, Miranda et al. (2011) showed that in a transwell system that immortalized BM-MSC could stimulate neurite outgrowth in a neurotrophin-dependent manner including BDNF secretion in Twitcher-derived DRG neurons, an *in vitro* model of Krabbe's disease. Moreover, the administration of exogenous BDNF to the sciatic nerve of Twitcher mice through BDNF-delivering osmotic pumps promoted morphometric ameliorations, suggesting BDNF as a promising candidate to be used in combination with BM-MSC treatment (Miranda et al., 2011). Nevertheless, simultaneous transplantation of BM-MSC and treatment with an antagonist of TrK receptors in a model of sciatic nerve crush completely reversed the inhibitory effect of the antagonist, suggesting that other mechanisms may be compensating the action of neurotrophins blocked by the antagonist.

A recent study has shown that CM-MSC promoted neuroprotective activities in *in vitro* PD models by increasing the viability of both rat and human dopaminergic neurons exposed to a neurotoxic insult (Parga et al., 2018). Additionally, the administration of MSC secretome through an injection into the substantia nigra and striatum in a PD mouse model led to an increase of dopaminergic neurons and neuronal terminals in those brain regions, which likely mediated the observed temporary motor performance improvements. Also, their findings suggest that the beneficial effects of human MSC secretome correlated with the increased presence of several neuroregulatory molecules, such as GDNF, BDNF, interleukin-6 (IL-6), VEGF, pigment epithelium-derived factor (PEDF), cystatin C, glia-derived nexin (GDN) and galectin-1 (Teixeira et al., 2017).

Alternatively, MSC potent anti-inflammatory and immunoregulatory effects might play a central role in tissue regeneration. For instance, MSC are suggested to inhibit T and B cell proliferation (Di Nicola et al., 2002; Franquesa et al., 2012), natural killer cytotoxicity (Spaggiari et al., 2008), and monocyte differentiation and maturation into dendritic cells (Ivanova-Todorova et al., 2009). These capacities have been linked to the release of molecules like indoleamine 2,3-dioxygenase (IDO), prostaglandin E2, interleukin-10 (IL-10), human leukocyte antigen-G, TGF-β, and HGF (Dorronsoro et al., 2013). It has been demonstrated that the systemic administration of CM derived from amniotic membrane MSC ameliorates motor dysfunctions, brain pathology, and decreased microglial activation in HD animal model (Giampà et al., 2019).

In summary, the bioactive molecules released by MSC can exert angiogenic, immunomodulatory, neurogenic, neuroprotective, and anti-apoptotic effects. Therefore, MSC's secretome seems to improve symptoms of several neurodegenerative disorders, including polyQs, in a paracrine-mediated manner, representing an alternative to cell therapies.

Extracellular Vesicles: Important Mediators of MSC Communication With Host Cells and Their Environment

MSC are known to release EVs, a group of heterogeneous membrane-limited structures secreted by almost all cell types. Exosomes are the most studied subpopulation of EVs that can are divided by intracellular origin, size, and cargo. EVs are recognized as important mediators of intercellular cross-talk communication, being involved in both physiological and pathological processes (Thery et al., 2002; Zöller, 2009; Colombo et al., 2014). In fact, after being released into the extracellular space, EVs can be internalized by recipient cells and transfer their content. Then, their cargo may either be degraded or mediate various signaling functions, modifying cellular fate, function, and/or plasticity (**Figure 3**; Mulcahy et al., 2014; Abels and Breakefield, 2016).

Depending on their subpopulation, the cell type of origin, and specific environmental challenges (for example stress conditions), EVs biomolecular content will change, bearing different lipids, proteins, and nucleic acids. To standardize the field, MISEV guidelines were created by the International Society of Extracellular vesicles (ISEV; Thery et al., 2018). Moreover, a harmonization criterion for MSC-EVs was created by four societies, ISEV, Society for Clinical Research and Translation of Extracellular Vesicles Singapore (SOCRATES), International Society for Cell and Gene Therapy (ISCT), and International Society of Blood Transfusion (ISBT), to boost MSC-EVs applications in clinical settings (Witwer et al., 2019).

Given EVs' physiological importance and role in brain disorders, their potential therapeutic use has been explored over the last decades. A comprehensive review can be found in Rufino-Ramos et al. (2017).

EVs Derived From MSC

The general morphological characteristics of EVs derived from MSC are common with EVs from other sources, yet they possess some distinctive features. Besides common surface markers, like CD9, CD63, and CD81, MSC-derived EVs also possess specific MSC integrins and adhesive molecules, including CD29, CD44, CD90, and CD73 (Lai et al., 2012). Their combination defines tropism to specific tissues.

EVs derived from MSC encapsulate mRNAs, miRNAs, and proteins related to multiple signaling pathways, namely cell cycle, proliferation, differentiation, and self-renewal (Lai et al., 2010; Kim et al., 2012). However, as aforementioned, extracellular conditions can affect their composition (Xin et al., 2014). Indeed, Lai et al. (2012) demonstrated that MSC-derived EVs obtained from CM from three different batches of the same cells had only 154 proteins in common. Furthermore, pre-conditioning MSC with other cells and/or exposure to disease insults may change exosomes' secretion profile (Kim et al., 2007; Sano et al., 2014; Cui et al., 2018).

Over the last decade, EVs derived from MSC have become promising therapeutic tools as researchers hypothesized that they may be the main paracrine effectors of MSC. In 2010, the therapeutic potential of MSC-derived EVs was tested in a mouse model of ischemia (Lai et al., 2010). Since then, EVs have been investigated in several other disease models, such as cardiovascular disorders (Wang N. et al., 2017), kidney injury (Aghajani Nargesi et al., 2017; Farzamfar et al., 2019), immune diseases (Lai et al., 2018, 2019), tumor growth (Wu et al., 2013; Zhang et al., 2017) and neurological diseases (Wang et al., 2018; Gorabi et al., 2019), obtaining encouraging results (Yu et al., 2014).

MSC-EVs have been shown to exert similar functions to those of the progenitor cells, participating in tissue repair and regeneration, immune system modulation, and inflammatory suppression in neurodegenerative disorders (Pashoutan Sarvar et al., 2016). Xin et al. (2013) performed a proof-of-concept study providing evidence that treatment with MSC-derived EVs alone could promote functional recovery, neurogenesis, and neurovascular remodeling in rats subjected to middle cerebral artery occlusion, a stroke model. Interestingly, the outcomes were similar to those obtained with systemic administration of MSC. In a previous study, the same group had found that MSC was able to promote neurite outgrowth and functional recovery partially by the transference of miRNA 133b (miR-133b) *via* exosomes (Xin et al., 2012). Levels of miR-133b were increased in the ipsilateral hemisphere of the stroke rat model after treatment with MSC. Plus, *in vitro* exposure of MSC to damaged brain tissue led to the enrichment of miR-133b in cells and their exosomes. Finally, when primary neurons co-cultured with astrocytes were treated with these exosomes, miR-133b levels were found to be increased. Moreover, this effect was lost in astrocytes treated with a miR-133b inhibitor. In this way, Xin et al. (2012) showed that exosomes play a role in MSC communication with brain parenchymal cells, mediating the transfer of bioactive molecules like miR-133b. In a different context, EVs isolated from adipose-derived stromal cells (AD-MSC) exerted a neuroprotective effect in an *in vitro* model of amyotrophic lateral sclerosis (ALS). Upon an oxidative insult, the motoneuron-like NSC-34 cells expressing ALS mutations were treated with exosomes, which showed the capacity to protect ALS motoneurons from oxidative stress, increasing their viability (Bonafede et al., 2016).

In line with those results, in another study, it was demonstrated that both MSC and MSC-derived EVs were able to protect hippocampal neurons from oxidative stress and synapse damage induced by amyloid-β peptide (AβO) upon IC delivery (de Godoy et al., 2018). These authors also unveiled that such neuroprotective action could be explained by MSC's ability to internalize and degrade AβO, secret EVs, as well as trophic and anti-inflammatory cytokines. EVs were shown to contain an antioxidant enzyme, catalase, which contributed to the decrease of oxidative stress. Recently, Reza-Zaldivar et al. (2019) also reported that both MSC and MSC-derived EVs were similarly able to promote neurogenesis and ameliorate cognitive impairments caused by β amyloid 1–42 aggregates, an established Alzheimer's disease animal model.

Hypoxia conditions were shown to increase mir-21 levels in MSC cells and their derived EVs. By systemically injecting EVs in Transgenic APP/PS1, it has been shown that EVs produced during hypoxia could reach the brain and increase the levels of mir-21. By increasing the levels of mir-21 in the brain, learning and memory capabilities were restored while plaque deposition and Aβ levels were decreased. These findings substantiate the therapeutic potential of MSC-EVs in Alzheimer's disease context (Cui et al., 2018).

MSC can ultimately be genetically modified to enhance EVs production that specifically targets the brain and promote cargo delivery in neurons. Kojima et al. (2018) engineered MSC to: (1) boost EVs production by acting in exosomes biogenesis; (2) package mRNA catalase inside EVs; and (3) express at EVs surface the RVG peptide that targets the brain and connexin 43 to enhance information transference in the target cells. This study showed that engineered cells can produce modified EVs able to deliver therapeutic cargo to the brain in a PD model (Kojima et al., 2018).

One of the major drawbacks of using MSC-EVs for brain therapy is the low number of vesicles that cross BBB and reach the brain. To overcome this, a recent study engineered MSC derived vesicles with magnetic properties that reach the brain after application of a magnetic field (Kim et al., 2020). Moreover, stimuli with magnetic particles increase growth factors expression in MSC-derived vesicles. These magnetic vesicles were administered through a systemic injection in middle-cerebral-artery-occlusion (MCAO)-induced rats promoting anti-inflammatory response, angiogenesis, and anti-apoptosis in the ischemic brain lesion leading to an improvement in the motor function.

Altogether, these studies support the idea that EVs derived from MSC may be an important player in the paracrine effect of MSC and, therefore, a potential therapeutic agent for the treatment of polyQ diseases.

HOW TO STANDARDIZE MSC OR MSC's EFFECTORS?

As aforementioned, multiple extrinsic factors including tissue source, culture methods, passage number, and oxygen concentration can significantly interfere with MSC innate functional potential (Hagmann et al., 2013; Lee et al., 2013; Heathman et al., 2015; Elahi et al., 2016). To comply with Good Manufacturing Practices (GMP) and current Good Tissue Practices, certain aspects of MSC or MSC's effector's production like: (1) the method of isolation; (2) culture medium, serum and supplements used; (3) cell seeding density; and (4) devices of expansion, must be optimized and standardized, thus ensuring consistent production, efficacy, and safety (Rojewski et al., 2013; Sensebé et al., 2013; Sharma et al., 2014).

Another important aspect that needs further clarification is the cryopreservation procedure as no consensus on whether the use of freshly cultured or freeze-thawed MSC is more advantageous. Multiple studies are showing that thawed MSC/MSC products have reduced therapeutic capabilities (François et al., 2012; Chinnadurai et al., 2016, 2018) and, contrarily, some studies demonstrate that both thawed and fresh MSC have similar functionalities (Cruz et al., 2015; Bárcia et al., 2017; Tan et al., 2019). One explanation for this disparity may be the temperature fluctuation under frozen conditions (Chabot et al., 2017).

Given the heterogeneity of practices and outcomes among different laboratories, both in pre-clinic and clinic grades, in 2016 ISCT issued a position statement urging the need to identify MSC functional markers of potency and implement assays for the measurement of such markers. This is a fundamental step that could help in predicting and improving the efficiency of MSC related therapies, meeting the requirements of Regulatory Authorities for extensive quality-control protocols for advanced clinical studies and registrations (Galipeau et al., 2016). Nevertheless, MSC has complex mechanisms of action that are not yet completely understood, which hinders the determination of which biological proprieties of MSC are more relevant to assess (Galipeau and Krampera, 2015). Ideally, multiple complementary assays (matrix assay) should be developed to characterize several attributes of MSC/MSC's effectors that may be important for each therapeutic purpose (Galipeau et al., 2016).

Regarding studies that use MSC/MSC's effectors as a therapeutic agent for neurodegenerative disorders, in addition to current characterization criteria, which include the presence and/or absence of specific markers and their differentiation potency (Dominici et al., 2006), it is essential to establish assays for the analysis of their immunomodulatory capability and senescence status, especially in clinical settings. A summary of the assays that could be used to standardize MSC/MSC's effectors is given in **Table 4**.

It has been demonstrated that after multiple passages, MSC enters senescence, and cells show loss of differentiation potential, different secretion profiles, and telomere shortening (Wagner et al., 2008). Thus, sensitive evaluation of key genes and/or molecules involved in senescence pathways such as myc, p21,

TABLE 4 | Summary of assays that can be used to standardize MSC/MSC's effectors.

Study	Assayed Parameters	Assayed description/outcomes
Tarte et al. (2010)	MSC senescence	Evaluation of the expression of key genes of senescence pathways (myc, p21, p53, $p16^{ink4}$, hTERT) to guarantee MSC fitness in later passages.
Boregowda et al. (2016)	Angiogenic, anti-inflammatory and immunomodulatory potential	Clinical Indications Prediction scale that assesses MSC potential in accordance to the expression levels of the transfection factor TWIST1: ↓ levels of TWIST1–↑ anti-inflammatory and immunomodulatory potential; ↑ levels of TWIST1–↑ angiogenic potential
Chinnadurai et al. (2019)	Immunomodulatory potential	Assay matrix approach that assesses MSC immunoinhibitory potential upon of MSC-PBMC co-culture: ↓ TNF-α, IFNγ, IL-13, IL-5, IL-2R, CCL3, and CCL4 cytokine levels in the secretome–↑ suppression of T-cells; ↑ VEGF, IFNα, CXCL10, GCSF, CXCL9, IL-7, and CCL2 levels in the secretome–↑ suppression of T-cells
Chinnadurai et al. (2019)	Immunomodulatory potential	Assay matrix approach that assesses MSC immunoinhibitory potential based on the phosphorylation of STAT in MSC after contact with the secretome of MSC-PBMC co-culture: ↓ phosphorylation of STAT1 and STAT3–↑ suppression of T-cells

List of abbreviations: CCL2, monocyte chemoattractant protein-1; CCL3, macrophage inflammatory protein-1 α; CCL4, Chemokine C-C motif ligands 4; CXCL, C-X-C Motif Chemokine Ligand; IL, interleukin; IFN, Interferon; GCSF, Granulocyte colony-stimulating factor; MSC, Mesenchymal stromal cells; PBMC, peripheral blood mononuclear cells; STAT, signal transducer and activator of transcription; TNF-α, Tumor Necrosis Factor-α; VEGF, Vascular endothelial growth factor.

p53, p16ink4, hTERT is essential to guarantee the genomic stability of MSC in long-term culture (Tarte et al., 2010).

In 2018, Chinnadurai et al. (2018) proposed for the first time a dual assay matrix approach that combined secretome analysis and quantitative RNA-based array to characterize MSC crosstalk interaction with peripheral blood mononuclear cells (PBMC) upon co-culture. In this study the authors identified cytokines and chemokines whose increased or decreased expression directly correlated with MSC ability to suppress PMBC proliferation, reporting on MSC functionality to modulate the secretome immune response (Chinnadurai et al., 2018). Recently, the same group developed another assay matrix approach based on the phosphorylation of signal transducer and activator of transcription (STAT) of MSC after interaction with MSC-PBMC co-culture secretome. This informs on MSC functionality to sense and modulate the environment. The authors reported that STAT1 and STAT3 phosphorylation correlated with MSC immunoinhibitory ability (Chinnadurai et al., 2019). Nevertheless, depending on the disease, instead of PBMCs, the use of immune effector cells directly involved in the pathogenic mechanisms of the diseases may be more suitable and preferable.

Furthermore, Boregowda et al. (2016) developed a Clinical Indications Prediction (CLIP) scale for different diseases that predicts the therapeutic efficacy of populations of MSC from different donors, based on the expression levels of the transfection factor TWIST1. MSC expressing high levels of TWIST1 have greater angiogenic potential whereas low TWIST1 expressing populations are predicted to have higher anti-inflammatory and immunomodulatory capabilities.

Nonetheless, these potency assays still bear limitations, not providing complete predictive guidance. To complement these strategies in the future, it would also be important to study patient clinical parameters, such as the stage of the disease, co-morbidities, and age, among others, which could help predict patient responsiveness to treatment. Additionally, donors' age, gender, and clinical status should also be carefully considered. As different studies offer contradictory information on the impact of such parameters (Stolzing et al., 2008; Andrzejewska et al., 2019), a greater understanding of how these characteristics may correlate with MSC proliferation, multipotency, and efficiency is needed.

We expect that further research may shed light on MSC heterogeneity and their highly complex mechanisms of action, thus facilitating the optimization and standardization of MSC/MSC effector processes, leading to higher quality-control protocols along with better and more consistent therapeutic outcomes.

CONCLUSION: MSC—DO THEY HAVE A FUTURE IN PolyQ CLINICAL THERAPIES?

MSC have gathered considerable interest among the scientific community and their application to neurodegenerative disorders has not been an exception. It has been reported that the therapeutic application of MSC can produce encouraging results as they can promote tissue regeneration and cell turnover by: (i) increasing cell survival; (ii) inducing neurogenesis; (iii) inhibiting apoptosis; and (iv) modulating inflammation, among others, mainly through paracrine mechanisms. Further supporting their therapeutic potential, clinical trials have demonstrated that MSC are safe and may produce beneficial results in different neurodegenerative disorders. In polyQ disorders, patients enrolled in clinical trials showed a delay in the progression of disease symptoms. However, a considerable percentage of patients were reported to regress to their status before transplantation, indicating that MSC beneficial effects must be transient. On the other hand, the lack of appropriate animal models that could better mimic these human diseases is another limiting issue, as this hardens result interpretation.

Meanwhile, several strategies have emerged that aim at increasing either MSC survival or their efficacy *in vivo*. Pre-conditioning with hypoxic conditions or biomolecules, a combination of MSC with other pharmacological agents, or genetic modification of cells are some of these possible strategies. On the other hand, due to the strong paracrine properties of MSC, cell-free approaches using MSC secretome or MSC-derived EVs have naturally become a motif of interest in the case of neurological disorders, as they take advantages of the paracrine neuroprotective effect of MSC. Finally, it would be important to not only standardize procedures regarding MSC-derived therapies but also to uncover the role that MSC exert in polyQ models. Pinpointing possible MSC effects will certainly open avenues for triggering novel therapeutic targets for this group of neurodegenerative disorders.

As a conclusion, the final goal may now be to implement more standardized studies to discover a promising cell or cell-free based therapeutic strategy amenable to systemic delivery and enabling stably reaching the brain. In our point of view, this strategy would mean an add-on value in the clinics as it can be easily implemented due to its safety and minimally invasive character.

AUTHOR CONTRIBUTIONS

IB: discussion and article writing, and figure design. AM, TS, and JB: article writing. DR-R: article writing and figure design. LP: article discussion and reviewing. CM: article outline, discussion and article writing, and reviewing. All authors contributed to the article and approved the submitted version.

FUNDING

Our group is supported by the European Regional Development Fund through the Regional Operational Program Center2020, Competitiveness Factors Operational Program (COMPETE 2020), and National Funds through Foundation for Science and Technology (FCT): BrainHealth2020 projects (CENTRO-01-0145-FEDER-000008), ViraVector (CENTRO-01-0145-FEDER-022095), CortaCAGs (POCI-01-0145-FEDER-016719), SpreadSilencing POCI-01-0145-FEDER-029716, Imagene POCI-01-0145-FEDER-016807, CancelStem POCI-01-0145-FEDER-016390, as well as UID4950/2020 (to CM), BIN—National Brain Imaging Network (CENTRO-01-0145-FEDER-022118) and the Association Française Contre les Myopathies—Téléthon no. 21163 and the SynSpread, European SCA3/MJD Initiative and ModelPolyQ under the EU Joint Program, the last two co-funded by the European Union H2020 program, GA No. 643417; by National Ataxia Foundation, the American Portuguese Biomedical Research Fund, and the Richard Chin and Lily Lock Machado-Joseph Disease Research Fund. IB was supported by FCT fellowship (SFRH/BD/148877/2019).

REFERENCES

Abels, E. R., and Breakefield, X. O. (2016). Introduction to extracellular vesicles: biogenesis, RNA cargo selection, content, release and uptake. *Cell. Mol. Neurobiol.* 36, 301–312. doi: 10.1007/s10571-016-0366-z

Aghajani Nargesi, A., Lerman, L. O., and Eirin, A. (2017). Mesenchymal stem cell-derived extracellular vesicles for kidney repair: current status and looming challenges. *Stem Cell Res. Ther.* 8:273. doi: 10.1186/s13287-017-0727-7

Andrzejewska, A., Catar, R., Schoon, J., Qazi, T. H., Sass, F. A., Jacobi, D., et al. (2019). Multi-parameter analysis of biobanked human bone marrow stromal cells shows little influence for donor age and mild comorbidities on phenotypic and functional properties. *Front. Immunol.* 10:2474. doi: 10.3389/fimmu.2019.02474

Bader, A. M., Klose, K., Bieback, K., Korinth, D., Schneider, M., Seifert, M., et al. (2015). Hypoxic preconditioning increases survival and pro-angiogenic capacity of human cord blood mesenchymal stromal cells *in vitro*. *PLoS One* 10:e0138477. doi: 10.1371/journal.pone.0138477

Bai, L., Lennon, D. P., Caplan, A. I., DeChant, A., Hecker, J., Kranso, J., et al. (2012). Hepatocyte growth factor mediates mesenchymal stem cell-induced recovery in multiple sclerosis models. *Nat. Neurosci.* 15, 862–870. doi: 10.1038/nn.3109

Bantubungi, K., Blum, D., Cuvelier, L., Wislet-Gendebien, S., Rogister, B., Brouillet, E., et al. (2008). Stem cell factor and mesenchymal and neural stem cell transplantation in a rat model of Huntington's disease. *Mol. Cell. Neurosci.* 37, 454–470. doi: 10.1016/j.mcn.2007.11.001

Bárcia, R. N., Santos, J. M., Teixeira, M., Filipe, M., Pereira, A. R. S., Ministro, A., et al. (2017). Umbilical cord tissue-derived mesenchymal stromal cells maintain immunomodulatory and angiogenic potencies after cryopreservation and subsequent thawing. *Cytotherapy* 19, 360–370. doi: 10.1016/j.jcyt.2016.11.008

Bettencourt, C., and Lima, M. (2011). Machado-Joseph disease: from first descriptions to new perspectives. *Orphanet J. Rare Dis* 6:35. doi: 10.1186/1750-1172-6-35

Bibb, J. A., Yan, Z., Svenningsson, P., Snyder, G. L., Pieribone, V. A., Horiuchi, A., et al. (2000). Severe deficiencies in dopamine signaling in presymptomatic Huntington's disease mice. *Proc. Natl. Acad. Sci. U S A* 97, 6809–6814. doi: 10.1073/pnas.120166397

Bonafede, R., Scambi, I., Peroni, D., Potrich, V., Boschi, F., Benati, D., et al. (2016). Exosome derived from murine adipose-derived stromal cells: neuroprotective effect on *in vitro* model of amyotrophic lateral sclerosis. *Exp. Cell Res.* 340, 150–158. doi: 10.1016/j.yexcr.2015.12.009

Boregowda, S. V., Krishnappa, V., Haga, C. L., Ortiz, L. A., and Phinney, D. G. (2016). A clinical indications prediction scale based on TWIST1 for human mesenchymal stem cells. *EBioMedicine* 4, 62–73. doi: 10.1016/j.ebiom.2015.12.020

Caplan, A. I. (1991). Mesenchymal stem cells. *J. Orthop. Res.* 9, 641–650.

Cemal, C. K., Carroll, C. J., Lawrence, L., Lowrie, M. B., Ruddle, P., Al-Mahdawi, S., et al. (2002). YAC transgenic mice carrying pathological alleles of the MJD1 locus exhibit a mild and slowly progressive cerebellar deficit. *Hum. Mol. Genet.* 11, 1075–1094. doi: 10.1093/hmg/11.9.1075

Chabot, D., Tremblay, T., Paré, I., Bazin, R., and Loubaki, L. (2017). Transient warming events occurring after freezing impairs umbilical cord-derived mesenchymal stromal cells functionality. *Cytotherapy* 19, 978–989. doi: 10.1016/j.jcyt.2017.04.005

Chamberlain, G., Fox, J., Ashton, B., and Middleton, J. (2007). Concise review: mesenchymal stem cells: their phenotype, differentiation capacity, immunological features and potential for homing. *Stem Cells* 25, 2739–2749. doi: 10.1634/stemcells.2007-0197

Chang, Y. K., Chen, M. H., Chiang, Y. H., Chen, Y. F., Ma, W. H., Tseng, C. Y., et al. (2011). Mesenchymal stem cell transplantation ameliorates motor function deterioration of spinocerebellar ataxia by rescuing cerebellar Purkinje cells. *J. Biomed. Sci.* 18:54. doi: 10.1186/1423-0127-18-54

Chao, T. K., Hu, J., and Pringsheim, T. (2017). Risk factors for the onset and progression of Huntington disease. *Neurotoxicology* 61, 79–99. doi: 10.1016/j.neuro.2017.01.005

Chen, Q., Long, Y., Yuan, X., Zou, L., Sun, J., Chen, S., et al. (2005). Protective effects of bone marrow stromal cell transplantation in injured rodent brain: synthesis of neurotrophic factors. *J. Neurosci. Res.* 80, 611–619. doi: 10.1002/jnr.20494

Chinnadurai, R., Copland, I. B., Garcia, M. A., Petersen, C. T., Lewis, C. N., Waller, E. K., et al. (2016). Cryopreserved mesenchymal stromal cells are susceptible to T-cell mediated apoptosis which is partly rescued by IFNγ licensing. *Stem Cells* 34, 2429–2442. doi: 10.1002/stem.2415

Chinnadurai, R., Rajakumar, A., Schneider, A. J., Bushman, W. A., Hematti, P., and Galipeau, J. (2019). Potency analysis of mesenchymal stromal cells using a phospho-STAT matrix loop analytical approach. *Stem Cells* 37, 1119–1125. doi: 10.1002/stem.3035

Chinnadurai, R., Rajan, D., Qayed, M., Arafat, D., Garcia, M., Liu, Y., et al. (2018). Potency analysis of mesenchymal stromal cells using a combinatorial assay matrix approach. *Cell Rep.* 22, 2504–2517. doi: 10.1016/j.celrep.2018.02.013

Chiu, C. T., Wang, Z., Hunsberger, J. G., and Chuang, D. M. (2013). Therapeutic potential of mood stabilizers lithium and valproic acid: beyond bipolar disorder. *Pharmacol. Rev.* 65, 105–142. doi: 10.1124/pr.111.005512

Chopp, M., Zhang, X. H., Li, Y., Wang, L., Chen, J., Lu, D., et al. (2000). Spinal cord injury in rat: treatment with bone marrow stromal cell transplantation. *Neuroreport* 11, 3001–3005. doi: 10.1097/00001756-200009110-00035

Colombo, M., Raposo, G., and Théry, C. (2014). Biogenesis, secretion and intercellular interactions of exosomes and other extracellular vesicles. *Annu. Rev. Cell Dev. Biol.* 30, 255–289. doi: 10.1146/annurev-cellbio-101512-122326

Crigler, L., Robey, R. C., Asawachaicharn, A., Gaupp, D., and Phinney, D. G. (2006). Human mesenchymal stem cell subpopulations express a variety of neuro-regulatory molecules and promote neuronal cell survival and neuritogenesis. *Exp. Neurol.* 198, 54–64. doi: 10.1016/j.expneurol.2005.10.029

Cruz, F. F., Borg, Z. D., Goodwin, M., Sokocevic, D., Wagner, D., McKenna, D. H., et al. (2015). Freshly thawed and continuously cultured human bone marrow-derived mesenchymal stromal cells comparably ameliorate allergic airways inflammation in immunocompetent mice. *Stem Cells Transl. Med.* 4, 615–624. doi: 10.5966/sctm.2014-0268

Cui, G. H., Wu, J., Mou, F. F., Xie, W. H., Wang, F. B., Wang, Q. L., et al. (2018). Exosomes derived from hypoxia-preconditioned mesenchymal stromal cells ameliorate cognitive decline by rescuing synaptic dysfunction and regulating inflammatory responses in APP/PS1 mice. *FASEB J.* 32, 654–668. doi: 10.1096/fj.201700600r

Das, R., Jahr, H., van Osch, G. J., and Farrell, E. (2010). The role of hypoxia in bone marrow-derived mesenchymal stem cells: considerations for regenerative medicine approaches. *Tissue Eng. Part B Rev.* 16, 159–168. doi: 10.1089/ten.teb.2009.0296

de Godoy, M. A., Saraiva, L. M., de Carvalho, L. R. P., Vasconcelos-Dos-Santos, A., Beiral, H. J. V., Ramos, A. B., et al. (2018). Mesenchymal stem cells and cell-derived extracellular vesicles protect hippocampal neurons from oxidative stress and synapse damage induced by amyloid-β oligomers. *J. Biol. Chem.* 293, 1957–1975. doi: 10.1074/jbc.M117.807180

Deng, P., Anderson, J. D., Yu, A. S., Annett, G., Fink, K. D., and Nolta, J. A. (2016). Engineered BDNF producing cells as a potential treatment for neurologic disease. *Expert Opin. Biol. Ther.* 16, 1025–1033. doi: 10.1080/14712598.2016.1183641

Deschepper, M., Oudina, K., David, B., Myrtil, V., Collet, C., Bensidhoum, M., et al. (2011). Survival and function of mesenchymal stem cells (MSCs) depend on glucose to overcome exposure to long-term, severe and continuous hypoxia. *J. Cell. Mol. Med.* 15, 1505–1514. doi: 10.1111/j.1582-4934.2010.01138.x

Dey, N. D., Bombard, M. C., Roland, B. P., Davidson, S., Lu, M., Rossignol, J., et al. (2010). Genetically engineered mesenchymal stem cells reduce behavioral deficits in the YAC 128 mouse model of Huntington's disease. *Behav. Brain Res.* 214, 193–200. doi: 10.1016/j.bbr.2010.05.023

Di Nicola, M., Carlo-Stella, C., Magni, M., Milanesi, M., Longoni, P. D., Matteucci, P., et al. (2002). Human bone marrow stromal cells suppress T-lymphocyte proliferation induced by cellular or nonspecific mitogenic stimuli. *Blood* 99, 3838–3843. doi: 10.1182/blood.v99.10.3838

Dierks, C., Grbic, J., Zirlik, K., Beigi, R., Englund, N. P., Guo, G. R., et al. (2007). Essential role of stromally induced hedgehog signaling in B-cell malignancies. *Nat. Med.* 13, 944–951. doi: 10.1038/nm1614

Dominici, M., Le Blanc, K., Mueller, I., Slaper-Cortenbach, I., Marini, F. C., Krause, D. S., et al. (2006). Minimal criteria for defining multipotent mesenchymal stromal cells. The International society for cellular therapy position statement. *Cytotherapy* 8, 315–317. doi: 10.1080/14653240600855905

Dongmei, H., Jing, L., Mei, X., Ling, Z., Hongmin, Y., Zhidong, W., et al. (2011). Clinical analysis of the treatment of spinocerebellar ataxia and multiple system atrophy-cerebellar type with umbilical cord mesenchymal stromal cells. *Cytotherapy* 13, 913–917. doi: 10.3109/14653249.2011.579958

Dorronsoro, A., Fernandez-Rueda, J., Fechter, K., Ferrin, I., Salcedo, J. M., Jakobsson, E., et al. (2013). Human mesenchymal stromal cell-mediated immunoregulation: mechanisms of action and clinical applications. *Bone Marrow Res.* 2013:203643. doi: 10.1155/2013/203643

Drouin-Ouellet, J., Sawiak, S. J., Cisbani, G., Lagace, M., Kuan, W. L., Saint-Pierre, M., et al. (2015). Cerebrovascular and blood-brain barrier impairments in Huntington's disease: potential implications for its pathophysiology. *Ann. Neurol.* 78, 160–177. doi: 10.1002/ana.24406

Drüsedau, M., Dreesen, J. C., De Die-Smulders, C., Hardy, K., Bras, M., Dumoulin, J. C., et al. (2004). Preimplantation genetic diagnosis of spinocerebellar ataxia 3 by (CAG)(n) repeat detection. *Mol. Hum. Reprod.* 10, 71–75. doi: 10.1093/molehr/gah008

Dürr, A., Stevanin, G., Cancel, G., Duyckaerts, C., Abbas, N., Didierjean, O., et al. (1996). Spinocerebellar ataxia 3 and Machado-Joseph disease: clinical, molecular, and neuropathological features. *Ann. Neurol.* 39, 490–499. doi: 10.1002/ana.410390411

Ebrahimi, M. J., Aliaghaei, A., Boroujeni, M. E., Khodagholi, F., Meftahi, G., Abdollahifar, M. A., et al. (2018). Human umbilical cord matrix stem cells reverse oxidative stress-induced cell death and ameliorate motor function and striatal atrophy in rat model of Huntington disease. *Neurotox. Res.* 34, 273–284. doi: 10.1007/s12640-018-9884-4

Edalatmanesh, M. A., Bahrami, A. R., Hosseini, E., Hosseini, M., and Khatamsaz, S. (2011). Bone marrow derived mesenchymal stem cell transplantation in cerebellar degeneration: a behavioral study. *Behav. Brain Res.* 225, 63–70. doi: 10.1016/j.bbr.2011.06.030

Elahi, K. C., Klein, G., Avci-Adali, M., Sievert, K. D., MacNeil, S., and Aicher, W. K. (2016). Human mesenchymal stromal cells from different sources diverge in their expression of cell surface proteins and display distinct differentiation patterns. *Stem Cells Int.* 2016:5646384. doi: 10.1155/2016/5646384

Elbaz, E. M., Helmy, H. S., El-Sahar, A. E., Saad, M. A., and Sayed, R. H. (2019). Lercanidipine boosts the efficacy of mesenchymal stem cell therapy in 3-NP-induced Huntington's disease model rats *via* modulation of the calcium/calcineurin/NFATc4 and Wnt/β-catenin signalling pathways. *Neurochem. Int.* 131:104548. doi: 10.1016/j.neuint.2019.104548

Esteves, S., Duarte-Silva, S., Naia, L., Neves-Carvalho, A., Teixeira-Castro, A., Rego, A. C., et al. (2015). Limited effect of chronic valproic acid treatment in a mouse model of Machado-Joseph disease. *PLoS One* 10:e0141610. doi: 10.1371/journal.pone.0141610

Evans, S. J., Douglas, I., Rawlins, M. D., Wexler, N. S., Tabrizi, S. J., and Smeeth, L. (2013). Prevalence of adult Huntington's disease in the UK based on diagnoses recorded in general practice records. *J. Neurol. Neurosurg. Psychiatry* 84, 1156–1160. doi: 10.1136/jnnp-2012-304636

Farzamfar, S., Hasanpour, A., Nazeri, N., Razavi, H., Salehi, M., Shafei, S., et al. (2019). Extracellular micro/nanovesicles rescue kidney from ischemia-reperfusion injury. *J. Cell. Physiol.* 234, 12290–12300. doi: 10.1002/jcp.27998

Ferreira, J. R., Teixeira, G. Q., Santos, S. G., Barbosa, M. A., Almeida-Porada, G., and Goncalves, R. M. (2018). Mesenchymal stromal cell secretome: influencing therapeutic potential by cellular pre-conditioning. *Front. Immunol.* 9:2837. doi: 10.3389/fimmu.2018.02837

Fink, K. D., Rossignol, J., Crane, A. T., Davis, K. K., Bombard, M. C., Bavar, A. M., et al. (2013). Transplantation of umbilical cord-derived mesenchymal stem cells into the striata of R6/2 mice: behavioral and neuropathological analysis. *Stem Cell Res. Ther.* 4:130. doi: 10.1186/scrt341

Fisher, E. R., and Hayden, M. R. (2014). Multisource ascertainment of Huntington disease in Canada: prevalence and population at risk. *Mov. Disord.* 29, 105–114. doi: 10.1002/mds.25717

François, M., Romieu-Mourez, R., Li, M., and Galipeau, J. (2012). Human MSC suppression correlates with cytokine induction of indoleamine 2,3-dioxygenase and bystander M2 macrophage differentiation. *Mol. Ther.* 20, 187–195. doi: 10.1038/mt.2011.189

Franquesa, M., Hoogduijn, M. J., Bestard, O., and Grinyo, J. M. (2012). Immunomodulatory effect of mesenchymal stem cells on B cells. *Front. Immunol.* 3:212. doi: 10.3389/fimmu.2012.00212

Friedenstein, A. J., Chailakhjan, R. K., and Lalykina, K. S. (1970). The development of fibroblast colonies in monolayer cultures of guinea-pig bone marrow and spleen cells. *Cell Tissue Kinet.* 3, 393–403. doi: 10.1111/j.1365-2184.1970.tb00347.x

Friedenstein, A. J., Deriglasova, U. F., Kulagina, N. N., Panasuk, A. F., Rudakowa, S. F., Luria, E. A., et al. (1974). Precursors for fibroblasts in different populations of hematopoietic cells as detected by the *in vitro* colony assay method. *Exp. Hematol.* 2, 83–92.

Galipeau, J., and Krampera, M. (2015). The challenge of defining mesenchymal stromal cell potency assays and their potential use as release criteria. *Cytotherapy* 17, 125–127. doi: 10.1016/j.jcyt.2014.12.008

Galipeau, J., Krampera, M., Barrett, J., Dazzi, F., Deans, R. J., DeBruijn, J., et al. (2016). International Society for Cellular Therapy perspective on immune functional assays for mesenchymal stromal cells as potency release criterion for advanced phase clinical trials. *Cytotherapy* 18, 151–159. doi: 10.1016/j.jcyt.2015.11.008

Gao, J., Dennis, J. E., Muzic, R. F., Lundberg, M., and Caplan, A. I. (2001). The dynamic *in vivo* distribution of bone marrow-derived mesenchymal stem cells after infusion. *Cells Tissues Organs* 169, 12–20. doi: 10.1159/000047856

Giampà, C., Alvino, A., Magatti, M., Silini, A. R., Cardinale, A., Paldino, E., et al. (2019). Conditioned medium from amniotic cells protects striatal degeneration and ameliorates motor deficits in the R6/2 mouse model of Huntington's disease. *J. Cell. Mol. Med.* 23, 1581–1592. doi: 10.1111/jcmm.14113

Gorabi, A. M., Kiaie, N., Barreto, G. E., Read, M. I., Tafti, H. A., and Sahebkar, A. (2019). The therapeutic potential of mesenchymal stem cell-derived exosomes in treatment of neurodegenerative diseases. *Mol. Neurobiol.* 56, 8157–8167. doi: 10.1007/s12035-019-01663-0

Hagmann, S., Moradi, B., Frank, S., Dreher, T., Kammerer, P. W., Richter, W., et al. (2013). Different culture media affect growth characteristics, surface marker distribution and chondrogenic differentiation of human bone marrow-derived mesenchymal stromal cells. *BMC Musculoskelet. Disord.* 14:223. doi: 10.1186/1471-2474-14-223

Hao, P., Liang, Z., Piao, H., Ji, X., Wang, Y., Liu, Y., et al. (2014). Conditioned medium of human adipose-derived mesenchymal stem cells mediates protection in neurons following glutamate excitotoxicity by regulating energy metabolism and GAP-43 expression. *Metab. Brain Dis.* 29, 193–205. doi: 10.1007/s11011-014-9490-y

Harper, S. Q., Staber, P. D., He, X., Eliason, S. L., Martins, I. H., Mao, Q., et al. (2005). RNA interference improves motor and neuropathological abnormalities in a Huntington's disease mouse model. *Proc. Natl. Acad. Sci. U S A* 102, 5820–5825. doi: 10.1073/pnas.0501507102

Hassel, B., Tessler, S., Faull, R. L., and Emson, P. C. (2008). Glutamate uptake is reduced in prefrontal cortex in Huntington's disease. *Neurochem. Res.* 33, 232–237. doi: 10.1007/s11064-007-9463-1

Havel, L. S., Li, S., and Li, X. J. (2009). Nuclear accumulation of polyglutamine disease proteins and neuropathology. *Mol. Brain* 2:21. doi: 10.1186/1756-6606-2-21

Heathman, T. R., Glyn, V. A., Picken, A., Rafiq, Q. A., Coopman, K., Nienow, A. W., et al. (2015). Expansion, harvest and cryopreservation of human mesenchymal stem cells in a serum-free microcarrier process. *Biotechnol. Bioeng.* 112, 1696–1707. doi: 10.1002/bit.25582

Hidalgo-Bastida, L. A., and Cartmell, S. H. (2010). Mesenchymal stem cells, osteoblasts and extracellular matrix proteins: enhancing cell adhesion and differentiation for bone tissue engineering. *Tissue Eng. Part B Rev.* 16, 405–412. doi: 10.1089/ten.teb.2009.0714

Hoban, D. B., Howard, L., and Dowd, E. (2015). GDNF-secreting mesenchymal stem cells provide localized neuroprotection in an inflammation-driven rat model of Parkinson's disease. *Neuroscience* 303, 402–411. doi: 10.1016/j.neuroscience.2015.07.014

Hosseini, M., Moghadas, M., Edalatmanesh, M. A., and Hashemzadeh, M. R. (2015). Xenotransplantation of human adipose derived mesenchymal stem cells in a rodent model of Huntington's disease: motor and non-motor outcomes. *Neurol. Res.* 37, 309–319. doi: 10.1179/1743132814y.0000000456

Huang, B., Tabata, Y., and Gao, J. Q. (2012). Mesenchymal stem cells as therapeutic agents and potential targeted gene delivery vehicle for brain diseases. *J. Control Release* 162, 464–473. doi: 10.1016/j.jconrel.2012.07.034

Huda, F., Fan, Y., Suzuki, M., Konno, A., Matsuzaki, Y., Takahashi, N., et al. (2016). Fusion of human fetal mesenchymal stem cells with "degenerating" cerebellar neurons in Spinocerebellar ataxia type 1 model mice. *PLoS One* 11:e0164202. doi: 10.1371/journal.pone.0164202

Im, W., Lee, S. T., Park, J. E., Oh, H. J., Shim, J., Lim, J., et al. (2010). Transplantation of patient-derived adipose stem cells in YAC128 Huntington's disease transgenic mice. *PLoS Curr.* 2:RRN1183. doi: 10.1371/currents.rrn1183

Ivanova-Todorova, E., Bochev, I., Mourdjeva, M., Dimitrov, R., Bukarev, D., Kyurkchiev, S., et al. (2009). Adipose tissue-derived mesenchymal stem cells are more potent suppressors of dendritic cells differentiation compared to bone marrow-derived mesenchymal stem cells. *Immunol. Lett.* 126, 37–42. doi: 10.1016/j.imlet.2009.07.010

Jin, J. L., Liu, Z., Lu, Z. J., Guan, D. N., Wang, C., Chen, Z. B., et al. (2013). Safety and efficacy of umbilical cord mesenchymal stem cell therapy in hereditary spinocerebellar ataxia. *Curr. Neurovasc. Res.* 10, 11–20. doi: 10.2174/156720213804805936

Katona, R. L., Sinko, I., Hollo, G., Szucs, K. S., Praznovszky, T., Kereso, J., et al. (2008). A combined artificial chromosome-stem cell therapy method in a model experiment aimed at the treatment of Krabbe's disease in the Twitcher mouse. *Cell Mol. Life Sci.* 65, 3830–3838. doi: 10.1007/s00018-008-8442-2

Kim, H. S., Choi, D. Y., Yun, S. J., Choi, S. M., Kang, J. W., Jung, J. W., et al. (2012). Proteomic analysis of microvesicles derived from human mesenchymal stem cells. *J. Proteome Res.* 11, 839–849. doi: 10.1021/pr200682z

Kim, H. Y., Kim, T. J., Kang, L., Kim, Y. J., Kang, M. K., Kim, J., et al. (2020). Mesenchymal stem cell-derived magnetic extracellular nanovesicles for targeting and treatment of ischemic stroke. *Biomaterials* 243:119942. doi: 10.1016/j.biomaterials.2020.119942

Kim, S. H., Bianco, N. R., Shufesky, W. J., Morelli, A. E., and Robbins, P. D. (2007). Effective treatment of inflammatory disease models with exosomes derived from dendritic cells genetically modified to express IL-4. *J. Immunol.* 179, 2242–2249. doi: 10.4049/jimmunol.179.4.2242

Klockgether, T., Mariotti, C., and Paulson, H. L. (2019). Spinocerebellar ataxia. *Nat. Rev. Dis. Primers* 5:24. doi: 10.1038/s41572-019-0074-3

Kojima, R., Bojar, D., Rizzi, G., Hamri, G. C., El-Baba, M. D., Saxena, P., et al. (2018). Designer exosomes produced by implanted cells intracerebrally deliver therapeutic cargo for Parkinson's disease treatment. *Nat. Commun.* 9:1305. doi: 10.1038/s41467-018-03733-8

Kunter, U., Rong, S., Boor, P., Eitner, F., Muller-Newen, G., Djuric, Z., et al. (2007). Mesenchymal stem cells prevent progressive experimental renal failure but maldifferentiate into glomerular adipocytes. *J. Am. Soc. Nephrol.* 18, 1754–1764. doi: 10.1681/asn.2007010044

Labbadia, J., and Morimoto, R. I. (2013). Huntington's disease: underlying molecular mechanisms and emerging concepts. *Trends Biochem. Sci.* 38, 378–385. doi: 10.1016/j.tibs.2013.05.003

Lai, R. C., Arslan, F., Lee, M. M., Sze, N. S., Choo, A., Chen, T. S., et al. (2010). Exosome secreted by MSC reduces myocardial ischemia/reperfusion injury. *Stem Cell Res.* 4, 214–222. doi: 10.1016/j.scr.2009.12.003

Lai, P., Chen, X., Guo, L., Wang, Y., Liu, X., Liu, Y., et al. (2018). A potent immunomodulatory role of exosomes derived from mesenchymal stromal cells in preventing cGVHD. *J. Hematol. Oncol.* 11:135. doi: 10.1186/s13045-018-0680-7

Lai, R. C., Tan, S. S., Teh, B. J., Sze, S. K., Arslan, F., de Kleijn, D. P., et al. (2012). Proteolytic potential of the msc exosome proteome: implications for an exosome-mediated delivery of therapeutic proteasome. *Int. J. Proteomics* 2012:971907. doi: 10.1155/2012/971907

Lai, P., Weng, J., Guo, L., Chen, X., and Du, X. (2019). Novel insights into MSC-EVs therapy for immune diseases. *Biomark. Res.* 7:6. doi: 10.1186/s40364-019-0156-0

Lee, K. S., Cha, S. H., Kang, H. W., Song, J. Y., Lee, K. W., Ko, K. B., et al. (2013). Effects of serial passage on the characteristics and chondrogenic differentiation of canine umbilical cord matrix derived mesenchymal stem cells. *Asian-Australas. J. Anim. Sci.* 26, 588–595. doi: 10.5713/ajas.2012.12488

Lee, S. T., Chu, K., Jung, K. H., Im, W. S., Park, J. E., Lim, H. C., et al. (2009). Slowed progression in models of Huntington disease by adipose stem cell transplantation. *Ann. Neurol.* 66, 671–681. doi: 10.1002/ana.21788

Lee, J. H., Yoon, Y. M., and Lee, S. H. (2017). Hypoxic preconditioning promotes the bioactivities of mesenchymal stem cells *via* the HIF-1α-GRP78-Akt axis. *Int. J. Mol. Sci.* 18:1320. doi: 10.3390/ijms18061320

Lei, L. F., Yang, G. P., Wang, J. L., Chuang, D. M., Song, W. H., Tang, B. S., et al. (2016). Safety and efficacy of valproic acid treatment in SCA3/MJD patients. *Parkinsonism Relat. Disord.* 26, 55–61. doi: 10.1016/j.parkreldis.2016.03.005

Li, L. Y., Li, J. T., Wu, Q. Y., Li, J., Feng, Z. T., Liu, S., et al. (2008). Transplantation of NGF-gene-modified bone marrow stromal cells into a rat model of Alzheimer' disease. *J. Mol. Neurosci.* 34, 157–163. doi: 10.1007/s12031-007-9022-x

Li, T., Liu, Y., Yu, L., Lao, J., Zhang, M., Jin, J., et al. (2018). Human umbilical cord mesenchymal stem cells protect against SCA3 by modulating the level of 70 kD heat shock protein. *Cell. Mol. Neurobiol.* 38, 641–655. doi: 10.1007/s10571-017-0513-1

Lin, Y. T., Chern, Y., Shen, C. K., Wen, H. L., Chang, Y. C., Li, H., et al. (2011). Human mesenchymal stem cells prolong survival and ameliorate motor deficit through trophic support in Huntington's disease mouse models. *PLoS One* 6:e22924. doi: 10.1371/journal.pone.0022924

Linares, G. R., Chiu, C. T., Scheuing, L., Leng, Y., Liao, H. M., Maric, D., et al. (2016). Preconditioning mesenchymal stem cells with the mood stabilizers lithium and valproic acid enhances therapeutic efficacy in a mouse model of Huntington's disease. *Exp. Neurol.* 281, 81–92. doi: 10.1016/j.expneurol.2016.04.003

Lobo, D., Nobre, R. J., Miranda, C. O., Pereira, D., Castelhano, J., Sereno, J., et al. (2020). The blood-brain barrier is disrupted in Machado-Joseph disease | Spinocerebellar Ataxia type 3: Evidence from transgenic mice and human post-mortem samples. *Acta Neuropathol. Commun.* 8:152. doi: 10.1186/s40478-020-00955-0

Lopes-Ramos, C. M., Pereira, T. C., Dogini, D. B., Gilioli, R., and Lopes-Cendes, I. (2016). Lithium carbonate and coenzyme Q10 reduce cell death in a cell model of Machado-Joseph disease. *Braz. J. Med. Biol. Res.* 49:e5805. doi: 10.1590/1414-431x20165805

Maciel, P., Gaspar, C., DeStefano, A. L., Silveira, I., Coutinho, P., Radvany, J., et al. (1995). Correlation between CAG repeat length and clinical features in Machado-Joseph disease. *Am. J. Hum. Genet.* 57, 54–61.

Maltman, D. J., Hardy, S. A., and Przyborski, S. A. (2011). Role of mesenchymal stem cells in neurogenesis and nervous system repair. *Neurochem. Int.* 59, 347–356. doi: 10.1016/j.neuint.2011.06.008

Matos, C. A., Carmona, V., Vijayakumar, U. G., Lopes, S., Albuquerque, P., Conceicao, M., et al. (2018). Gene therapies for polyglutamine diseases. *Adv. Exp. Med. Biol.* 1049, 395–438. doi: 10.1007/978-3-319-71779-1_20

Matsuura, S., Shuvaev, A. N., Iizuka, A., Nakamura, K., and Hirai, H. (2014). Mesenchymal stem cells ameliorate cerebellar pathology in a mouse model of spinocerebellar ataxia type 1. *Cerebellum* 13, 323–330. doi: 10.1007/s12311-013-0536-1

Mazzini, L., Fagioli, F., Boccaletti, R., Mareschi, K., Oliveri, G., Olivieri, C., et al. (2003). Stem cell therapy in amyotrophic lateral sclerosis: a methodological approach in humans. *Amyotroph. Lateral Scler. Other Motor Neuron Disord.* 4, 158–161. doi: 10.1080/14660820310014653

Mendonça, L. S., Onofre, I., Miranda, C. O., Perfeito, R., Nobrega, C., and de Almeida, L. P. (2018). Stem cell-based therapies for polyglutamine diseases. *Adv. Exp. Med. Biol.* 1049, 439–466. doi: 10.1007/978-3-319-71779-1_21

Miao, X., Wu, X., and Shi, W. (2015). Umbilical cord mesenchymal stem cells in neurological disorders: a clinical study. *Indian J. Biochem. Biophys.* 52, 140–146.

Mieda, T., Suto, N., Iizuka, A., Matsuura, S., Iizuka, H., Takagishi, K., et al. (2016). Mesenchymal stem cells attenuate peripheral neuronal degeneration in spinocerebellar ataxia type 1 knockin mice. *J. Neurosci. Res.* 94, 246–252. doi: 10.1002/jnr.23698

Millard, S. M., and Fisk, N. M. (2013). Mesenchymal stem cells for systemic therapy: shotgun approach or magic bullets? *Bioessays* 35, 173–182. doi: 10.1002/bies.201200087

Miranda, C. O., Teixeira, C. A., Liz, M. A., Sousa, V. F., Franquinho, F., Forte, G., et al. (2011). Systemic delivery of bone marrow-derived mesenchymal stromal cells diminishes neuropathology in a mouse model of Krabbe's disease. *Stem Cells* 29, 1738–1751. doi: 10.1002/stem.724

Moloney, T. C., Rooney, G. E., Barry, F. P., Howard, L., and Dowd, E. (2010). Potential of rat bone marrow-derived mesenchymal stem cells as vehicles for delivery of neurotrophins to the Parkinsonian rat brain. *Brain Res.* 1359, 33–43. doi: 10.1016/j.brainres.2010.08.040

Moon, G. J., Cho, Y. H., Kim, D. H., Sung, J. H., Son, J. P., Kim, S., et al. (2018). Serum-mediated activation of bone marrow-derived mesenchymal stem cells in ischemic stroke patients: a novel preconditioning method. *Cell Transplant.* 27, 485–500. doi: 10.1177/0963689718755404

Moya, A., Paquet, J., Deschepper, M., Larochette, N., Oudina, K., Denoeud, C., et al. (2018). Human mesenchymal stem cell failure to adapt to glucose shortage and rapidly use intracellular energy reserves through glycolysis explains poor cell survival after implantation. *Stem Cells* 36, 363–376. doi: 10.1002/stem.2763

Mulcahy, L. A., Pink, R. C., and Carter, D. R. (2014). Routes and mechanisms of extracellular vesicle uptake. *J. Extracell. Vesicles* 3:24641. doi: 10.3402/jev.v3.24641

Nakano, N., Nakai, Y., Seo, T. B., Yamada, Y., Ohno, T., Yamanaka, A., et al. (2010). Characterization of conditioned medium of cultured bone marrow stromal cells. *Neurosci. Lett.* 483, 57–61. doi: 10.1016/j.neulet.2010.07.062

Ning, H., Yang, F., Jiang, M., Hu, L., Feng, K., Zhang, J., et al. (2008). The correlation between cotransplantation of mesenchymal stem cells and higher recurrence rate in hematologic malignancy patients: outcome of a pilot clinical study. *Leukemia* 22, 593–599. doi: 10.1038/sj.leu.2405090

Noronha, N. C., Mizukami, A., Caliari-Oliveira, C., Cominal, J. G., Rocha, J. L. M., Covas, D. T., et al. (2019). Priming approaches to improve the efficacy of mesenchymal stromal cell-based therapies. *Stem Cell Res. Ther.* 10:131. doi: 10.1186/s13287-019-1224-y

Oliveira Miranda, C., Marcelo, A., Silva, T. P., Barata, J., Vasconcelos-Ferreira, A., Pereira, D., et al. (2018). Repeated mesenchymal stromal cell treatment sustainably alleviates machado-joseph disease. *Mol. Ther.* 26, 2131–2151. doi: 10.1016/j.ymthe.2018.07.007

Onda, T., Honmou, O., Harada, K., Houkin, K., Hamada, H., and Kocsis, J. D. (2008). Therapeutic benefits by human mesenchymal stem cells (hMSCs) and Ang-1 gene-modified hMSCs after cerebral ischemia. *J. Cereb. Blood Flow Metab.* 28, 329–340. doi: 10.1038/sj.jcbfm.9600527

Orr, H. T., and Zoghbi, H. Y. (2007). Trinucleotide repeat disorders. *Annu. Rev. Neurosci.* 30, 575–621. doi: 10.1016/B978-0-12-802395-2.00027-4

Owen, M., and Friedenstein, A. J. (1988). "Stromal stem cells: marrow-derived osteogenic precursors," in *Ciba Foundation Symposium 136 - Cell and Molecular Biology of Vertebrate Hard Tissues: Cell and Molecular Biology of Vertebrate Hard Tissues: Ciba Foundation Symposium 136*, eds D. Evered and S. Harnett (UK: John Wiley & Sons), 42–60. doi: 10.1002/9780470513637.ch4

Parga, J. A., Garcia-Garrote, M., Martinez, S., Raya, A., Labandeira-Garcia, J. L., and Rodriguez-Pallares, J. (2018). Prostaglandin EP2 receptors mediate mesenchymal stromal cell-neuroprotective effects on dopaminergic neurons. *Mol. Neurobiol.* 55, 4763–4776. doi: 10.1007/s12035-017-0681-5

Park, J. S., Suryaprakash, S., Lao, Y. H., and Leong, K. W. (2015). Engineering mesenchymal stem cells for regenerative medicine and drug delivery. *Methods* 84, 3–16. doi: 10.1016/j.ymeth.2015.03.002

Pashoutan Sarvar, D., Shamsasenjan, K., and Akbarzadehlaleh, P. (2016). Mesenchymal stem cell-derived exosomes: new opportunity in cell-free therapy. *Adv. Pharm. Bull.* 6, 293–299. doi: 10.15171/apb.2016.041

Paul, G., and Anisimov, S. V. (2013). The secretome of mesenchymal stem cells: potential implications for neuroregeneration. *Biochimie* 95, 2246–2256. doi: 10.1016/j.biochi.2013.07.013

Paulson, H. L. (2007). Dominantly inherited ataxias: lessons learned from Machado-Joseph disease/spinocerebellar ataxia type 3. *Semin. Neurol.* 27, 133–142. doi: 10.1055/s-2007-971172

Paulson, H. L. (2009). The spinocerebellar ataxias. *J. Neuroophthalmol.* 29, 227–237. doi: 10.1097/WNO0b013e3181b416de

Pittenger, M. F., Mackay, A. M., Beck, S. C., Jaiswal, R. K., Douglas, R., Mosca, J. D., et al. (1999). Multilineage potential of adult human mesenchymal stem cells. *Science* 284, 143–147. doi: 10.1126/science.284.5411.143

Pollock, K., Dahlenburg, H., Nelson, H., Fink, K. D., Cary, W., Hendrix, K., et al. (2016). Human mesenchymal stem cells genetically engineered to overexpress brain-derived neurotrophic factor improve outcomes in Huntington's disease mouse models. *Mol. Ther.* 24, 965–977. doi: 10.1038/mt.2016.12

Prockop, D. J. (1997). Marrow stromal cells as stem cells for nonhematopoietic tissues. *Science* 276, 71–74. doi: 10.1126/science.276.5309.71

Ranum, L. P., Lundgren, J. K., Schut, L. J., Ahrens, M. J., Perlman, S., Aita, J., et al. (1995). Spinocerebellar ataxia type 1 and Machado-Joseph disease: incidence of CAG expansions among adult-onset ataxia patients from 311 families with dominant, recessive, or sporadic ataxia. *Am. J. Hum. Genet.* 57, 603–608.

Ren, Z., Wang, J., Wang, S., Zou, C., Li, X., Guan, Y., et al. (2013). Autologous transplantation of GDNF-expressing mesenchymal stem cells protects against MPTP-induced damage in cynomolgus monkeys. *Sci. Rep.* 3:2786. doi: 10.1038/srep02786

Reza-Zaldivar, E. E., Hernández-Sapiéns, M. A., Gutiérrez-Mercado, Y. K., Sandoval-Avila, S., Gomez-Pinedo, U., Marquez-Aguirre, A. L., et al. (2019). Mesenchymal stem cell-derived exosomes promote neurogenesis and cognitive function recovery in a mouse model of Alzheimer's disease. *Neural Regen. Res.* 14, 1626–1634. doi: 10.4103/1673-5374.255978

Rojewski, M. T., Fekete, N., Baila, S., Nguyen, K., Furst, D., Antwiler, D., et al. (2013). GMP-compliant isolation and expansion of bone marrow-derived MSCs in the closed, automated device quantum cell expansion system. *Cell Transplant.* 22, 1981–2000. doi: 10.3727/096368912x657990

Ross, C. A., Poirier, M. A., Wanker, E. E., and Amzel, M. (2003). Polyglutamine fibrillogenesis: the pathway unfolds. *Proc. Natl. Acad. Sci. U S A* 100, 1–3. doi: 10.1073/pnas.0237018100

Rossignol, J., Boyer, C., Lévèque, X., Fink, K. D., Thinard, R., Blanchard, F., et al. (2011). Mesenchymal stem cell transplantation and DMEM administration in a 3NP rat model of Huntington's disease: morphological and behavioral outcomes. *Behav. Brain Res.* 217, 369–378. doi: 10.1016/j.bbr.2010.11.006

Rossignol, J., Fink, K. D., Crane, A. T., Davis, K. K., Bombard, M. C., Clerc, S., et al. (2015). Reductions in behavioral deficits and neuropathology in the R6/2 mouse model of Huntington's disease following transplantation of bone-marrow-derived mesenchymal stem cells is dependent on passage number. *Stem Cell Res. Ther.* 6:9. doi: 10.1186/scrt545

Rufino-Ramos, D., Albuquerque, P. R., Carmona, V., Perfeito, R., Nobre, R. J., and Pereira de Almeida, L. (2017). Extracellular vesicles: Novel promising delivery systems for therapy of brain diseases. *J. Control Release* 262, 247–258. doi: 10.1016/j.jconrel.2017.07.001

Sadan, O., Bahat-Stromza, M., Barhum, Y., Levy, Y. S., Pisnevsky, A., Peretz, H., et al. (2009). Protective effects of neurotrophic factor-secreting cells in a 6-OHDA rat model of Parkinson disease. *Stem Cells Dev.* 18, 1179–1190. doi: 10.1089/scd.2008.0411

Sadan, O., Shemesh, N., Barzilay, R., Bahat-Stromza, M., Melamed, E., Cohen, Y., et al. (2008). Migration of neurotrophic factors-secreting mesenchymal stem cells toward a quinolinic acid lesion as viewed by magnetic resonance imaging. *Stem Cells* 26, 2542–2551. doi: 10.1634/stemcells.2008-0240

Sadan, O., Shemesh, N., Barzilay, R., Dadon-Nahum, M., Blumenfeld-Katzir, T., Assaf, Y., et al. (2012). Mesenchymal stem cells induced to secrete neurotrophic factors attenuate quinolinic acid toxicity: a potential therapy for Huntington's disease. *Exp. Neurol.* 234, 417–427. doi: 10.1016/j.expneurol.2011.12.045

Sano, S., Izumi, Y., Yamaguchi, T., Yamazaki, T., Tanaka, M., Shiota, M., et al. (2014). Lipid synthesis is promoted by hypoxic adipocyte-derived exosomes in 3T3-L1 cells. *Biochem. Biophys. Res. Commun.* 445, 327–333. doi: 10.1016/j.bbrc.2014.01.183

Saute, J. A., de Castilhos, R. M., Monte, T. L., Schumacher-Schuh, A. F., Donis, K. C., D'Avila, R., et al. (2014). A randomized, phase 2 clinical trial of lithium carbonate in Machado-Joseph disease. *Mov. Disord.* 29, 568–573. doi: 10.1002/mds.25803

Schuler-Faccini, L., Osorio, C. M., Romariz, F., Paneque, M., Sequeiros, J., and Jardim, L. B. (2014). Genetic counseling and presymptomatic testing programs for machado-joseph disease: lessons from brazil and portugal. *Genet. Mol. Biol.* 37, 263–270. doi: 10.1590/s1415-47572014000200012

Sensebé, L., Gadelorge, M., and Fleury-Cappellesso, S. (2013). Production of mesenchymal stromal/stem cells according to good manufacturing practices: a review. *Stem Cell Res. Ther.* 4:66. doi: 10.1186/scrt217

Sequeiros, J., Maciel, P., Taborda, F., Ledo, S., Rocha, J. C., Lopes, A., et al. (1998). Prenatal diagnosis of Machado-Joseph disease by direct mutation analysis. *Prenat. Diagn.* 18, 611–617. doi: 10.1002/(sici)1097-0223(199806)18:6<611::aid-pd289>3.0.co;2-y

Shao, J., and Diamond, M. I. (2007). Polyglutamine diseases: emerging concepts in pathogenesis and therapy. *Hum. Mol. Genet.* 16, R115–R123. doi: 10.1093/hmg/ddm213

Sharma, R. R., Pollock, K., Hubel, A., and McKenna, D. (2014). Mesenchymal stem or stromal cells: a review of clinical applications and manufacturing practices. *Transfusion* 54, 1418–1437. doi: 10.1111/trf.12421

Snyder, B. R., Chiu, A. M., Prockop, D. J., and Chan, A. W. (2010). Human multipotent stromal cells (MSCs) increase neurogenesis and decrease atrophy of the striatum in a transgenic mouse model for Huntington's disease. *PLoS One* 5:e9347. doi: 10.1371/journal.pone.0009347

Sompol, P., and Norris, C. M. (2018). Ca^{2+}, astrocyte activation and calcineurin/NFAT signaling in age-related neurodegenerative diseases. *Front. Aging Neurosci.* 10:199. doi: 10.3389/fnagi.2018.00199

Spaggiari, G. M., Capobianco, A., Abdelrazik, H., Becchetti, F., Mingari, M. C., and Moretta, L. (2008). Mesenchymal stem cells inhibit natural killer-cell proliferation, cytotoxicity and cytokine production: role of indoleamine 2,3-dioxygenase and prostaglandin E2. *Blood* 111, 1327–1333. doi: 10.1182/blood-2007-02-074997

Stolzing, A., Jones, E., McGonagle, D., and Scutt, A. (2008). Age-related changes in human bone marrow-derived mesenchymal stem cells: consequences for cell therapies. *Mech. Ageing Dev.* 129, 163–173. doi: 10.1016/j.mad.2007.12.002

Sullivan, R., Duncan, K., Dailey, T., Kaneko, Y., Tajiri, N., and Borlongan, C. V. (2015). A possible new focus for stroke treatment—migrating stem cells. *Expert Opin. Biol. Ther.* 15, 949–958. doi: 10.1517/14712598.2015.1043264

Sullivan, R., Yau, W. Y., O'Connor, E., and Houlden, H. (2019). Spinocerebellar ataxia: an update. *J. Neurol.* 266, 533–544. doi: 10.1007/s00415-018-9076-4

Sundin, M., Orvell, C., Rasmusson, I., Sundberg, B., Ringden, O., and Le Blanc, K. (2006). Mesenchymal stem cells are susceptible to human herpesviruses, but viral DNA cannot be detected in the healthy seropositive individual. *Bone Marrow Transplant.* 37, 1051–1059. doi: 10.1038/sj.bmt.1705368

Suto, N., Mieda, T., Iizuka, A., Nakamura, K., and Hirai, H. (2016). Morphological and functional attenuation of degeneration of peripheral neurons by mesenchymal stem cell-conditioned medium in spinocerebellar ataxia type 1-knock-in mice. *CNS Neurosci. Ther.* 22, 670–676. doi: 10.1111/cns.12560

Tajiri, N., Kaneko, Y., Shinozuka, K., Ishikawa, H., Yankee, E., McGrogan, M., et al. (2013). Stem cell recruitment of newly formed host cells *via* a successful seduction? Filling the gap between neurogenic niche and injured brain site. *PLoS One* 8:e74857. doi: 10.1371/journal.pone.0074857

Tan, Y., Salkhordeh, M., Wang, J. P., McRae, A., Souza-Moreira, L., McIntyre, L., et al. (2019). Thawed mesenchymal stem cell product shows comparable immunomodulatory potency to cultured cells *in vitro* and in polymicrobial septic animals. *Sci. Rep.* 9:18078. doi: 10.1038/s41598-019-54462-x

Tarte, K., Gaillard, J., Lataillade, J. J., Fouillard, L., Becker, M., Mossafa, H., et al. (2010). Clinical-grade production of human mesenchymal stromal cells: occurrence of aneuploidy without transformation. *Blood* 115, 1549–1553. doi: 10.1182/blood-2009-05-219907

Teixeira, F. G., Carvalho, M. M., Panchalingam, K. M., Rodrigues, A. J., Mendes-Pinheiro, B., Anjo, S., et al. (2017). Impact of the secretome of human mesenchymal stem cells on brain structure and animal behavior in a rat model of Parkinson's disease. *Stem Cells Transl. Med.* 6, 634–646. doi: 10.5966/sctm.2016-0071

Thery, C., Witwer, K. W., Aikawa, E., Alcaraz, M. J., Anderson, J. D., Andriantsitohaina, R., et al. (2018). Minimal information for studies of extracellular vesicles 2018 (MISEV2018): a position statement of the International Society for Extracellular Vesicles and update of the MISEV2014 guidelines. *J. Extracell. Vesicles* 7:1535750. doi: 10.1080/20013078.2018.1535744

Thery, C., Zitvogel, L., and Amigorena, S. (2002). Exosomes: composition, biogenesis and function. *Nat. Rev. Immunol.* 2, 569–579. doi: 10.1038/nri855

Torashima, T., Koyama, C., Iizuka, A., Mitsumura, K., Takayama, K., Yanagi, S., et al. (2008). Lentivector-mediated rescue from cerebellar ataxia in a mouse model of spinocerebellar ataxia. *EMBO Rep.* 9, 393–399. doi: 10.1038/embor.2008.31

Tremain, N., Korkko, J., Ibberson, D., Kopen, G. C., DiGirolamo, C., and Phinney, D. G. (2001). MicroSAGE analysis of 2,353 expressed genes in a single cell-derived colony of undifferentiated human mesenchymal stem cells reveals mRNAs of multiple cell lineages. *Stem Cells* 19, 408–418. doi: 10.1634/stemcells.19-5-408

Tsai, Y. A., Liu, R. S., Lirng, J. F., Yang, B. H., Chang, C. H., Wang, Y. C., et al. (2017). Treatment of spinocerebellar ataxia with mesenchymal stem cells: a phase I/IIa clinical study. *Cell Transplant.* 26, 503–512. doi: 10.3727/096368916x694373

Tuite, P. J., Rogaeva, E. A., St George-Hyslop, P. H., and Lang, A. E. (1995). Dopa-responsive parkinsonism phenotype of Machado-Joseph disease: confirmation of 14q CAG expansion. *Ann. Neurol.* 38, 684–687. doi: 10.1002/ana.410380422

Wagner, W., Horn, P., Castoldi, M., Diehlmann, A., Bork, S., Saffrich, R., et al. (2008). Replicative senescence of mesenchymal stem cells: a continuous and organized process. *PLoS One* 3:e2213. doi: 10.1371/journal.pone.0002213

Wakitani, S., Saito, T., and Caplan, A. I. (1995). Myogenic cells derived from rat bone marrow mesenchymal stem cells exposed to 5-azacytidine. *Muscle Nerve* 18, 1417–1426. doi: 10.1002/mus.880181212

Wang, N., Chen, C., Yang, D., Liao, Q., Luo, H., Wang, X., et al. (2017). Mesenchymal stem cells-derived extracellular vesicles, *via* miR-210, improve infarcted cardiac function by promotion of angiogenesis. *Biochim. Biophys. Acta Mol. Basis Dis.* 1863, 2085–2092. doi: 10.1016/j.bbadis.2017.02.023

Wang, S. S., Jia, J., and Wang, Z. (2018). Mesenchymal stem cell-derived extracellular vesicles suppresses inos expression and ameliorates neural impairment in Alzheimer's disease mice. *J. Alzheimers Dis.* 61, 1005–1013. doi: 10.3233/jad-170848

Wang, J. W., Qiu, Y. R., Fu, Y., Liu, J., He, Z. J., and Huang, Z. T. (2017). Transplantation with hypoxia-preconditioned mesenchymal stem cells suppresses brain injury caused by cardiac arrest-induced global cerebral ischemia in rats. *J. Neurosci. Res.* 95, 2059–2070. doi: 10.1002/jnr.24025

Witwer, K. W., Van Balkom, B. W. M., Bruno, S., Choo, A., Dominici, M., Gimona, M., et al. (2019). Defining mesenchymal stromal cell (MSC)-derived small extracellular vesicles for therapeutic applications. *J. Extracell. Vesicles* 8:1609206. doi: 10.1080/20013078.2019.1609206

Wu, S., Ju, G. Q., Du, T., Zhu, Y. J., and Liu, G. H. (2013). Microvesicles derived from human umbilical cord Wharton's jelly mesenchymal stem cells attenuate bladder tumor cell growth *in vitro* and *in vivo*. *PLoS One* 8:e61366. doi: 10.1371/journal.pone.0061366

Xin, H., Li, Y., Buller, B., Katakowski, M., Zhang, Y., Wang, X., et al. (2012). Exosome-mediated transfer of miR-133b from multipotent mesenchymal stromal cells to neural cells contributes to neurite outgrowth. *Stem Cells* 30, 1556–1564. doi: 10.1002/stem.1129

Xin, H., Li, Y., and Chopp, M. (2014). Exosomes/miRNAs as mediating cell-based therapy of stroke. *Front. Cell. Neurosci.* 8:377. doi: 10.3389/fncel.2014.00377

Xin, H., Li, Y., Cui, Y., Yang, J. J., Zhang, Z. G., and Chopp, M. (2013). Systemic administration of exosomes released from mesenchymal stromal cells promote functional recovery and neurovascular plasticity after stroke in rats. *J. Cereb. Blood Flow Metab.* 33, 1711–1715. doi: 10.1038/jcbfm.2013.152

Yan, S., Tu, Z., Liu, Z., Fan, N., Yang, H., Yang, H., et al. (2018). A huntingtin knockin pig model recapitulates features of selective neurodegeneration in Huntington's disease. *Cell* 173, 989.e13–1002.e13. doi: 10.1016/j.cell.2018.03.005

Yu, B., Zhang, X., and Li, X. (2014). Exosomes derived from mesenchymal stem cells. *Int. J. Mol. Sci.* 15, 4142–4157. doi: 10.3390/ijms15034142

Yu-Taeger, L., Stricker-Shaver, J., Arnold, K., Bambynek-Dziuk, P., Novati, A., Singer, E., et al. (2019). Intranasal administration of mesenchymal stem cells ameliorates the abnormal dopamine transmission system and inflammatory reaction in the R6/2 mouse model of huntington disease. *Cells* 8:595. doi: 10.3390/cells8060595

Zhang, X., Tu, H., Yang, Y., Fang, L., Wu, Q., and Li, J. (2017). Mesenchymal stem cell-derived extracellular vesicles: roles in tumor growth, progression, and drug resistance. *Stem Cells Int.* 2017:1758139. doi: 10.1155/2017/1758139

Zhilai, Z., Biling, M., Sujun, Q., Chao, D., Benchao, S., Shuai, H., et al. (2016). Preconditioning in lowered oxygen enhances the therapeutic potential of human umbilical mesenchymal stem cells in a rat model of spinal cord injury. *Brain Res.* 1642, 426–435. doi: 10.1016/j.brainres.2016.04.025

Zoghbi, H. Y., and Orr, H. T. (2000). Glutamine repeats and neurodegeneration. *Annu. Rev. Neurosci.* 23, 217–247. doi: 10.1146/annurev.neuro.23.1.217

Zöller, M. (2009). Tetraspanins: push and pull in suppressing and promoting metastasis. *Nat. Rev. Cancer* 9, 40–55. doi: 10.1038/nrc2543

4

Sensorineural Hearing Loss and Mitochondrial Apoptosis of Cochlear Spiral Ganglion Neurons in Fibroblast Growth Factor 13 Knockout Mice

Yulou Yu[1,2†], Jing Yang[3†], Feng Luan[4], Guoqiang Gu[5], Ran Zhao[1,2], Qiong Wang[1,2], Zishan Dong[1,2], Junming Tang[6], Wei Wang[3], Jinpeng Sun[7], Ping Lv[1,2], Hailin Zhang[1,2*] and Chuan Wang[1,2*]*

[1] The Key Laboratory of Neural and Vascular Biology, Ministry of Education, Hebei Medical University, Shijiazhuang, China, [2] The Key Laboratory of New Drug Pharmacology and Toxicology, Department of Pharmacology, Hebei Medical University, Shijiazhuang, China, [3] Department of Physiology, Hebei Medical University, Shijiazhuang, China, [4] Department of Otolaryngology, The Third Hospital of Hebei Medical University, Shijiazhuang, China, [5] Department of Cardiology, The Second Hospital of Hebei Medical University, Shijiazhuang, China, [6] Hubei Key Laboratory of Embryonic Stem Cell Research, Hubei University of Medicine, Shiyan, China, [7] Key Laboratory Experimental Teratology of the Ministry of Education, Department of Biochemistry and Molecular Biology, School of Basic Medical Sciences, Cheeloo College of Medicine, Shandong University, Jinan, China

***Correspondence:**
Chuan Wang
wangchuan@hebmu.edu.cn
Hailin Zhang
zhanghl@hebmu.edu.cn
Ping Lv
lping77@hotmail.com

†These authors have contributed equally to this work

Deafness is known to occur in more than 400 syndromes and accounts for almost 30% of hereditary hearing loss. The molecular mechanisms underlying such syndromic deafness remain unclear. Furthermore, deafness has been a common feature in patients with three main syndromes, the BÖrjeson-Forssman-Lehmann syndrome, Wildervanck syndrome, and Congenital Generalized Hirsutism, all of which are characterized by loss-of-function mutations in the *Fgf13* gene. Whether the pathogenesis of deafness in these syndromes is associated with the *Fgf13* mutation is not known. To elucidate its role in auditory function, we generated a mouse line with conditional knockout of the *Fgf13* gene in the inner ear (*Fgf13* cKO). FGF13 is expressed predominantly in the organ of Corti, spiral ganglion neurons (SGNs), stria vascularis, and the supporting cells. Conditional knockout of the gene in the inner ear led to sensorineural deafness with low amplitude and increased latency of wave I in the auditory brainstem response test but had a normal distortion product otoacoustic emission threshold. *Fgf13* deficiency resulted in decreased SGN density from the apical to the basal region without significant morphological changes and those in the number of hair cells. TUNEL and caspase-3 immunocytochemistry assays showed that apoptotic cell death mediated the loss of SGNs. Further detection of apoptotic factors through qRT-PCR suggested the activation of the mitochondrial apoptotic pathway in SGNs. Together, this study reveals a novel role for *Fgf13* in auditory function, and indicates that the gene could be a potential candidate for understanding deafness. These findings may provide new perspectives on the molecular mechanisms and novel therapeutic targets for treatment deafness.

Keywords: fibroblast growth factor 13, deafness, spiral ganglion neuron, apoptosis, mitochondria, syndrome

INTRODUCTION

Hearing loss is one of the most common sensory deficits that can occur in newborns and has been linked to many environmental and genetic factors (Morton and Nance, 2006; Ideura et al., 2019). The mammalian cochlea (the peripheral organ for hearing) is a small, yet complicated snail-shaped organ that consists of heterologous cell types arranged precisely to detect and process sounds. Hair cells (HCs) convert the physical vibrations generated by sound stimuli into chemical signals, which are then transmitted by spiral ganglion neurons (SGNs) to the central nervous system via ribbon synapses (Coate and Kelley, 2013; Li et al., 2017). Both conductive and sensorineural impairments have been known to cause hearing loss. Damage to HCs or SGNs often results in irreversible, permanent sensorineural hearing loss due to their limited self-regenerating capacity (Kwon et al., 2014; Wong and Ryan, 2015).

Hearing loss, that is not associated with any other sign and symptom, is designated as non-syndromic hearing loss (NSHL) (Dror and Avraham, 2010), mainly caused due to mutations in the genes that are involved in auditory functions. More than 120 genes have been associated with NSHL[1]. In contrast, hearing loss is known to occur more frequently in syndromes that affect various other systems of the body, referred to as syndromic hearing loss (SHL) (Chen et al., 2016). Currently, there are over 400 syndromes that result in hearing loss, corresponding to 30% of inherited deafness cases (Ideura et al., 2019). Meanwhile, the molecular mechanisms that underlie the pathogenesis of most SHL have not yet been determined, presenting a great challenge to its clinical treatment. Previous studies have shown deafness to be a common feature in patients with the BÖrjeson-Forssman-Lehmann syndrome (BFLS; OMIM #301900), Wildervanck syndrome (WS; OMIM %314600), and Congenital Generalized Hirsutism (CGH; OMIM #307150) (Gecz et al., 1999; DeStefano et al., 2013; Abu-Amero et al., 2014). Interestingly, patients with these syndromes are characterized by a loss-of-function mutation in the *Fgf13* gene. Whether the pathogenesis of deafness in these syndromic patients is associated with the *Fgf13* mutation is unclear.

Fibroblast growth factor (FGF) 13 is a multifunctional non-secretory protein that belongs to the FGF homologous factor (FHF) subfamily, which includes four genes *Fhf1-4* (with the corresponding name *Fgf11-Fgf14*) based on distinct alternative sequences (Goldfarb, 2005; Wang et al., 2011, 2017; Yang et al., 2016; Wei et al., 2017). FHFs are involved in modulating voltage-dependent sodium channels (Liu et al., 2003; Lou et al., 2005; Goetz et al., 2009; Wang et al., 2011) and microtubule-stabilizing proteins (Wu et al., 2012). FHFs are widely expressed in the brain and are known to play crucial roles in the development and functioning of the nervous system (Goldfarb, 2005). Disruption of *Fgf13* causes genetic epilepsy and febrile seizures plus ($GEFS^+$) syndrome (Puranam et al., 2015), as well as results in impaired learning and memory (Wu et al., 2012). Furthermore, the *Fgf14* mutation has been known to induce inherited ataxia (Dalski et al., 2005), while the role of *Fgf13* in the auditory system has not yet been determined. Increased FGF13 expression has also been linked to cancer progression (Okada et al., 2013; Liu et al., 2018; Song and Li, 2019), and depletion of *Fgf13* induces apoptosis in cancer cells (Bublik et al., 2016). But its role in apoptotic pathways remains unknown.

In this study, we hypothesized *Fgf13* loss of function to be associated with hearing loss. We examined the localization of FGF13 in the murine cochlear tissue. Furthermore, a transgenic mouse line with a conditional knockout of the gene in the inner ear area (*Fgf13* cKO) was generated and characterized using ethological tests. We investigated the morphological structural changes of the cochlea and the potential underlying mechanisms in these mice. We found that *Fgf13* deficiency caused sensorineural deafness with activated mitochondrial apoptotic cell death in SGNs from the apical to the basal region, but there were no significant morphological and number changes in the HCs. Our study revealed a novel role for *Fgf13* in auditory function, where it regulates the survival of SGNs in the inner ear. Our results indicate that the gene could be a potential candidate for deafness, thus providing new insights into understanding its pathogenesis and creating novel therapeutic targets.

MATERIALS AND METHODS

Ethics Statement

All animal experiments were approved by the Laboratory Animal Ethical and Welfare Committee of Hebei Medical University (Shijiazhuang, China, Approval No. IACUC- Hebmu-PD-201720). All procedures were carried out in accordance with the National Institutes of Health Guide for the Care and Use of Laboratory Animals (Tan et al., 2018).

Animals

All genetically modified mice were maintained on a C57BL/6J genetic background. We generated a mouse line with *Fgf13* conditional knockout in the inner ear via *Cre/loxP*-mediated recombination by mating *Fgf13-loxP* mice (*Fgf13*$^{fl/fl}$ or *Fgf13*$^{fl/Y}$) with Atoh1-cre mice (Tg (Atoh1-cre) 1Bfri, also named Math1-cre, MGI #011104 from the Jackson Laboratory). *Fgf13-loxP* mice were generated in collaboration with Beijing Biocytogen, Co., Ltd. (Beijing, China) by flanking exon 3 of the mouse *Fgf13* gene with two *loxP* sites as described previously (Wang et al., 2017). Genotypes were verified by PCR. Specifically, Genomic DNA was extracted from the mouse tail tip using the following specific primers:

PCR for identification of *Fgf13-loxP* gene fragment: 5′-TAGTTCCATCTAACAGGGCTCATG (forward) and 5′- AGA CTTTGGTGGGAGCATCCTG (reverse). PCR for identification of *Fgf13 Frt* gene fragment: 5′- AGTTCGACAGACAGTGCCA

Abbreviations:ABR, auditory brainstem response; BFLS, BÖrjeson-Forssman-Lehmann syndrome; CC, Claudius' cells; CGH, congenital generalized hypertrichosis; DPOAE, distortion product otoacoustic emission; FGF, fibroblast growth factor; FHF, FGF homologous factors; $GEFS^+$, genetic epilepsy and febrile seizures plus; HCs, hair cells; IS, inner sulcus; Li, spiral limbus; NSHL, non-syndromic hearing loss; OC, organ of Corti; qRT-PCR, quantitative reverse transcriptase PCR; SGNs, spiral ganglion neurons; SHL, syndromic hearing loss; SPL, sound pressure level; SV, stria vascularis; WS, wildervanck syndrome; WT, wild-type.

[1]https://hereditaryhearingloss.org/

TTG (forward) and 5′- TCTGAACAGATTAGTAATGAACACA GATG (reverse). PCR for identification of *Atoh1-Cre* gene fragment: 5′- CCGGCAGAGTTTACAGAAGC (forward) and 5′- CTAGGCCACAGAATTGAAAGATCT (reverse). PCR for identification of *Atoh1-Cre* control gene fragment: 5′-GTAGGTGGAAATTCTAGCATCATCC (forward) and 5′-ATGTTTAGCTGG CCCAAATG (reverse).

The sizes of amplified PCR products were 241 bp for the *loxP* allele or 183 bp for the wild-type (WT) allele; 324 bp for the *Cre* control allele and 450 bp for the *Cre* allele. Homozygous mice *Fgf13*$^{fl/Y}$; *Atoh1-cre* (Fgf13$^{-/Y}$) and *Fgf13*$^{-/-}$; *Atoh1-cre* (*Fgf13*$^{-/-}$) were denoted knockout (*Fgf13* cKO) mice and *Fgf13*$^{-/+}$; *Atoh1-cre* (*Fgf13*$^{-/+}$) were heterozygous knockout mice. WT and Atoh1-cre mice were used as the control mice. All animals (males and females) used in our experiments were adult C57BL/6J mice aged 8–12 weeks.

Assessment of Auditory Functions

Auditory brainstem recording (ABR) and distortion product otoacoustic emission (DPOAE) recordings were performed in WT, Atoh1-cre and *Fgf13* cKO mice in a soundproof room. Mice were anesthetized with intraperitoneal injections of 10% chloral hydrate (0.04 mL/10 g body weight, dissolved in saline solution) before recordings. The depth of anesthesia was periodically verified by the lack of foot-pinch response. For ABR recordings (Guarch et al., 2018), three needle electrodes were inserted subdermally at the cranial vertex (active), the external ear (reference), and the subcutaneous hind leg (ground), respectively. ABR click stimuli of different intensities (100 μsec duration) and tone pips (1 ms rise-fall time with 3 ms plateau) of 4, 8, 12, 16, 20, 24, 28, and 32 kHz frequencies were delivered using a Tucker Davis Technologies (TDT) workstation running SigGen32 software (Fetoni et al., 2018). Auditory function was tested by decreasing the sound intensities from 90 to 20 dB sound pressure level (SPL) in 5 dB SPL steps. The hearing threshold was determined by the lowest intensities at which reproducible electrical response waves could be recognized.

Distortion product otoacoustic emission response thresholds were tested as described previously (Men et al., 2015). The DPOAE responses at diction frequency 2f1-f2 were measured with two primary tone frequencies (f1 and f2 with f2/f1 = 1.2) to predict the auditory thresholds. DPOAE response thresholds were recorded at 8–32 kHz frequencies and intensities ranging from 90 to 20 dB SPL in 5 dB SPL decrements, the same as the ABR test protocol by the acoustic microphone probe and TDT system.

Quantitative Real-Time Reverse Transcription PCR (qRT-PCR)

qRT-PCR was performed as previously described (Wang et al., 2017). Total RNA was extracted at 4°C using Trizol RNA isolating reagent (Thermo Fisher Scientific, Waltham, MA, United States) according to the established procedures. Total RNA (1000 ng) was reversely transcribed using PrimeScript™ RT reagent kit with gDNA Eraser (perfect real time) kit (Takara, Japan) for the synthesis of a single-stranded cDNA library according to the manufacturer's introduction. Gene-specific mRNA analyses were performed using the standard protocol of SYBR Premix Ex Taq™ II (TliRnaseH plus) kit (Takara, Japan). Relative quantification was performed using the comparative threshold method (ΔΔCT) and *Gapdh* gene was used as a reference to normalize the specific gene mRNA expression. After amplification, each qPCR product was sequenced using electrophoresis to ensure the specificity. The primers used were listed in **Table 1**.

Tissue Preparation

Mice were transcardially perfused with 4% paraformaldehyde (PFA) under terminal anesthesia (10% chloral hydrate, 0.04 mL/10 g body weight). Cochleae were quickly dissected, post-fixed at 4°C for 12 h, decalcified in 10% EDTA for 24–48 h and incubated in 10% and 30% sucrose solution, respectively, at 4°C for 24 h. The samples were then embedded in OCT, frozen, and cryosectioned in 10 μm-thick sections (von Bartheld et al., 2016). All middle sections of cochleae were used. For whole-mount immunostaining, cochleae were exposed to the sensory epithelium and dissected into basal, middle, and apical sections as previously described (Montgomery and Cox, 2016).

Immunostaining Analysis and Confocal Imaging

After being washed by 10 mM PBS, specimens were permeabilized with 3% bovine serum albumin (BSA) and 0.3% Triton X-100 (Sigma, MO, United States) solution at 37°C for 1 h and blocked with 10% goat serum (Solarbio, China) at 37°C for 30 min and at RT for 30 min, respectively. For whole-mount staining, the cochleae were washed with 10 mM

TABLE 1 | List of qRT-PCR primer sequences.

Gene name	Primer sequence (5′-3′)	Length (bp)
Fgf13	F-CAGCCGACAAGGCTACCAC R-GTTCCGAGGTGTACAAGTATCC	184
Fgf12	F-GGAGAGCAAAGAACCCCAG R-CACCACACGCAGTCCTACAG	159
Caspase 3	F-GGAGCAGCTTTGTGTGTGTG R-CTTTCCAGTCAGACTCCGGC	131
Caspase 9	F-GGACCGTGACAAACTTGAGC R-TCTCCATCAAAGCCGTGACC	101
Caspase 12	F-CTGGCTCTCATCATCTGCAACAA R-CGGCCAGCAAACTGCATTAAC	173
Cytochrome C	F-GAGGCAAGCATAAGACTGGA R-TACTCCATCAGGGTATCCTC	133
P53	F-CCCGAGTATCTGGAAGACAG R-ATAGGTCGGCGGTTCAT	146
Bcl-2	F-TGACTTCTCTCGTCGCTACCG R-GTGAAGGGCGTCAGGTGCAG	69
Bcl-xl	F-GAGGCAGGCGATGAGTT R-ACGATGCGACCCCAGTTT	149
Bak	F-AAACCTCTCTCCCTACCCCA R-AGGATGGGGTTCAGTAGCAC	162
Gapdh	F-TGTCAGCAATGCATCCTGCA R-CCGTTCAGCTCTGGGATGAC	240

PBS and blocked (1% BSA, 1% Triton X-100 and 5% goat serum) at RT for 1 h. The sections or whole mounts were incubated with different primary antibodies: rabbit anti-FGF13 (1:200; Yenzym), mouse anti-tubulin βIII (Tuj1; 1:200; GTX631836, GeneTex, United States), mouse anti-Myosin VIIa (1:200; sc-74516, Santa Cruz Biotechnology, United States), rabbit anti-cleaved caspase-3 (1:250; #9664, Cell Signaling, United States) and rabbit anti-cytochrome C (1:100; 10993-1-AP, Proteintech, United States) overnight at 4°C. The secondary antibodies were Fluorescein (FITC)-conjugated AffiniPure goat-anti-mouse IgG (1:300; 115-095-166, Jackson ImmunoResearch, United States) and cy^{TM}3-conjugated AffiniPure goat-anti-rabbit IgG (1:200; 111-165-144, Jackson Immunoresearch, United States) diluted in the solution (1% BSA and 0.1% Triton X-100 for sections; 1% BSA, 0.1% Triton X-100 and 5% goat serum for whole mounts) at RT for 90 min. As controls for specificity, the primary antibody was co-incubated with the peptide used for immunization or samples were incubated with secondary antibody only. Nuclei were stained with 4, 6-diamidino-2-phenylindole (DAPI; 1:200; Southern Biotech, United States). The samples were observed under a laser scanning confocal microscope (Leica, Model: SP5, Wetzlar, Germany). For quantification of immunofluorescent staining, uniform microscope settings were kept during all image capture sections.

Hematoxylin-Eosin Staining

The hematoxylin-eosin (H&E) staining was performed as described previously (Fetoni et al., 2018). Frozen-sections at a thickness of 10 μm were stained with hematoxylin and eosin for the histological assessment of cochlear morphological damage. A standard H&E protocol was followed with 3 min incubation in hematoxylin and 2.5 min staining in eosin, then mounted with neutral balsam (Solarbo, China). Only one middle section of a mouse cochlea was used to quantify the SGNs loss. SGNs were counted in the apical, middle, and basal regions of the cochlear sections using a × 20 objective as previously described (Someya et al., 2009). Type I and type II neurons were not differentiated, and viable neurons with a clear round nucleus and homogeneous cytoplasm were counted. The corresponding area of Rosenthal canal was measured on digital photomicrographs of each canal profile. The perimeter of the canal was traced with a cursor using ImageJ software (National Institutes of Health). The computer then calculated the area within the outline. The SGNs density was calculated as the number of SGNs per mm^2.

TUNEL Assay

TUNEL assay (in situ cell death detection kit, Roche) was used to examine DNA fragmentation in the nuclei of apoptotic cells in SGNs of the cochlea. The assay was performed on cochlear cryosections according to the manufacturer's instructions. Specimens were permeabilized with 0.3% Triton X-100 and 3% BSA solution at 37°C for 1 h and incubated with freshly prepared working solution at 37°C for 2 h. After rinsing in 10 mM PBS, specimens were coverslipped. Nuclei of TUNEL-positive cells intensely labeled by lilac plus green were identified as apoptotic cells.

Statistical Analysis

All data analyses were carried out with IBM SPSS 21 Statistics software and image processing with Origin 8, Adobe Illustrator 10 and GraphPad Prism 6.0. Data were presented as mean ± standard deviation ($\bar{x}$ ± SD). The comparisons between two groups were performed with the Student's *t*-test and differences among groups were analyzed by One-Way ANOVA followed by Scheffe (C) and Bonferroni (B) analyses (SPSS). $P < 0.05$ was considered statistically significant.

RESULTS

Localization of FGF13 in the Cochlea

We first examined the localization of FGF13 in murine cochlear tissue using immunohistochemistry (IHC) (**Figure 1**). FGF13 was located primarily in the organ of Corti (OC), SGN, stria vascularis (SV), spiral limbus (Li), inner sulcus (IS), and the Claudius' cells (CC) from the apical to basal regions of the cochlea tissue (**Figure 1A**). We further investigated the expression patterns of FGF13 in cultured SGNs and whole-mount HC staining. The results showed expression in the cytoplasm, membrane, and neurites of the SGN (**Figure 1B**). In HCs, staining was also observed in the cytoplasm and membrane, with more prominence in the inner rather than the outer regions (**Figure 1C**). We did not observe FGF13 expression in the nucleus of either the SGNs or HCs (**Figures 1B,C**). We did not find any significant difference between the post-natal expression levels in the mouse cochlea at P0, P7, P14, P30, and P60 days (**Supplementary Figure 1**).

Generation of a Mouse Line With Conditional Knockout of *Fgf13* in the Inner Ear

We generated a mouse line with a selectively deleted *Fgf13* in the inner ear using the *loxP/Cre* system, to investigate its role in auditory function. The gene is located on the X chromosome and comprises 72.3 kb of the genomic DNA in mice. Homozygous mice with floxed exon 3 of *Fgf13* (*Fgf13*$^{fl/Y}$ or *Fgf13*$^{fl/fl}$) were generated as described previously (Wang et al., 2017). To achieve inner ear-specific *Fgf13* deletion, *Fgf13*$^{fl/Y}$ or *Fgf13*$^{fl/fl}$ mice were crossed with Atoh1-cre mice (Maţei et al., 2005). The latter consisted of a 1.5 kb *Atoh1* enhancer fragment that drove Cre expression. *Cre* mediated beta-galactosidase identified the *loxP* site, and was expressed in all HCs and SGNs, as well as some supporting cells of both the cochlea and the vestibule (Matei et al., 2005)[2], resulting in the specific knockout of *Fgf13*. The pups were genotyped using PCR analysis (**Figure 2A**). To further confirm *Fgf13* specific deletion in the cochlea, quantitative reverse transcriptase PCR (qRT-PCR) was performed to determine the levels of cochlear mRNA in WT, Atoh1-cre, and *Fgf13* cKO mice. The gene transcript level was reduced to ~33.8% in *Fgf13* cKO mice compared to levels in WT and Atoh1-cre control mice (**Figure 2B**). While FGF12 mRNA levels were unaltered

[2]https://www.jax.org/strain/011104

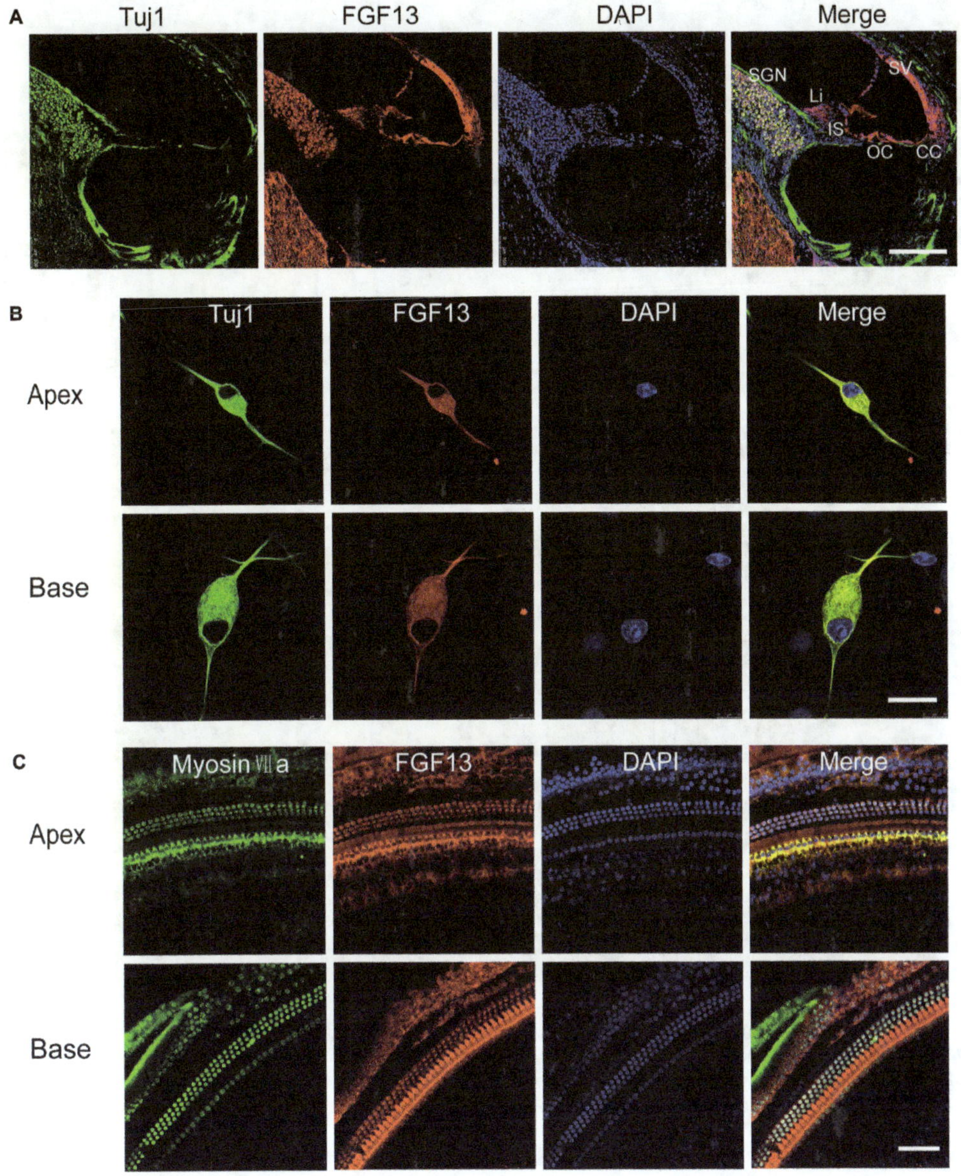

FIGURE 1 | FGF13 is highly expressed in SGNs and HCs of the cochlea. **(A)** Confocal images of cryosections from the base of the cochlea show immunodetection of FGF13 (red), Tuj1 (green), and DAPI (blue). Tuj1 was used as a neuron marker. SV, stria vascularis; OC, organ of Corti; CC, Claudius's cells; IS, inner sulcus; Li, spiral limbus; SGN, spiral ganglion neurons; Scale bar = 200 μm. **(B)** Immunofluorescence staining of cultured SGNs at a cellular level. Scale bar = 20 μm. **(C)** Confocal whole-mount images of FGF13 in hair cells. Myosin VIIa (green) served as a hair cell marker. Scale bar = 50 μm.

in *Fgf13* cKO mice, confirming the specificity, efficacy as well as a lack of FGF12 compensation in the knockout (**Figure 2B**). Furthermore, IHC staining of the cochlea showed significantly decreased levels of FGF13 in *Fgf13* cKO mice (**Figure 2C**). Together, these data demonstrate the efficient deletion of *Fgf13* in the cochlea of the cKO mice.

Elevated ABR Thresholds but Not DPOAE in *Fgf13* cKO Mice

We next sought to determine whether *Fgf13* cKO mice exhibited any hearing impairment-based phenotypes. The auditory brainstem response (ABR) test was performed to evaluate the functional integrity of the auditory system (**Figure 3**). The test consisted of a click stimulus in the 20–90 dB SPL range and frequency-specific stimuli for the tone test at 4–32 kHz. Consistent with our hypothesis, the *Fgf13* cKO mice displayed significantly higher ABR thresholds in response to the click stimulus and tone test across the entire auditory spectrum compared to WT and Atoh1-cre control mice, indicating the hearing impairment in cKO mice (**Figures 3A–C**). It is worth a mention that only the homozygous knockout mice ($Fgf13^{-/-}$ or $Fgf13^{-/Y}$) exhibited impaired auditory function, while the heterozygous mice ($Fgf13^{+/-}$) showed a normal

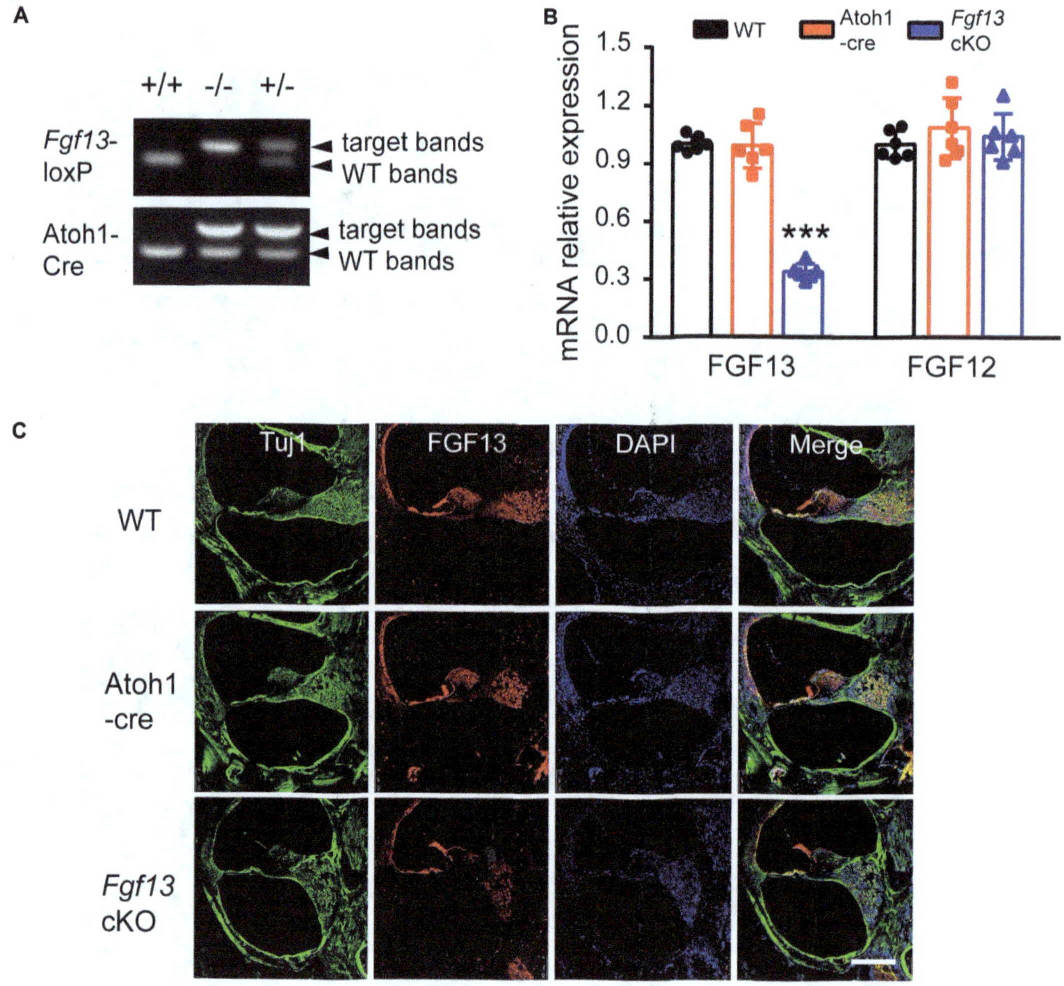

FIGURE 2 | Generation of inner ear conditional knockout *Fgf13* mice using the *loxP-Cre* system. **(A)** PCR genotyping of wild-type (+/+), homozygote (–/Y or–/–), and hemizygous (+/–) *Fgf13* knockout mice using genomic DNA prepared from tail biopsies. **(B)** qRT-PCR of FGF13 and FGF12B. Expression levels were normalized to endogenous GAPDH. Each expression level was standardized to WT and given a value of 1. $^{***}P < 0.001$ compared to WT as determined by One-Way ANOVA. **(C)** Confocal images of FGF13 (red) and Tuj1 (green) in the base of the cochlea. Scale bar = 200 μm.

hearing threshold (**Supplementary Figure 2**). Further wave analysis in response to ABR click stimuli of 80 dB SPL revealed lower amplitude and increased latency of wave I (but not wave II-IV) in the *Fgf13* cKO mice compared to that of control groups (**Figures 3D–F**).

To test the function of the outer HCs, we measured the distortion product otoacoustic emission (DPOAE) test at frequencies of 8–32 kHz in the mice. Surprisingly, in comparison with WT and Atoh1-cre mice, no significant alteration of thresholds was found in *Fgf13* cKO mice at all tested frequencies, indicating the normal functioning of the outer HCs (**Figure 3G**).

Fgf13 Deficiency Reduced SGNs Density

The results for lower amplitude and increased latency of wave I in the ABR test suggested that *Fgf13* deficiency might cause damage to SGNs. Thus, we performed immunostaining on the SGNs to examine their morphology and any other changes in *Fgf13* cKO mice. Fluorescence intensity of FGF13 staining reduced significantly from the apex to the base in *Fgf13* cKO mice, confirming effective knockdown (**Figures 4A–C**). Importantly, when compared with controls, cell densities of type I SGNs (counts/mm^2 marked by Tuj1, a neuronal marker) significantly decreased from the apex to the base (with normal morphology) in *Fgf13* cKO mice, with a more significant loss of SGNs in the base of the cochlea (**Figures 4D–F**).

The gross histological features of the cochlear sections were also examined (**Figure 5A**). Hematoxylin-eosin staining data demonstrated that the cochlea of *Fgf13* cKO mice exhibited largely normal morphological structures of OC, SV, SL, and tectorial membrane with no obvious cell loss in either the inner or outer HCs (**Figure 5B**, upper panel). In contrast, loss of SGNs was observed in *Fgf13* cKO mice and was more significant in the basal region (**Figure 5B** lower panel, summary results in **Figures 5C–E**). Besides, whole-mount staining showed that the deletion of *Fgf13* did not affect the morphology and densities of both the inner and outer HCs from the apex to the

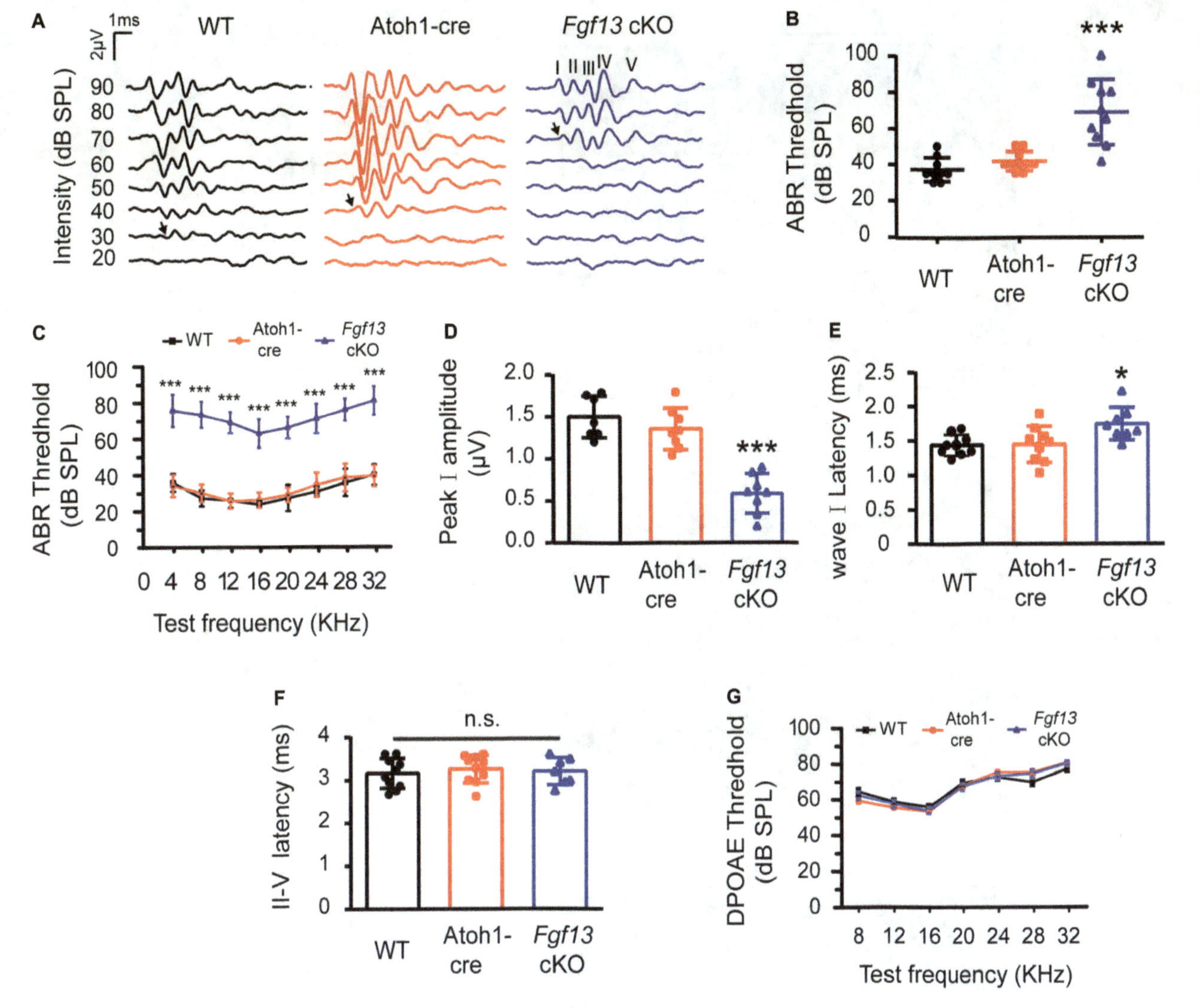

FIGURE 3 | *Fgf13* cKO mice exhibit sensorineural hearing loss. **(A)** A representative click-evoked ABR waveform at decreasing intensities from 90 to 20 dB SPL in WT, Atoh1-cre and *Fgf13* cKO mice. Arrows indicate threshold levels. **(B,C)** ABR measurements for click **(B)** and tone burst responses **(C)**. n = 9/WT (5 M, 4 F); n = 10/Atoh1-cre (6 M, 4 F); n = 10/*Fgf13* cKO (7 M, 3 F). **(D–F)** The peak I amplitude **(D)**, latency of wave I **(E)**, and II-V **(F)** in ABR test were analyzed. The traces analyzed in **(D–F)** were from click test at SPL of 80 dB. n = 7/WT (5 M, 2 F); n = 8/Atoh1-cre and *Fgf13* cKO (5 M, 3 F) in peak I amplitude. n = 9/WT and Atoh1-cre (5 M, 4 F); n = 8/*Fgf13* cKO (5 M, 3 F) in wave I latency. n = 10/WT and Atoh1-cre (6 M, 4 F); n = 7/*Fgf13* cKO (5 F, 2 M) in II-V wave latency. $^{*}P < 0.05$, $^{***}P < 0.001$ compared to controls as determined by One-Way ANOVA test. **(G)** DPOAE measurements in WT, Atoh1-cre and *Fgf13* cKO mice. n = 10/ WT (6 F, 4 M); n = 9/Atoh1-cre (5 M, 4 F); n = 4/*Fgf13* cKO (3 M, 1 F). F, female; M, male; n.s., no significance. $P > 0.05$ as determined by One-Way ANOVA test.

base (**Figure 6**). These results were consistent with the normal functioning of the outer HCs as demonstrated by the DPOAE test (**Figure 3G**).

Activated Mitochondrial Apoptosis Pathway in *Fgf13* cKO Mice

We further investigated whether the loss of SGNs in the cochlea was associated with apoptosis. We used the TUNEL assay to measure nuclear DNA fragmentation, a key feature of apoptosis (Martin et al., 2009; Someya et al., 2009). As shown in **Figure 7**, a significant number of TUNEL-positive SGNs were observed in *Fgf13* cKO mice from the apex to the base of the cochlea, with much fewer apoptotic cells detected in WT and Atoh1-cre mice. We further studied whether SGN apoptosis was caspase-dependent by examining caspase-3 activation through immunofluorescent staining with antibodies against cleaved-caspase-3 and Tuj1 (**Figure 8**). In agreement with the TUNEL results, the number of cleaved-caspase-3 positive cells was significantly higher in SGN of *Fgf13* cKO mice from the apex to the base compared to WT and Atoh1-cre controls (**Figures 8A–C**, summary results in **Figures 8D–F**), indicating the participation of caspases in the apoptotic pathway.

To further investigate which pathways participated in SGN apoptosis, we examined the expression of several key factors in both extrinsic and intrinsic apoptotic pathways by

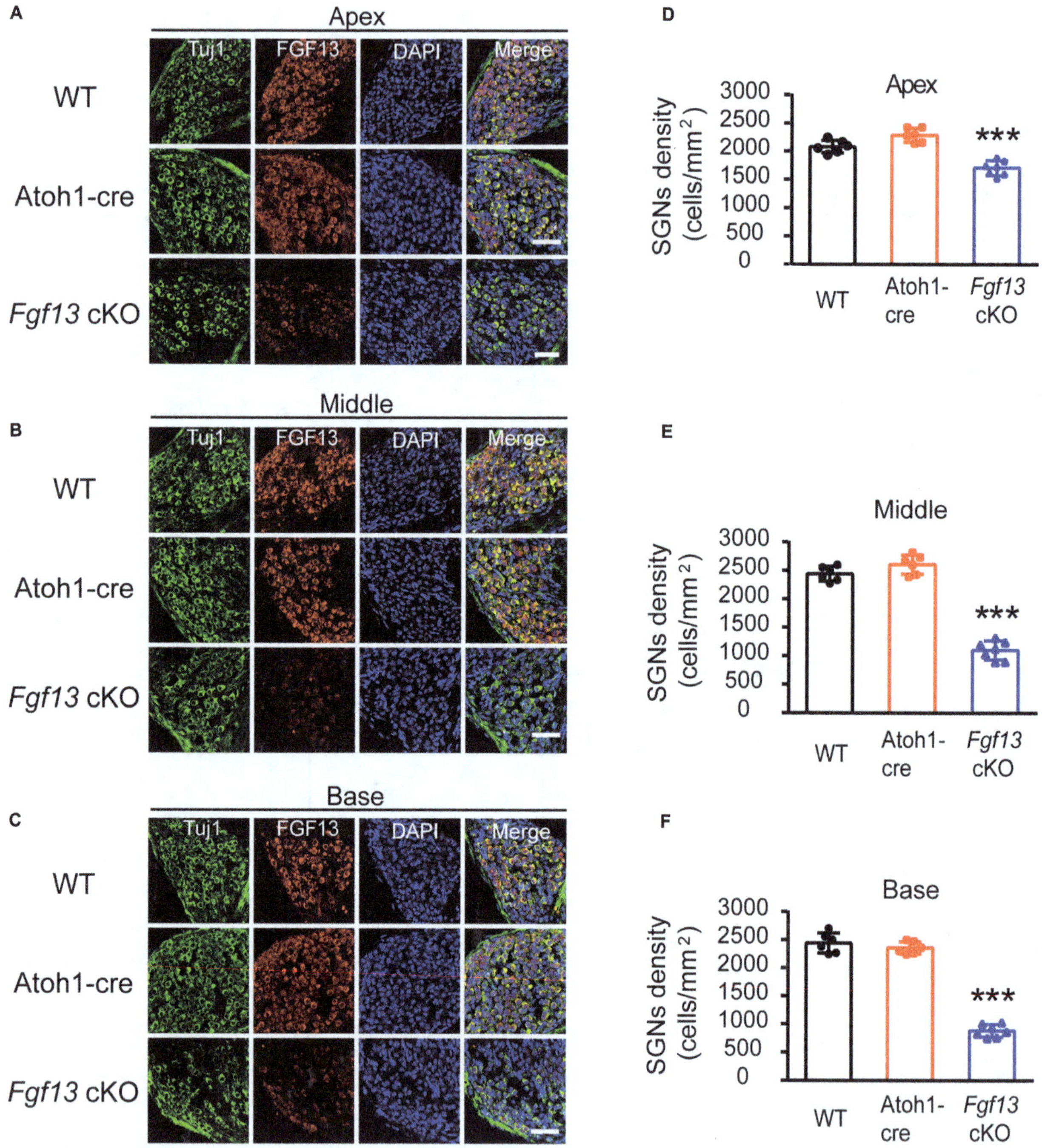

FIGURE 4 | Reduced SGN density in *Fgf13* cKO mice. **(A–C)** Immunostaining of FGF13 in SGNs marked by Tuj1 from apex **(A)**, middle **(B)**, and base **(C)** in WT, Atoh1-cre and *Fgf13* cKO mice. Scale bar = 50 μm. **(D–F)** Quantitative assessment of SGNs densities from the apex to the base. n = 6/WT (5 M, 1 F); n = 6/Atoh1-cre (4 M, 2 F); n = 7/*Fgf13* cKO (5 M, 2 F). F, female; M, male. ***$P < 0.001$ compared to controls as determined by One-Way ANOVA.

qRT-PCR. *Fgf13* cKO mice exhibited significantly higher levels of pro-apoptotic genes, including *caspase-3*, *caspase-9*, *caspase-12*, *P53*, *cytochrome C*, and *Bak*, compared to that of WT and Atoh1-cre controls (**Figures 9A–F**). Consistently, the levels of anti-apoptotic factors *Bcl-2* and *Bcl-xl* were significantly decreased (**Figures 9G,H**) accompanied by a ~53% FGF13 reduction in the cKO mice (**Figure 9I**). There were no significant alterations in *caspase-8*, *AIF*, *Bim*, and *Bax* among the three groups (**Supplementary Figure 3**), which indicates a potential activation of the mitochondrial apoptotic pathway (Wu and Bratton, 2013). Thus, we investigated the localization of cytochrome C, the release of which into the cytoplasm is a key feature for the activation of the mitochondrial apoptotic pathway. In both the WT and Atoh1-cre groups, cytochrome C was distributed

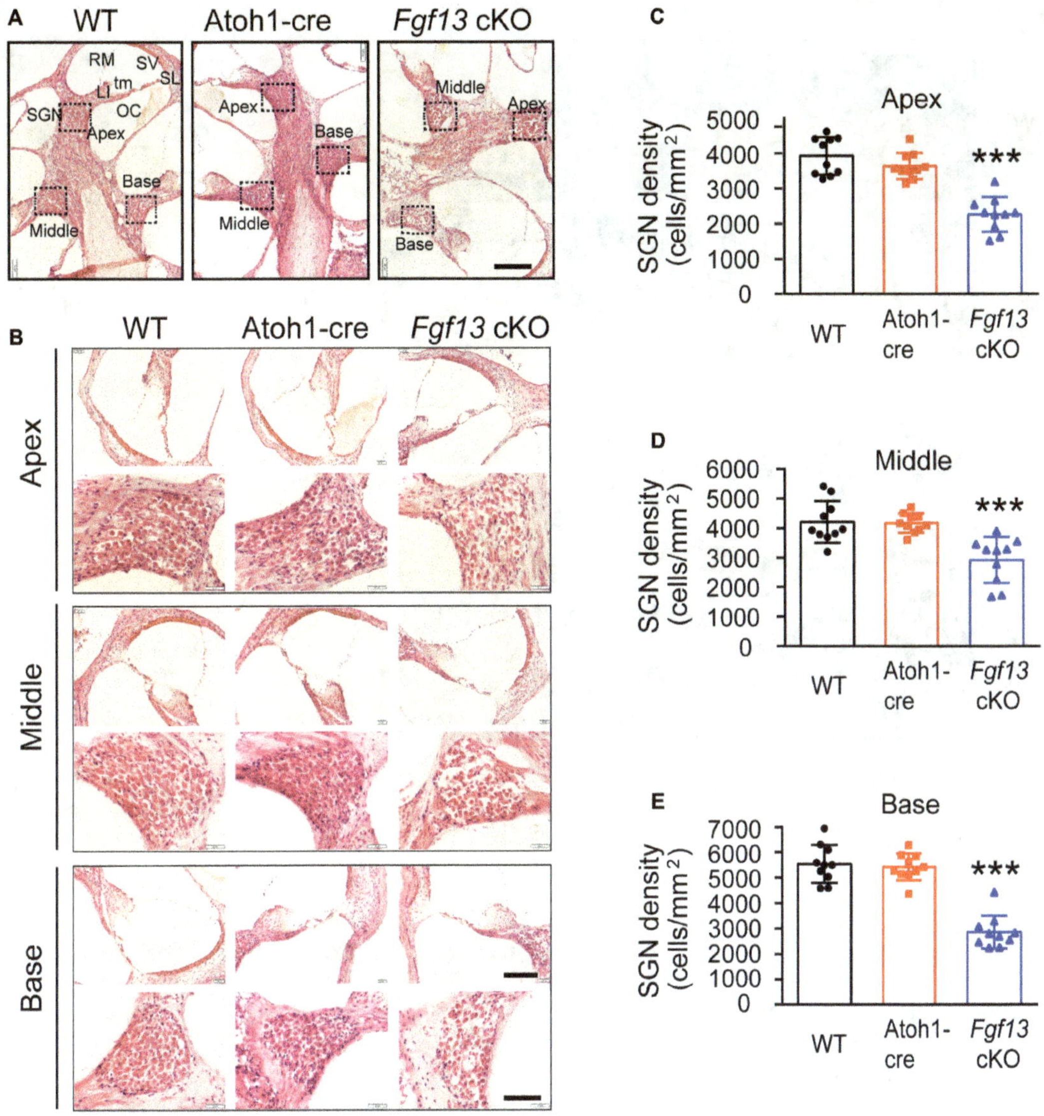

FIGURE 5 | *Fgf13* knockout causes the loss of SGNs with normal morphological structure of the cochlea in the mice. **(A)** Modiolar cochlear sections depicting the apex to the base stained by hematoxylin-eosin. SGN, spiral ganglion neuron; LI, spiral limbus; OC, organ of Corti; tm, tectorial membrane; RM, Reissner's membrane; SV, stria vascularis; SL, spiral ligament. **(B)** OC (top) and SGN (bottom) are shown at higher magnification. **(C–E)** Quantitative analysis of SGNs densities from the apex to the base. n = 10/WT (6 M, 4 F); n = 10/Atoh1-cre (7 M, 3 F), n = 10/*Fgf13* cKO (7 M, 3 F). Scale bar: G = 100 μm, Scale bar: H = 80 μm (top), and Scale bar: H = 40 μm (bottom). F, female; M, male. ***$P < 0.001$ compared to controls as determined by One-Way ANOVA.

uniformly as puncta in SGNs. In contrast, in *Fgf13* cKO mice, there was an uneven distribution as well as plaque aggregation accompanied by an increased expression in the cytoplasm of the neurons. These data indicate that knockout of *Fgf13* induced release of cytochrome C into the cytoplasm and hence, activated the mitochondrial apoptotic pathway in SGNs (**Figure 10**).

DISCUSSION

Multiple genes are involved in the development and functioning of the cochlea. They participate in various cellular functions and about 30% of congenital hereditary deafness has been reported to be SHL caused by genetic mutations (Ideura et al., 2019). For example, *Tbx1* gene mutations are associated with the human DiGeorge syndrome accompanied by a deficiency in hearing function (Chen et al., 2016). Moreover, functional null mutations of *Hars2* (encoding mitochondrial histidyl tRNA synthetase) cause Perrault syndrome, which is characterized by ovarian dysgenesis and sensorineural hearing loss (Pierce et al., 2011). Hearing loss is a common feature found in patients with BFLS, WS and CGH syndromes, all of which are characterized by genetic loss-of-function mutations in the *Fgf13* gene (Gecz et al., 1999; DeStefano et al., 2013; Abu-Amero et al., 2014). Whether the deafness in these patients is caused by the loss function of *Fgf13* and whether this gene plays any role in the auditory system

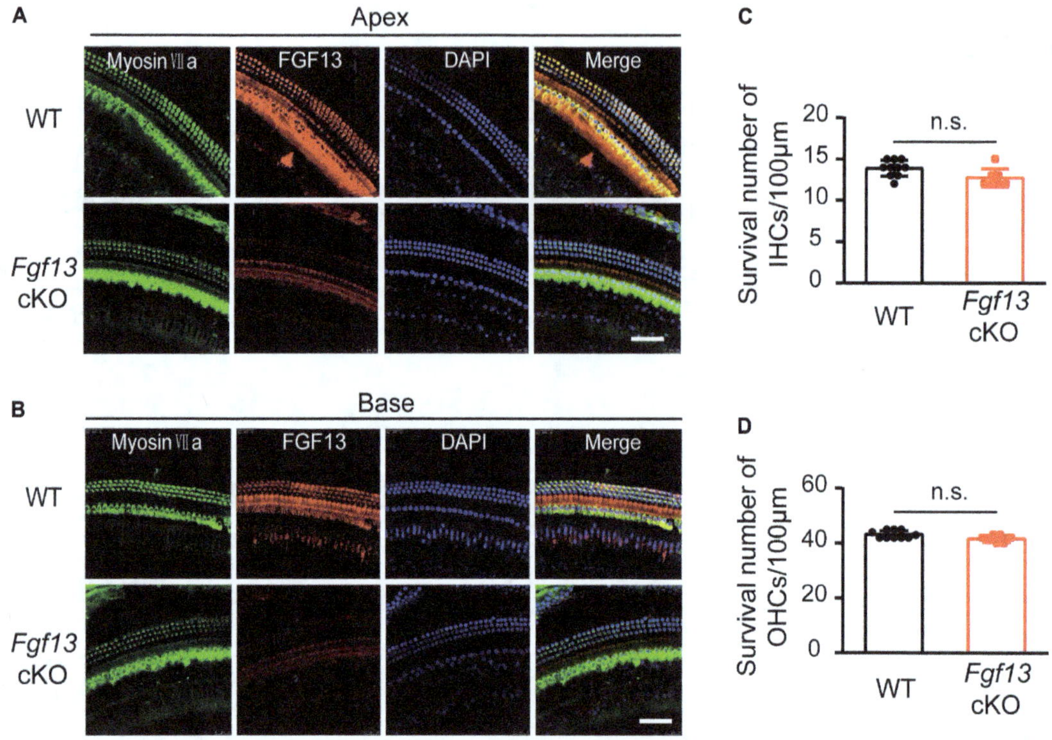

FIGURE 6 | *Fgf13* cKO unalters the morphology and densities of HCs. **(A,B)** Confocal whole-mount images of cochlea HCs in WT and *Fgf13* cKO mice form the apex **(A)** to the base **(B)**. **(C,D)** Quantitative analysis of the survived inner HCs **(C)** and outer HCs **(D)**. There are no obvious abnormalities of morphology **(A,B)** and densities **(C,D)** for both inner HCs and outer HCs in *Fgf13* cKO mice. $n = 10$/WT (7 males, 3 females); $n = 7$/*Fgf13* cKO (4 males, 3 females). n.s., no significance. Scale bar = 50 μm; $P > 0.05$ as determined by One-Way ANOVA.

is unknown. In this study, we found that selective deletion of *Fgf13* in the inner ear of mice caused sensorineural deafness. It also increased apoptotic cell loss of SGNs associated with the mitochondrial apoptotic pathway in the mouse cochlea. Our data revealed a novel role for *Fgf13* in the auditory system and suggest that the gene could be a potential candidate for understanding deafness.

In the current study, we investigated the role of *Fgf13* in hearing using the ABR and DPOAE tests. The ethological results showed that *Fgf13* cKO mice displayed impaired auditory function (**Figures 3A–C**) and gene deletion affected the wave I component but not the wave II-V in the click-evoked ABR test (**Figures 3D–F**). These results suggest that *Fgf13* deletion selectively causes damage to SGN but not to the central auditory pathway, since the amplitude and peak latency of wave I reflect the function of the peripheral SGNs, whereas waves II-V indicate a function of the ascending auditory pathway (Zuccotti et al., 2013; Eggermont, 2019). This observation was further supported by two results, abundant expression of FGF13 in SGNs of the inner ear (**Figure 1**) as well as that the cKO mice functioned normally in the DPOAE test (**Figure 3G**), indicating an intact function for outer HCs. Both SGNs and HCs play crucial roles in the production and transmission of sound. Although the investigation of the electrophysiological function of inner HCs and SGNs deserves further study, *Fgf13* cKO mice displayed normal morphology and densities of both inner and outer HCs, but showed significant cell loss in SGNs, indicating a contribution to hearing deficit in the knockout mice.

In previous studies, C57BL/6J mice developed a completely mature auditory pathway from P21 to P30 (Li et al., 2017), and showed a slow decline in auditory function, in a time-dependent manner, with age (Someya et al., 2009). Therefore, only mice aged 8–12 weeks were included in this study. In our study, 2-month-old *Fgf13* cKO mice showed sensorineural deafness and mitochondrial apoptosis associated with SGN loss, but the temporal aspects of the loss were unclear. RNA-seq results (Elkon et al., 2015) and online gEAR database showed a high expression level of FGF13 at E16, which reduced quickly until P0, and then remained stable at a low level at P7. The qRT-PCR results in our study also showed that the expression levels remained stable and no significant changes were observed between P0 and P60 in the mice (**Supplementary Figure 1**). This particular trend of FGF13 expression indicates a potentially important function in SGN development during the pre-natal stage. Atoh1-cre mediated beta-galactosidase caused *Fgf13* deletion in the otocyst as early as E10.5 (Matei et al., 2005). Moreover, *Fgf13* knockout only caused 30–47% SGN loss in 8–12-week-old mice, which might be increased gradually with age (**Figures 4, 5**). Due to this, we cannot exclude the possibility that *Fgf13* knockout could affect SGN development and display a congenital progressive hearing loss in mice, which would be an interesting topic for future studies (Ma et al., 2000; Pirvola et al., 2000).

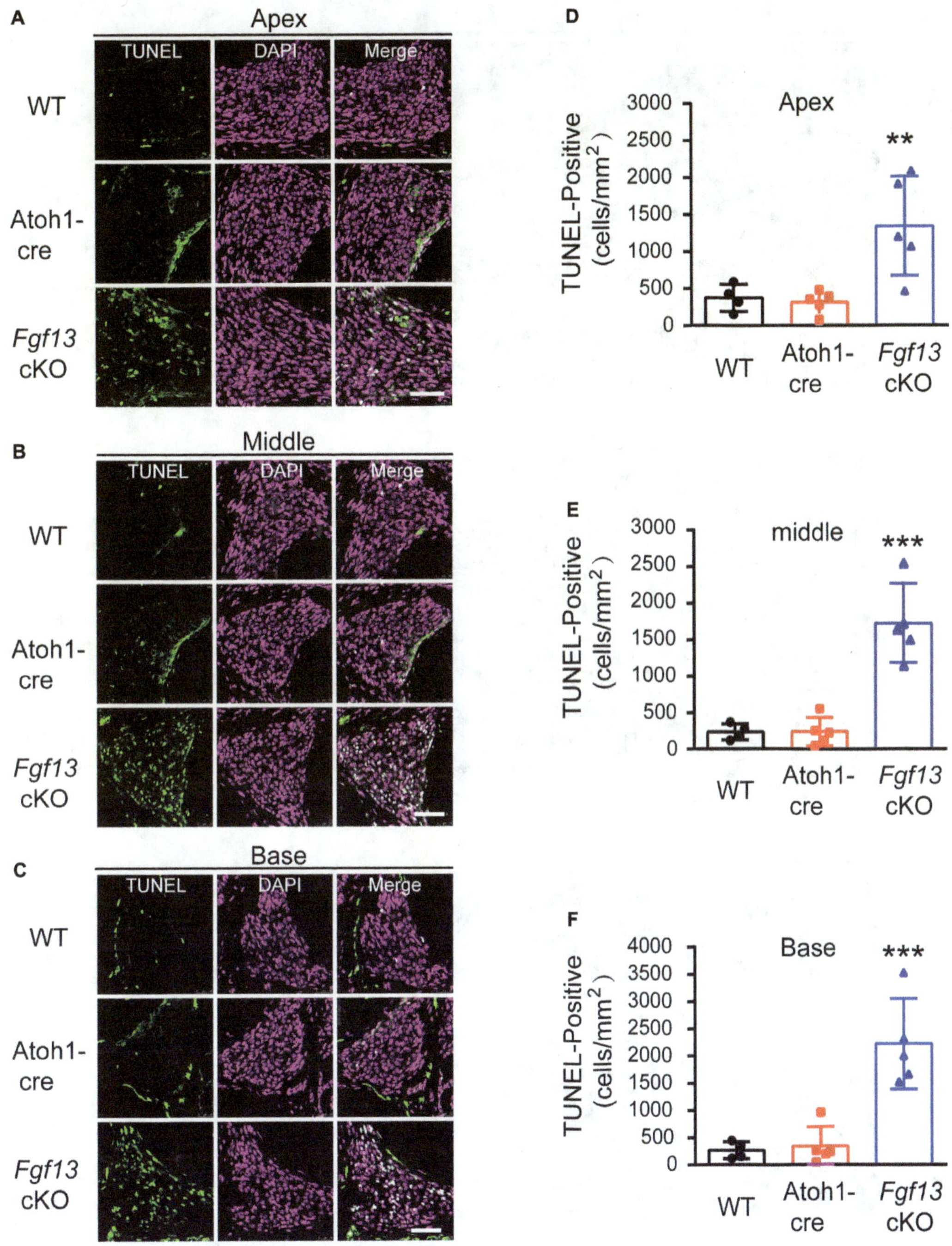

FIGURE 7 | *Fgf13* cKO mice show accelerated SGNs apoptosis detected by TUNEL staining. **(A–C)** TUNEL staining of cochlea in WT, Atoh1-cre and *Fgf13* cKO mice. Scale bar = 50 μm. **(D–F)** Quantitative analysis of TUNEL-positive cells in apex **(D)**, middle **(E)** and base **(F)**. n = 4/WT (3 males, 1 females); n = 5/Atoh1-cre and *Fgf13* cKO (3 males, 2 females). $^{**}P < 0.01$, $^{***}P < 0.001$ compared to controls as determined by One-Way ANOVA test.

SGN development is regulated by complex networks of transcription factors and signaling molecules, which have been proven in various transgenic mouse models (Kim et al., 2001; Puligilla et al., 2010; Yang et al., 2011; Dvorakova et al., 2016; Macova et al., 2019; Filova et al., 2020; Pavlinkova, 2020; Chizhikov et al., 2021). For example, *Neurod1* (the basic helix-loop-helix gene) is an essential gene for SGN development. *Neurod1* null mutants, aged 2–3 months, eventually lose most of their sensory neurons, cause projection defects of SGNs (Kim et al., 2001; Macova et al., 2019) and show the elevation at ∼35 dB of SPL at 4–38 kHz frequencies in the ABR test (Macova et al., 2019). In our study, *Fgf13* knockout mice aged 8–12 weeks showed 30–47% SGN loss and elevated ∼40 dB of SPL at 4–32 kHz frequencies in the ABR test, which might provide new

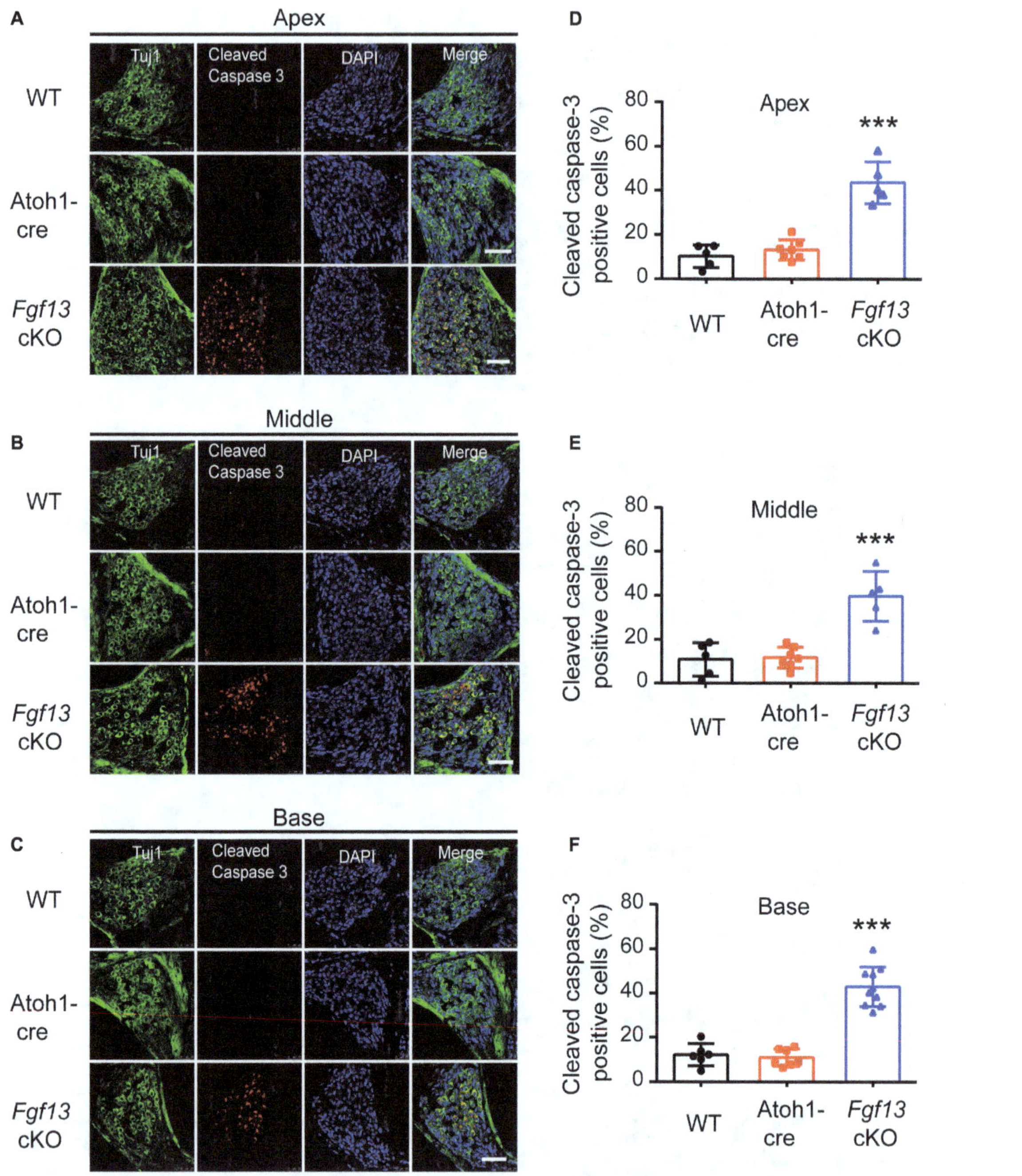

FIGURE 8 | *Fgf13* cKO induces activation of cleaved-caspase-3 in SGNs. **(A–C)** The expression of cleaved-caspase-3 in cochlea SGNs from the apex to the base in WT, Atoh1-cre and *Fgf13* cKO mice. Scale bar = 50 μm. **(D–F)** Quantitative analysis of cleaved-caspase-3 positive cells in apex **(D)**, middle **(E)**, and base **(F)** for WT, Atoh1-cre and *Fgf13* cKO mice. Apex and Middle: n = 5/WT (4 M and 1 F); n = 7/Atoh1-cre (4 M, 3 F); n = 5/*Fgf13* cKO (3 M, 2 F). Base: n = 6 (5 M and 1 F) for WT, n = 7/Atoh1-cre (4 M, 3 F); n = 10/*Fgf13* cKO (7 M, 3 F). F, female, M, male. $^{***}P < 0.001$ compared with controls as detected by One-Way ANOVA test.

perspectives on the molecular mechanisms underlying SGN loss and deafness. FGF13 is known to regulate neuronal polarization and migration by stabilizing microtubules. Loss of function results in an increase in the branching of axons and leading processes of cortical neurons (Wu et al., 2012). Therefore, we also cannot exclude the possibility that *Fgf13* knockout in the inner ear could affect the projection of SGNs, which may require future study to completely understand the role (Puligilla et al., 2007, 2010; Jahan et al., 2018).

Apoptosis is a major type of programmed cell death, where caspase activation plays a key role in the execution of the pathway. Caspases can be activated by two apoptotic pathways: the extrinsic (death receptor) pathway and the intrinsic (mitochondrial and ER stress) pathways (Li et al., 2017). We

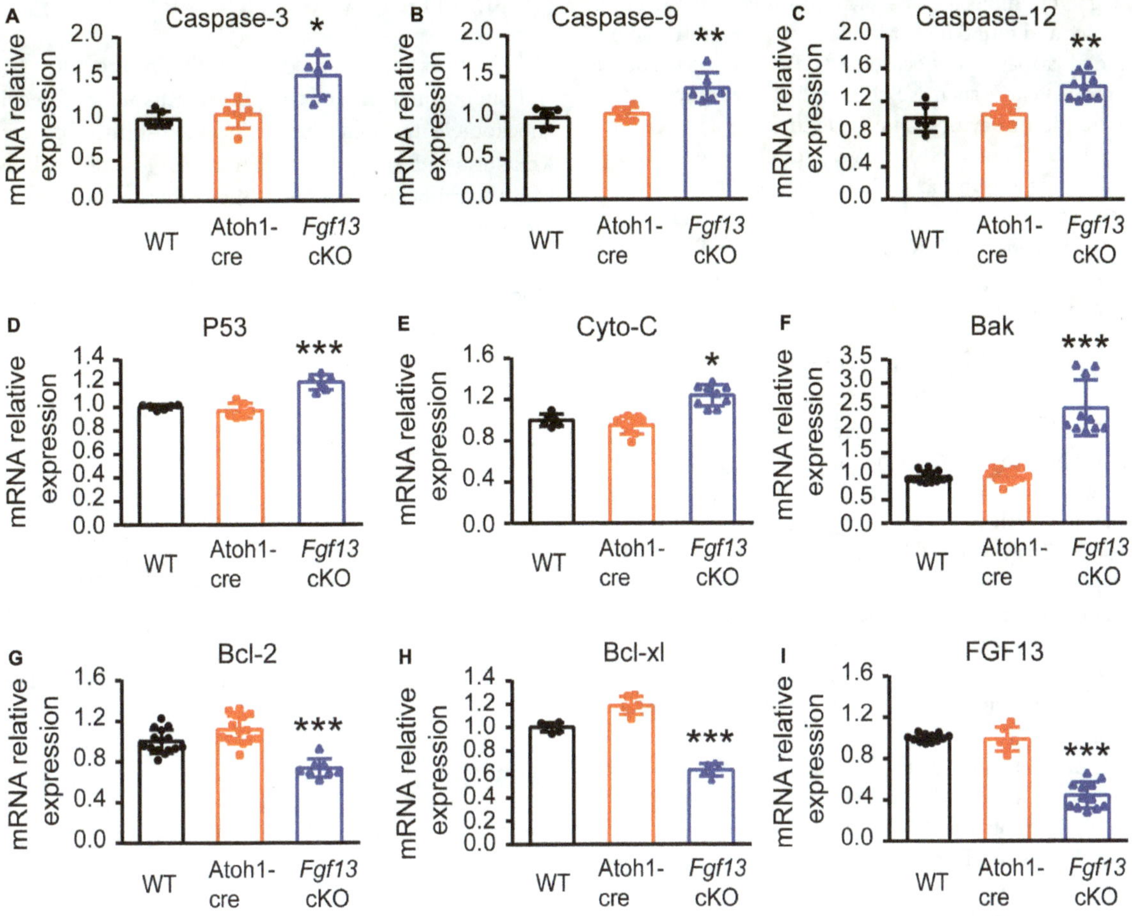

FIGURE 9 | *Fgf13* deficiency changes apoptosis related gene expression through qRT-PCR. qRT-PCR data show significantly higher mRNA levels of caspase-3, caspase-9, caspase-12, P53, cytochrome C and Bak **(A–F),** and lower levels of Bcl-2 and Bcl-xl **(G,H)** accompanied by ~53% reduction of FGF13 **(I)** in *Fgf13* cKO mice compared to the controls. Gene expression was calculated using the $2^{-\Delta\Delta Ct}$ method. Ct values were corrected for GAPDH and normalized to the WT group. $^{*}P < 0.05$, $^{**}P < 0.01$, and $^{***}P < 0.001$ compared with controls as detected by One-way ANOVA test.

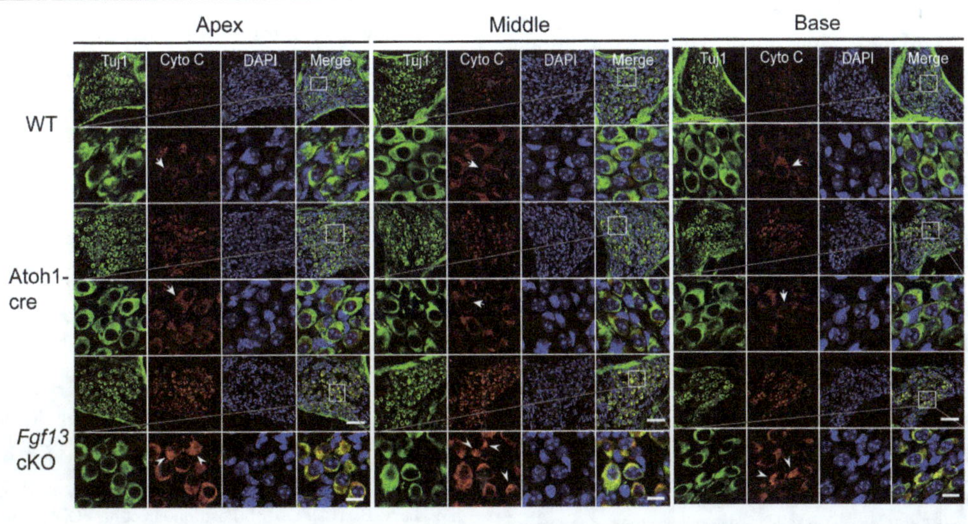

FIGURE 10 | *Fgf13* cKO alters cytochrome C localization in SGNs. Immunofluorescence shows uniformly distributed cytochrome C (red) as puncta (arrows) from the apex to the base in SGNs of WT and Atoh1-cre mice. However, the distribution of cytochrome C was no longer uniform in the cytoplasm and are aggregated as a plaque (arrowhead) with increased expression in the cytoplasm of SGN in *Fgf13* cKO mice. Scale bar = 50 μm (top) and 10 μm (bottom).

tested key signaling molecules associated with both pathways in *Fgf13* cKO mice and found higher levels of pro-apoptotic factors, which included caspase-9, cytochrome C, Bak, and lower levels of anti-apoptotic factors such as Bcl-2 and Bcl-xl (**Figure 9**). The uneven distribution of cytochrome C (**Figure 10**) and activated cleaved-caspase-3 (**Figure 8**) in *Fgf13* cKO mice indicated the involvement of the mitochondrial apoptotic pathway. The increased levels of caspase-12 suggested that ER stress may also be activated in the SGN of *Fgf13* cKO mice (**Figure 9C**). The detailed mechanisms underlying the activation of apoptosis require further study. A previous study showed that the downregulation of FGF13 elicited cell apoptosis directly through P53 interaction in cancer cells (Bublik et al., 2016). P53, a tumor suppressor, acts primarily as a transcription factor that regulates cell fate decisions, including cellular senescence, cell death, DNA repair, and metabolic homeostasis (Levine and Oren, 2009; Vousden and Prives, 2009). The activation of P53 can trigger apoptosis in a wide range of cell types, including neurons (Culmsee and Mattson, 2005). Expression of FGF13/miR-504 is repressed by P53, and depletion of FGF13 induces cell death by increasing P53 expression in cancer cells (Bublik et al., 2016). In our study, the higher expression of P53 mRNA in *Fgf13* cKO mice indicated the potential role of P53 in apoptotic SGN. It would be interesting to investigate whether knockout of *Fgf13* in SGN activates apoptosis through P53.

Bax and Bak belong to the multidomain Bcl-2 family of proteins, both of which are critically involved in the mitochondrial apoptotic pathway. In the present study, we observed higher levels of Bak mRNA with unaltered Bax mRNA levels in the cochlea of *Fgf13* cKO mice (**Figure 8F** and **Supplementary Figure 3**), indicating the participation of Bak in mitochondrial apoptosis but not Bax with suppression of *Fgf13*. These findings are also consistent with previous studies showing that Bak and Bax play different roles in mitochondrial apoptosis (Brooks et al., 2007; Fei et al., 2008; Someya et al., 2009). For example, Someya et al. (2009) showed that Bak-dependent (Bax-independent) mitochondrial apoptosis mediates age-related hearing loss in C57BL/6J mice. Thus, our data provides additional evidence for the specific function of *Bak* in mitochondrial apoptosis in the adult mouse cochlea.

CONCLUSION

FGF13 is expressed predominantly in the OC, SGNs, SV, and supporting cells of the cochlear tissue. *Fgf13* conditional knockout in the inner ear induces sensorineural hearing loss, while also increasing the apoptotic cell loss of SGNs associated with the mitochondrial apoptotic pathway in the mice cochlea. Moreover, these findings reveal a novel role for *Fgf13* in hearing function and suggest that it could be a potentially novel candidate gene for understanding deafness. Thus, this study may provide new perspectives on the molecular mechanisms and novel therapeutic targets for deafness.

AUTHOR CONTRIBUTIONS

CW, YY, and JY designed the experiments. YY, JY, PL, RZ, QW, and ZD performed the research. WW, JT, GG, JS, and HZ contributed new reagents and analytic tools. YY, JY, and FL interpreted the data. YY, HZ, PL, and CW wrote and reviewed the manuscript. CW, HZ, and PL supervised the project. All authors contributed to the article and approved the submitted version.

FUNDING

This work was supported by the National Natural Science Foundation of China (Nos. 81770407 and 31171097 to CW, No. 81900249 to JY, and No. 81670939 to PL), the Natural Science Foundation of Hebei Province (No. H2017206262 to CW, Nos. C2018206277 and H2020206003 to JY, and No. H2018206265 to PL), Key Project of Precision Medicine Joint Fund of Natural Science Foundation of Hebei Province (No. H20202064), Embryonic Stem Cell Research of The First Open Subject of Hubei Key Laboratory (No. 2020ESOF003), and High-Level Talent Support Project of Hebei Province (No. A2017005070 to CW and No. A201901032 to JY).

REFERENCES

Abu-Amero, K. K., Kondkar, A. A., Alorainy, I. A., Khan, A. O., Al-Enazy, L. A., Oystreck, D. T., et al. (2014). Xq26.3 microdeletion in a male with Wildervanck syndrome. *Ophthalmic Genet.* 35, 18–24. doi: 10.3109/13816810.2013.766218

Brooks, C., Wei, Q., Feng, L., Dong, G., Tao, Y., Mei, L., et al. (2007). Bak regulates mitochondrial morphology and pathology during apoptosis by interacting with mitofusins. *Proc. Natl. Acad. Sci. U. S. A.* 104, 11649–11654.

Bublik, D. R., Bursać, S., Sheffer, M., Oršolić, I., Shalit, T., Tarcic, O., et al. (2016). Regulatory module involving FGF13, miR-504, and p53 regulates ribosomal biogenesis and supports cancer cell survival. *Proc. Natl. Acad. Sci. U. S. A.* 114, E496–E505. doi: 10.1073/pnas.1614876114

Chen, J., Zhang, X., Li, J., Song, C., Jia, Y., and Xiong, W. (2016). Identification of a novel ENU-induced mutation in mouse Tbx1 linked to human DiGeorge syndrome. *Neural Plast.* 2016, 5836143. doi: 10.1155/2016/5836143

Chizhikov, V. V., Iskusnykh, I. Y., Fattakhov, N., and Fritzsch, B. (2021). Lmx1a and Lmx1b are Redundantly Required for the Development of Multiple Components of the Mammalian Auditory System. *Neuroscience.* 452, 247–264. doi: 10.1016/j.neuroscience.2020.11.013

Coate, T. M., and Kelley, M. W. (2013). Making connections in the inner ear: recent insights into the development of spiral ganglion neurons and their connectivity with sensory hair cells. *Semin. Cell Dev. Biol.* 24, 460–469. doi: 10.1016/j.semcdb.2013.04.003

Culmsee, C., and Mattson, M. P. (2005). p53 in neuronal apoptosis. *Biochem. Biophys. Res. Commun.* 331, 761–777. doi: 10.1016/j.bbrc.2005.03.149

Dalski, A., Atici, J., Kreuz, F. R., Hellenbroich, Y., Schwinger, E., and Zühlke, C. (2005). Mutation analysis in the fibroblast growth factor 14 gene: frameshift mutation and polymorphisms in patients with inherited ataxias. *Eur. J. Hum. Genet.* 13, 118–120. doi: 10.1038/sj.ejhg.5201286

DeStefano, G. M., Fantauzzo, K. A., Petukhova, L., Kurban, M., Tadin-Strapps, M., Levy, B., et al. (2013). Position effect on FGF13 associated with X-linked congenital generalized hypertrichosis. *Proc. Natl. Acad. Sci. U. S. A.* 110, 7790–7795. doi: 10.1073/pnas.1216412110

Dror, A. A., and Avraham, K. B. (2010). Hearing impairment: a panoply of genes and functions. *Neuron* 68, 293–308. doi: 10.1016/j.neuron.2010.10.011

Dvorakova, M., Jahan, I., Macova, I., Chumak, T., Bohuslavova, R., Syka, J., et al. (2016). Incomplete and delayed Sox2 deletion defines residual ear neurosensory development and maintenance. *Sci. Rep.* 6, 38253. doi: 10.1038/srep38253

Eggermont, J. J. (2019). Auditory brainstem response. *Handb. Clin. Neurol.* 160, 451–464. doi: 10.1016/B978-0-444-64032-1.00030-8

Elkon, R., Milon, B., Morrison, L., Shah, M., Vijayakumar, S., Racherla, M., et al. (2015). RFX transcription factors are essential for hearing in mice. *Nat. Commun.* 15, 8549. doi: 10.1038/ncomms9549

Fei, Q., McCormack, A. L., Monte, D. A. D., and Ethell, D. W. (2008). Paraquat neurotoxicity is mediated by a Bak-dependent mechanism. *J. Biol. Chem.* 283, 3357–3364. doi: 10.1074/jbc.M708451200

Fetoni, A. R., Zorzi, V., Paciello, F., Ziraldo, G., Peres, C., Raspa, M., et al. (2018). Cx26 partial loss causes accelerated presbycusis by redox imbalance and dysregulation of Nfr2 pathway. *Redox Biol.* 19, 301–317. doi: 10.1016/j.redox.2018.08.002

Filova, I., Dvorakova, M., Bohuslavova, R., Pavlinek, A., Elliott, K. L., Vochyanova, S., et al. (2020). Combined Atoh1 and Neurod1 Deletion Reveals Autonomous Growth of Auditory Nerve Fibers. *Mol. Neurobiol.* 57, 5307–5323. doi: 10.1007/s12035-020-02092-0

Gecz, J., Baker, E., Donnelly, A., Ming, J. E., McDonald-McGinn, D. M., Spinner, N. B., et al. (1999). Fibroblast growth factor homologous factor 2 (FHF2): gene structure, expression and mapping to the Börjeson-Forssman-Lehmann syndrome region in Xq26 delineated by a duplication breakpoint in a BFLS-like patient. *Hum. Genet.* 104, 56–63. doi: 10.1007/s004390050910

Goetz, R., Dover, K., Laezza, F., Shtraizent, N., Huang, X., Tchetchik, D., et al. (2009). Crystal structure of a fibroblast growth factor homologous factor (FHF) defines a conserved surface on FHFs for binding and modulation of voltage-gated sodium channels. *J. Biol. Chem.* 284, 17883–17896. doi: 10.1074/jbc.M109.001842

Goldfarb, M. (2005). Fibroblast growth factor homologous factors: evolution, structure, and function. *Cytokine Growth Factor Rev.* 16, 215–220. doi: 10.1016/j.cytogfr.2005.02.002

Guarch, M. E., Font-Llitjós, M., Murillo-Cuesta, S., Errasti-Murugarren, E., Celaya, A. M., Girotto, G., et al. (2018). Mutations in L-type amino acid transporter-2 support SLC7A8 as a novel gene involved in age-related hearing loss. *Elife* 7, e31511. doi: 10.7554/eLife.31511

Ideura, M., Nishio, S. Y., Moteki, H., Takumi, Y., Miyagawa, M., Sato, T., et al. (2019). Comprehensive analysis of syndromic hearing loss patients in Japan. *Sci. Rep.* 9, 11976. doi: 10.1038/s41598-019-47141-4

Jahan, I., Elliott, K. L., and Fritzsch, B. (2018). Understanding Molecular Evolution and Development of the Organ of Corti Can Provide Clues for Hearing Restoration. *Integr. Comp. Biol.* 58, 351–365. doi: 10.1093/icb/icy019

Kim, W. Y., Fritzsch, B., Serls, A., Bakel, L. A., Huang, E. J., Reichardt, L. F., et al. (2001). NeuroD-null mice are deaf due to a severe loss of the inner ear sensory neurons during development. *Development* 3, 417–426.

Kwon, T. J., Cho, H. J., Kim, U. K., Lee, E., Oh, S. K., Bok, J., et al. (2014). Methionine sulfoxide reductase B3 deficiency causes hearing loss due to stereocilia degeneration and apoptotic cell death in cochlear hair cells. *Hum. Mol. Genet.* 23, 1591–1601. doi: 10.1093/hmg/ddt549

Levine, A. J., and Oren, M. (2009). The first 30 years of p53: growing ever more complex. *Nat. Rev. Cancer* 9, 749–758. doi: 10.1038/nrc2723

Li, X., Xu, L., Sun, G., Wu, X., Bai, X., Li, J., et al. (2017). Spag6 mutant mice have defects in development and function of spiral ganglion neurons, apoptosis, and higher sensitivity to paclitaxel. *Sci. Rep.* 7, 8638. doi: 10.1038/s41598-017-08739-8

Liu, C. J., Dib-Hajj, S. D., Renganathan, M., Cummins, T. R., and Waxman, S. G. (2003). Modulation of the cardiac sodium channel Nav1.5 by fibroblast growth factor homologous factor 1B. *J. Biol. Chem.* 278, 1029–1036. doi: 10.1074/jbc.M207074200

Liu, Y., Li, S., Tao, T., Li, X., Zhu, Q., Liao, Y., et al. (2018). Intrafollicular fibroblast growth factor 13 in polycystic ovary syndrome: relationship with androgen levels and oocyte developmental competence. *J. Ovarian Res.* 11, 87. doi: 10.1186/s13048-018-0455-3

Lou, J. Y., Laezza, F., Gerber, B. R., Xiao, M., Yamada, K. A., Hartmann, H., et al. (2005). Fibroblast growth factor 14 is an intracellular modulator of voltage-gated sodium channels. *J. Physiol.* 569, 179–193. doi: 10.1113/jphysiol.2005.097220

Ma, Q., Anderson, D. J., and Fritzsch, B. (2000). Neurogenin 1 null mutant ears develop fewer, morphologically normal hair cells in smaller sensory epithelia devoid of innervation. *J. Assoc. Res. Otolaryngol.* 1, 129–143. doi: 10.1007/s101620010017

Macova, I., Pysanenko, K., Chumak, T., Dvorakova, M., Bohuslavova, R., Syka, J., et al. (2019). Neurod1 Is Essential for the Primary Tonotopic Organization and Related Auditory Information Processing in the Midbrain. *J. Neurosci.* 39, 984–1004. doi: 10.1523/JNEUROSCI.2557-18.2018

Martin, L. J., Liu, Z., Pipino, J., Chestnut, B., and Landek, M. A. (2009). Molecular regulation of DNA damage-induced apoptosis in neurons of cerebral cortex. *Cereb. Cortex.* 19, 1273–1293. doi: 10.1093/cercor/bhn167

Matei, V., Pauley, S., Kaing, S., Rowitch, D., Beisel, K. W., Morris, K., et al. (2005). Smaller inner ear sensory epithelia in Neurog 1 null mice are related to earlier hair cell cycle exit. *Dev. Dyn.* 234, 633–650. doi: 10.1002/dvdy.20551

Men, Y., Zhang, A., Li, H., Zhang, T., Jin, Y., Li, H., et al. (2015). LKB1 is required for the development and maintenance of stereocilia in inner ear hair cells in mice. *PLoS One* 10:e0135841. doi: 10.1371/journal.pone.0135841

Montgomery, S. C., and Cox, B. C. (2016). Whole mount dissection and immunofluorescence of the adult mouse cochlea. *J. Vis. Exp.* 16, 53561. doi: 10.3791/53561

Morton, C. C., and Nance, W. E. (2006). Newborn hearing screening-a silent revolution. *N. Engl. J. Med.* 354, 2151–2164. doi: 10.1056/NEJMra050700

Okada, T., Murata, K., Hirose, R., Matsuda, C., Komatsu, T., Ikekita, M., et al. (2013). Upregulated expression of FGF13/FHF2 mediates resistance to platinum drugs in cervical cancer cells. *Sci. Rep.* 3, 2899. doi: 10.1038/srep02899

Pavlinkova, G. (2020). Molecular Aspects of the Development and Function of Auditory Neurons. *Int. J. Mol. Sci.* 22, 131. doi: 10.3390/ijms22010131

Pierce, S. B., Chisholm, K. M., Lynch, E. D., Lee, M. K., Walsh, T., Opitz, J. M., et al. (2011). Mutations in mitochondrial histidyl tRNA synthetase HARS2 cause ovarian dysgenesis and sensorineural hearing loss of Perrault syndrome. *Proc. Natl. Acad. Sci. U. S. A.* 108, 6543–6548. doi: 10.1073/pnas.1103471108

Pirvola, U., Spencer-Dene, B., Xing-Qun, L., Kettunen, P., Thesleff, I., Fritzsch, B., et al. (2000). FGF/FGFR-2(IIIb) signaling is essential for inner ear morphogenesis. *J. Neurosci.* 20, 6125–6134. doi: 10.1523/JNEUROSCI

Puligilla, C., Dabdoub, A., Brenowitz, S. D., and Kelley, M. W. (2010). Sox2 induces neuronal formation in the developing mammalian cochlea. *J. Neurosci.* 30, 714–722. doi: 10.1523/JNEUROSCI.3852-09.2010

Puligilla, C., Feng, F., Ishikawa, K., Bertuzzi, S., Dabdoub, A., Griffith, A. J., et al. (2007). Disruption of fibroblast growth factor receptor 3 signaling results in defects in cellular differentiation, neuronal patterning, and hearing impairment. *Dev. Dyn.* 236, 1905–1917. doi: 10.1002/dvdy.21192

Puranam, R. S., He, X. P., Yao, L., Le, T., Jang, W., Rehder, C. W., et al. (2015). Disruption of Fgf13 causes synaptic excitatory-inhibitory imbalance and genetic epilepsy and febrile seizures plus. *J. Neurosci.* 35, 8866–8881. doi: 10.1523/JNEUROSCI.3470-14.2015

Someya, S., Xu, J., Kondo, K., Ding, D., Salvi, R. J., Yamasoba, T., et al. (2009). Age-related hearing loss in C57BL/6J mice is mediated by Bak-dependent mitochondrial apoptosis. *Proc. Natl. Acad. Sci. U. S. A.* 106, 19432–19437. doi: 10.1073/pnas.0908786106

Song, J. J., and Li, W. (2019). MiR-10b suppresses the growth and metastasis of colorectal cancer cell by targeting FGF13. *Eur. Rev. Med. Pharmacol. Sci.* 23, 576–587. doi: 10.26355/eurrev_201901_16870

Tan, X., Jahan, I., Xu, Y., Stock, S., Kwan, C. C., Soriano, C., et al. (2018). Auditory Neural Activity in Congenitally Deaf Mice Induced by Infrared Neural Stimulation. *Sci. Rep.* 8, 388. doi: 10.1038/s41598-017-18814-9

von Bartheld, C. S., Bahney, J., and Herculano-Houzel, S. (2016). The search for true numbers of neurons and glial cells in the human brain: A review of 150 years of cell counting. *J. Comp. Neurol.* 524, 3865–3895. doi: 10.1002/cne.24040

Vousden, K. H., and Prives, C. (2009). Blinded by the Light: The Growing Complexity of p53. *Cell* 137, 413–431. doi: 10.1016/j.cell.2009.04.037

Wang, C., Hennessey, J. A., Kirkton, R. D., Wang, C., Graham, V., Puranam, R. S., et al. (2011). Fibroblast growth factor homologous factor 13 regulates Na+ channels and conduction velocity in murine hearts. *Circ. Res.* 109, 775–782. doi: 10.1161/CIRCRESAHA.111.247957

Wang, X., Tang, H., Wei, E. Q., Wang, Z., Yang, J., Yang, R., et al. (2017). Conditional knockout of Fgf13 in murine hearts increases arrhythmia susceptibility and reveals novel ion channel modulatory roles. *J. Mol. Cell. Cardiol.* 104, 63–74. doi: 10.1016/j.yjmcc.2017.01.009

Wei, E. Q., Sinden, D. S., Mao, L., Zhang, H., Wang, C., and Pitt, G. S. (2017). Inducible Fgf13 ablation enhances caveolae-mediated cardioprotection during cardiac pressure overload. *Proc. Natl. Acad. Sci. U. S. A.* 114, E4010–E4019. doi: 10.1073/pnas.1616393114

Wong, A. C. Y., and Ryan, A. F. (2015). Mechanisms of sensorineural cell damage, death and survival in the cochlea. *Front. Aging Neurosci.* 7:58. doi: 10.3389/fnagi.2015.00058

Wu, C. C., and Bratton, S. B. (2013). Regulation of the intrinsic apoptosis pathway by reactive oxygen species. *Antioxid. Redox Signal.* 19, 546–558. doi: 10.1089/ars.2012.4905

Wu, Q. F., Yang, L., Li, S., Wang, Q., Yuan, X. B., Gao, X., et al. (2012). Fibroblast growth factor 13 is a microtubule-stabilizing protein regulating neuronal polarization and migration. *Cell* 149, 1549–1564. doi: 10.1016/j.cell.2012.04.046

Yang, J., Wang, Z., Sinden, D. S., Wang, X., Shan, B., Yu, X., et al. (2016). FGF13 modulates the gating properties of the cardiac sodium channel Na(v)1.5 in an isoform-specific manner. *Channels (Austin)* 10, 410–420. doi: 10.1080/19336950.2016

Yang, T., Kersigo, J., Jahan, I., Pan, N., and Fritzsch, B. (2011). The molecular basis of making spiral ganglion neurons and connecting them to hair cells of the organ of Corti. *Hear. Res.* 278, 21–33. doi: 10.1016/j.heares.2011.03.002

Zuccotti, A., Lee, S. C., Campanelli, D., Singer, W., Satheesh, S. V., Patriarchi, T., et al. (2013). L-type CaV1.2 deletion in the cochlea but not in the brainstem reduces noise vulnerability: implication for CaV1.2-mediated control of cochlear BDNF expression. *Front. Mol. Neurosci.* 6:20. doi: 10.3389/fnmol.2013.00020

5

Modeling Retinitis Pigmentosa: Retinal Organoids Generated from the iPSCs of a Patient with the USH2A Mutation Show Early Developmental Abnormalities

Yonglong Guo[1,2], Peiyuan Wang[1], Jacey Hongjie Ma[3,4], Zekai Cui[3,5], Quan Yu[6], Shiwei Liu[1], Yunxia Xue[7], Deliang Zhu[2], Jixing Cao[1], Zhijie Li[7], Shibo Tang[3,5] and Jiansu Chen[1,2,3,5,7]**

[1] Ophthalmology Department, The First Affiliated Hospital of Jinan University, Guangzhou, China, [2] Key Laboratory for Regenerative Medicine of Ministry of Education, Jinan University, Guangzhou, China, [3] Aier School of Ophthalmology, Central South University, Changsha, China, [4] Shenzhen Aier Eye Hospital, Shenzhen, China, [5] Aier Eye Institute, Changsha, China, [6] Centric Laboratory, Medical College, Jinan University, Guangzhou, China, [7] Institute of Ophthalmology, Medical College, Jinan University, Guangzhou, China

***Correspondence:**
Shibo Tang
tangshibo@vip.163.com
Jiansu Chen
chenjiansu2000@163.com*

Retinitis pigmentosa (RP) represents a group of inherited retinopathies with early-onset nyctalopia followed by progressive photoreceptor degeneration causing irreversible vision loss. Mutations in USH2A are the most common cause of non-syndromic RP. Here, we reprogrammed induced pluripotent stem cells (iPSCs) from a RP patient with a mutation in *USH2A* (c.8559-2A > G/c.9127_9129delTCC). Then, multilayer retinal organoids including neural retina (NR) and retinal pigment epithelium (RPE) were generated by three-step "induction-reversal culture." The early retinal organoids derived from the RP patient with the USH2A mutation exhibited significant defects in terms of morphology, immunofluorescence staining and transcriptional profiling. To the best of our knowledge, the pathogenic mutation (c.9127_9129delTCC) in *USH2A* has not been reported previously among RP patients. Notably, the expression of laminin in the USH2A mutation organoids was significantly lower than in the iPSCs derived from healthy, age- and sex-matched controls during the retinal organogenesis. We also observed that abnormal retinal neuroepithelium differentiation and polarization caused defective retinal progenitor cell development and retinal layer formation, disordered organization of NRs in the presence of the USH2A mutation. Furthermore, the USH2A mutation bearing RPE cells presented abnormal morphology, lacking pigmented foci and showing an apoptotic trend and reduced expression of specific makers, such as MITF, *PEDF*, and RPE65. In addition, the USH2A mutation organoids had lower expression of cilium-associated (especially *CFAP43*, *PIFO*) and dopaminergic synapse-related genes (including *DLGAP1*, *GRIK1*, *SLC17A7*, and *SLC17A8*), while there was higher expression of neuron apoptotic process-related genes (especially *HIF1A*, *ADARB1*, and *CASP3*). This study may provide essential assistance in the molecular diagnosis and screening of RP. This work recapitulates the pathogenesis of USH2A using patient-specific organoids and demonstrated that alterations in USH2A function due to mutations may lead to cellular and molecular abnormalities.

Keywords: retinitis pigmentosa, USH2A, iPSCs, organoid, RPE, basement membrane

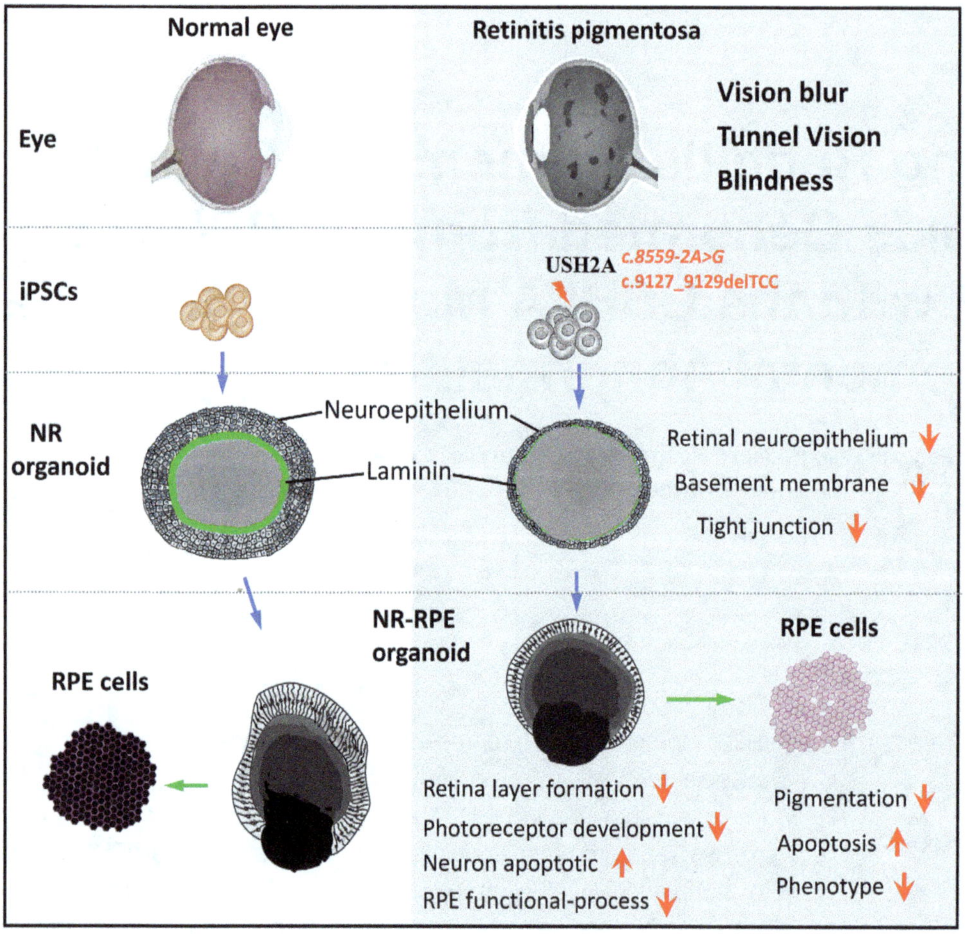

GRAPHICAL ABSTRACT | A model of USH2A mutation-associated 3D retinal developmental abnormalities.

INTRODUCTION

Retinitis pigmentosa (RP) is a group of inherited and progressive eye diseases that cause vision loss (Hartong et al., 2006). Unfortunately, RP disease is still incurable although RP patients have benefited from progress in research (Campochiaro and Mir, 2018; Dona et al., 2018). Initial RP symptoms include difficulty seeing at night and decreased peripheral vision due to rod- and cone-photoreceptor degeneration in the retinal structure (Shintani et al., 2009). As the disease progresses, the patient visual field (VF) gradually becomes only tunnel vision, eventually leading to dysfunction and blindness (Hartong et al., 2006). Most hereditary forms of RP are monogenic, and the inheritance patterns include autosomal dominant, autosomal recessive, X-linked, or maternal (mitochondria) (Rivolta et al., 2002). Currently, more than 95 RP pathogenic genes have been mapped and identified, among which 65 are implicated in the autosomal recessive form[1]. These known genes account for only approximately 60–70% of all RP cases and the other 40% have not been identified (Chen et al., 2018). Mutations of the *USH2A* gene in RP cases were discovered two decades ago (Weston et al., 2000). Over 600 different mutations in USH2A have been identified[2] and mutations in this gene are by far the most frequent cause of autosomal recessive non-syndromic RP (12–25%) (Slijkerman et al., 2018).

The *USH2A* gene is located at chromosome 1q41 and contains 72 exons ranging in length from 127 bp to 78 kb, which encodes the protein USHERIN (van Wijk et al., 2004). The *USH2A* gene has two isoforms, A and B. Isoform A consists of 21 exons encoding ~170 kDa of an extracellular protein (Weston et al., 2000), and isoform B is the full-length encoding an ~580 kDa complex transmembrane protein (Liquori et al., 2016). USHERIN contains laminin EGF motifs, a pentraxin domain, a short intracellular region with a PDZ-binding motif, some fibronectin type III motifs and so on (Dona et al., 2018). It is generally considered that USHERIN is an important stabilizing component in the centrosome-cilium interface region of the photoreceptor, where it is fixed by interactions with HARMONIN, SANS and WHIRLIN (Chen et al., 2014; Sorusch et al., 2017). It has also been proposed that USHERIN might play an important role in vesicle trafficking between the inner and outer segments of the photoreceptor based on its protein interactions and

[1] https://sph.uth.edu/retnet/sum-dis.htm

[2] USHbases; http://www.lovd.nl/USH2A

localization (Reiners et al., 2006; Maerker et al., 2007). On the other hand, USHERIN may also play a key role in structural maintenance of the apical inner segment, the basal outer segment, and the connecting cilium (Insinna and Besharse, 2008; Nachury et al., 2010). Huang et al. demonstrated that photoreceptors synthesized the *USH2A* protein and selectively deposited it into the interphotoreceptor cell matrix (IPM). The USH2A protein also displays homology to laminin and other extracellular matrix (ECM) proteins containing laminin epidermal growth factor (LE) and fibronectin type III (F3) motifs. Therefore, USH2A could perform ECM functions, including providing an environment for mechanical and physiological support of the surrounding cells, attachment of cells to the underlying epithelium, and/or signals for differentiation (Huang et al., 2002).

Animal models are useful for understanding the biological function of the *USH2A* gene and the pathogenic mechanisms underlying USH2A's role in RP. Liu et al. (2007) generated an *Ush2a* knockout mouse model in 2007. Their data showed that the ~580 kDa long USHERIN isoform is the predominant form in photoreceptor cells and USHERIN holds the apical inner segment recess that wraps around the connecting cilia. They also found that more than one-half of the photoreceptors degenerated and the outer segments of photoreceptors became very short and disorganized in *Ush2a* knockout mice by 20 months. Recently, there have been two reports that found that defects or the absence of USHERIN produced early onset retinal dysfunction in zebrafish models, which were specifically represented by reductions in a-wave and b-wave amplitudes in zebrafish model larvae compared to wild-type larvae. Dona et al. (2018) showed that mutation of *ush2a* led to decreases in WHIRLIN and ADGRV1 levels at the periciliary region of the photoreceptor and increased the apoptosis of photoreceptors. Han et al. (2018) revealed that the photoreceptors progressively degenerated and rod degeneration occurred prior to cone degeneration in *ush2a* knockout zebrafish. The studies of these *USH2A* knockout or mutation animal models provide some clues to uncovering the function of the *USH2A* gene. However, solid evidence supporting the function of *USH2A* is still difficult to obtain for the human retina.

Induced pluripotent stem cells (iPSCs) are a favorable tool in modeling inherited retinal disease. Cells derived from skin, eye, blood, or urine of patients can be reprogrammed to become iPSCs and then differentiated into retinal cell types. A disease model is created by compared the phenotypes and genotype between diseased retinal cells and normal cells. The recent development of iPSCs has enabled researchers to recapitulate the retinal structure, physiology, functionality, pathological changes and mechanisms *in vitro* (Jin et al., 2011). Researchers have successfully revealed evidence of pathogenic mechanisms by using patient specific iPSCs. Megaw et al. (2017) showed that iPSCs-derived photoreceptors from *RPGR* mutation patients exhibited increased actin polymerization compared to the control, which was due to a disruption of cell signaling pathways regulating actin turnover. Schwarz et al. (2017) validated that iPSC-derived 3D optic cups from a patient with the *RP2* mutation (p. R120X) develop normally, but the photoreceptors in the optic cups displayed reduced Kif7 staining at their cilia tips compared with controls. Jin et al. (2012) revealed that diffuse distribution of RHO protein is associated with endoplasmic reticulum stress in an iPSCs disease model derived from a RP patient with a CHO mutation.

Remarkably, three-dimensional (3D) culture technology allows embryonic stem cells (ESCs)/iPSCs in dishes to utilize self-assembly characteristics to generate retinal organoids, which reflect major structural and functional properties of real organs. Organoid technology is therefore, conducive to increasing insight into human retinal development, providing new avenues for drug screening, and serving as disease models (Li and Izpisua Belmonte, 2019). Eiraku et al. (2011) generated dynamic optic cup structures from mouse ESCs by 3D culture reminiscent of retinal development *in vivo*. Then, they demonstrated that optic cup organoids can be produced by self-organization of human ESCs, and the optic cup organoids have the capacity to form multilayered neural retina (NR) containing both rods and cones (Nakano et al., 2012). There are some differences between human ESCs and mouse ESCs derived optic cup organoids in structure and morphogenetics. More recently, a method of optic cup organoids produced by iPSCs has also become feasible. Meyer et al. found that optic vesicle-like structures arose at the appropriate time from human iPSCs by 3D culture of retinogenesis, and the vesicle-like structures from iPSCs are indistinguishable from those derived from human ESCs (Meyer et al., 2011). Zhong et al. (2014) showed that human iPSCs enable the generation of 3D retinal cups that contain mature photoreceptors with outer-segment-disc and photosensitivity. Sharma et al. (2017) reported that iPSCs and iPSC-derived retinal organoids carrying the *TRNT1* mutation exhibited a deficit in autophagy and inefficient expression of full-length TRNT1 protein. Deng et al. (2018) also demonstrated that urine derived from a RP patient with a RPGR mutation could be reprogrammed into iPSCs, which enabled differentiation into RPE cells and retinal organoids with shorted cilium. Thus, this cutting-edge technology provides a possibility for understanding the effect of mutations in *USH2A* on the development of human retinal organogenesis.

In the present study, we identified a novel pathogenic mutation in the *USH2A* gene (c.8559-2A > G/c.9127_9129delTCC) that lead to non-syndrome RP. The iPSCs were generated from the urine cells (UCs) of a RP patient with a *USH2A* mutation, and 3D retinal organoids were generated and differentiated into retinal pigment epithelium (RPE) cells to recapitulate the disease *in vitro*. We compared the 3D retinal organoids formation of iPSCs with the USH2A mutation with those derived from age- and sex-matched normal iPSCs by integrating the morphology and phenotype of retinal differentiation data with transcriptome profiling. We were able to demonstrate abnormal developmental features of the USH2A mutation in 3D retina and RPE cells.

MATERIALS AND METHODS

Clinical Diagnosis of RP

The proband and his family members were evaluated at Shenzhen Aier Eye Hospital (Shenzhen, China). The proband underwent a slit-lamp examination, visual acuity testing, fundus photography, spectral domain optical coherence tomography (SD-OCT), and full-field electrophysiological testing (ERG). The diagnosis of RP was established based on the appearance of abnormal pigmentation in fundus examination, and an extinguished aptitude on full-field rod ERG. Computerized testing of the VFs. The proband signed informed consent for participation in this study. The study was approved by the ethical committee of Aier Eye Institution and adhered to the tenets of the Declaration of Helsinki.

The Isolation of Urine Cells From the RP Patient

Urine cells were collected from 400 mL of fresh urine that was centrifuged, then the pellet obtained was washed with PBS, resuspended in urine cell isolation medium (UCI; Cellapy Biotechnology, Beijing, China), and plated onto 12-well plates. Six to seven days after cell seeding, the medium was replaced with urine cell expansion medium (UCE; Cellapy Biotechnology).

Generation of RPiPS Cells From UCs

Induced pluripotent stem cells were produced by the integration-free CytoTune-iPS 2.0 Sendai Reprogramming Kit as previously reported (Guo et al., 2018). Prior to viral transduction, 2×10^5 UCs were seeded onto one well of a Matrigel-coated 12-well plate. After 24 h, the medium was discarded and we added fresh UCE medium containing 2.5 ng/ml of basic FGF2. After 72 h of transduction, the medium was discarded and we added a fresh 1:1 (v/v) mix of UCE (no FGF2) and E6 (10 ng/ml FGF2) media. At the fourth day post transduction, the medium was changed to E6 medium (10 ng/ml FGF2) until colonies appeared. At that time the medium was changed to E8 medium. It should be noted that the culture medium before and after cell passage was consistent in reprogramming induction. At least three patient-derived iPSCs clones were used for subsequent experiments.

Generation of Retinal Organoid and RPE Cells

3D retinal organoids were generated from iPSCs using a method previously reported (Kuwahara et al., 2015). Briefly, iPSCs were digested into single cells and reseeded in low-cell-adhesion 96-well plates with V-bottomed conical wells at a density of 12,000 cells per well in NR induction medium supplemented with 20 μM Y-27632 under 5% CO_2 conditions. On day 6, recombinant human BMP4 was added into a culture well at a concentration of 1.5 nM, and its concentration was diluted into half by half medium change every third day. On day 18, the NR containing aggregates were transferred onto low-cell-adhesion 6-well plates (6 aggregates per well) in RPE-induction medium supplement with CHI99021 (3 μM) and SU5402 (5 μM) under 5% CO_2 conditions for 6 days culture. On day 24, the aggregates with RPE-like were cultured in NR and RPE induction medium under 40% O_2, 5% CO_2 conditions for 6 days culture for long term culture. RPE cells were differentiated from 3D retinal organoids following the method previously published with some modifications (Reichman et al., 2017). On day 34, identified pigmented patches were cut around from retinal organoid aggregate and cultured onto 6-well plates coated with 0.1% gelatin with RPE medium containing DMEM/F12, 1% MEM non-essential amino acids, 1% N2 supplement, 100 U ml-1 penicillin and 100 μg ml−1 streptomycin. The medium was changed every 2–3 day.

RNA-Seq Analysis

The RNA-seq analysis was performed as previously described (Cui et al., 2018). Briefly, total RNA of organoids in RPiPSCs and NiPSCs was extracted. Each group had three repetitions. The realization of the RNASeq was subcontracted to Chi Biotech Co., Ltd. (Shenzhen, China). The RNA of samples was submitted for the library construction. After the library sequencing, RPKM (Reads Per Kilo bases per Million reads) was calculated to obtain normalized gene expression levels. FANSe2 was used to map the original RNA-seq to the reference transcriptome sequence. The correlation coefficients between gene expression levels were calculated and plotted as a correlation heatmap. The screening threshold for the differentially expressed genes (DEGs) was set to: $|\log 2\ (\text{FoldChange})| > 1$ and $P\text{-Value} < 0.05$. Gene ontology (GO) analysis was performed using TopGO software (version 2.18.0). Pathway enrichment analysis was primarily based on the Kyoto Encyclopedia of Genes and Genomes (KEGG) database. KOBAS software (kobas2.0-20150126) was used, and comparisons between the two groups were made using the hypergeometric test.

QPCR

The qPCR protocol was performed as described in our previous report (Guo et al., 2017). Total RNA was extracted using a Tissue RNA Miniprep Kit. RNA (1 mg) was reverse transcribed into cDNA by using a ReverTra Ace qPCR RT Kit (TOYOBO, Japan). A qPCR reaction in a 20 μl total volume mixture containing 2 × SYBR, 250 nM of forward and reverse primers, and 10 ng of cDNA. The primers used are shown in **Table 1**.

Immunofluorescence Staining

The immunofluorescence staining was performed as described in our previous report (Guo et al., 2015). Samples were fixed in paraformaldehyde and permeabilized with 0.1% Triton X-100 in PBS. The samples were incubated with the primary antibody overnight and then incubated with the secondary antibodies. Finally, the samples were incubated with DAPI for nuclear staining. The antibodies used are shown in **Table 2**.

Data Analysis

All experiments were performed with at least 3 different iPSCs clones from patient and healthy control, 6–8 retinal organoids were required in each experiment. All data are presented as the averages ± standard deviations from 3–6 independent experiments (3–6 iPSCs clones). Statistical analyses were conducted using a two-sided unpaired Student's t-test to compare differences between two groups. $p < 0.05$ was considered statistically significant.

RESULTS

Identification of a Novel Pathogenic Mutation of the *USH2A* Gene in RP

The proband (**Figure 1A**, arrow) is a man who was 27 years old at the time of diagnosis. The patient exhibited typical clinical features of RP, including initial nyctalopia and visual acuity impairment. Best corrective visual acuity (decimal) of the proband was 20/25 (0.8, feet equivalent) in his both eyes. There was VF constriction and patient left eye presented more severe VF loss. His clinical phenotype in both eyes showed an extensive leopard spot-like pattern with a granular appearance that had individual pigment mottling under fundus images (**Figure 1B**).

SD-OCT images revealed that the outer limiting membrane, ellipsoidal zone and interdigitation zone had disappeared from almost all of the retina, except for a residual in the fovea. The thickness of the retinal neuroepithelial and outer nuclear layers (ONLs) was significantly decreased in both eyes (**Figure 1B**). The ERG results showed that decreased or distinguished b wave amplitude in scotopic ERG, which suggested the loss of rod function. The cone function was also impaired in photopic ERG testing (**Supplementary Figure S1**). His vestibular and hearing function were normal. Genomic sequencing data confirmed that the patient carries two compound heterozygous mutants in the *USH2A* gene. One was c.8559-2A > G inherited from the patient' father, which has been reported to be pathogenic when in combination with a missense mutation c.11806A > C

TABLE 1 | List of primers.

Primers		Sequences (5′to 3′)	Size (bp)	GeneBank Accession Number
GAPDH	Sense	GGTCGGAGTCAACGGATTTG	219	BC059110
GAPDH	Antisense	TGGAAGATGGTGATGGGATT		
PAX6	Sense	ACATCTGGCTCCATGTTGGG	184	NM_013435
PAX6	Antisense	ATAACTCCGCCCATTCACCG		
RAX	Sense	CTCCTCTCCGTCTCCAAAGC	275	NM_003106.
RAX	Antisense	TCCCGTCGTCCTTGGTAAAC		
SIX6	Sense	GGTTCAAAAACCGCCGACAA	165	NM_007374
SIX6	Antisense	CCTTGCTGGATAGACTGGCG		
CHX10	Sense	TTCAACGAAGCCCACTACCC	167	NM_182894
CHX10	Antisense	TAGAGCCCATACTCCGCCAT		
CASP3	Sense	TTGGAACCAAAGATCATACATGGAA	178	NM_004346
CASP3	Antisense	TGAGGTTTGCTGCATCGACA		
MSH2	Sense	GACTTCTATACGGCGCACGG	105	NM_000251
MSH2	Antisense	CAGATTCTTTGCTCCTGCCG		
ADARB1	Sense	GGCATGGAGAGCTTAAGGGG	179	NM_015833
ADARB1	Antisense	CTTGACTGGCGGAGACTGTT		
HIF1A	Sense	GCCAGACGATCATGCAGCTA	171	NM_001530
HIF1A	Antisense	CTGGTCAGCTGTGGTAATCCA		
PEDF	Sense	AGTGTGCAGGCTTAGAGGGAC	107	NM_002615
PEDF	Antisense	CCCGAGGAGGGCTCCAATG		
RPE65	Sense	CCACCTGTTTGATGGGCAAG	162	NM_000329
RPE65	Antisense	CAGGGATCTGGGAAAGCACA		
TJP1	Sense	AGCCATTCCCGAAGGAGTTG	175	NM_003257
TJP1	Antisense	ATCACAGTGTGGTAAGCGCA		
MITF	Sense	AGCTTGTATCTCAGTTCCGC	200	NM_198177
MITF	Antisense	ATGGCTGGTGTCTGACTCAC		
OTX2	Sense	CCTCACTCGCCACATCTACT	196	NM_021728
OTX2	Antisense	GTTTGGAGGTGCAAAGTCGG		
COL4A6	Sense	ATTTTGGTCGGTGCTCCCTG	166	NM_001847.3
COL4A6	Antisense	CTCGGTCAGGCACAACGTAA		
LAMb1	Sense	TTCAGTTTCTTAGCCCTGTGC	175	NM_002291.3
LAMb1	Antisense	CGATACAGTAGGGTTCGGGC		
TNC	Sense	CCAAAACATTTCTGGACAGTACCT	182	NM_002160.4
TNC	Antisense	CCCCTCCCACTTGACCACTA		
CLDN4	Sense	AGCCTTCCAGGTCCTCAACT	167	NM_001305.4
CLDN4	Antisense	GCGAGGTGACAATGTTGCTG		
CLDN19	Sense	AACCCAAGCACACCTGTCAA	189	NM_148960.2
CLDN19	Antisense	AACAACTGGTTCTCGGGCAG		

TABLE 2 | List of antibodies.

Antigen	Host species	Company or provider	Cat. no.
OCT4	Rb	Proteintech	11263-1-AP
SSEA4	MS	Cell Signaling	4755s
TRA-160	MS	Abcam	ab16288
SOX2	Rb	Proteintech	11064
AFP	Rb	Abclonal Technology	A0200
SMA	Rb	ThermoFish	PA5-16697
Nestin	Ms	Santa Cruz Biotechnology	SC23927
RAX	Ms	Antibodies	AA 104-206
N-cadherin	Ms	R&D	MAB13881
CDH2	Sheep	R&D	AF6426
CHX10	Ms	Santa	sc-365519
PAX6	Rb	Abcam	ab154253
aPKC	Ms	Abcam	ab11723
Ki67	Rb	Abcam	ab15580
RPE65	Ms	Abcam	ab13826
TYR	Rb	Abcam	ab180753
MITF	Rb	Abcam	ab122982
ZO-1	Ms	ThermoFish	Cat # 33-9100
Laminin	Rb	Abcam	ab11575

(p.T3936P) (Dai et al., 2008). The other was c.9127_9129delTCC inherited from the patient' mother, which is a novel deletion mutation in *USH2A* (**Supplementary Figure S2**). However, the patients' parents had no disease phenotypes despite carrying these mutations. Taken together, the patient was diagnosed as the proband with non-syndrome RP disease by clinical phenotype and genotype demonstration.

Reprogramming Patient-Specific Urine Cells

RP patient-specific and control (age- and sex- matched healthy donor) UCs were harvested from donated urine samples and reprogrammed to induced pluripotent stem cell (iPS) lines by using an integration-free CytoTune^{TM}-iPS 2.0 Sendai Reprogramming Kit (**Figure 2A**). The morphology of the RPiPSCs clones resembled that of typical iPS clones (**Figure 2B**). Immunofluorescence staining showed that RPiPSCs at passage 7 positively expressed pluripotency markers Oct4, Sox2, SSEA4, and TRA-160 (**Figure 2C**). Immunofluorescence staining also revealed that the RPiPSCs had the ability to form embryoid bodies and spontaneous differentiated into three germ layer cells, which were positive for Nestin, SMA, and AFP (**Figure 2D**). The RPiPSCs clones showed a normal karyotype (46, XY) and the USH2A mutations were confirmed (**Figures 2D,E**).

Association of the USH2A Mutations With Abnormal Organoid Induction and Increased Apoptosis

To establish the disease model of RP with mutated USH2A *in vitro*, we generated NR – RPE conversion organoids from RPiPSCs and normal iPSCs (**Figure 3A**). The organoids displayed

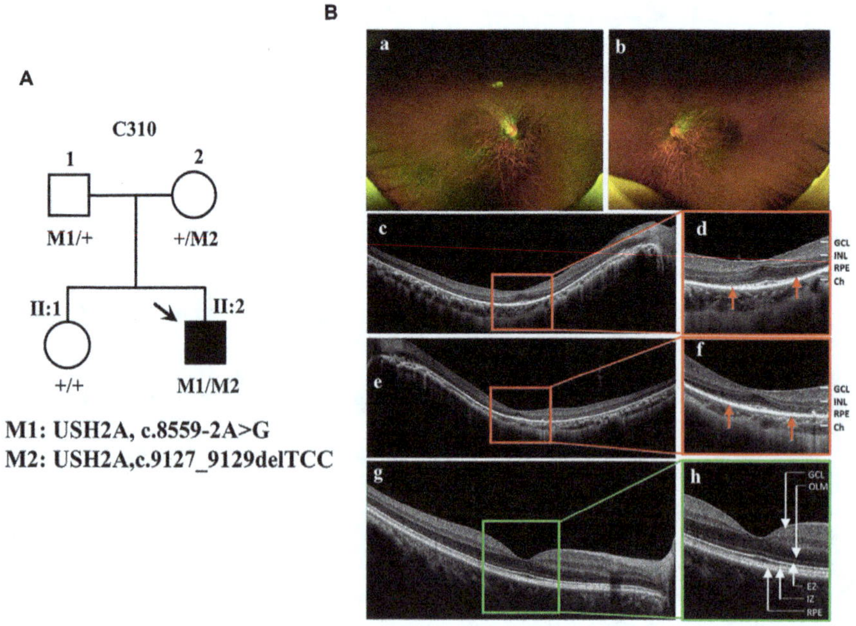

FIGURE 1 | Identification of a novel pathogenic mutation of the USH2A gene in RP. **(A)** Pedigrees and mutations segregating in family C310. Males and females are represented by squares and circles, respectively. The symbols for affected family member are filled. The symbols for proband with black arrow. The genotype of each evaluated individual is shown below the individuals' symbol and identification number. Abbreviations: Wild type (+); c.8559-2A > G (M1); c.9127_9129delTCC (M2). **(B)** Fundus images of the proband and health control. Fundus photographs showed the abnormal pigmentation in the peripheral area in both eye of the proband **(a,b)**. Loss of outer limiting membrane (OLM), ellipsoid zone (EZ) and interdigitation zone (IZ) was visualized by SD-OCT in the proband **(c–f)**, when comparing to the normal eye **(g,h)**.

NR and RPE features after 34 days of NR-RPE conversion culture in the RP and normal group (**Figure 3B**). Temporal analysis revealed a delay (approximately 4–6 days) in the self-forming of NRs for USH2A mutation iPSCs compared to controls (**Figure 3C**). Quantification data showed that fewer NR organoids were induced from the RPiPSCs than those in the normal group (**Figure 3D**). The organoid diameters among those with the USH2A mutation were smaller than the controls at days 8–20 (**Figure 3E**). Additionally, the most peripheral cells of the organoids were more likely to fallen off from the USH2A mutation organoids during the cultivation (**Supplementary Figure S3A**). Furthermore, qPCR analysis revealed that the expression levels of apoptosis related genes, such as CASP3, MSH2, ADARB1, and HIF1A, were significantly increased in the USH2A mutation group at days 34 compared to the control (**Figure 3F**). According to the results of DEGs and GO enrichment of RNA-Seq analysis, the neuronal apoptosis-related genes were significantly upregulated in the RP groups compared with the control group at day 34 (**Figure 3H**). Additionally, a comparison of cell proliferation revealed fewer Ki67 positive cells in the USH2A mutation versus the control group (**Figure 3G** and **Supplementary Figure S3B**). These results collectively suggest that the USH2A mutation in iPSCs is associated with abnormal NR organoids induction, which is accompanied by decreased proliferation.

Association of the USH2A Mutation With Defective Retinal Neuroepithelial and Photoreceptor Development

The thickness of the retinal neuroepithelium in the RP-organoids was significantly reduced compared with the control group at days 15–22 (**Figures 4A,B**). To determine whether there was an adverse effect of the USH2A mutation on retinal organogenesis and neuroepithelium, we examined the expression of factors that regulate eye development growth (RAX, PAX6 and CHX10) and polarized guidance (N-cadherin and aPKC). Coimmunostaining analysis of the organoids presented a significantly reduced proportion of RAX^+ and N-cadherin$^+$ cells in the USH2A mutation group than in the controls at day 10 (**Figure 4C** and **Supplementary Figure S4A**). Over time, we observed that significantly fewer $CHX10^+$, $PAX6^+$, N-cadherin$^+$, and $aPKC^+$

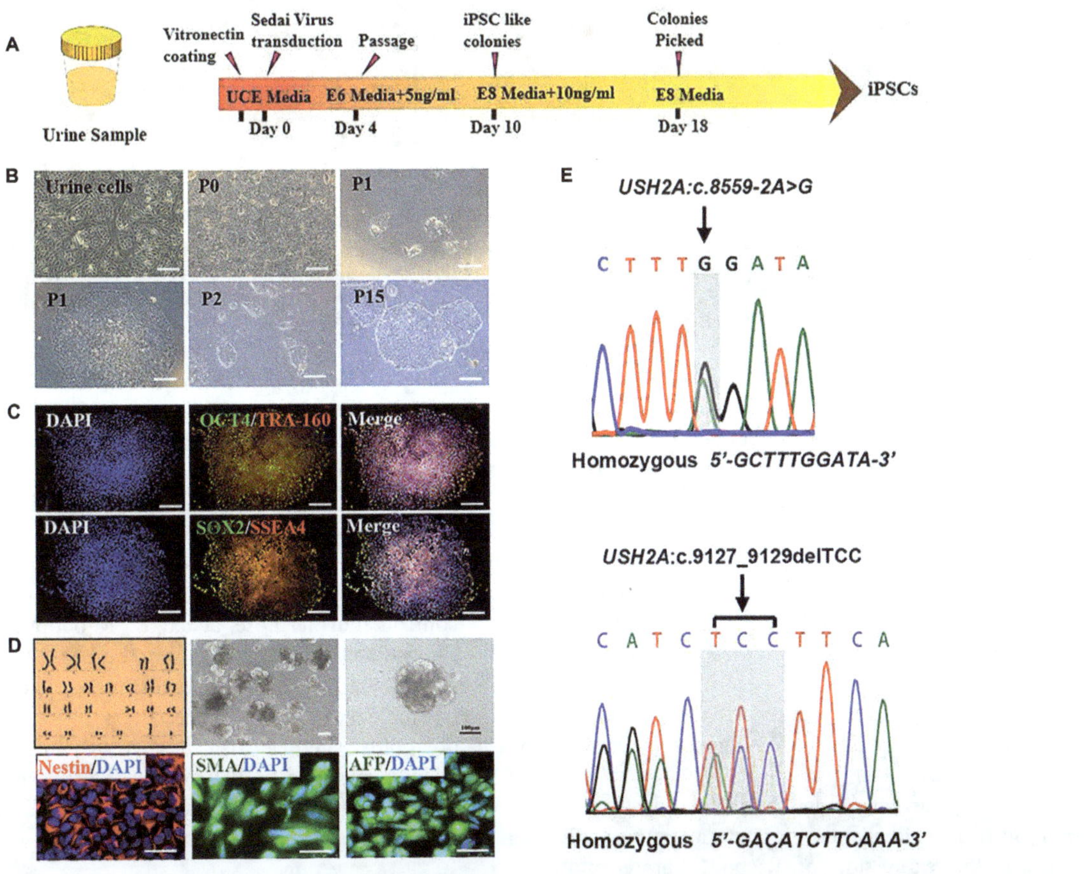

FIGURE 2 | Characteristics of iPSCs derived from RP patient urine cells. **(A)** A schematic of RP patient-specific iPSCs generated by use of an integration-free Sendai Reprogramming Kit. **(B)** Morphology of RP patient-specific urine cells **(a)** and iPSCs in different passages **(b–f)**. **(C)** Immunofluorescence staining analysis results of the pluripotency markers OCT4, TRA-160, SOX2, and SSEA4 in RP patient-specific iPSCs. **(D)** Normal karyotype (46, XY), embryoid body formation, and spontaneous differentiation experiments show that the RP patient-specific iPSCs had the capacity to form three germ layers, which was verified by immunofluorescence staining with Nestin, SMA, and AFP. **(E)** The USH2A mutation was confirmed in RP patient-specific iPSCs (scale bar = 100 μm).

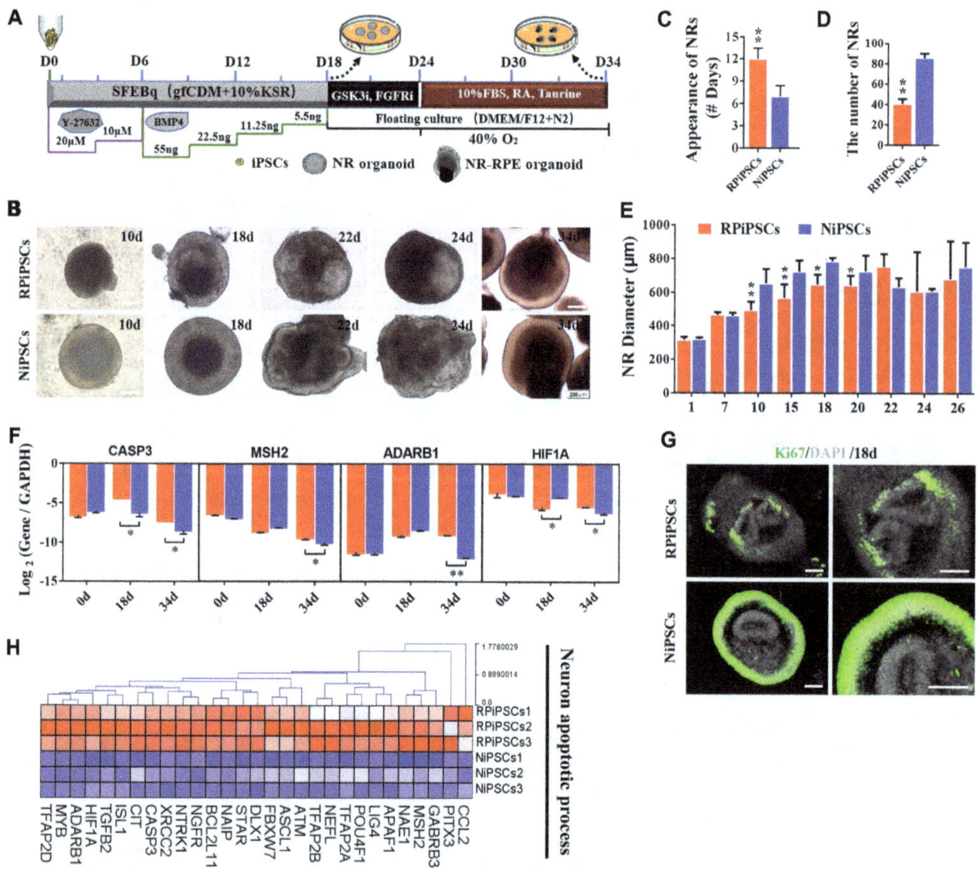

FIGURE 3 | USH2A mutation neural retina (NR) organoids have abnormal organoid induction and increased apoptosis. **(A)** A schematic representation of patient-specific iPSC differentiation along the NR organoid lineage. Brightfield images of developing NR over time **(B)** and USH2A mutation iPSCs generated NR at a slower pace **(C)**, less efficiently **(D)**, and with a smaller diameter **(E)** versus controls. **(F)** qPCR analysis revealed mRNA levels of neuron apoptosis cell specific markers *CASP3*, *MSH2*, *ADARB1*, and *HIF1A* (*GAPDH* gene as a control). **(G)** Immunostaining of retinal organoids showing the expression of cell-proliferation marker Ki67. **(H)** Neuron apoptotic process related DEGs at day 34 were normalized by *Z*-scores. Red represents upregulated expression. Blue represents downregulated expression, Bar is represents *Z*-score ($n = 3$ independent experiments; each experiment need 2–4 organoids; scale bar = 200 μm).

cells expressed immunoreactivity corresponding to these factors in the USH2A mutation groups compared with controls at day 18 (**Figures 4D–F** and **Supplementary Figures S4B–E**). qPCR analysis also showed that the mRNA expression level of the RPCs genes, PAX6, RAX, SIX6, and CHX10, were significantly lower in the USH2A mutation group than the control at day 18 (**Figure 4G**). However, at day 34, the mRNA expression levels of these genes were expressed higher in the USH2A mutation group than the control group, which is consistent with the significantly upregulated expression of retinal development related- and RPCs related-genes detected by RNA-seq analysis (**Figures 4G–I**).

Moreover, to further understand the effect of USH2A mutation on neuroepithelial develop into a stratified NR tissue, we analyzed the expression of Reep6 (photoreceptors), GS (Müller glia), bHLHB5 (amacrine cells) by immunofluorescence. After 86 days of retina differentiated culture, the NR-RPE organoids displayed NR features and positive for bHLHB5, GS, Reep6 staining in the RP and normal group. However, the stratified NR tissue contain major retinal cell types were disordered organization in USH2A mutation group, while the normal organoids differentiated retinal cells, such as photoreceptors, Müller cells, and Amacrine cells all arranged in their proper layers (**Figure 5**). Meanwhile, there was a significantly decreased proportion of bHLHB5^{+}, GS^{+}, and Reep6^{+} cells in the USH2A mutation groups compared with controls. Furthermore, at day 86 Rhodopsin immunoreactivity showed negatively expression in the USH2A mutation organoids and instead revealed weak expression at the apical surface where developing photoreceptors were located in normal organoids (**Figure 5**). These results demonstrate that the USH2A mutation in iPSCs is associated with aberrant organoids polarization, defective neuroepithelium, and abnormal RPCs and photoceptors differentiation.

Association of USH2A Mutations With Abnormal RPE Development and Growth

To better understand the effect of the USH2A mutation on RPE cells differentiated from RPiPSCs, we detected the

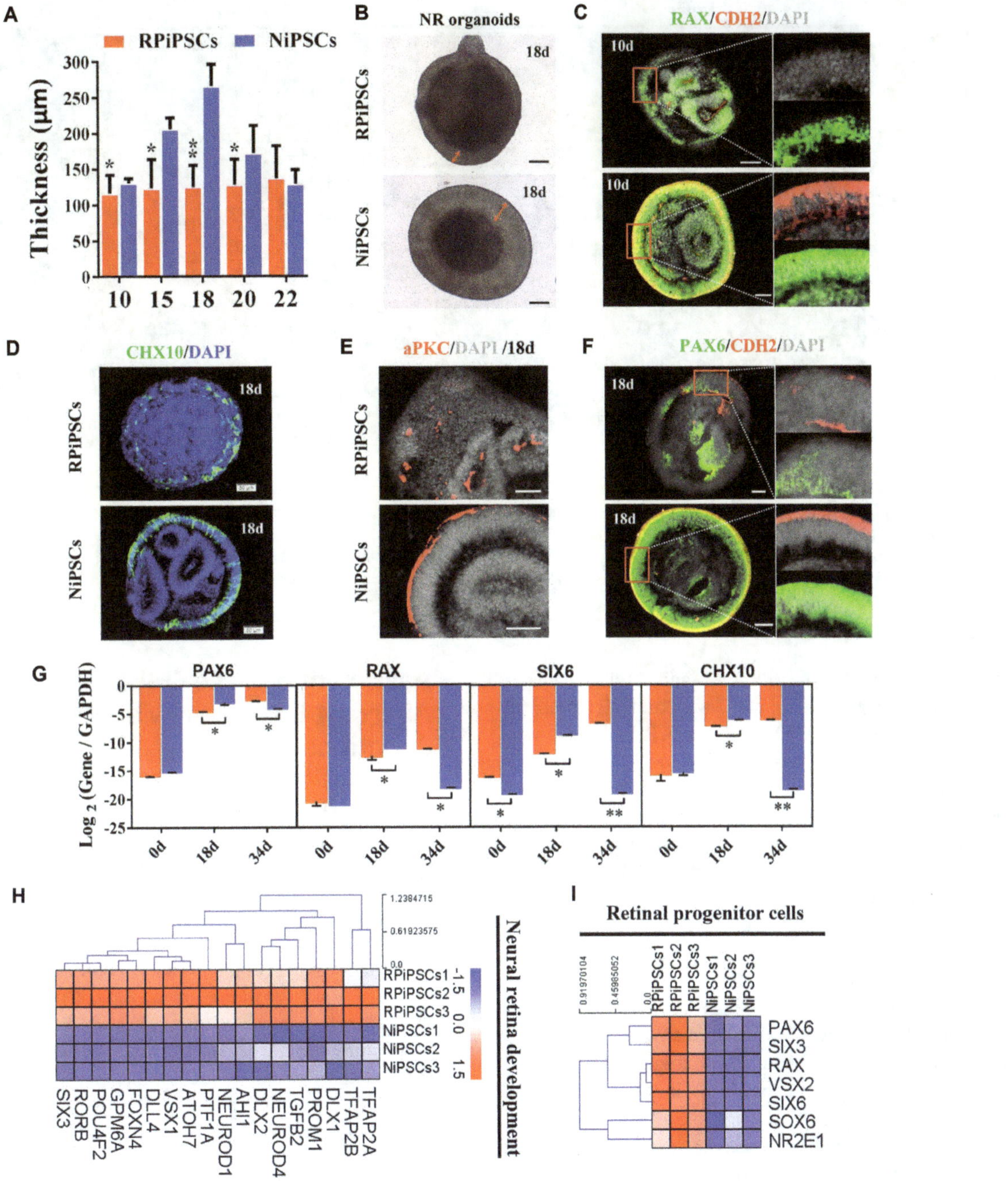

FIGURE 4 | USH2A mutation iPSCs generate retinal organoids with defective retinal neuroepithelial and RPC development. **(A)** The thickness of retinal neuroepithelium in RP-organoids at days 15–22. **(B)** Brightfield images of developing retinal neuroepithelium at day 18. **(C)** Immunofluorescence staining shows the cells in retinal organoids co-expressing RAX and CDH2 (N-cadherin) immunoreactivity at day 10. **(D)** Immunofluorescence images shows the cells in NR organoids expressing CHX10 immunoreactivity at day 18. **(E,F)** Immunostaining of NR organoids showing the expression of specific markers PAX6, N-cadherin, and aPKC at day 18. **(G)** qPCR analysis reveals mRNA levels of RPCs specific markers *PAX6*, *RAX*, *SIX6*, and *CHX10* (*GAPDH* gene as a control). **(H,I)** Neural retina development and RPCs related DEGs at day 34 were normalized by *Z*-scores. Red represents upregulated expression. Blue represents downregulated expression, Bar is represents *Z*-score (n = 3 independent experiments; each experiment need 2–4 organoids; scale bar = 200 μm).

morphology, phenotype and transcriptome of RPE. After NR-RPE conversion culture, the NR-RPE organoids showed RPE features as distinct pigmented foci in the RP and normal group at day 34. However, the pigmented foci in the RP organoids were significantly less than those in the control (**Figure 6A**). The immunofluorescence staining data showed the NR-RPE organoids were positive for RPE cell markers RPE65 and MITF in both groups at day 26. However, there was a significantly

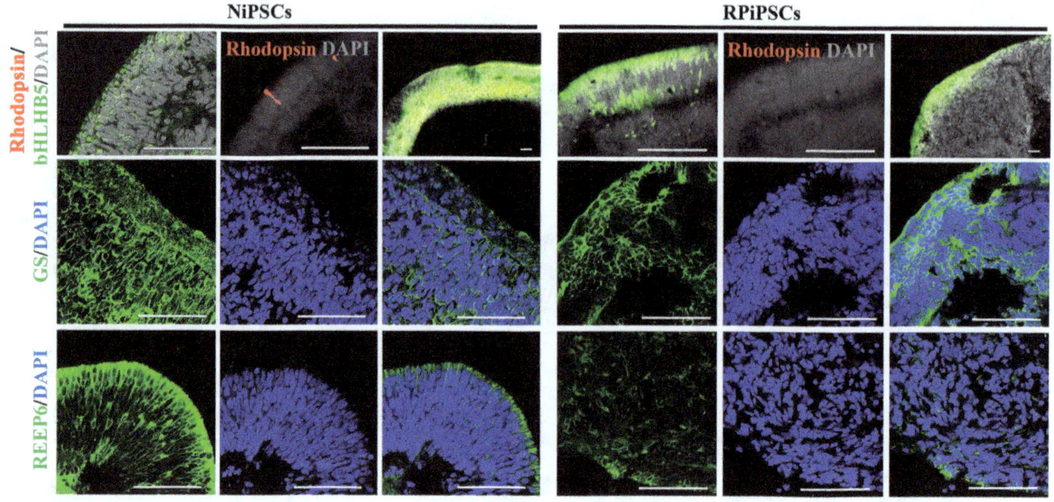

FIGURE 5 | Differentiation of neural retinas in organoids at day 86. bHLHB5, glutamine synthetase (GS) and Reep6 are markers of amacrine cells, Müller cells, and rod photoreceptors, respectively. Rhodopsin immunoreactivity is most intense in the rudimentary outer segments. The stratified neural retinas differentiated from organoids were disordered organization and negative for Rhodopsin in USH2A mutation group, while the normal organoids differentiated retinal cells, such as photoreceptors, Müller cells, and Amacrine cells all arranged in their proper layers and weak expressed Rhodopsin. The bHLHB5, GS, Reep6 were positively expressed in the RP and normal organoids. Nuclei were stained with DAPI (blue) (n = 2 independent experiments; each experiment need 2 organoids; scale bar = 100 μm).

decreased proportion of RPE 65^{+} and MITF^{+} cells in the USH2A mutation groups compared with controls (**Figure 6B** and **Supplementary Figures S5A,B**). The qPCR analysis showed that the mRNA expression levels of OTX2 and MITF were lower in USH2A mutation organoids than those in the control at days 18–34 (**Supplementary Figure S5C**). Furthermore, RNA-seq data at day 34 showed that the major upregulated genes involved in the RPE functional process included ALDH1A3 and RDH10, while the major downregulated genes were RPE65, RDH5, and DHRS3 (**Figure 6C**). Melanin metabolism-related genes were all downregulated, especially the markers TYR, OCA2, RARG, and GPR143 at day 34 (**Figure 6D**).

Moreover, the adherent RPE cells at passage 2 differentiated from control organoids expressed many more RPE features than those in the USH2A mutation organoids. We observed abnormal morphology and absent melanin in RPE cells derived from the USH2A mutation organoids, while the RPE cells in the control presented a classic pigmented, cobblestone morphology (**Figure 6E**). Immunofluorescence staining assay revealed that there were significantly fewer MITF and ZO-1 positive cells in the USH2A mutation group compared to the control (**Figures 6E,F**). qPCR analysis also revealed that the RPE cells in the USH2A mutation group expressed lower levels of transcripts corresponding to RPE65, TJP1, and a mature marker, PEDF, compared to control (**Figures 6G–I**). Additionally, the RPE cells differentiated from USH2A mutation organoids displayed atrophic trends after passage in culture. Together, these observations demonstrate that the USH2A mutation is associated with abnormal RPE development and phenotype expression as well as increased cell death in *de novo* generated RPE cells.

Association of the USH2A Mutation With Aberrant Basement Membrane and Tight Junctions

We found that approximately 40% of the organoids in the USH2A mutation group degraded at day 19 of organoid induction. The most peripheral cells of the organoids began to shed and underwent apoptosis by day 26 (**Figure 7A**). Notably, fluorescent staining showed that the expression of the basement membrane marker Laminin in the USH2A mutation group was significantly lower than that in the control group at day 18 (**Figure 7B** and **Supplementary Figure S6A**). The Laminin^{+} expression was also lower in the USH2A mutation group than that in the control at day 26 and day 86 (**Figures 7C,D** and **Supplementary Figure S6B**). Likewise, the qPCR results showed that basement membrane markers COL4A6, LAMb1, and TNC were downregulated at day 15 and day 34 in the USH2A mutation group compared to the control organoids. Moreover, we also found that the tight junction interactions related transcription factors, such as CLDN4, CLDN19, and TJP1, were also significantly downregulated at day 15 and day 34 in the USH2A mutation group compared to the control organoids (**Figures 7E,F**). Furthermore, the RNA-seq data showed that the expression levels of basement membrane-associated mRNAs in the USH2A mutation group were significantly lower than those in the control group, such as COL4A6, CST3, and LAMA3 (**Figure 7H**). Meanwhile, it was also found that the tight junction interactions related transcription factors also showed significant downregulation, such as 4, 3, and 7 of the CLDN family (**Figure 7G**). Together, these observations suggest that the USH2A mutation is associated with degeneration of the basement membrane, tight junction and other intercellular conjunctions.

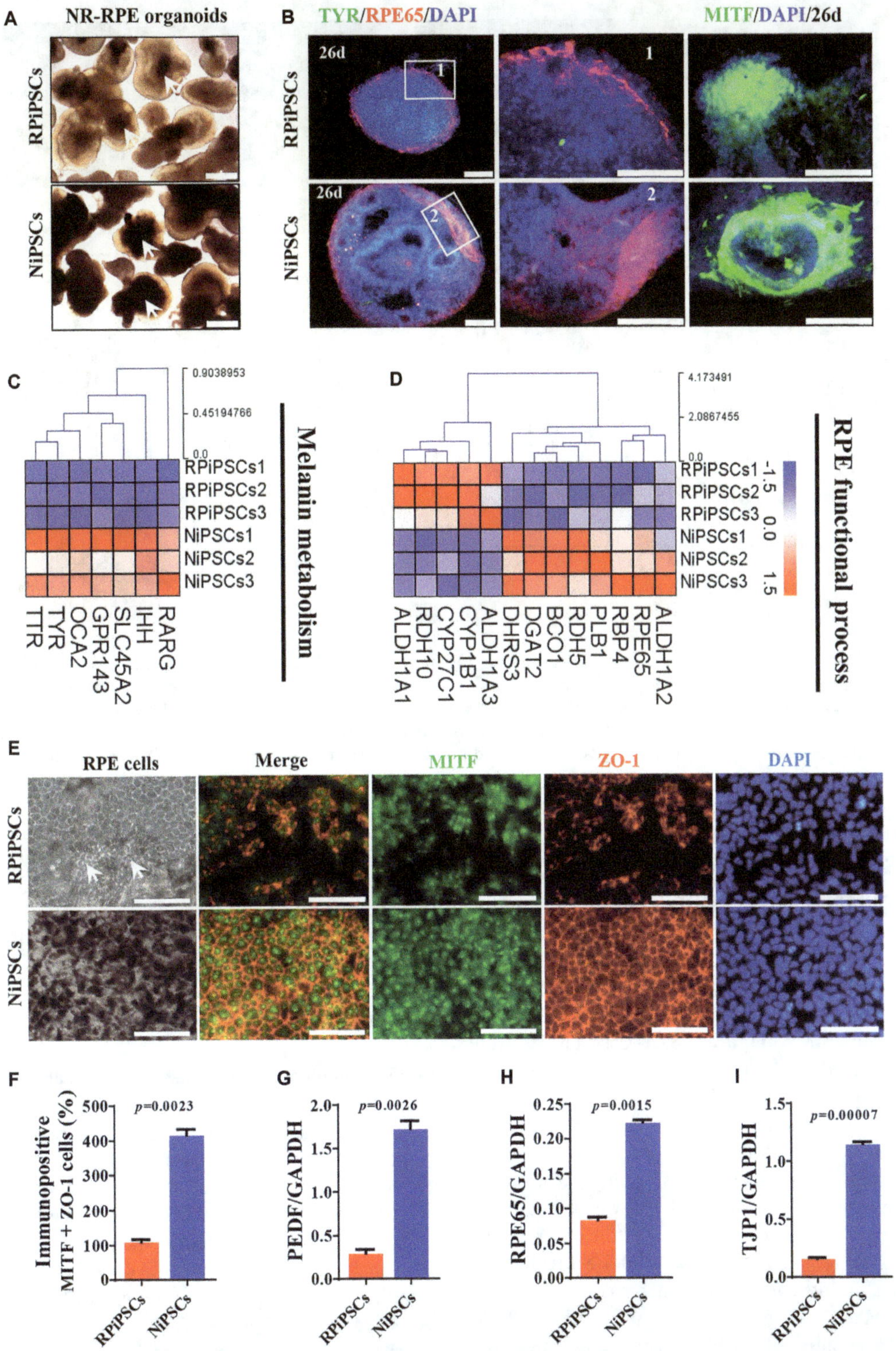

FIGURE 6 | USH2A mutation iPSCs generate abnormal RPE development and growth. **(A)** Brightfield images of developing NR-RPE organoids at day 34. **(B)** Immunostaining of NR-RPE organoids showing the expression of RPE cell specific markers TYR, RPE65, and MITF. **(C,D)** Melanin metabolism and RPE functional process related DEGs at day 34 were picked and normalized by Z-scores. Red represents upregulated expression, blue represents downregulated expression, Bar is represents *Z*-score. **(E)** Immunostaining of RPE cells derived from NR-RPE organoids showing the expression of RPE cell specific markers MITF and ZO-1. **(F)** A separate bar graph shows the percentage of MIFT^{+} ZO-1^{+} cells for immunostaining in a merged image. **(G–I)** qPCR analysis revealed mRNA levels of RPE cell specific markers *PEDF*, *RPE65*, and *TJP1* (*GAPDH* gene as a control) (*n* = 3 independent experiments; each experiment need 2–4 organoids. Scar bar = 100 μm).

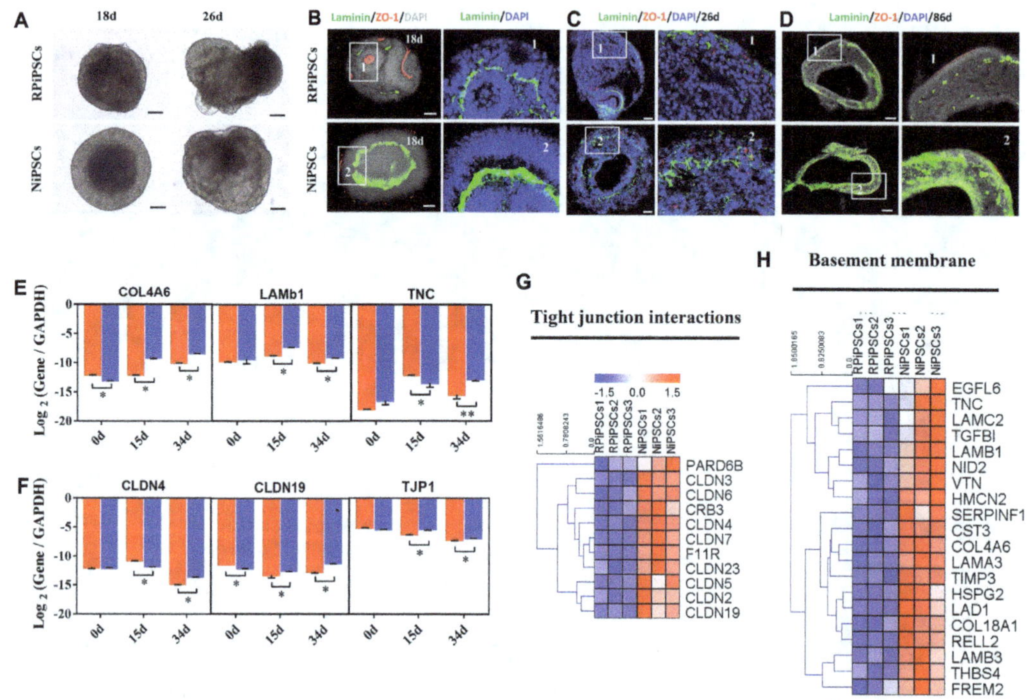

FIGURE 7 | USH2A mutation iPSCs generated retinal organoids have aberrant basement membrane and tight junctions. **(A)** Brightfield images of developing retinal organoids at day 18 and 26. **(B–D)** Immunostaining of retinal organoids showing the expression of Laminin and ZO-1 at day 18, 26, and 86. **(E,F)** qPCR analysis reveals mRNA levels of basement membrane and tight junction regulators, *COL4A6*, *LAMb1*, *TNC*, *CLDN4*, *CLDN19*, and *TJP1* (*GAPDH* gene as a control). **(G,H)** Basement membrane and tight junction interactions associated DEGs at day 34 were picked and normalized by *Z*-scores. Red represents upregulated expression, blue represents downregulated expression, Bar is represents *Z*-score ($n = 5$ independent experiments; each experiment need 4–6 organoids; scale bar = 200 μm).

Association of the USH2A Mutation With Abnormal Gene Expression and Pathways by RNA-Seq Analysis

To understand the mechanism underlying the USH2A mutation affected genes and pathways, we examined RNA-seq expression of both groups' organoids at day 34. Transcriptome data revealed significant differences between the control and RP organoids (**Supplementary Figures S7A–C**). There were 1853 DEGs upregulated and 1808 DEGs were downregulated (**Supplementary Figure S7D**). Pathway enrichment analysis showed that the USH2A mutation primarily affected the cell adhesion molecules pathway, followed by the MAPK signaling pathway, glutamatergic synapse, the Wnt signaling pathway and to a lesser extent, the calcium signaling pathway (**Figure 8A** and **Supplementary Figure S7G**). Remarkably, the differentially expressed gene ontology (GO) terms were those involved in the USH2 complex and its components, such as downregulated PDZD7, WHRN, MYO7A, and USH1C, and upregulated VEZT, CLRN1 and LCA5 (**Figure 8B**). qPCR analysis confirmed that the gene expression level of CLRN1 was abnormal in the USH2A mutation group compared to the control (**Supplementary Figure S7E**). Additionally, the cilium related genes were all downregulated, especially CFAP126, PIFO, and CFAP161 (**Figure 8C**). We also observed that "calcium signaling," "retinal layer formation," "dopaminergic synapse," and "vesical transport" were among the top-ranking categories identified at the NR-RPE stage by GO analysis (**Figures 8D–G**). Together, the trend in these DEGs indicate that the USH2A mutation may adversely affect cell adhesion molecules, the glutamatergic synapse pathway and cilium, calcium signaling, dopaminergic synapse, and vesical transport related gene ontology terms in the early organoids.

DISCUSSION

Retinitis pigmentosa patients first gradually lose night vision and side vision, and then experience late central vision loss because of progressive loss of rod and cone photoreceptor cells (National Eye Institute, 2004). Mutations of the *USH2A* gene are the most common cause of RP and are found in around 10–15% of recessive RP and 30–40% of Usher syndrome type 2 cases (Sun et al., 2016; Huang et al., 2018). Here, the modeling RP retinal organoid was studied. We generated retinal organoids from the iPSCs of a patient with the USH2A mutation of c.8559-2A > G and c.9127_9129delTCC. The c.9127_9129delTCC was inherited from the patient's mother, which is a novel deletion mutation in *USH2A* and may change the amino acid sequence and affect the protein features as predicted by Mutation Taster[3]. c.8559-2A > G was inherited from the patient' father, which has previously been reported to be pathogenic when in combination with the missense mutation c.11806A > C (p.T3936P) (Dai et al.,

[3]http://www.mutationtaster.org/

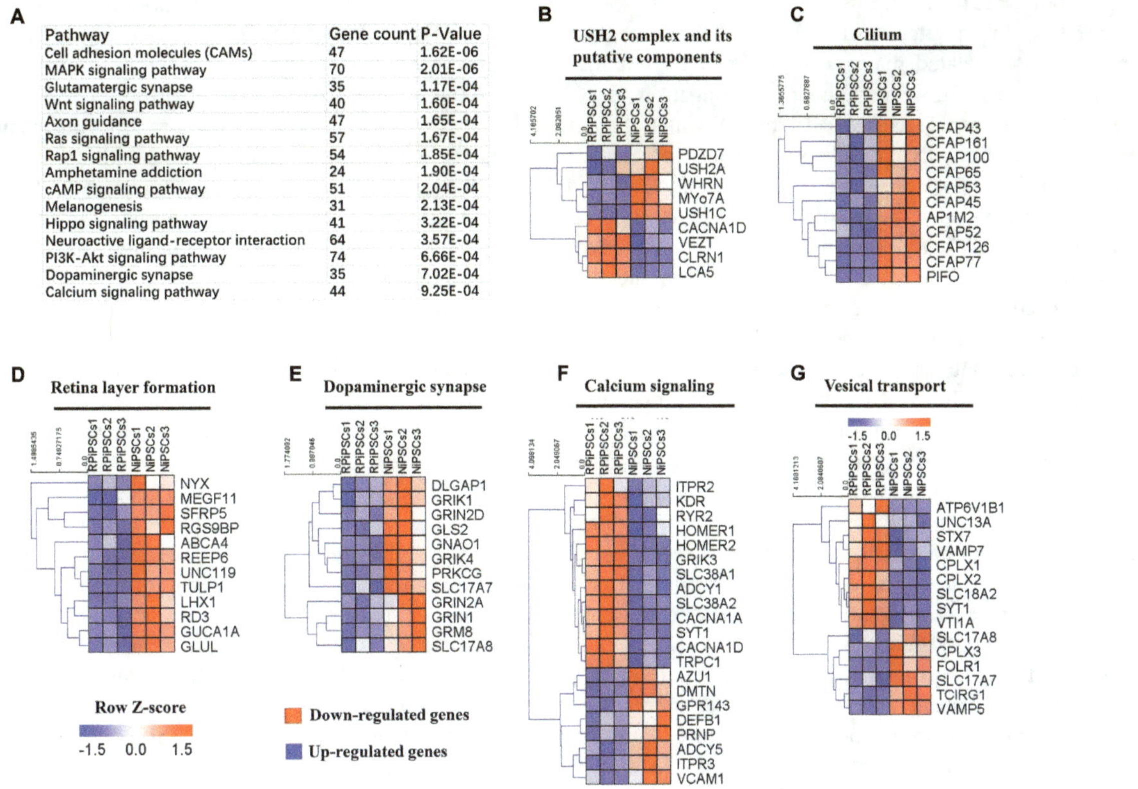

FIGURE 8 | KEGG pathway and heatmaps of other major types of genes. **(A)** GO enrichment for KEGG pathways of the differentially expressed genes. USH2A complex **(B)**, cilium **(C)**, retina layer formation **(D)**, dopaminergic synapse **(E)**, calcium signaling **(F)**, and vesical transport **(G)** related DEGs were picked and normalized by Z-scores. Red represents upregulated expression. Blue represents upregulated expression. Bar is represents Z-score.

2008). Sun et al. found that mutation c.8559-2A > G was the most frequent point mutation, which was only detected in Chinese and Japanese patients, accounting for approximately 19.1% of the identified USH2A mutations in 67 probands (Sun et al., 2018).

In this study, the proband is 27 years old at the time of diagnosis. This patient exhibited obviously visual acuity impairment, seriously decreased retinal neuroepithelial layers of SD-OCT, and severely abnormal ERG testing. From these clinical phenotype, we can reason that the patient has suffered from RP disease for quite a while and at an advanced stage of the pathology. On the other hand, we demonstrated for the first time that this RP patient iPSCs were able to form NR organoids but revealed abnormal early developmental features, including delayed self-organization, thinned retinal neuroepithelium, and disordered organization of retinal cells within a specific time window. Our results also revealed that the expression levels of neuron apoptosis related genes were up regulated in the USH2A mutation group compared to the control. These defection features in USH2A mutation organoid are corresponds to the disease characteristics of the RP patient.

USH2A-associated retinal degeneration in humans has a slow and progressive pathology. However, USH2A mutant retinal degeneration might also undergo some early pathological changes. Dona et al. reported that zebrafish ush2a mutant's models presented with early-onset retinal dysfunction, as evidenced by significantly reduced ERG a- and b-wave responses recorded at 5 days post fertilization (Dona et al., 2018). Ush2a knockdown experiments in zebrafish produced moderate levels of photoreceptor cell death in larvae. This cell death was restricted to photoreceptors and the retinas were morphologically normal (Aller et al., 2013). Photoreceptor degeneration in the Ush2a-/- mice is slowly progressive, similar to RP in human patients. However, GFAP was found to be up-regulated from as early as 2 months of age and remained so at the older ages in the Ush2a-/- mice. GFAP up-regulation is a non-specific indicator of photoreceptor degeneration and typically precedes overt cell loss (Liu et al., 2007). Our study also confirmed that retinal organoids generated from a RP patient with the USH2A mutation did exhibit early retinal developmental abnormalities.

According to the procedures of "induction-reversal" organoid culture in our study, firstly, the NR containing aggregates were cultured in NR induction medium for 18 days. Then, they were transferred to culture plates in RPE-induction medium from day 18 to day 24. Finally, they became NR-RPE like tissue in NR and RPE induction medium after day 24. Interestingly, HIF1A as well as RPCs related genes, such as PAX6, RAX, SIX6, and CHX10, at D18 in NR induction significantly higher expressed in controls and then at D34 in USH2A mutant organoids. Such changes in HIF1A may be related to the different oxygen concentration during the organoid culture. The oxygen concentration was 20% before 18 days, while we used 40% oxygen concentration during later culture. Additionally, corresponding changes in RPCs related markers might involve compensation growth and regeneration mechanism. McGuigan et al. (2017) reported that autosomal recessive RP with EYS Mutations showed decreased retinal and ONL as well as apparently increased inner nuclear layer (INL). Such observations of thickening INL coupled with thinning ONL have been noted in other inherited retinal degenerations (IRDs). The explanation has not been provided whereas this is only one of many exceedingly complicated and continuous retinal changes after photoreceptor loss, it could be a marker for a stage of remodeling (McGuigan et al., 2017). It was reported that acute damage to the retina triggers the reprogramming of Müller glial cells into retinal progenitor cells that are able to differentiate into all major types of retinal neurons including photoreceptors (Wan and Goldman, 2016). Dona et al. uncovered that mutant ush2a zebrafish models showed no progressive loss of retinal photoreceptor cells under normal light conditions, although increased levels of photoreceptor apoptosis and impaired visual function were observed within the first week of life when larvae were challenged by exposure to constant illumination with 3000 lux of white light (Dona et al., 2018). The observed lack of progressive retinal degeneration in these mutants suggests that the rate of photoreceptor apoptosis might be compensated by the rate of photoreceptor regeneration when fish are raised in low intensity light. Furthermore, after RPE-induction, the pigmented domain of RPE is gradually increasing according to the procedures of "induction-reversal" organoid culture in our study. And NR-RPE like tissue exhibited a large semispherical domain of continuous NR epithelium with a small pigmented domain of thin and winding RPE in an adjacent location (Kuwahara et al., 2015). We suppose that the development of RPE cells helps the growth and differentiation of RPC, which is manifested by the increase in the thickness of neuroepithelial. The RPE cells are responsible for producing some of the molecular signals that influence RPC differentiation and for providing metabolic support for the photoreceptors in the mature retina (Lu and Barnstable, 2019). Zhu et al. (2011) showed that the hESCs derived RPE cells secrete a high level of PEDF, which can promote RPC proliferation and survival. But in first NR induction medium before day 18 (including day 20 just beginning RPE growth), USH2A mutations indeed induced defective and reduced thickness of the neuroepithelium. Therefore, in our study, RPCs in USH2A mutant organoids firstly underwent apoptosis and decreased expression, which triggers later remodeling and compensatory increased expression and by regenerative pathway and further intensified with RPE action.

In addition, a significantly decreased expression of the basement membrane markers Laminin, *COL4A6*, *LAMB1*, *TNC*, *LAMA3*, and *LAMC2* were confirmed by immunofluorescence staining, qPCR and RNA-Seq, which were connected with abnormal retinal organoids with USH2A mutant in this study. Firstly, decreased cell growth and increased cell apoptosis happen when there is improper basement membrane. We showed that there was a significantly smaller proportion of Ki67^{+} cells and more apoptosis in the USH2A mutant organoids, as well as upregulated neuro apoptotic process related genes including *CASP3*, *MSH2*, *ADARB1*, and *HIF1A*. Our results are broadly consistent with other perspectives. Urbano et al. (2009) demonstrated that mutation of the *Lamb1* gene not only prevented basement membrane from forming, but it also generated abnormal organogenesis with defective adhesion and cell migration in Drosophila. Laperle et al. (2015) found that knockdown of the *LAMA5* gene significantly increased human ESCs and iPSCs apoptosis but did not affect their pluripotency. Secondly, abnormal cell polarization can be affected in a situation with an abnormal basement membrane. we also demonstrated that the USH2A mutation organoids displayed sluggish self-assembly and chaotic polarization, which is reflected in their dramatic decrease in the expression level of aPKC, CDH2, and ZO-1 compared to normal organoids. Similarly, Huang et al. (2003) showed that laminin is essential for the formation of cell-anchored polymers, which are required for basement membrane assembly and epithelial cell polarization. Ivanovitch et al. (2013) revealed that the absence of *laminin1* caused misoriented AB polarity and pseudostratified neuroepithelium in zebrafish. Thirdly, stem cell differentiation and adherence may be affected by defective basement membrane. We observed that the retinal development and RPC differentiation were delayed in USH2A mutation organoids at days 0–18 compared to healthy organoids, and the specific manifestation was that the RPC makers *PAX6*, *RAX*, *SIX6*, and *CHX10* were downregulated in the thinned retinal neuroepithelium. Serjanov et al. (2018) showed that deletion of *Lamb2* exhibited a loss of RPC basal processes that lead to decreased RPC proliferation and altered cellular composition of the retina in mice. Gopalan et al. showed that RPCs adhered to the inner limiting membrane, which consists of laminins, such as α1, α5, β1, β2, γ1, γ2, and γ3, during the development of the retina (Varshney et al., 2015).

CONCLUSION

By combining iPSCs and 3D retinal organoids technologies with panel sequencing, we were able to demonstrate the two disease-causing mutations in a patient with non-syndromic USH2A-associated RP were associated with abnormal NR and RPE development. We identified a novel pathogenic mutation (c.8559-2A > G/c.9127_9129delTCC) in USH2A that has not been reported previously. Furthermore, we revealed that UCs cultured from a patient can be reprogrammed into iPSCs and further form a multilayered 3D retina organoid structure. Moreover, the resulting 3D retinas in the early stages display abnormal structure and function, including reduced organoid diameter and thickness, reduced laminin, increased apoptosis, and dysregulated RPC gene expression, defective photoreceptors and RPE cell phenotype. Abnormal retinal organoids with USH2A mutant correspond to RP disease characteristics such as atrophic pigment mottling on the fundus, decreased neuroepithelial and ONLs from OCT images, and reduced ERG a-wave and b-wave amplitude.

The retinal organoids enable to recapitulate human retinal development and disease that are not easily, early and accurately modeled in animals (Li and Izpisua Belmonte, 2019). Such kind of patients can immensely benefit, in terms of early prophylaxis, from the early pinpointing molecular diagnosis. The greatest benefit of the study genotype - phenotype - organoids correlations will allow the early molecular diagnosis as well as target based pathogenic mechanisms and intervention treatment. Being able to analyze and predict the course of the disease early in the process is highly desirable and becomes paramount (Pierrache et al., 2016; Jouret et al., 2019). Although iPS derived organoids can provide a unique platform for understanding diseases, it is important to keep in mind that the current organoids are very simplistic and immature (Qian et al., 2019). Organoids are semi-physiologic models because they are no vascularization and immune system (Li and Belmonte, 2019). Specially, the human retinal organoids require longer time to matured development (Lancaster and Knoblich, 2014). Our study is the first to investigate the effect of USH2A gene mutation on RP using 3D retinal organoid technology based on the related literature search. The limitation of this study is to involve early development in USH2A-associated organoids. Long-term culture of retinal organoid may recapitulate the phenotype more adequately. Further development of technologies, such as accelerating functional maturation and incorporating glia, microglia and vessel cells, will push our study toward more comprehensive and faithful RP models. Therefore, to overcome the technical challenges retinal organoid culture with USH2A mutation, for example easily disintegrated organoids, will be our future efforts. Additionally, our study just relies on comparison of one patient derived iPSC lined versus one healthy control. Next, we will screen more USH2A mutant RP cases including the proband pedigree by sequencing. Furthermore, applying patient derived iPSC-organoids system after CRISPR/Cas9-mediated correction of the mutation in treatment of RP is attractive, and the replacement of retinal tissue may inspire potential therapies (Jin et al., 2018).

AUTHOR CONTRIBUTIONS

JC and YG did the background research and study design. JM collected the clinical data. ZC, DZ, YX, and QY analyzed and interpreted the data. SL, JC, and PW carried out the immunostaining analysis. YG drafted the manuscript. ZL and JC critically reviewed and revised the manuscript for intellectual content. JC and ST supervised the study and provided mentorship. All authors contributed to the important intellectual content during manuscript drafting or revision, and accepted the accountability for the overall work.

REFERENCES

Aller, E., Sanchezsanchez, A. V., Chicote, J. U., Garciagarcia, G., Udaondo, P., Cavalle, L., et al. (2013). Analysis of the Ush2a gene in medaka fish (Oryzias latipes). *PLoS One* 8:e74995. doi: 10.1371/journal.pone.0074995

Campochiaro, P. A., and Mir, T. A. (2018). The mechanism of cone cell death in Retinitis Pigmentosa. *Prog. Retin. Eye Res.* 62, 24–37. doi: 10.1016/j.preteyeres.2017.08.004

Chen, Q., Zou, J., Shen, Z., Zhang, W., and Yang, J. (2014). Whirlin and PDZ domain containing 7 (PDZD7) proteins are both required to form the quaternary protein complex associated with Usher syndrome type 2. *J. Biol. Chem.* 289, 36070–36088. doi: 10.1074/jbc.M114.610535

Chen, X., Jiang, C., Yang, D., Sun, R., Wang, M., Sun, H., et al. (2018). CRB2 mutation causes autosomal recessive retinitis pigmentosa. *Exp. Eye Res.* 180, 164–173. doi: 10.1016/j.exer.2018.12.018

Cui, Z., Zeng, Q., Liu, S., Zhang, Y., Zhu, D., Guo, Y., et al. (2018). Cell-Laden and orthogonal-multilayer tissue-engineered corneal stroma induced by a mechanical collagen microenvironment and transplantation in a rabbit model. *Acta Biomater.* 75, 183–199. doi: 10.1016/j.actbio.2018.06.005

Dai, H., Zhang, X., Zhao, X., Deng, T., Dong, B., Wang, J., et al. (2008). Identification of five novel mutations in the long isoform of the USH2A gene in Chinese families with usher syndrome type II. *Mol. Vis.* 14:2067.

Deng, W. L., Gao, M. L., Lei, X. L., Lv, J. N., Zhao, H., He, K. W., et al. (2018). Gene correction reverses ciliopathy and photoreceptor loss in iPSC-derived retinal organoids from retinitis pigmentosa patients. *Stem Cell Rep.* 10, 1267–1281. doi: 10.1016/j.stemcr.2018.02.003

Dona, M., Slijkerman, R., Lerner, K., Broekman, S., Wegner, J., Howat, T., et al. (2018). Usherin defects lead to early-onset retinal dysfunction in zebrafish. *Exp. Eye Res.* 173, 148–159. doi: 10.1016/j.exer.2018.05.015

Eiraku, M., Takata, N., Ishibashi, H., Kawada, M., Sakakura, E., Okuda, S., et al. (2011). Self-organizing optic-cup morphogenesis in three-dimensional culture. *Nature* 472, 51–56. doi: 10.1038/nature09941

Guo, Y., Liu, Q., Yang, Y., Guo, X., Lian, R., Li, S., et al. (2015). The effects of ROCK inhibitor Y-27632 on injectable spheroids of bovine corneal endothelial cells. *Cell. Reprogram.* 17, 77–87. doi: 10.1089/cell.2014.0070

Guo, Y., Yu, Q., Mathew, S., Lian, R., Xue, Y., Cui, Z., et al. (2017). Cocktail of chemical compounds and recombinant proteins robustly promote the stemness of adipose-derived stem cells. *Cell. Reprogram.* 19, 363–371. doi: 10.1089/cell.2017.0022

Guo, Y., Zeng, Q., Liu, S., Yu, Q., Wang, P., Ma, H., et al. (2018). Generation of an iPS cell line via a non-integrative method using urine-derived cells from a patient with USH2A-associated retinitis pigmentosa. *Stem Cell Res.* 29, 139–142. doi: 10.1016/j.scr.2018.03.022

Han, S., Liu, X., Xie, S., Gao, M., Liu, F., Yu, S., et al. (2018). Knockout of ush2a gene in zebrafish causes hearing impairment and late onset rod-cone dystrophy. *Hum. Genet.* 137, 779–794. doi: 10.1007/s00439-018-1936-6

Hartong, D. T., Berson, E. L., and Dryja, T. P. (2006). Retinitis pigmentosa. *Lancet* 368, 1795–1809.

Huang, C. C., Hall, D. H., Hedgecock, E. M., Kao, G., Karantza, V., Vogel, B. E., et al. (2003). Laminin alpha subunits and their role in C. elegans development. *Development* 130, 3343–3358. doi: 10.1242/dev.00481

Huang, D., Eudy, J. D., Uzvolgyi, E., Davis, J. R., Talmadge, C. B., Pretto, D., et al. (2002). Identification of the mouse and rat orthologs of the gene mutated in Usher syndrome type IIA and the cellular source of USH2A mRNA in retina, a target tissue of the disease. *Genomics* 80, 195–203. doi: 10.1006/geno.2002.6823

Huang, L., Mao, Y., Yang, J., Li, Y., Li, Y., and Yang, Z. (2018). Mutation screening of the USH2A gene in retinitis pigmentosa and USHER patients in a Han Chinese population. *Eye* 32, 1608–1614. doi: 10.1038/s41433-018-0130-3

Insinna, C., and Besharse, J. C. (2008). Intraflagellar transport and the sensory outer segment of vertebrate photoreceptors. *Dev. Dyn.* 237, 1982–1992. doi: 10.1002/dvdy.21554

Ivanovitch, K., Cavodeassi, F., and Wilson, S. W. (2013). Precocious acquisition of neuroepithelial character in the eye field underlies the onset of eye morphogenesis. *Dev. Cell* 27, 293–305. doi: 10.1016/j.devcel.2013.09.023

Jin, Z. B., Gao, M. L., Deng, W. L., Wu, K. C., Sugita, S., Mandai, M., et al. (2018). Stemming retinal regeneration with pluripotent stem cells. *Prog. Retin. Eye Res.* 69, 38–56. doi: 10.1016/j.preteyeres.2018.11.003

Jin, Z. B., Okamoto, S., Osakada, F., Homma, K., Assawachananont, J., Hirami, Y., et al. (2011). Modeling retinal degeneration using patient-specific induced pluripotent stem cells. *PLoS One* 6:e17084. doi: 10.1371/journal.pone.0017084

Jin, Z. B., Okamoto, S., Xiang, P., and Takahashi, M. (2012). Integration-free induced pluripotent stem cells derived from retinitis pigmentosa patient for disease modeling. *Stem Cells. Med.* 1, 503–509. doi: 10.5966/sctm.2012-0005

Jouret, G., Poirsier, C., Spodenkiewicz, M., Jaquin, C., Gouy, E., Arndt, C., et al. (2019). Genetics of usher syndrome: new insights from a meta-analysis. *Otol. Neurotol.* 40, 121–129. doi: 10.1097/MAO.0000000000002054

Kuwahara, A., Ozone, C., Nakano, T., Saito, K., Eiraku, M., and Sasai, Y. (2015). Generation of a ciliary margin-like stem cell niche from self-organizing human retinal tissue. *Nat. Commun.* 6:6286. doi: 10.1038/ncomms7286

Lancaster, M. A., and Knoblich, J. A. (2014). Organogenesis in a dish: modeling development and disease using organoid technologies. *Science* 345:1247125. doi: 10.1126/science.1247125

Laperle, A., Hsiao, C., Lampe, M., Mortier, J., Saha, K., Palecek, S. P., et al. (2015). alpha-5 laminin synthesized by human pluripotent stem cells promotes self-renewal. *Stem Cell Rep.* 5, 195–206. doi: 10.1016/j.stemcr.2015.06.009

Li, M., and Belmonte, J. C. I. (2019). Organoids — preclinical models of human disease. *N. Engl. J. Med.* 380, 569–579. doi: 10.1056/nejmra1806175

Li, M., and Izpisua Belmonte, J. C. (2019). Organoids - preclinical models of human disease. *N. Engl. J. Med.* 380, 569–579. doi: 10.1056/nejmra1806175

Liquori, A., Vaché, C., Baux, D., Blanchet, C., Hamel, C., Malcolm, S., et al. (2016). Whole USH2A gene sequencing identifies several new deep intronic mutations. *Hum. Mutat.* 37, 184–193. doi: 10.1002/humu.22926

Liu, X., Bulgakov, O. V., Darrow, K. N., Pawlyk, B., Adamian, M., Liberman, M. C., et al. (2007). Usherin is required for maintenance of retinal photoreceptors and normal development of cochlear hair cells. *Proc. Natl. Acad. Sci. U.S.A.* 104, 4413–4418. doi: 10.1073/pnas.0610950104

Lu, A. Q., and Barnstable, C. J. (2019). Pluripotent stem cells as models of retina development. *Mol. Neurobiol.* [Epub ahead of print].

Maerker, T., van Wijk, E., Overlack, N., Kersten, F. F., McGee, J., Goldmann, T., et al. (2007). A novel usher protein network at the periciliary reloading point between molecular transport machineries in vertebrate photoreceptor cells. *Hum. Mol. Genet.* 17, 71–86. doi: 10.1093/hmg/ddm285

McGuigan, D. B., Heon, E., Cideciyan, A. V., Ratnapriya, R., Lu, M., Sumaroka, A., et al. (2017). EYS mutations causing autosomal recessive retinitis pigmentosa: changes of retinal structure and function with disease progression. *Genes* 8:E178. doi: 10.3390/genes8070178

Megaw, R., Abu-Arafeh, H., Jungnickel, M., Mellough, C., Gurniak, C., Witke, W., et al. (2017). Gelsolin dysfunction causes photoreceptor loss in induced pluripotent cell and animal retinitis pigmentosa models. *Nat. Commun.* 8:271. doi: 10.1038/s41467-017-00111-8

Meyer, J. S., Howden, S. E., Wallace, K. A., Verhoeven, A. D., Wright, L. S., Capowski, E. E., et al. (2011). Optic vesicle-like structures derived from human pluripotent stem cells facilitate a customized approach to retinal disease treatment. *Stem Cells* 29, 1206–1218. doi: 10.1002/stem.674

Nachury, M. V., Seeley, E. S., and Jin, H. (2010). Trafficking to the ciliary membrane: how to get across the periciliary diffusion barrier? *Annu. Rev. Cell Dev. Biol.* 26, 59–87. doi: 10.1146/annurev.cellbio.042308.113337

Nakano, T., Ando, S., Takata, N., Kawada, M., Muguruma, K., Sekiguchi, K., et al. (2012). Self-formation of optic cups and storable stratified neural retina from human ESCs. *Cell Stem Cell* 10, 771–785. doi: 10.1016/j.stem.2012.05.009

National Eye Institute, (2004). *Facts About Retinitis Pigmentosa.* Bethesda, MD: National Eye Institute.

Pierrache, L., Hartel, B. P., Van Wijk, E., Meestersmoor, M. A., Cremers, F. P. M., De Baere, E., et al. (2016). Visual prognosis in USH2A-associated retinitis pigmentosa is worse for patients with usher syndrome type iia than for those with nonsyndromic retinitis pigmentosa. *Ophthalmology* 123, 1151–1160. doi: 10.1016/j.ophtha.2016.01.021

Qian, X., Song, H., and Ming, G.-L. (2019). Brain organoids: advances, applications and challenges. *Development* 146:dev166074. doi: 10.1242/dev.166074

Reichman, S., Slembrouck, A., Gagliardi, G., Chaffiol, A., Terray, A., Nanteau, C., et al. (2017). Generation of storable retinal organoids and retinal pigmented epithelium from adherent human iPS Cells in Xeno-free and feeder-free conditions. *Stem Cells* 35, 1176–1188. doi: 10.1002/stem.2586

Reiners, J., Nagel-Wolfrum, K., Jürgens, K., Märker, T., and Wolfrum, U. (2006). Molecular basis of human Usher syndrome: deciphering the meshes of the Usher protein network provides insights into the pathomechanisms of the Usher disease. *Exp. Eye Res.* 83, 97–119. doi: 10.1016/j.exer.2005.11.010

Rivolta, C., Sharon, D., DeAngelis, M. M., and Dryja, T. P. (2002). Retinitis pigmentosa and allied diseases: numerous diseases, genes, and inheritance patterns. *Hum. Mol. Genet.* 11, 1219–1227. doi: 10.1093/hmg/11.10.1219

Schwarz, N., Lane, A., Jovanovic, K., Parfitt, D. A., Aguila, M., Thompson, C. L., et al. (2017). Arl3 and RP2 regulate the trafficking of ciliary tip kinesins. *Hum. Mol. Genet.* 26, 2480–2492. doi: 10.1093/hmg/ddx143

Serjanov, D., Bachay, G., Hunter, D. D., and Brunken, W. J. (2018). Laminin beta2 chain regulates retinal progenitor cell mitotic spindle orientation via dystroglycan. *J. Neurosci.* 38, 5996–6010. doi: 10.1523/JNEUROSCI.0551-18.2018

Sharma, T. P., Wiley, L. A., Whitmore, S. S., Anfinson, K. R., Cranston, C. M., Oppedal, D. J., et al. (2017). Patient-specific induced pluripotent stem cells to evaluate the pathophysiology of TRNT1-associated Retinitis pigmentosa. *Stem Cell Res.* 21, 58–70. doi: 10.1016/j.scr.2017.03.005

Shintani, K., Shechtman, D. L., and Gurwood, A. S. (2009). Review and update: current treatment trends for patients with retinitis pigmentosa. *Optometry* 80, 384–401. doi: 10.1016/j.optm.2008.01.026

Slijkerman, R., Goloborodko, A., Broekman, S., de Vrieze, E., Hetterschijt, L., Peters, T., et al. (2018). Poor splice-site recognition in a humanized zebrafish knockin model for the recurrent deep-intronic c.7595-2144A>G Mutation in USH2A. *Zebrafish* 15, 597–609. doi: 10.1089/zeb.2018.1613

Sorusch, N., Bauß, K., Plutniok, J., Samanta, A., Knapp, B., Nagel-Wolfrum, K., et al. (2017). Characterization of the ternary Usher syndrome SANS/ush2a/whirlin protein complex. *Hum. Mol. Genet.* 26, 1157–1172. doi: 10.1093/hmg/ddx027

Sun, L. W., Johnson, R. D., Langlo, C. S., Cooper, R. F., Razeen, M. M., Russillo, M. C., et al. (2016). Assessing photoreceptor structure in retinitis pigmentosa and Usher syndrome. *Investig. Ophthalmol. Visual Sci.* 57, 2428–2442. doi: 10.1167/iovs.15-18246

Sun, T., Xu, K., Ren, Y., Xie, Y., Zhang, X., Tian, L., et al. (2018). Comprehensive molecular screening in chinese usher syndrome patients. *Investig. Ophthalmol. Visual Sci.* 59, 1229–1237.

Urbano, J. M., Torgler, C. N., Molnar, C., Tepass, U., Lopez-Varea, A., Brown, N. H., et al. (2009). Drosophila laminins act as key regulators of basement membrane assembly and morphogenesis. *Development* 136, 4165–4176. doi: 10.1242/dev.044263

van Wijk, E., Pennings, R. J., te Brinke, H., Claassen, A., Yntema, H. G., Hoefsloot, L. H., et al. (2004). Identification of 51 novel exons of the Usher syndrome type 2A (USH2A) gene that encode multiple conserved functional domains and that are mutated in patients with Usher syndrome type II. *Am. J. Hum. Genet.* 74, 738–744. doi: 10.1086/383096

Varshney, S., Hunter, D. D., and Brunken, W. J. (2015). Extracellular matrix components regulate cellular polarity and tissue structure in the developing and mature retina. *J. Ophthalmic Vision Res.* 10, 329–339. doi: 10.4103/2008-322X.170354

Wan, J., and Goldman, D. (2016). Retina regeneration in zebrafish. *Curr. Opin. Genet. Dev.* 40, 41–47. doi: 10.1016/j.gde.2016.05.009

Weston, M., Eudy, J., Fujita, S., Yao, S.-F., Usami, S., Cremers, C., et al. (2000). Genomic structure and identification of novel mutations in usherin, the gene responsible for Usher syndrome type IIa. *Am. J. Hum. Genet.* 66, 1199–1210. doi: 10.1086/302855

Zhong, X., Gutierrez, C., Xue, T., Hampton, C., Vergara, M. N., Cao, L.-H., et al. (2014). Generation of three-dimensional retinal tissue with functional photoreceptors from human iPSCs. *Nat. Commun.* 5:4047. doi: 10.1038/ncomms5047

Zhu, D., Deng, X., Spee, C., Sonoda, S., Hsieh, C. L., Barron, E., et al. (2011). Polarized secretion of PEDF from human embryonic stem cell-derived RPE promotes retinal progenitor cell survival. *Invest. Ophthalmol. Vis. Sci.* 52, 1573–1585. doi: 10.1167/iovs.10-6413

6

Studying Abnormal Chromosomal Diseases using Patient-Derived Induced Pluripotent Stem Cells

Yohei Hayashi, Miho Takami and Mami Matsuo-Takasaki*

iPS Advanced Characterization and Development Team, RIKEN BioResource Research Center, Tsukuba, Japan

***Correspondence:**
Yohei Hayashi
yohei.hayashi@riken.jp

Chromosomal abnormality causes congenital and acquired intractable diseases. In general, there are no fundamental treatments for these diseases. To establish platforms to develop therapeutics for these diseases, patient-derived induced pluripotent stem cells (iPSCs) are highly beneficial. To study abnormal chromosomal diseases, it is often hard to apply animal disease models because the chromosomal structures are variable among species. It is also difficult to apply simple genome editing technology in cells or individuals for abnormal chromosomes. Thus, these patient-derived iPSCs have advantages for developing disease models with multiple cell and tissue types, which are typically seen in the symptoms of abnormal chromosomal diseases. Here we review the studies of patient-derived iPSCs carrying abnormal chromosomes, focusing on pluripotent state and neural lineages. We also discuss the technological advances in chromosomal manipulations toward establishing experimental models and future therapeutics. Patient-derived iPSCs carrying chromosomal abnormality are valuable as cellular bioresources since they can indefinitely proliferate and provide various cell types. Also, these findings and technologies are important for future studies on elucidating pathogenesis, drug development, regenerative medicine, and gene therapy for abnormal chromosomal diseases.

Keywords: abnormal chromosome, iPSCs, chromosomal deletion, trisomy, monosomy

INTRODUCTION

The development of induced pluripotent stem cells (iPSCs) offers unprecedented opportunities for life sciences, drug development, and cell therapy (Takahashi and Yamanaka, 2006; Takahashi et al., 2007). Human iPSCs have been established from somatic cells in many patients who suffered from various genetic diseases (Park et al., 2008; Hayashi, 2017). These patient-derived iPSCs have been widely utilized for recapitulating pathogenesis *in vitro*, thereby contributing to disease modeling and drug development (Matsumoto et al., 2013; Hayashi et al., 2016).

Utilizing patient-derived iPSCs for abnormal chromosomal diseases has several advantages. First, animal disease modeling is often inappropriate to evaluate abnormal chromosomes due to the structural difference among species. Second, it is hard to apply simple genome editing technology in cells or individuals for abnormal chromosomes. Third, since iPSCs are pluripotent, patient-derived iPSCs can develop disease models with multiple cell and tissue types, which are typically seen in the symptoms of chromosomal diseases.

Thus, patient-derived iPSCs carrying chromosomal abnormality are valuable as cellular bioresources. In this review, we introduce the studies of abnormal chromosomal diseases using patient-derived iPSCs, focusing on pluripotent state and neural lineages. We also discuss the technological advances in chromosomal manipulations toward establishing experimental models and future therapeutic methods.

STUDIES OF ABNORMAL CHROMOSOMAL DISEASES USING PATIENT-SPECIFIC iPSCs

We show the summary list of studies using abnormal chromosomal disease-specific iPSC lines in **Table 1**.

Down Syndrome (Chromosome 21 Trisomy)

Down syndrome (DS) is a genetic disease caused by a third copy of chromosome 21 and is the most frequent chromosome abnormality, occurring in about 1 in 1,000 (Mai et al., 2019). DS leads to physical growth delays, intellectual disability, characteristic facial features, poor immune function, congenital heart diseases, epilepsy, leukemia, thyroid diseases, and mental illnesses. The elucidation of the pathological mechanisms has been hampered by the lack of experimental models that contain the chromosome 21 trisomy in various cell types. To overcome this limitation, many groups have generated iPSCs and ESCs from DS patients.

These DS-iPSCs were firstly examined in their pluripotent state. In the case of a pair of maternal chromosomes in DS-iPSCs, the homologous copies of chromosomes tended to form an adjacent pair and to locate relatively inside in a nucleus (Omori et al., 2017). Transcriptional profiling of these DS-iPSCs showed the specific effects of the pair of maternal chromosomes in trisomy 21 on gene expression patterns. These results suggested the pathological phenotypes of DS might be contributed by topological interaction between these homologous chromosomes as well as the increased number of human chromosome 21. DS-iPSCs showed the misexpression of neural genes, such as Microtubule-Associated Protein 2 (*MAP2*), Glutamate [NMDA] receptor subunit 3A (*GRIN3A*), Gamma-aminobutyric acid receptor subunit alpha-2 (GABRA2), and Stathmin 2 (*STMN2*) as well as the increased gene expression produced from chromosome 21 trisomy (Gonzales et al., 2018). These results suggested that the chromosome 21 trisomy in iPSCs might disturb the maintenance of pluripotency, but not intrinsically limit neuronal differentiation.

When DS-ESCs were differentiated into neural progenitor cells (NPCs), these NPCs displayed increased apoptosis and downregulation of forehead developmental genes (Halevy et al., 2016). It was also found that *RUNX1* is a key transcription regulator in DS-ESC-NPCs for neural differentiation (Halevy et al., 2016). In these teratomas generated from DS-iPSCs injected intramuscularly into immunodeficient mice, ectodermal tissues were largely absent. When DS-iPSCs were differentiated into neural lineages *in vitro*, the architecture and density of neurons, astroglia, and oligodendrocytes were abnormal together with the misexpression of neuronal genes (Hibaoui et al., 2014). Excitatory and inhibitory synapses in DS-iPSCs-cortical neurons displayed reduced synaptic activity (Weick et al., 2013). Transplantation of DS-iPS-derived cortical neurons into the adult mouse cortex showed increased synaptic stability and reduced oscillation in transplants (Real et al., 2018). DS is associated with an increased risk of Alzheimer's disease (AD). DS-neurons showed up-regulated expression of the *APP* (Amyloid precursor protein) gene and increased secretion and accumulation of amyloid-β (Aβ) granules made of Abeta42 pathological isoform (Dashinimaev et al., 2017). Additionally, expression levels of AD-associated genes, such as *BACE2*, *RCAN1*, *ETS2*, and *TMED10*, were increased. These findings suggested DS-neurons might recapitulate cellular signs of AD and could be useful models for studying this DS-associated AD subtype. Astroglia from DS-iPSCs produced decreased levels of synaptogenic molecules, thrombospondins 1 and 2 (TSP-1 and TSP-2), and increased levels of reactive oxygen species. Conditioned medium collected from the culture of DS-astroglia invoked neural cell toxicity and failed to promote the maturation of voltage-gated sodium and potassium ion channels and synapse formation in normal and DS neurons. Transplanted DS-astroglia also did not promote neural development in the developing brain of immunodeficient mice (Chen et al., 2014). DS-astroglia also exhibited abnormal synaptogenesis and neuronal excitability with transcriptomic and epigenetic changes in genes associated with neurodevelopmental, cell adhesion and extracellular matrix functions (Araujo et al., 2018; Mizuno et al., 2018; Ponroy Bally et al., 2020). Together, DS-iPSCs and their differentiated neural derivatives recapitulated various cellular defects that were consistent with various symptoms in DS patients and might enable the discovery of the underlying pathology and the development of treatments for DS.

Aneuploidy of Sex Chromosomes: Turner Syndrome and Klinefelter Syndrome

X chromosome monosomy is the most frequent genetic aberration in about 2% of human conceptions; however, 99% of the conceptions are spontaneously aborted. Those who survive after birth, as called Turner syndrome (TS) patients, are likely to be suffered from specific dysmorphic stigmata, short stature, hypogonadism, and renal dysfunctions, cardiac diseases, skeletal defects, endocrine failure, and metabolic deficiency. Several studies have successfully generated TS-iPSCs (Li et al., 2012; Parveen et al., 2017; Luo et al., 2018; Lu et al., 2019). TS-iPSCs were examined their global gene expression patterns (Zhang et al., 2013). They differentiated into various somatic cells in embryoid bodies (EBs) but displayed lower expression of placental genes, *ASMTL*, *PPP2R3B*, and *CSF2RA* in the genomic pseudoautosomal region. These findings suggested that abnormal organogenesis and embryonic lethality in TS might not be caused by an abnormal tissue-specific differentiation capacity, but might be caused by other abnormalities including impaired placental development (Li et al., 2012). TS-iPSCs were also able to form germ-cell-like cells *in vivo* through xenotransplantation into mice (Dominguez et al., 2014). These results suggested that

TABLE 1 | Summary list of studies using abnormal chromosomal disease-specific iPSC lines.

Disease name	Genomic features	Major symptoms	Major cellular phenotypes revealed by studies using disease-specific iPSCs	References
Down Syndrome	Trisomy 21	Physical growth delays, intellectual disability, characteristic facial features, poor immune function, congenital heart diseases, epilepsy, leukemia, thyroid diseases, and mental illnesses.	Abnormal chromosome location in a nucleus; abnormal synaptogenesis and neuronal excitability; increased apotosis and downregulation of forehead developmental genes in neural progenitor cells; an abnormal neural differentiation; reduced synaptic activity; decreased levels of synaptogenic molecules and increased levels of reactive oxygen species in astroglia;	Weick et al. (2013), Chen et al. (2014), Hibaoui et al. (2014), Halevy et al. (2016), Dashinimaev et al. (2017), Omori et al. (2017), Araujo et al. (2018), Gonzales et al. (2018), Mizuno et al. (2018), Real et al. (2018) and Ponroy Bally et al. (2020)
Turner Syndrome	Monosomy X	Specific dysmorphic stigmata, short stature, hypogonadism, and renal dysfunctions, cardiac diseases, skeletal defects, endocrine failure, and metabolic deficiency	Lower expression of placental genes in the genomic pseudoautosomal region	Li et al. (2012), Zhang et al. (2013), Dominguez et al. (2014), Parveen et al. (2017), Luo et al. (2018) and Lu et al. (2019)
Klinefelter Syndrome	Sex chromosome trisomy	Tall stature, reduced muscle tone, and hypogonadism	Abnormal gene expression associated with KS symptoms	Ma et al. (2012), Shimizu et al. (2016) and Panula et al. (2019)
Angelman Syndrome	Maternal dysfunction of chromosome 15q11-q13 imprinted region	A small head and a specific facial appearance, severe intellectual disability, developmental disability, speaking problems, balance and movement problems, seizures, and sleep problems	Abnormal DNA methylation status and neuronal maturation defects	Chamberlain et al. (2010), Stanurova et al. (2016), Fink et al. (2017), Takahashi et al. (2017), Neureiter et al. (2018), Pólvora-Brandão et al. (2018) and Niki et al. (2019)
Prader-Willi Syndrome	Paternal dysfunction of chromosome 15q11-q13 imprinted region	Developmental delay, obesity and type 2 diabetes, intellectual impairment and behavioral problems, a narrow forehead, small hands and feet, short height, light skin and hair, infertility	High DNA methylation levels in the imprinting center of the maternal allele; abnormal adherens junctions and apical-basal polarity in neural progenitors	Chamberlain et al. (2010), Yang et al. (2010), Martins-Taylor et al. (2014), Burnett et al. (2016, 2017), Okuno et al. (2017), Langouët et al. (2018) and Soeda et al. (2019)
DiGeorge Syndrome	Heterozygous 22q11.2 deletion syndrome	Congenital heart problems, characteristic facial features, poor immune function, developmental delay, learning problems, cleft palate, kidney dysfunctions, and hearing loss	Reduced size of neurospheres; lower neural differentiation capacity, poor neurite outgrowth, poor neural migration, and an abnormal transition in the neurogenic-to-gliogenic competence	Zhao et al. (2015), Lin et al. (2016) and Toyoshima et al. (2016)
Miller-Dieker Syndrome	Heterozygous deletion of chromosome 17p13.3	Lissencephaly	Neural cell migration defect, reduced size, and mitotic defects associated with a switch from symmetric to asymmetric cell division of ventricular zone radial glia	Bershteyn et al. (2017) and Iefremova et al. (2017)
Phelan-McDermid Syndrome	Heterozygous deletion of chromosome 22q13.3	Developmental delay, intellectual disability, and an increased risk of autism spectrum disorders (ASDs)	Decreased SHANK3 expression and defects in excitatory synaptic transmission	Shcheglovitov et al. (2013)
16p11.2 CNV change	Deletion or duplication of 16p11.2	Neurodevelopmental diseases, including macrocephaly, ASD and schizophrenia	Changes in cell body size and dendrite length, decreased synaptic density	Deshpande et al. (2017)
Williams Syndrome	Deletion of chromosome 7q11.23 region	Intellectual disability, specific personality traits, characteristic facial features, and cardiovascular problems	Abnormal neurogenic commitment from WS-iPSCs	Chailangkarn et al. (2016) and Lalli et al. (2016)

two intact X chromosomes might not be essential for human germ cell differentiation at its initial phase.

Klinefelter syndrome (KS), or two or more X chromosomes in males (XXY), occurs in about 1 in 1,000 (Bojesen et al., 2003). Because X chromosome inactivation (XCI) occurs in KS patients heterogeneously, they develop a variety of clinical symptoms, including tall stature, reduced muscle tone, and hypogonadism. The KS pathophysiology remains elusive due to the lack of experimental models. Some studies reported the successful generation of KS-iPSCs with a karyotype of 47, XXY (Ma et al., 2012; Shimizu et al., 2016; Panula et al., 2019). Although XCI occurs in KS-iPSCs, their transcriptome profile identified abnormally expressed genes associated with KS symptoms. These KS-iPSCs should be useful in revealing the mechanisms of human XCI and fertility.

Angelman Syndrome (AS) and Prader-Willi Syndrome (PWS)

Angelman Syndrome (AS) and Prader-Willi Syndrome (PWS) are two distinct neurological diseases caused by the dysfunction of the genes located on chromosome 15q11-q13 imprinted region. AS arises from dysfunction of the ubiquitin-protein ligase E3A (UBE3A), while the responsible genetic defects in PWS remain elusive. Several research groups have successfully generated AS-iPSCs and demonstrated abnormal DNA methylation status and neuronal maturation defects (Chamberlain et al., 2010; Sakurai et al., 2016; Fink et al., 2017; Takahashi et al., 2017; Neureiter et al., 2018; Pólvora-Brandão et al., 2018; Niki et al., 2019). These phenotypes could be reversed by recovering paternal *UBE3A* expression by the treatment with a topoisomerase inhibitor, topotecan (Fink et al., 2017). Some research groups have also generated iPSCs from PWS patients (Chamberlain et al., 2010; Yang et al., 2010; Martins-Taylor et al., 2014; Burnett et al., 2016, 2017; Okuno et al., 2017; Langouët et al., 2018; Soeda et al., 2019). These PWS-iPSCs showed high DNA methylation levels in the imprinting center of the maternal allele. Also, maternally expressed genes in the *DLK1-DIO3* imprinting loci on chromosome 14, which was regulated by IPW, a long non-coding RNA located in chromosome 15q11-q13 imprinted region, was up-regulated (Stelzer et al., 2014). These results suggested that a part of PWS phenotypes might be caused by the dysregulation of imprinted loci distinct from the 15q11-q13 imprinted region. Another study showed that PWS-iPSC-derived neural progenitors exhibited abnormal adherens junctions and apical-basal polarity (Yoon et al., 2014). These findings suggested that AS-and PWS-iPSCs were useful tools to study genetic imprinting diseases.

DiGeorge Syndrome (DGS)

DiGeorge syndrome (DGS), also known as 22q11.2 deletion syndrome or CATCH 22, causes congenital heart problems, characteristic facial features, poor immune function, developmental delay, learning problems, cleft palate, kidney dysfunctions, and hearing loss. Several groups have generated DGS patient-derived iPSCs carrying 22q11.2 deletion. Neurospheres, a culture system composed of floating clusters of neural stem cells, derived from the DGS-iPSCs showed reduced size, lower neural differentiation capacity, poor neurite outgrowth, poor neural migration, and an abnormal transition in the neurogenic-to-gliogenic competence. Transcriptomic profile in DGS-neurons also revealed significant differences in many genes outside of the deleted region as well as gene expression reduction in the 22q11.2 region. Key pathways and gene ontology, such as apoptosis, cell cycle and survival, and MAPK signaling, were uncovered by functional enrichment and network analysis on the differentially expressed genes (Lin et al., 2016). The global miRNA profiling in these neurospheres showed decreased expression of the miR-17/92 cluster and miR-106a/b, which were known to control cell proliferation (Toyoshima et al., 2016). Many differentially expressed miRNAs were also detected, including miRNAs located in the 22q11.2 region. Genes involved in neurological diseases were predicted to be the targets for the differentially expressed miRNAs by the pathway and gene ontology enrichment analysis (Zhao et al., 2015).

Neurodevelopmental Diseases Associated With Chromosomal Abnormalities

Miller-Dieker syndrome (MDS) is caused by a heterozygous deletion of chromosome 17p13.3 and leads to lissencephaly (meaning "smooth brain"), a lack of development of brain folds and grooves because of defective neuronal migration. Patient-specific forebrain-type organoids were generated from MDS-iPSCs to investigate pathological changes associated with MDS (Bershteyn et al., 2017; Iefremova et al., 2017). MDS-organoids showed cell migration defect, reduced size, and mitotic defects associated with a switch from symmetric to asymmetric cell division of ventricular zone radial glia. The cell migration defect was rescued when the chromosomal deletion in MDS-iPSCs was corrected as uniparental disomy (UPD; the details of the methods will be explained later; Bershteyn et al., 2014). Mitotic defects in outer radial glia, which could not be seen in the experimental rodent model of lissencephaly, was observed (Bershteyn et al., 2017). In another study, the treatment of a WNT beta-catenin signaling activator, CHIR99021, rescued cell division modes, and organoid growth defects (Iefremova et al., 2017).

Phelan-McDermid syndrome (PMDS) is caused by a heterozygous deletion of chromosome 22q13.3 and leads to developmental delay, intellectual disability, and an increased risk of autism spectrum disorders (ASDs). Among the genes in the deleted region, *SHANK3*, which encodes a multidomain scaffold protein in the postsynaptic density, is the candidate responsible gene for the neurological symptoms. One study reported the generation of PMDS-iPSCs to recapitulate neuronal symptoms. PMDS-neurons exhibited decreased SHANK3 expression and defects in excitatory α-amino-3-hydroxy-5-methyl-4-isoxazolepropionic acid (AMPA)-and N-methyl-D-aspartic acid (NMDA)-mediated synaptic transmission. These defects were rescued by reactivating SHANK3 expression or by treating with insulin-like growth factor 1 (IGF1), which had been previously reported to increase synaptic transmission (Shcheglovitov et al., 2013).

A change of copy number variation (CNV) in the 16p11.2 region, which is caused by both deletion (16pdel) and duplication (16pdup), is associated with neurodevelopmental diseases, including macrocephaly, ASD and schizophrenia. One research group generated iPSCs from 16pdel and 16pdup patients and differentiated them into neurons to identify causal cellular phenotypes underlying neurological symptoms. 16pdel neurons showed increased cell body size and dendrite length, while 16pdup neurons showed decreased cell body size and dendrite length. Notably, both 16pdel and 16pdup neurons showed decreased synaptic density (Deshpande et al., 2017). These results suggested the 16p11.2 region might regulate brain size and neuronal connectivity distinctively.

Williams syndrome (WS) is caused by a deletion of chromosome 7q11.23 region, which contains approximately 28 genes. WS patients are suffered from intellectual disability, specific personality traits, characteristic facial features, and cardiovascular problems. Two research groups generated patient-derived WS-iPSCs and differentiation them into neurons (Chailangkarn et al., 2016; Lalli et al., 2016). Both studies showed that the global transcriptional profile of WS-iPSC-derived neurons confirmed the expression changes of the deleted genes and that NPCs from WS-iPSCs exhibited abnormal neurogenic commitment; however, the rate of cell cycle and apoptosis in these cells was somewhat inconsistent between these studies. The first study focused on Bromodomain Adjacent To Zinc Finger Domain 1B (*BAZ1B*) gene, which is deleted in WS and encodes an ATP-dependent chromatin remodeling protein. Knocking down BAZ1B mimicked transcriptional and neural differentiation defects as seen in WS-derived cells. These defects were rescued by antagonizing Wnt signaling because this signaling pathway was identified in target genes of BAZ1B revealed by ChIP-seq analysis (Lalli et al., 2016). The second study focused on Frizzled-9 (*FZD9*), which was also deleted in WS and encodes transmembrane receptors for Wnt proteins. Reduced cell viability was rescued by a GSK3 inhibitor, which acts as a Wnt signaling activator. Although these studies seemingly demonstrated contradictory findings, both emphasized the role of Wnt signaling in WS pathology.

DISCUSSION: CHROMOSOMAL MANIPULATION TECHNOLOGIES

So far, we have reviewed how patient-derived iPSCs have been used as valuable tools to recapitulate abnormal chromosomal diseases to elucidate disease mechanisms and develop potential therapies. To enhance, the technological advances of chromosomal manipulation are keys to improved usability of patient-derived iPSCs carrying abnormal chromosomes and the development of disease models and future therapeutic methods. In this chapter, we discuss current technologies for chromosomal manipulation (illustrated in a scheme in **Figure 1**).

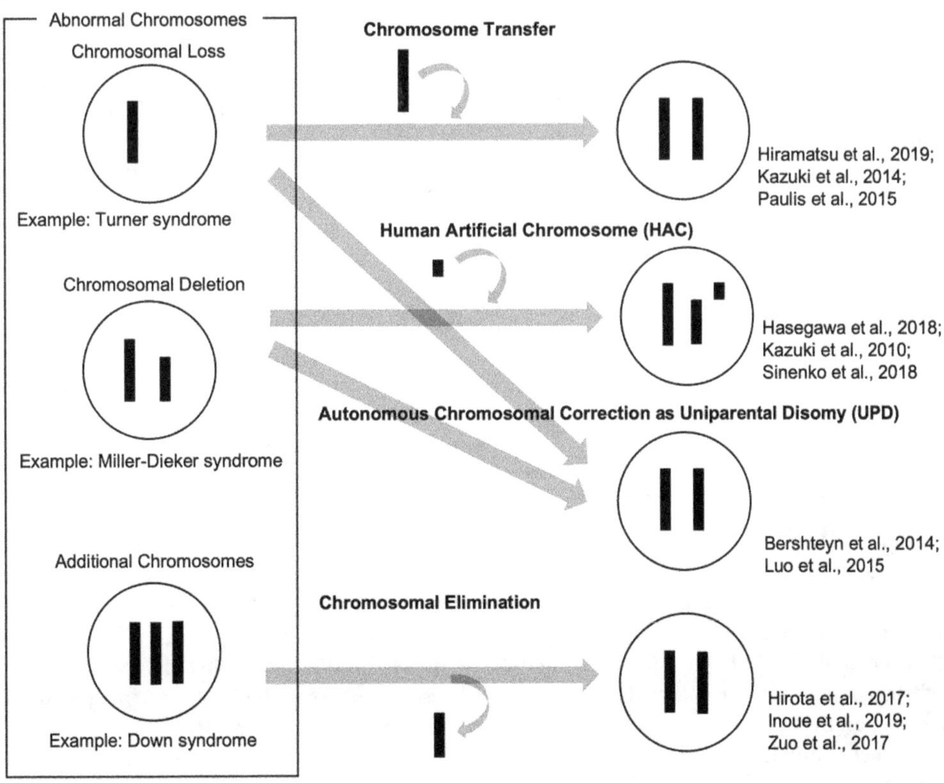

FIGURE 1 | Schemes of current technologies for chromosomal manipulation.

Chromosome Transfer

Chromosome transfer can be achieved by extracting a chromosome from a cell and inserting it into another cell. Human pluripotent stem cells carrying trisomy 8, 13, 18, and 21 were generated by a single chromosome insertion *via* microcell-mediated chromosome transfer. Global trisomic expression levels were confirmed by microarray analysis in each cell line (Kazuki et al., 2014; Hiramatsu et al., 2019). Another study reported successful chromosome transfers by replacing an X chromosome carrying a mutation in the *Hprt* gene with a normal one without the mutation in mouse embryonic stem cells (mESCs), which enables HAT (hypoxanthine-aminopterin-thymidine) selection. They also transferred a Y chromosome to rescue the defective Hprt gene, which is common in sex chromosomes. These mESC clones, which were transferred with a sex chromosome, maintained their pluripotency and genomic integrity and contributed to chimera formation (Paulis et al., 2015). Although it has not been directly applied, these studies suggested that intact chromosomes could be transferred into patient-derived iPSCs carrying abnormal chromosomes. This approach might be used to cure various disorders with chromosomal loss or deletions.

Human Artificial Chromosome (HAC)

Human artificial chromosome (HAC) vectors hold the potential to rescue diseases associated with chromosomal deletion and loss since HAC could hold DNA inserts of any size in principle. Several studies showed the successful integration of HAC vectors into iPSCs (Kazuki et al., 2010; Hasegawa et al., 2018; Sinenko et al., 2018). One study showed that a HAC vector carrying whole dystrophin (*DMD*) genomic sequence rescued dystrophin expression of Duchenne muscular dystrophy (DMD) in patient-specific iPSCs (Kazuki et al., 2010). These results suggested that the HAC vectors containing defective genes could be valuable tools to rescue the phenotypes in patient-specific iPSCs carrying specific gene deletions. Although there has been no direct evidence that HAC vectors can be used for abnormal chromosomal diseases to date, this strategy could be applied to achieve gene and cell therapies for abnormal chromosomal diseases. However, it is challenging that cloning desired long regions of the human genome into HAC and transferring them into iPSCs and other human somatic cells.

Autonomous Chromosomal Correction as Uniparental Disomy (UPD)

Several studies demonstrated that abnormal chromosomes were autonomously corrected in iPSCs. A Ring chromosome is an aberrant chromosome whose ends have fused to form a ring. We previously generated iPSCs from patients' somatic cells carrying ring chromosomes with terminal deletions and identified that these reprogrammed iPSCs lost ring chromosomes and duplicated the intact homologous chromosome. The resulting iPSCs carried a pair of identical chromosomes, UPD (Bershteyn et al., 2014). These karyotypically normal UPD iPSCs proliferated at a better/improved rate over co-existing subpopulations carrying ring chromosomes or monosomy. These patient-derived iPSCs free from the original chromosomal abnormality were efficiently isolated. Another study reported that some iPSCs generated from somatic cells in TS patients, carrying X monosomy acquired XX UPD karyotype. The XX-UPD iPSCs showed XCI, better pluripotent stem cell morphology, and higher mitotic rate than uncorrected ones (Luo et al., 2015). These studies demonstrated human iPSCs were a useful model for the investigation of mechanisms that control the number and behaviors of chromosomes during development/differentiation. Although the UPD phenomenon occurs spontaneously, it could serve as a model strategy for the development of innovative methods aimed at targeting large-scale chromosomal deletions or entire chromosome loss.

Chromosome Elimination

Some studies demonstrated spontaneous or targeted chromosomal elimination in iPSCs that carry extra chromosomes. A study showed that iPSCs from XXY and XYY trisomy mice lost an extra sex chromosome during the reprogramming process (Hirota et al., 2017). Resulting disomic XY iPSCs successfully differentiated into male germ cell lineages and functional sperms which produced fertile offspring with normal chromosomes. Other research groups demonstrated to eliminate Y chromosomes in male mESCs using genome editing technology by removing the centromere or shredding the chromosome arm (Adikusuma et al., 2017; Zuo et al., 2017). Also, several studies demonstrated to produce a targeted autosome loss in human TS-iPSCs with trisomy 21 and KS-iPSCs with sex chromosome trisomy (Hirota et al., 2017; Zuo et al., 2017; Inoue et al., 2019). Another method was reported to use X-Inactive Specific Transcript (XIST) RNA, which was normally located in the X chromosome and acts as a major effector of the X-inactivation process. Knocking-in XIST locus in the extra chromosome 21 of DS patient-derived cells could induce chromosome-wide silencing of the targeted chromosome (Jiang et al., 2013; Chiang et al., 2018; Czermiński and Lawrence, 2020). Thus, spontaneous and targeted chromosome elimination or silencing can offer new approaches to enhance the usability of disease models with abnormal chromosomes and to provide potential therapeutic methods for diseases involving additional chromosomes.

CONCLUSIONS

We have discussed the usability of patient-derived iPSCs carrying chromosomal abnormality. To establish platforms to develop therapeutics for these diseases related to chromosomal abnormality, patient-derived iPSCs are highly beneficial. Disease-specific iPSCs in major cell banks worldwide can be searched for using ICSCB (Integrated Collection of Stem Cell Bank) data[1] by MIACARM (Minimum Information About a Cellular Assay for Regenerative Medicine; Sakurai et al., 2016). We might apologize for not referring to some

[1] https://icscb.stemcellinformatics.org/

studies using patient-derived iPSCs on rare chromosomal abnormalities due to the word limit. Although many of these studies are still immature, they have generated patient-derived iPSCs that will serve as valuable bioresources/models for the investigation into the pathogenesis of abnormal chromosome disease with the potential of developing treatment strategies. We have also introduced current technologies in chromosomal manipulation. For the future clinical application of these technologies, we need to discuss adequately to reach social consensus on ethical conflict/controversy with such genetic manipulations.

AUTHOR CONTRIBUTIONS

YH contributed to the conception and design of the study. YH, MT, and MM-T wrote sections of the manuscript. All authors contributed to the article and approved the submitted version.

FUNDING

This research was supported in part by the grants from a JSPS KAKENHI Grant-in-Aid for Young Scientists (A; 17H05063) to YH, Grants for Regenerative Medicine, Japan Agency for Medical Research and Development (AMED) to YH, Grants for RARE/INTRACTABLE DISEASE PROJECT OF JAPAN, AMED, to YH, a Kowa Life Science Foundation Research Grant to YH, Takeda Science Foundation to YH, Uehara Memorial Foundation to YH, Mochida Foundation to YH, Mother and Child Health Foundation to YH, The Tokyo Biochemical Research Foundation to YH.

ACKNOWLEDGMENTS

We would like to express our sincere gratitude to all our coworkers and collaborators and Kumiko Omori for her administrative support.

REFERENCES

Adikusuma, F., Williams, N., Grutzner, F., Hughes, J., and Thomas, P. (2017). Targeted deletion of an entire chromosome using CRISPR/Cas9. *Mol. Ther.* 25, 1736–1738. doi: 10.1016/j.ymthe.2017.05.021

Araujo, B. H. S., Kaid, C., De Souza, J. S., Gomes da Silva, S., Goulart, E., Caires, L. C. J., et al. (2018). Down syndrome iPSC-derived astrocytes impair neuronal synaptogenesis and the mTOR pathway *in vitro*. *Mol. Neurobiol.* 55, 5962–5975. doi: 10.1007/s12035-017-0818-6

Bershteyn, M., Hayashi, Y., Desachy, G., Hsiao, E. C., Sami, S., Tsang, K. M., et al. (2014). Cell-autonomous correction of ring chromosomes in human induced pluripotent stem cells. *Nature* 507, 99–103. doi: 10.1038/nature12923

Bershteyn, M., Nowakowski, T. J., Pollen, A. A., Di Lullo, E., Nene, A., Wynshaw-Boris, A., et al. (2017). Human iPSC-derived cerebral organoids model cellular features of lissencephaly and reveal prolonged mitosis of outer radial glia. *Cell Stem Cell* 20, 435.e4–449.e4. doi: 10.1016/j.stem.2016.12.007

Bojesen, A., Juul, S., and Gravholt, C. H. (2003). Prenatal and postnatal prevalence of Klinefelter syndrome: a national registry study. *J. Clin. Endocrinol. Metab.* 88, 622–626. doi: 10.1210/jc.2002-021491

Burnett, L. C., LeDuc, C. A., Sulsona, C. R., Paull, D., Eddiry, S., Levy, B., et al. (2016). Induced pluripotent stem cells (iPSC) created from skin fibroblasts of patients with Prader-Willi syndrome (PWS) retain the molecular signature of PWS. *Stem Cell Res.* 17, 526–530. doi: 10.1016/j.scr.2016.08.008

Burnett, L. C., LeDuc, C. A., Sulsona, C. R., Paull, D., Rausch, R., Eddiry, S., et al. (2017). Deficiency in prohormone convertase PC1 impairs prohormone processing in Prader-Willi syndrome. *J. Clin. Invest.* 127, 293–305. doi: 10.1172/JCI88648

Chailangkarn, T., Trujillo, C. A., Freitas, B. C., Hrvoj-Mihic, B., Herai, R. H., Yu, D. X., et al. (2016). A human neurodevelopmental model for Williams syndrome. *Nature* 536, 338–343. doi: 10.1038/nature19067

Chamberlain, S. J., Chen, P.-F., Ng, K. Y., Bourgois-Rocha, F., Lemtiri-Chlieh, F., Levine, E. S., et al. (2010). Induced pluripotent stem cell models of the genomic imprinting disorders Angelman and Prader-Willi syndromes. *Proc. Natl. Acad. Sci. U S A* 107, 17668–17673. doi: 10.1073/pnas.1004487107

Chen, C., Jiang, P., Xue, H., Peterson, S. E., Tran, H. T., McCann, A. E., et al. (2014). Role of astroglia in Down's syndrome revealed by patient-derived human-induced pluripotent stem cells. *Nat. Commun.* 5:4430. doi: 10.1038/ncomms5430

Chiang, J.-C., Jiang, J., Newburger, P. E., and Lawrence, J. B. (2018). Trisomy silencing by XIST normalizes Down syndrome cell pathogenesis demonstrated for hematopoietic defects *in vitro*. *Nat. Commun.* 9:5180. doi: 10.1038/s41467-018-07630-y

Czermiński, J. T., and Lawrence, J. B. (2020). Silencing trisomy 21 with XIST in neural stem cells promotes neuronal differentiation. *Dev. Cell* 52, 294.e3–308.e3. doi: 10.1016/j.devcel.2019.12.015

Dashinimaev, E. B., Artyuhov, A. S., Bolshakov, A. P., Vorotelyak, E. A., and Vasiliev, A. V. (2017). Neurons derived from induced pluripotent stem cells of patients with down syndrome reproduce early stages of Alzheimer's disease type pathology *in vitro*. *J. Alzheimers Dis.* 56, 835–847. doi: 10.3233/jad-160945

Deshpande, A., Yadav, S., Dao, D. Q., Wu, Z.-Y., Hokanson, K. C., Cahill, M. K., et al. (2017). Cellular phenotypes in human iPSC-derived neurons from a genetic model of autism spectrum disorder. *Cell Rep.* 21, 2678–2687. doi: 10.1016/j.celrep.2017.11.037

Dominguez, A. A., Chiang, H. R., Sukhwani, M., Orwig, K. E., and Reijo Pera, R. A. (2014). Human germ cell formation in xenotransplants of induced pluripotent stem cells carrying X chromosome aneuploidies. *Sci. Rep.* 4:6432. doi: 10.1038/srep06432

Fink, J. J., Robinson, T. M., Germain, N. D., Sirois, C. L., Bolduc, K. A., Ward, A. J., et al. (2017). Disrupted neuronal maturation in Angelman syndrome-derived induced pluripotent stem cells. *Nat. Commun.* 8:15038. doi: 10.1038/ncomms15038

Gonzales, P. K., Roberts, C. M., Fonte, V., Jacobsen, C., Stein, G. H., and Link, C. D. (2018). Transcriptome analysis of genetically matched human induced pluripotent stem cells disomic or trisomic for chromosome 21. *PLoS One* 13:e0194581. doi: 10.1371/journal.pone.0194581

Halevy, T., Biancotti, J.-C., Yanuka, O., Golan-Lev, T., and Benvenisty, N. (2016). Molecular characterization of down syndrome embryonic stem cells reveals a role for RUNX1 in neural differentiation. *Stem Cell Reports* 7, 777–786. doi: 10.1016/j.stemcr.2016.08.003

Hasegawa, Y., Ikeno, M., Suzuki, N., Nakayama, M., and Ohara, O. (2018). Improving the efficiency of gene insertion in a human artificial chromosome vector and its transfer in human-induced pluripotent stem cells. *Biol. Methods Protoc.* 3:bpy013. doi: 10.1093/biomethods/bpy013

Hayashi, Y. (2017). Human mutations affecting reprogramming into induced pluripotent stem cellsle. *AIMS Cell Tissue Eng.* 1, 31–46. doi: 10.3934/celltissue.2017.1.31

Hayashi, Y., Hsiao, E. C., Sami, S., Lancero, M., Schlieve, C. R., Nguyen, T., et al. (2016). BMP-SMAD-ID promotes reprogramming to pluripotency by inhibiting p16/INK4A-dependent senescence. *Proc. Natl. Acad. Sci. U S A* 113, 13057–13062. doi: 10.1073/pnas.1603668113

Hibaoui, Y., Grad, I., Letourneau, A., Sailani, M. R., Dahoun, S., Santoni, F. A., et al. (2014). Modelling and rescuing neurodevelopmental defect of Down syndrome using induced pluripotent stem cells from monozygotic twins discordant for trisomy 21. *EMBO Mol. Med.* 6, 259–277. doi: 10.1002/emmm.201302848

Hiramatsu, K., Abe, S., Kazuki, K., Osaki, M., Kajitani, N., Yakura, Y., et al. (2019). Generation of a novel isogenic trisomy panel in human embryonic stem cells *via* microcell-mediated chromosome transfer. *Biochem. Biophys. Res. Commun.* 508, 603–607. doi: 10.1016/j.bbrc.2018.11.138

Hirota, T., Ohta, H., Powell, B. E., Mahadevaiah, S. K., Ojarikre, O. A., Saitou, M., et al. (2017). Fertile offspring from sterile sex chromosome trisomic mice. *Science* 357, 932–935. doi: 10.1126/science.aam9046.

Iefremova, V., Manikakis, G., Krefft, O., Jabali, A., Weynans, K., Wilkens, R., et al. (2017). An organoid-based model of cortical development identifies non-cell-autonomous defects in wnt signaling contributing to miller-dieker syndrome. *Cell Rep.* 19, 50–59. doi: 10.1016/j.celrep.2017.03.047

Inoue, M., Kajiwara, K., Yamaguchi, A., Kiyono, T., Samura, O., Akutsu, H., et al. (2019). Autonomous trisomic rescue of Down syndrome cells. *Lab. Investig.* 99, 885–897. doi: 10.1038/s41374-019-0230-0

Jiang, J., Jing, Y., Cost, G. J., Chiang, J.-C., Kolpa, H. J., Cotton, A. M., et al. (2013). Translating dosage compensation to trisomy 21. *Nature* 500, 296–300. doi: 10.1038/nature12394

Kazuki, Y., Hiratsuka, M., Takiguchi, M., Osaki, M., Kajitani, N., Hoshiya, H., et al. (2010). Complete genetic correction of iPS cells from duchenne muscular dystrophy. *Mol. Ther.* 18, 386–393. doi: 10.1038/mt.2009.274

Kazuki, Y., Yakura, Y., Abe, S., Osaki, M., Kajitani, N., Kazuki, K., et al. (2014). Down syndrome-associated haematopoiesis abnormalities created by chromosome transfer and genome editing technologies. *Sci. Rep.* 4:6136. doi: 10.1038/srep06136

Lalli, M. A., Jang, J., Park, J.-H. C., Wang, Y., Guzman, E., Zhou, H., et al. (2016). Haploinsufficiency of BAZ1B contributes to Williams syndrome through transcriptional dysregulation of neurodevelopmental pathways. *Hum. Mol. Genet.* 25, 1294–1306. doi: 10.1093/hmg/ddw010

Langouët, M., Glatt-Deeley, H. R., Chung, M. S., Dupont-Thibert, C. M., Mathieux, E., Banda, E. C., et al. (2018). Zinc finger protein 274 regulates imprinted expression of transcripts in Prader-Willi syndrome neurons. *Hum. Mol. Genet.* 27, 505–515. doi: 10.1093/hmg/ddx420

Li, W., Wang, X., Fan, W., Zhao, P., Chan, Y.-C., Chen, S., et al. (2012). Modeling abnormal early development with induced pluripotent stem cells from aneuploid syndromes. *Hum. Mol. Genet.* 21, 32–45. doi: 10.1093/hmg/ddr435

Lin, M., Pedrosa, E., Hrabovsky, A., Chen, J., Puliafito, B. R., Gilbert, S. R., et al. (2016). Integrative transcriptome network analysis of iPSC-derived neurons from schizophrenia and schizoaffective disorder patients with 22q11.2 deletion. *BMC Syst. Biol.* 10:105. doi: 10.1186/s12918-016-0366-0

Lu, C. Y., Chen, Y. A., Syu, S. H., Lu, H. E., Ho, H. N., and Chen, H. F. (2019). Generation of induced pluripotent stem cell line-NTUHi001-A from a premature ovarian failure patient with Turner's syndrome mosaicism. *Stem Cell Res.* 37:101422. doi: 10.1016/j.scr.2019.101422

Luo, Y., Zhu, D., Du, R., Gong, Y., Xie, C., Xu, X., et al. (2015). Uniparental disomy of the entire X chromosome in Turner syndrome patient-specific induced pluripotent stem cells. *Cell Discov.* 1:15022. doi: 10.1038/celldisc.2015.22

Luo, Y., Zhu, D., Xu, X., Ge, L., Sun, X., Chen, G., et al. (2018). Generation of an induced pluripotent stem cell line from an adult male with 45,X/46,XY mosaicism. *Stem Cell Res.* 27, 42–45. doi: 10.1016/j.scr.2018.01.003

Ma, Y., Li, C., Gu, J., Tang, F., Li, C., Li, P., et al. (2012). Aberrant gene expression profiles in pluripotent stem cells induced from fibroblasts of a Klinefelter syndrome patient. *J. Biol. Chem.* 287, 38970–38979. doi: 10.1074/jbc.m112.380204

Mai, C. T., Isenburg, J. L., Canfield, M. A., Meyer, R. E., Correa, A., Alverson, C. J., et al. (2019). National population-based estimates for major birth defects, 2010–2014. *Birth Defects Res.* 111, 1420–1435. doi: 10.1002/bdr2.1589

Martins-Taylor, K., Hsiao, J. S., Chen, P.-F., Glatt-Deeley, H., De Smith, A. J., Blakemore, A. I. F., et al. (2014). Imprinted expression of UBE3A in non-neuronal cells from a Prader-Willi syndrome patient with an atypical deletion. *Hum. Mol. Genet.* 23, 2364–2373. doi: 10.1093/hmg/ddt628

Matsumoto, Y., Hayashi, Y., Schlieve, C. R., Ikeya, M., Kim, H., Nguyen, T. D., et al. (2013). Induced pluripotent stem cells from patients with human fibrodysplasia ossificans progressiva show increased mineralization and cartilage formation. *Orphanet J. Rare Dis.* 8:190. doi: 10.1186/1750-1172-8-190

Mizuno, G. O., Wang, Y., Shi, G., Wang, Y., Sun, J., Papadopoulos, S., et al. (2018). Aberrant calcium signaling in astrocytes inhibits neuronal excitability in a human down syndrome stem cell model. *Cell Rep.* 24, 355–365. doi: 10.1016/j.celrep.2018.06.033

Neureiter, A., Brändl, B., Hiber, M., Tandon, R., Müller, F.-J., and Steenpass, L. (2018). Generation of an iPSC line of a patient with Angelman syndrome due to an imprinting defect. *Stem Cell Res.* 33, 20–24. doi: 10.1016/j.scr.2018.09.015

Niki, T., Imamura, K., Enami, T., Kinoshita, M., and Inoue, H. (2019). Establishment of human induced pluripotent stem cell line from a patient with Angelman syndrome carrying the deletion of maternal chromosome 15q11.2–q13. *Stem Cell Res.* 34:101363. doi: 10.1016/j.scr.2018.101363

Okuno, H., Nakabayashi, K., Abe, K., Ando, T., Sanosaka, T., Kohyama, J., et al. (2017). Changeability of the fully methylated status of the 15q11.2 region in induced pluripotent stem cells derived from a patient with Prader-Willi syndrome. *Congenit. Anom.* 57, 96–103. doi: 10.1111/cga.12206

Omori, S., Tanabe, H., Banno, K., Tsuji, A., Nawa, N., Hirata, K., et al. (2017). A pair of maternal chromosomes derived from meiotic nondisjunction in trisomy 21 affects nuclear architecture and transcriptional regulation. *Sci. Rep.* 7:764. doi: 10.1038/s41598-017-00714-7

Panula, S., Kurek, M., Kumar, P., Albalushi, H., Sánchez, S. P., Damdimopoulou, P., et al. (2019). Human induced pluripotent stem cells from two azoospermic patients with Klinefelter syndrome show similar X chromosome inactivation behavior to female pluripotent stem cells. *Hum. Reprod.* 34, 2297–2310. doi: 10.1093/humrep/dez134

Park, I. H., Arora, N., Huo, H., Maherali, N., Ahfeldt, T., Shimamura, A., et al. (2008). Disease-specific induced pluripotent stem cells. *Cell* 134, 877–886. doi: 10.1016/j.cell.2008.07.041

Parveen, S., Panicker, M. M., and Gupta, P. K. (2017). Generation of an induced pluripotent stem cell line from chorionic villi of a Turner syndrome spontaneous abortion. *Stem Cell Res.* 19, 12–16. doi: 10.1016/j.scr.2016.12.016

Paulis, M., Castelli, A., Susani, L., Lizier, M., Lagutina, I., Focarelli, M. L., et al. (2015). Chromosome transplantation as a novel approach for correcting complex genomic disorders. *Oncotarget* 6, 35218–35230. doi: 10.18632/oncotarget.6143

Pólvora-Brandão, D., Joaquim, M., Godinho, I., Aprile, D., Álvaro, A. R., Onofre, I., et al. (2018). Loss of hierarchical imprinting regulation at the Prader-Willi/Angelman syndrome locus in human iPSCs. *Hum. Mol. Genet.* 27, 3999–4011. doi: 10.1093/hmg/ddy274

Ponroy Bally, B., Farmer, W. T., Jones, E. V., Jessa, S., Kacerovsky, J. B., Mayran, A., et al. (2020). Human iPSC-derived Down syndrome astrocytes display genome-wide perturbations in gene expression, an altered adhesion profile, and increased cellular dynamics. *Hum. Mol. Genet.* 29, 785–802. doi: 10.1093/hmg/ddaa003

Real, R., Peter, M., Trabalza, A., Khan, S., Smith, M. A., Dopp, J., et al. (2018). *In vivo* modeling of human neuron dynamics and Down syndrome. *Science* 362:eaau1810. doi: 10.1126/science.aau1810

Sakurai, K., Kurtz, A., Stacey, G., Sheldon, M., and Fujibuchi, W. (2016). First proposal of minimum information about a cellular assay for regenerative medicine. *Stem Cells Transl. Med.* 5, 1345–1361. doi: 10.5966/sctm.2015-0393

Shcheglovitov, A., Shcheglovitova, O., Yazawa, M., Portmann, T., Shu, R., Sebastiano, V., et al. (2013). SHANK3 and IGF1 restore synaptic deficits in neurons from 22q13 deletion syndrome patients. *Nature* 503, 267–271. doi: 10.1038/nature12618

Shimizu, T., Shiohara, M., Tai, T., Nagao, K., Nakajima, K., and Kobayashi, H. (2016). Derivation of integration-free iPSCs from a Klinefelter syndrome patient. *Reprod. Med. Biol.* 15, 35–43. doi: 10.1007/s12522-015-0213-9

Sinenko, S., Skvortsova, E., Liskovykh, M., Ponomartsev, S., Kuzmin, A., Khudiakov, A., et al. (2018). Transfer of synthetic human chromosome into human induced pluripotent stem cells for biomedical applications. *Cells* 7:261. doi: 10.3390/cells7120261

Soeda, S., Saito, R., Fujita, N., Fukuta, K., and Taniura, H. (2019). Neuronal differentiation defects in induced pluripotent stem cells derived from a Prader-Willi syndrome patient. *Neurosci. Lett.* 703, 162–167. doi: 10.1016/j.neulet.2019.03.029

Stanurova, J., Neureiter, A., Hiber, M., De Oliveira Kessler, H., Stolp, K., Goetzke, R., et al. (2016). Angelman syndrome-derived neurons display late onset of paternal UBE3A silencing. *Sci. Rep.* 6:30792. doi: 10.1038/srep30792

Stelzer, Y., Sagi, I., Yanuka, O., Eiges, R., and Benvenisty, N. (2014). The noncoding RNA IPW regulates the imprinted DLK1-DIO3 locus in an induced pluripotent stem cell model of Prader-Willi syndrome. *Nat. Genet.* 46, 551–557. doi: 10.1038/ng.2968

Takahashi, K., Tanabe, K., Ohnuki, M., Narita, M., Ichisaka, T., Tomoda, K., et al. (2007). Induction of pluripotent stem cells from adult human fibroblasts by defined factors. *Cell* 131, 861–872. doi: 10.1016/j.cell.2007.11.019

Takahashi, Y., Wu, J., Suzuki, K., Martinez-Redondo, P., Li, M., Liao, H.-K., et al. (2017). Integration of CpG-free DNA induces de novo methylation of CpG islands in pluripotent stem cells. *Science* 356, 503–508. doi: 10.1126/science.aag3260

Takahashi, K., and Yamanaka, S. (2006). Induction of pluripotent stem cells from mouse embryonic and adult fibroblast cultures by defined factors. *Cell* 126, 663–676. doi: 10.1016/j.cell.2006.07.024

Toyoshima, M., Akamatsu, W., Okada, Y., Ohnishi, T., Balan, S., Hisano, Y., et al. (2016). Analysis of induced pluripotent stem cells carrying 22q11.2 deletion. *Transl. Psychiatry* 6:e934. doi: 10.1038/tp.2016.206

Weick, J. P., Held, D. L., Bonadurer, G. F. III., Doers, M. E., Liu, Y., Maguire, C., et al. (2013). Deficits in human trisomy 21 iPSCs and neurons. *Proc. Natl. Acad. Sci. U S A* 110, 9962–9967. doi: 10.1073/pnas.1216575110

Yang, J., Cai, J., Zhang, Y., Wang, X., Li, W., Xu, J., et al. (2010). Induced pluripotent stem cells can be used to model the genomic imprinting disorder Prader-Willi syndrome. *J. Biol. Chem.* 285, 40303–40311. doi: 10.1074/jbc.m110.183392

Yoon, K.-J., Nguyen, H. N., Ursini, G., Zhang, F., Kim, N.-S., Wen, Z., et al. (2014). Modeling a genetic risk for schizophrenia in iPSCs and mice reveals neural stem cell deficits associated with adherens junctions and polarity. *Cell Stem Cell* 15, 79–91. doi: 10.1016/j.stem.2014.05.003

Zhang, R., Hao, L., Wang, L., Chen, M., Li, W., Li, R., et al. (2013). Gene expression analysis of induced pluripotent stem cells from aneuploid chromosomal syndromes. *BMC Genomics* 14:S8. doi: 10.1186/1471-2164-14-s5-s8

Zhao, D., Lin, M., Chen, J., Pedrosa, E., Hrabovsky, A., Fourcade, H. M., et al. (2015). MicroRNA profiling of neurons generated using induced pluripotent stem cells derived from patients with schizophrenia and schizoaffective disorder and 22q11.2 del. *PLoS One* 10:e0132387. doi: 10.1371/journal.pone.0132387

Zuo, E., Huo, X., Yao, X., Hu, X., Sun, Y., Yin, J., et al. (2017). CRISPR/Cas9-mediated targeted chromosome elimination. *Genome Biol.* 18:224. doi: 10.1186/s13059-017-1354-4

7

Olfactory Mucosa Mesenchymal Stem Cells Ameliorate Cerebral Ischemic/Reperfusion Injury Through Modulation of UBIAD1 Expression

Jianyang Liu[1], Yan Huang[1], Jialin He[1], Yi Zhuo[2], Wei Chen[2], Lite Ge[1], Da Duan[2], Ming Lu[2,3] and Zhiping Hu[1]**

[1] *Department of Neurology, The Second Xiangya Hospital, Central South University, Changsha, China,* [2] *Developmental Biology of Ministry of Education, College of Life Sciences, Hunan Normal University, Changsha, China,* [3] *Hunan Provincial Key Laboratory of Neurorestoratology, Second Affiliated Hospital of Hunan Normal University, Changsha, China*

***Correspondence:**
Zhiping Hu
zhipinghu@csu.edu.cn
Ming Lu
lumingcs163@163.com

Mesenchymal stem cells (MSCs) have presented a promising neuroprotective effect in cerebral ischemia/reperfusion (I/R). Olfactory mucosa MSCs (OM-MSCs), a novel source of MSCs located in the human nasal cavity, are easy to obtain and situated for autologous transplantation. The present study was designed to evaluate the neuroprotective effects of OM-MSCs on cerebral I/R injury and the possible mechanisms. In the transient middle cerebral artery occlusion (t-MCAO) model, excessive oxidative stress and increased swollen mitochondria were observed in the peri-infarct cortex. Intravenous injection of OM-MSCs ameliorated mitochondrial damage and restored oxidant/antioxidant imbalance. Using the oxygen glucose deprivation/reperfusion (OGD/R) model *in vitro*, we discovered that the exposure of mouse neuroblastoma N2a cells to OGD/R triggers excessive reactive oxygen species (ROS) generation and induces mitochondrial deterioration with decreased mitochondrial membrane potential and reduces ATP content. OM-MSC transwell coculture attenuated the above perturbations accompanied with increased UbiA prenyltransferase domain-containing 1 (UBIAD1) expression, whereas these protective effects of OM-MSCs were blocked when UBIAD1 was knocked down. UBIAD1-specific small interfering RNA (siRNA) reversed the increased membrane potential and ATP content promoted by OM-MSCs. Additionally, UBIAD1-specific siRNA blocked the oxidant/antioxidant balance treated by OM-MSCs. Overall, our results suggested that OM-MSCs exert neuroprotective effects in cerebral I/R injury by attenuating mitochondrial dysfunction and enhancing antioxidation via upregulation of UBIAD1.

Keywords: cerebral ischemic/reperfusion injury, mesenchymal stem cell, neuroprotection, oxidative stress, mitochondria, UBIAD1

INTRODUCTION

Stroke is the third leading cause of death according to the systematic analysis for the Global Burden of Disease Study 2017 (GBD 2017 Causes of Death Collaborators, 2018), and someone has a stroke approximately every 40 s in the United States (Virani et al., 2020). Strokes occur under interruption of cerebral blood flow. Approximately 87% are ischemic strokes (Virani et al., 2020). The only FDA-approved therapy with proven efficacy for ischemic stroke was alteplase for dissolving the thrombus and increasing cerebral blood flow (Hacke et al., 2008). Alteplase intravenously injected within 3–4.5 hof ischemic stroke onset improves functional recovery and survival (Tsivgoulis et al., 2020). Although restoration of cerebral blood flow by mechanical or chemical therapies is essential to prevent irreversible brain damage, reestablishing blood flow paradoxically amplifies the initial brain tissue damage. This phenomenon is termed as cerebral ischemia/reperfusion (I/R) injury (Al-Mufti et al., 2018). It can be defined as a deterioration of ischemic brain tissue that reverses the benefits of endovascular recanalization (Jung et al., 2010). Ischemic stroke outcome in the forms of moderate to severe neurological deficits and mortality mainly results from cerebral I/R injury (Al-Mufti et al., 2018). Multiple biomechanisms play a role in the pathology of this injury, including oxidative stress, leukocyte infiltration, inflammation, and apoptosis. Oxidative stress, the result of reactive oxygen species (ROS) overproduction, is regarded as the primary event in cerebral I/R injury (Janardhan and Qureshi, 2004; Granger and Kvietys, 2015). ROS trigger many cellular and molecular events, which leads to the oxidation of proteins and lipids and eventually induces neuronal death (Sugawara and Chan, 2003). Mitochondria are the predominant organelle responsible for the generation of ROS (Marchi et al., 2012). Previous studies have suggested that cerebral I/R produces oxygen free radicals, mostly secreted by the mitochondria, thereby resulting in excessive oxidative damage in neurons (Christophe and Nicolas, 2006; Zhao et al., 2018). Hence, oxidant/antioxidant imbalance and mitochondrial dysfunction are fundamental triggers to neuronal injury in cerebral I/R.

Mesenchymal stem cell (MSC) transplantation therapy has shown promise for cerebral I/R injury. Various source tissues have been examined for therapies of ischemic strokes, such as adipose (Zhou et al., 2015), bone marrow (Liu et al., 2006), umbilical cord (Zuo et al., 2019), umbilical cord blood (Park et al., 2015), placenta (Kholodenko et al., 2012), and olfactory mucosa (Fan et al., 2018; Veron et al., 2018). The mechanism of MSCs in ischemic stroke therapy includes the promotion of angiogenesis, immunomodulation, secretion of neurotrophic factors, and enhancement of endogenous repair process (Eckert et al., 2013; Marei et al., 2018). Olfactory mucosa MSCs (OM-MSCs), localized in nasal lamina propria, are an attractive source of stem cells as they are relatively easy to obtain and ideally situated for autologous transplantation (Nivet et al., 2011). A previous study has demonstrated that the OM-MSC transplantation can restore cognitive abilities in global cerebral ischemia rats (Veron et al., 2018), but no study explores the mechanism of OM-MSC therapy in cerebral I/R injury.

UbiA prenyltransferase domain-containing 1 (UBIAD1) (aka TERE1) is an antioxidant enzyme catalyzing the biosynthesis of coenzyme Q10 and vitamin K2. The loss of UBIAD1 reduces the expression of the coenzyme Q10 and results in ROS-mediated lipid peroxidation (Mugoni et al., 2013). Mutations in *UBIAD1* were found to cause corneal cholesterol accumulation and induce the Schnyder corneal dystrophy (Nickerson et al., 2010). Vitamin K2 is involved in mitochondrial electron transport, ectopic UBIAD1 expression-elevated mitochondrial membrane potential, and ROS/RNS overproduction (Fredericks et al., 2013a). Silencing UBIAD1 in carcinoma cells causes morphological changes in the mitochondria (Morales et al., 2014). These studies of UBIAD1 emphasize its important role in oxidative/nitrosative stress, mitochondrial function, and cholesterol metabolism. Our previous study has demonstrated that UBIAD1 protects against I/R-induced mitochondrial dysfunction (Huang and Hu, 2018). Using the oxygen-glucose deprivation/reperfusion (OGD/R) model *in vitro* and transient middle cerebral artery occlusion (t-MCAO) model *in vivo*, the present study investigated the protective effects of OM-MSCs in cerebral I/R injury and whether OM-MSCs protect neurons by attenuating mitochondrial dysfunction and enhancing antioxidant activity via upregulation of UBIAD1.

MATERIALS AND METHODS

Isolation and Identification of OM-MSCs

Human OM-MSCs from healthy donors (two males, two females, 20–40 years old) were isolated from the surface interior of the concha nasalis media under otolaryngology endoscopy operation at the Department of Otolaryngologic Surgery, the Second Affiliated Hospital of Hunan Normal University (Changsha, China). Informed consent was given to each subject before the operations. The ethics committee of Hunan Normal University has approved this procedure protocol (Approved No. 2009163009).

Olfactory mucosa MSCs were isolated and cultured following a published protocol (Girard et al., 2011). The human olfactory mucosa tissues were immersed and washed with the antibiotic–antimycotic solution (Invitrogen, Carlsbad, CA, United States) for three times under 37°C. After washing, the tissues were cut into 1–2-mm^3 tissue pieces and cultured in Dulbecco's modified Eagle's medium: nutrient mixture F12 (DMEM/F12; Invitrogen) with 10% fetal bovine serum (FBS; Invitrogen, United States) at 37°C in 5% CO_2 atmosphere. OM-MSCs were chosen the fourth passage for use in this experiment.

After being incubated with 5 μL monoclonal PE-conjugated antibodies against specific membrane markers (CD105, CD90, CD73, CD44, CD146, CD133, CD34, and CD45; eBioscience, San Diego, CA, United States) for 30 min, fluorescence signals were evaluated by flow cytometry with a FACSCaliber instrument (Becton Dickinson, CA, United States).

In vivo Experimental Design

Animals

Male Sprague–Dawley (SD) rats (weighing 240–260 g) were purchased from the animal center of Hunan Normal University. The rats were housed in controlled conditions (standard lighting conditions, temperature of 20–25°C and humidity of 40–60%). All work for the animal study was approved by the Animal Care and Use Committee of Hunan Normal University (Approved No. 2020-164).

Transient Middle Cerebral Artery Occlusion (t-MCAO)

The cerebral I/R injury model *in vivo* was induced by t-MCAO as described previously (Cechetto et al., 1989). 100 male

SD rats were randomly allocated to three groups: (i) sham-operated group (n = 20), (ii) t-MCAO + saline group, and (iii) t-MCAO + MSCs group (**Figure 1A**). Briefly, the rats were anesthetized with 3.5% isoflurane and maintained with 1.0–2.0% isoflurane in 30% oxygen (0.3 L/min) and 70% nitrous oxide (0.7 L/min) mixture. A nylon filament was inserted in the right common carotid artery to block the right middle cerebral artery. After 2 h, the nylon filament was removed to enable reperfusion. Sham-operated rats underwent the same procedure without the insertion of the nylon filament. The operation period per rat did not exceed 15 min.

OM-MSC Transplantation Procedure

After surgery, rat mortality was approximately 15%. The survived 60 rats were randomly allocated to the MSC group (n = 30) or saline group (n = 30) 24 h post MCAO surgery. The rats were anesthetized as described above. Cell volume was set at 5 × 10^6 cells in 1 mL solution (saline) for intravenous transplantation. The infusion rate was approximately at 0.2 mL per minute.

The behavioral tests in rats were quantified at 3, 7, and 14 days post-MCAO, or before sacrifice. After being sacrificed, the brains were quickly removed to collect the peri-infarct cortex. Measures of brain infarction volume, neuronal apoptosis, level of oxidative stress, and mitochondrial function were tested at 3 and 7 days post-MCAO occlusion (2 and 6 days after MSC transplantation).

In vitro Experimental Design

Mouse Neuroblastoma (N2a) Cells

Mouse N2a neuroblastoma cells were purchased from the Cell Storage Center of Chinese Academy of Sciences (Shanghai, China). N2a neuroblastoma cells were cultured in Dulbecco's Modification of Eagle's medium (DMEM; Gibco) containing 10% FBS (Gibco) in 5% CO2 at 37°C.

Oxygen Glucose Deprivation/Reperfusion (OGD/R)

For the *in vitro* study, the cerebral I/R injury model was set up by OGD/R as described previously (Tang et al., 2016). N2a neuroblastoma cells were treated as follows: (i) Control, normal cell; (ii) OGD/R group; and (iii) OGD/R + MSC group (**Figure 3A**). For OGD/R group, 1 × 10^5 N2a cells were grown at the six-well culture plates and then placed into a modular incubator chamber (Billups Rothenberg, Inc., Del Mar, CA, United States) with a gas mixture of 5% CO_2 and 95% N_2. The culture medium was replaced with deoxygenated glucose-free Hanks' Balanced Salt Solution (Biological Industries) for 4 h. After OGD, the Hanks' Balanced Salt Solution was removed and the fresh culture medium (DMEM with 10% FBS) was added back for the re-oxygen and re-glucose.

Transwell Coculture

The coculture was set up by 0.4-μm pore size Transwell plates (Corning Incorporated, Wujiang, China) that allow the diffusion of soluble factors but not cells. For the OGD/R + MSC group,

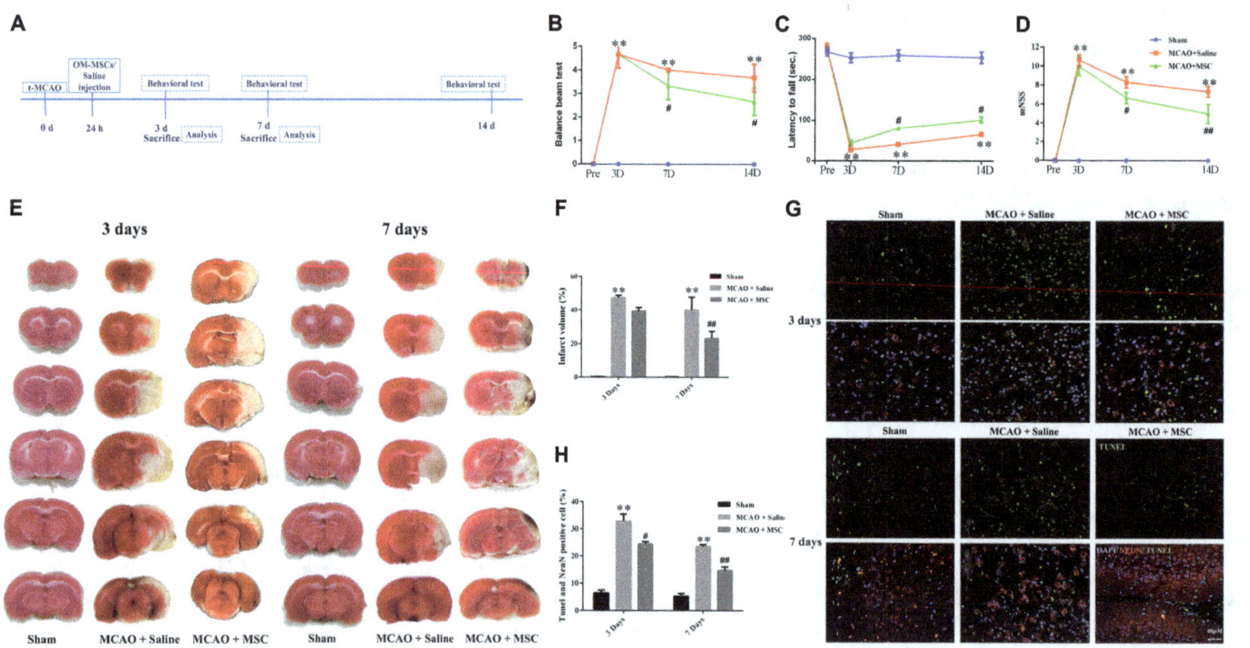

FIGURE 1 | OM-MSCs ameliorated neurological deficit and inhibited neuronal apoptosis in an MCAO animal model. **(A)** Schematic representation of the experimental design for *in vivo* experiments. **(B)** The balance beam test, **(C)** the rotarod test, and **(D)** the modified neurological severity score (mNSS) test were performed before MCAO and on 3, 7, and 14 days after MCAO (n = 7/group at 3 and 7 days; n = 5/group at 14 days). **(E,F)** The cerebral infarct volume, assessed by TTC staining of coronal brain sections after MCAO (n = 3). **(G)** Neuronal apoptosis in the ipsilateral cortex as detected by NeuN and TUNEL immunofluorescence staining after MCAO. **(H)** Quantification of NeuN and terminal transferase-mediated dUTP nick end labeling (TUNEL) double-stained cells (n = 3). All data are displayed as mean ± SEM. ($^*P$ < 0.05, $^{**}P$ < 0.01 vs. sham-operated, $^{\#}P$ < 0.05, $^{\#\#}P$ < 0.01 vs. MCAO + saline).

1×10^5 N2a cells were grown at the bottom of six-well culture plates and treated with OGD for 4 h. After OGD treatment, 1.5×10^5 OM-MSCs were grown on the upper chamber of transwell plate inserts with a pore-size of 0.4 μm. The transwell plates were then cultured in the normal incubator for 24 h. The relative measurements were performed after coculture for 24 h.

Cell Viability, Cell Apoptosis, and Lactate Dehydrogenase (LDH) Release Assay

The viability of N2a cells generally was detected using CCK-8 Assay Kit (Dojindo Molecular Technologies) according to the manufacturer's protocol.

The apoptosis of N2a cells was measured using the FITC Annexin V apoptosis detection kit (KGA108, KeyGen Biotech, Jiangsu, China) following the manufacturer's protocol. Apoptotic cells were analyzed via a flow cytometer.

To evaluate the integrity of the membrane and release of cellular contents, LDH activity from cultured cells into the supernatants was determined using a colorimetric assay kit (Nanjing Jiancheng Bioengineering Institute, Nanjing, China) according to the manufacturer's protocol.

Measurement of the Level of Oxidative Stress

Measurement of Intracellular ROS Generation

Change of intracellular ROS in N2a cells and brain tissue was measured using the semiquantitative dichlorofluorescein-diacetate (DCFH-DA, Beyotime Biotechnology) according to the manufacturer's instruction and then documented by a flow cytometer (Becton Dickinson, CA, United States).

Measurement of Total SOD, GSH-PX, MDA, LPO, and CAT Activity

The supernatant of N2a cells was used for the analysis of superoxide dismutase (SOD) activities, glutathione peroxidase (GSH-Px) activities, and malondialdehyde (MDA) levels. The level was determined with the commercial kits according to the manufacturer's instructions (Nanjing Jiancheng Biotech, Nanjing, China).

The peri-infarct cortex homogenates (10% wt/vol) were resuspended in cold saline. SOD activities, catalase (CAT) activities, MDA, and lipid peroxidase (LPO) levels were analyzed using commercial kits according to the manufacturer's instructions (Nanjing Jiancheng Biotech, Nanjing, China).

Measurement of Mitochondrial Function

Measurement of Mitochondrial Membrane Potential

For N2a neuroblastoma cells, the change in mitochondrial membrane potential ($\Delta\psi m$) was assessed using the JC-1 dye (Beyotime Institute of Biotechnology, China) following the manufacturer's instruction. Cells were harvested and analyzed on flow cytometry (Becton Dickinson, CA, United States). The ratio of red (aggregates)/green (monomers) fluorescence was calculated.

For brain tissue, the mitochondria were isolated from the penumbra cortex using the mitochondria isolation kit (Beyotime Institute of Biotechnology, China) and were treated with medium containing JC-1 dye. The fluorescence was detected using a flow cytometer.

ATP Measurement

ATP content was determined using an ATP Assay Kit (Nanjing Jiancheng Biotech, Nanjing, China) according to the manufacturer's protocol.

Infarct Volume Assessment

On 3 and 7 days post-MCAO, the brain was removed and sliced into 2-mm coronal sections and incubated with 2% 2,3-5-triphenyl-tetrazolium chloride (TTC; Sigma) in PBS at 37°C for 30 min. Brain sections were scanned using ImageJ. Then the infarct area (mm^2) of staining in each slice was multiplied by the slice thickness (2 mm) to examine the infarct volume (mm^3). The brain infarct volume was the summation of six individual section volumes.

TUNEL Assay

Brain sections were cut into 10-μm thickness, and the terminal deoxynucleotidyl transferase biotin-mediated dUTP Nick-end labeling (TUNEL) staining kit (DeadEnd Fluorometric TUNEL System, Promega, Madison, WI, United States) was used according to the manufacturer's instructions. Subsequently, the neuron was stained with NeuN (Sigma). Nuclei were stained with 6-diamidino-2-phenylindole (DAPI; Sigma). For each coverslip, five random fields were examined under a fluorescent microscope. The result was presented as a percentage of TUNEL and NeuN double-positive cells compared with all nuclei within 400× magnification fields.

Transmission Electron Microscope

A transmission electron microscope was used to observe the morphology of the mitochondrial ultrastructure in the fresh rat penumbra cortex. The observations were carried out using an electron microscope (Hitachi, HT7700, Japan). The percentage of abnormal mitochondria was evaluated by randomly selecting 20 micrographs per sample.

Behavioral Test

Functional behaviors in rats were tested at 3, 7, and 14 post-MCAO. All behavioral tests were estimated by two investigators who were blinded to the experimental groups. The 12-point modified neurologic severity scores (mNSS) were used to evaluate the sensorimotor integration of forelimbs. The rotarod test was used to test motor coordination (Bederson et al., 1986). Rats performed rotarod training for 3 days before MCAO. The balance beam test consisted of a beam that was placed 0.5 m above the ground. The motor performance was estimated using a five-point scale (Jiang et al., 2018).

Knockdown of UBIAD1 by Small Interfering RNA (siRNA)

The small interfering RNAs (siRNAs) for UBIAD1, along with control siRNA, were purchased from Genechem

(Shanghai, China). The sequences of UBIAD1 siRNA were forward, 5′-CACUUGGCUCUUAUCUACUdTdT-3′ and reverse, 5′-AGUAGAUAAGAGCCAAGUGdTdT-3′. The sequences of the NC were as follows: forward, 5′-UUCUCCGAACGUGUCACGUTT-3′ and reverse, 5′-ACGUGACACGUUCGGAGAATT-3′. Gene silencing was proved by the analysis of protein expression using Western blotting.

Western Blotting

Proteins were extracted from N2a cells or the peri-infarct cortex using a total protein extraction kit (Beyotime). The concentration was determined by a BCA Protein Assay Kit (Beyotime Biotechnology). Protein extracts were separated by SDS-PAGE and transferred to PVDF membranes. The membranes were blocked and incubated with indicated primary antibodies against UBIAD1 (1: 750, ab191691, Abcam) and mouse anti-actin antibodies (1: 5000; Proteintech) at 4°C overnight. After washing, the membranes were incubated with horseradish peroxidase-conjugated secondary antibody (1: 5000; anti-mouse or anti-rabbit IgG; Proteintech). The blots were visualized using an ECL detection kit (Bio-Rad, Munich, Germany).

Real-Time PCR Quantification

Total RNA was isolated from N2a cells using TriZol (Tiangen, Beijing, China). Reverse transcription was performed using a reverse transcription kit (Tiangen). The following qPCR primer sequences were used to generate specific fragments: 5′-GGCCATTCTCCATTCCAACA-3′ and 5′-GCCAGCCTCTC GGTCAGA-3′ for mouse UBIAD1 and 5′-GTCCCTGTATGCCTCTGGTC-3′ and 5′-GGTCTTTACGGATGTCAACG-3′ for mouse β -actin.

Statistical Analysis

All experiments were performed in at least three replicates. Data are expressed as mean ± SEM. Differences between groups were estimated using two-sided unpaired Student's *t*-test or two-sided ANOVA with the Bonferroni correction for the *post hoc* *t*-test as appropriate. Statistical analysis was conducted with GraphPad Prism 6 Software (La Jolla, CA, United States). Differences with the probability of $P < 0.05$ were considered significant.

RESULTS

Identification of OM-MSCs

Most OM-MSCs adhered to the culture plate surface and adopted a spindle-shaped morphology (**Supplementary Figure S1**). OM-MSCs were characterized by eight membrane markers using flow cytometry. Flow cytometry results revealed that OM-MSCs were uniformly positive for the MSC markers CD44, CD90, CD105, CD146, CD133, and CD73 and negative for CD45 and CD34 (**Supplementary Figure S1**), which indicated high MSC purity.

OM-MSCs Ameliorate Neurological Deficit *in vivo*

Rats were subjected to 2 h of MCAO and received saline or OM-MSCs intravenously at 24 h post-MCAO (**Figure 1A**). Three behavior tests were performed to test the protective effects of OM-MSCs on cerebral I/R injury, including the rotarod test, balance beam test, and mNSS. The MCAO + saline group exhibited severe neurological deficits at 3, 7, and 14 days post-stroke compared with the sham-operated group ($p < 0.01$). The behavioral tests showed that there was no significant difference between MSC-treated groups and saline groups at 3 days post-MCAO ($p > 0.05$, **Figures 1B–D**). However, at 7 or 14 days post-MCAO, the rotarod test, balance beam test, and mNSS were significantly improved in the OM-MSC-transplanted groups compared with that in the saline group ($p < 0.05$, **Figures 1B–D**). In total, these measurements provided evidence that OM-MSC treatment attenuated impairment on behavioral function post-MCAO.

OM-MSCs Reduce Lesion Volume and Neuron Apoptosis *in vivo*

2,3-5-Triphenyl-tetrazolium chloride staining was used to determine the brain infarction volume. The staining was performed at 3 and 7 days to evaluate whether OM-MSC administration decreased the infarction volume in animals post-MCAO. The representative pictures are shown in **Figure 1E**, showing normal brain tissue stains with TTC, but significant reduced TTC staining on the lesion side after MCAO surgery, which confirmed the success of the MCAO rat model. The infarction volume in the OM-MSC-treated rats was indistinguishable from the saline-treated group at 3 days ($p > 0.05$). At 7 days, the infarction area in the OM-MSC group was significantly reduced compared with that in the saline group after MCAO ($p < 0.01$, **Figure 1F**).

Neuronal apoptosis has been identified as a major determinant in cerebral I/R injury. The apoptosis rate was analyzed by counting the percentage of NeuN$^+$ TUNEL$^+$ cells in the peri-infarct region of MCAO rats. At 3 and 7 days after MCAO, NeuN and TUNEL double-positive cells remarkably increased. OM-MSC administration significantly abrogated the increased percentage of NeuN$^+$ TUNEL$^+$ cells at days 3 and 7 following MCAO in comparison to the MCAO + saline group ($p < 0.05$ and $p < 0.01$, respectively; **Figures 1G,H**), which demonstrated that OM-MSCs could suppress neuron apoptosis in cerebral I/R injury *in vivo*.

OM-MSCs Dampen Excessive Oxidative Stress and Mitochondrial Dysfunction *in vivo*

Oxidative stress plays a significant role in the pathological process associated with ischemic tissue. To evaluate the antioxidative effect of OM-MSCs in MCAO, we examined the effects of OM-MSCs on the ROS content and the levels of total SOD, CAT, MDA, and LPO in the peri-infarct cortex. As a result, the content of ROS was higher in the MCAO group compared with the sham-operated group as evidenced by flow cytometry-based evaluation. OM-MSC administration significantly reduced the content of ROS compared with the MCAO group at 3 and 7 days post-MCAO ($p < 0.05$ and $p < 0.01$, respectively; **Figures 2A,B**). Similarly, MCAO significantly reduced SOD and CAT activity and increased MDA and LPO levels ($p < 0.01$, **Figures 2C–F**). These effects were reversed by OM-MSC treatment at days 7 post-MCAO ($p < 0.01$, **Figures 2C–F**). At days 3, OM-MSC treatment could attenuate the level of SOD, CAT, and MDA activity ($p < 0.05$) but could not suppress the level of LPO activity ($p > 0.05$). In total, these data indicated that OM-MSCs could attenuate excessive oxidative stress induced by cerebral I/R.

By using JC-1 and a flow cytometer, **Figures 2G,H** demonstrate that the collapse of the membrane potential in the isolated mitochondria from MCAO rats was remarkably reversed by OM-MSC administration at 3 and 7 days ($p < 0.05$ and $p < 0.01$, respectively). Furthermore, transmission electron microscope analysis was carried out at 7 days to observe the shape and structure of mitochondria. **Figures 2I,J** demonstrated morphological aberrations like membrane ruptures, vacuole formation, and cristae swelling of mitochondria in neurons post-MCAO. However, with the administration of OM-MSCs, mitochondria displayed fewer abnormalities in morphology.

Overall, *in vivo* experiments, we observed that OM-MSCs can protect neurons from oxidative damage and mitochondria dysfunction at 3 and 7 days post-MCAO. OM-MSCs also ameliorate neurological deficit and lesion volume at 7 days post-MCAO. However, OM-MSC administration could not reduce the behavioral impairment and infarction volume at 3 days after cerebral I/R injury.

OM-MSCs Enhance the Survival of I/R-Induced Cell Injury *in vitro*

In the *in vitro* model, N2a cells were treated by OGD for 4 h and then transwell coculture with OM-MSCs for 24 h (**Figure 3A**). To verify the protective effects of OM-MSCs on the cultured N2a cells subjected to OGD/R, the cell viability of N2a cells was determined by CCK-8 assay, cell apoptosis was tested using Annexin V assay, and cell necrosis was evaluated by the LDH-release assay. The cell viability of N2a cells in the OGD/R group was reduced to 0.38 ± 0.01-fold of the control group. The detrimental effect of OGD/R was reversed by OM-MSC transwell coculture, and the cell viability was restored to 0.70 ± 0.01-fold of the control group (**Figure 3B**). The coculture of OGD/R-treated N2a cells with OM-MSCs led to a strong enhancement in cell viability and a further decrease in LDH release, compared with that in the OGD/R group (126.5 ± 5.97- vs. 153.6 ± 5.84-fold of the control group) ($p < 0.01$, **Figure 3C**). Meanwhile, OM-MSCs could inhibit the toxic effect of OGD/R on cell apoptosis in N2a cells. In the OGD/R + MSC group, the apoptotic cell population was markedly reduced in the Annexin V assay (**Figure 3E**). The apoptosis rates were 0.97, 16.43, and 10.28% in the control, OGD/R, and OGD/R + MSC groups, respectively ($p < 0.01$, **Figure 3D**). Thus, the above results suggest that OM-MSCs promote cell survival under cerebral I/R injury.

OM-MSCs Reduce OGD/R-Induced Oxidative Stress

Oxidative damage is considered to be an important contributor to cerebral I/R injury. Excessive oxidative stress could contribute directly to cell apoptosis and necrosis in cerebral I/R injury (Rodrigo et al., 2013). For this reason, we attempted to study whether OM-MSCs could attenuate OGD/R-induced ROS overproduction. In the present study, we found that the total intracellular ROS production increased by 4.19 ± 0.06-fold of the control group after OGD/R, which was reduced to 2.48 ± 0.06-fold with OM-MSC transwell coculture (**Figures 4A,B**). Besides, oxidative stress is closely related to the activity of oxidants and antioxidants, which is often a consequence of oxidant-induced ROS production. We revealed that OGD/R reduced total SOD and GSH-PX activity and increased MDA activity, which was neutralized by OM-MSC treatment (**Figures 4C–E**). These results indicate that OM-MSC treatment inhibits the ROS overproduction and can reverse the impaired oxidant/antioxidant balance.

OM-MSCs Suppress OGD/R Induced Mitochondrial Dysfunction

Mitochondria are primary consumers of oxygen and an important source of free radicals. Previously, we have demonstrated that OGD/R could lead to mitochondrial dysfunction in N2a cells (Huang and Hu, 2018). We asked whether OM-MSCs can protect against it in this study. ATP content, an indicator of mitochondrial function, plays an important role in energy transfer (Cadenas and Davies, 2000). In this study, the ATP content in the OGD/R group was significantly decreased. OM-MSC transwell cocultures attenuate the reduced ATP content from 32.18 ± 2.95-fold to 52.75 ± 2.28-fold that in the control group ($p < 0.01$, **Figure 4F**). Next, we evaluated mitochondrial membrane potential following OGD/R with or without OM-MSC treatment. OGD/R reduced mitochondrial membrane potential, which was presented as a decrease in red/green fluorescence. The mitochondrial membrane potential was significantly increased when N2a cells were treated with OM-MSCs (19.59 ± 2.14- vs. 32.05 ± 2.68-fold of the control group) ($p < 0.01$, **Figures 4G,H**). Collectively, the above results suggested that OM-MSCs limited ROS production in OGD/R-injured N2a cells, potentially via improving mitochondrial function.

UBIAD1 Is Essential for Neuroprotection of OM-MSCs in I/R-Induced Injury

In a previous study, we found that UBIAD1 plays a protective role in OGD/R-induced mitochondrial dysfunction (Huang and Hu, 2018). Therefore, we next tested whether the neuroprotection of OM-MSCs against oxidative stress and mitochondrial function involved the upregulation of UBIAD1. In the *in vivo* model, we found that the UBIAD1 protein levels were decreased after MCAO surgery but reversed by OM-MSC administration at 3 and 7 days ($p < 0.05$ and $p < 0.01$, respectively, **Figures 5A,B**). At 7 days post-MCAO, the expression of UBIAD1 was decreased to 0.38 ± 0.05-fold of the sham-operated group, which was

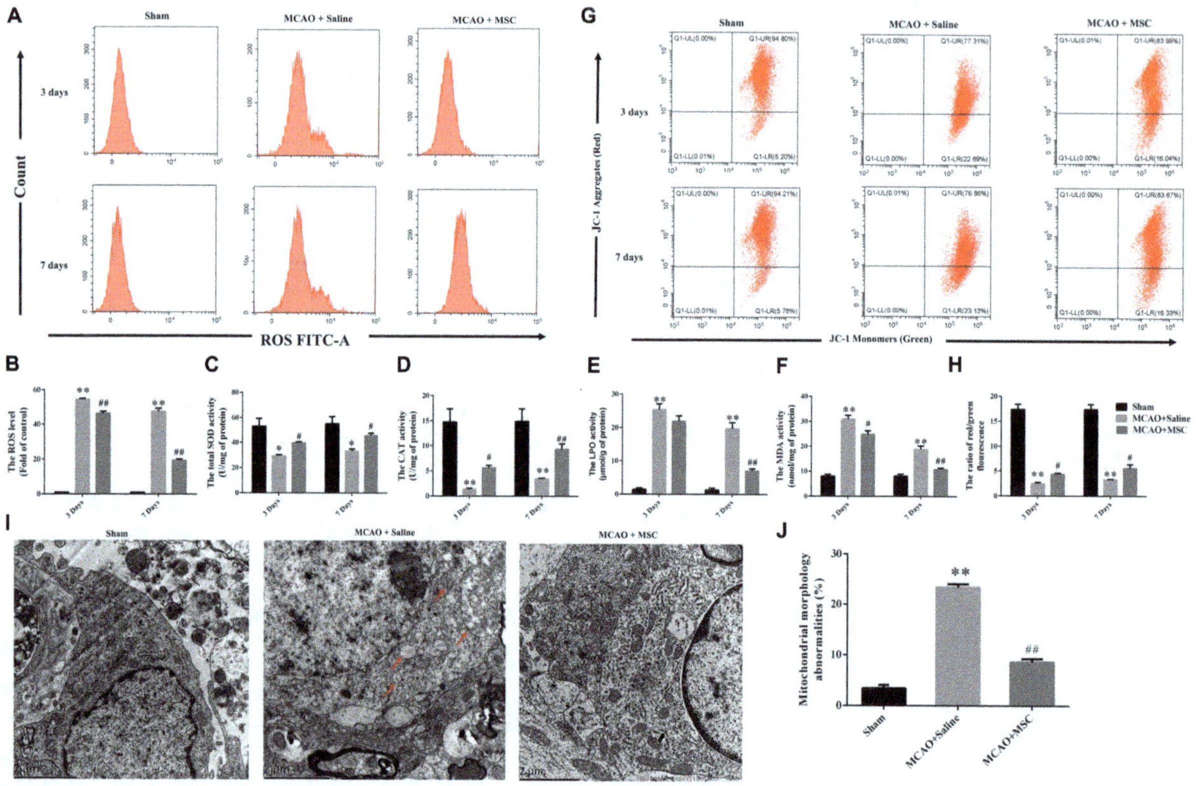

FIGURE 2 | OM-MSC administratio reduced oxidative stress level and alleviated mitochondrial dysfunction in an MCAO animal model. **(A,B)** ROS levels in the peri-infarct cortex were detected, and the determination was carried out using a flow cytometer. The levels were normalized by the fluorescence intensity of the control group (the value of control cells was presented as 1.0). **(C–F)** Measurement of antioxidative enzyme activities of CAT and SOD, and oxidative enzyme activities of LPO and MDA. **(G–H)** Mitochondrial membrane potential was measured using a JC-1 probe determined by flow cytometry. **(I)** Representative photographs of the transmission electron microscope showed ultrastructural changes in mitochondria morphology in the peri-infarct cortex and its reversal after OM-MSC treatment. In the MCAO + saline group, mitochondria with abnormal shapes, such as focal enlargement of the intermembrane space and loss of a normal cristae pattern (red arrows), were observed. **(J)** Based on representative photographs, quantification of aberrant mitochondria (membrane ruptures, vacuole formation) was performed. All data are displayed as mean ± SEM (n = 3). ($^{*}P < 0.05$, $^{**}P < 0.01$ vs. sham-operated, $^{\#}P < 0.05$, $^{\#\#}P < 0.01$ vs. MCAO + saline).

increased to 0.66 ± 0.04-fold of the sham-operated group with OM-MSC administration ($p < 0.01$). In the *in vitro* model, we found that the expression of UBIAD1 was decreased by 0.34 ± 0.06-fold of the control group after OGD/R, which was increased to 0.65 ± 0.06-fold of control with OM-MSC treatment, as demonstrated using western blot analysis ($p < 0.01$, **Figures 5C,D**). The same result was demonstrated by real-time PCR ($p < 0.05$, **Figure 5E**). For further exploration of whether UBIAD1 was essential for OM-MSCs in regulating mitochondrial function and oxidative stress, the N2a cells were treated with gene silencing. UBIAD1-specific siRNA was transfected into N2a cells to reduce the expression of UBIAD1. The protein and mRNA levels of UBIAD1 were reduced remarkably in the UBIAD1 RNAi group compared with the N2a cells transfected with control siRNA (**Figures 5F–H**).

We then demonstrated the contribution of UBAD1 to oxidant/antioxidant balance in OGD/R-induced N2a cells. UBIAD1 silencing blocked the ability of OM-MSCs to reverse the OGD/R-induced total intracellular ROS. There were no statistically significant differences in the ROS production between OGD/R with the control RNAi group and OGD/R + MSC with UBIAD1 RNAi group (**Figures 6A,B**). The same results were affirmed by the detection of SOD, GSH-PX, and MDA activity (**Figures 6C–E**). Collectively, OM-MSC treatment reduced the oxidative stress level in cells transfected with control siRNA but failed to do so in N2a cells transfected with UBIAD1 siRNA.

The improvement in mitochondrial function by OM-MSCs was also abolished by UBIAD1 silencing, evidenced by the mitochondrial membrane potential (**Figures 6F,G**) and the ATP production (**Figure 6H**). There were no statistically significant differences between OGD/R with the control RNAi group and OGD/R + MSC with the UBIAD1 RNAi group. OM-MSC treatment restored the damaged mitochondrial function in N2a cells transfected with control siRNA but failed to do so in cells transfected with UBIAD1 siRNA. Conclusively, UBIAD1 plays an important role in OM-MSC-driven suppression of oxidative injury and improvement of mitochondrial function. CCK-8 assay results showed that the levels of cell viability were decreased in the UBIAD1 RNAi group compared with the control siRNA group (**Figure 6I**). Cell apoptosis assay results showed that the apoptosis rates were increased in the UBIAD1 RNAi group compared with the control siRNA group (**Figures 6J,K**). Thus, these results

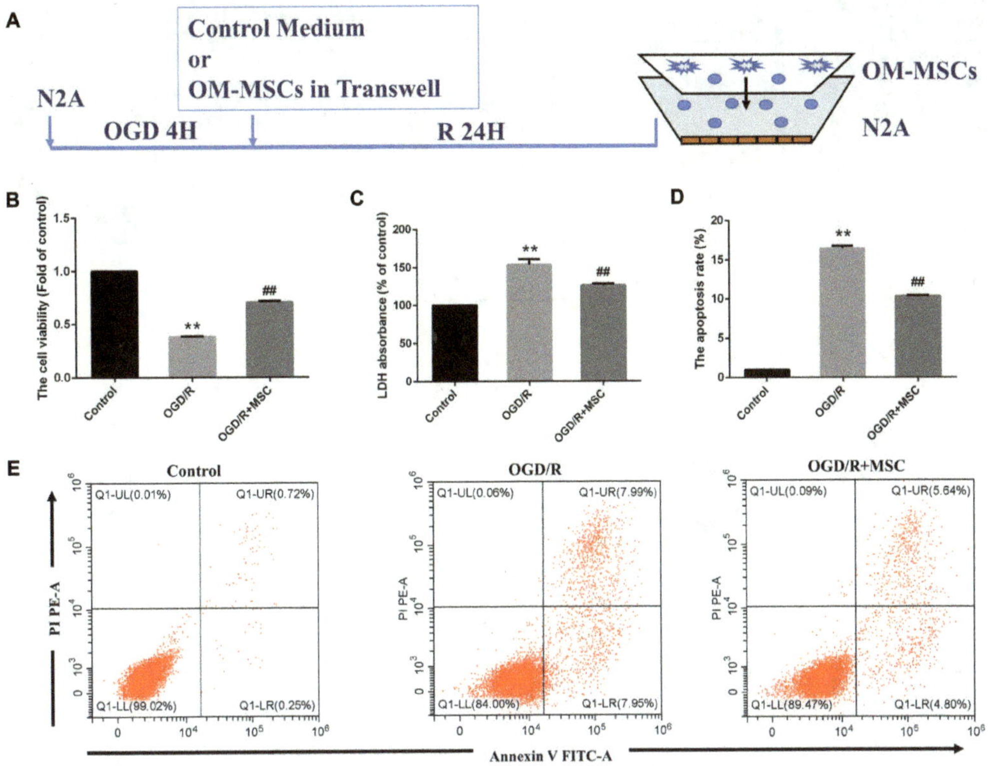

FIGURE 3 | OM-MSCs ameliorated OGD/R-induced N_2a injury. **(A)** Schematic representation of the experimental design and the transwell system used for *in vitro* experiments. **(B)** Cell viability was determined using the CCK-8 assay, and data are normalized by control cells (the value of control cells was presented as 1.0). **(C)** Cell necrosis was determined using the LDH assay, and data are expressed as a percentage of the control. **(D)** Apoptotic cells are represented as the percentage of Annexin-V single-positive plus Annexin-V/PI double-positive cells. **(E)** Representative plots of FACS by Annexin V–FITC/PI dual staining. All data are displayed as mean ± SEM (n = 3). ($^{**}P < 0.01$ vs. control; $^{\#\#}P < 0.01$ vs. OGD/R).

demonstrate that the intervention with OM-MSCs minimizes oxidative stress damage and mitochondrial dysfunction, at least in part, via upregulating the expression of UBIAD1.

DISCUSSION

Mitochondrial Dysfunction and ROS Overproduction in Cerebral I/R Injury

In the prognosis of ischemic stroke, cerebral I/R injury is the main cause of moderate to severe neurological deficits and mortality. This injury involves free radical damage, mitochondrial dysfunction, aberrant immune responses, apoptosis, and necrosis. Here, we discuss the relations between mitochondrial dysfunction and oxidative stress under I/R status. First, the mitochondrion is the major source of ROS. Under physiological status, ROS is a by-product of mitochondrial respiration (Chance et al., 1979). The mitochondrial respiratory chain is composed of four complexes, complex I (NADH-coenzyme Q reductase), complex II (succinate-coenzyme Q reductase or succinate dehydrogenase), complex III (ubiquinol cytochrome c reductase), and complex IV (cytochrome c oxidase) (Chance and Williams, 1955; Friedrich and Böttcher, 2004). Complexes I, III, and IV formed the electron transport chain. In the electron transport chain complex, the escape of electron during the electron transport to the electron acceptor leads their binding to molecular oxygen and is considered to be the major source of ROS. Upon I/R condition, various changes occur in the electron transport chain complex, which may result in succinate reoxidization and accumulation of intermediate succinate (Chouchani et al., 2014). Meanwhile, the I/R condition induces calcium overload, which results in calcium influx into mitochondria and mitochondrial permeability transition pore opening (**Figure 7A**) (Jang et al., 2017). Second, oxidative stress contributes to mitochondrial dysfunction in neurons. During I/R, an oxidative burst occurs, which is generated by succinate accumulation and NMDA receptor-mediated calcium influx into mitochondria. Excessive oxidative stress could induce mitochondrial permeability transition (Reddy et al., 2008), which is characterized by the mitochondrial permeability transition pore opening in the inner mitochondrial membrane. The increased permeability of the mitochondrial membrane leads to the collapse of mitochondrial membrane potential, decreased the ATP content, and further generated additional ROS (**Figure 7B**) (Zorov et al., 2006). Generally, mitochondrial dysfunction and ROS burst form a vicious circle in cerebral I/R injury and ultimately result in neuron apoptosis and necrosis.

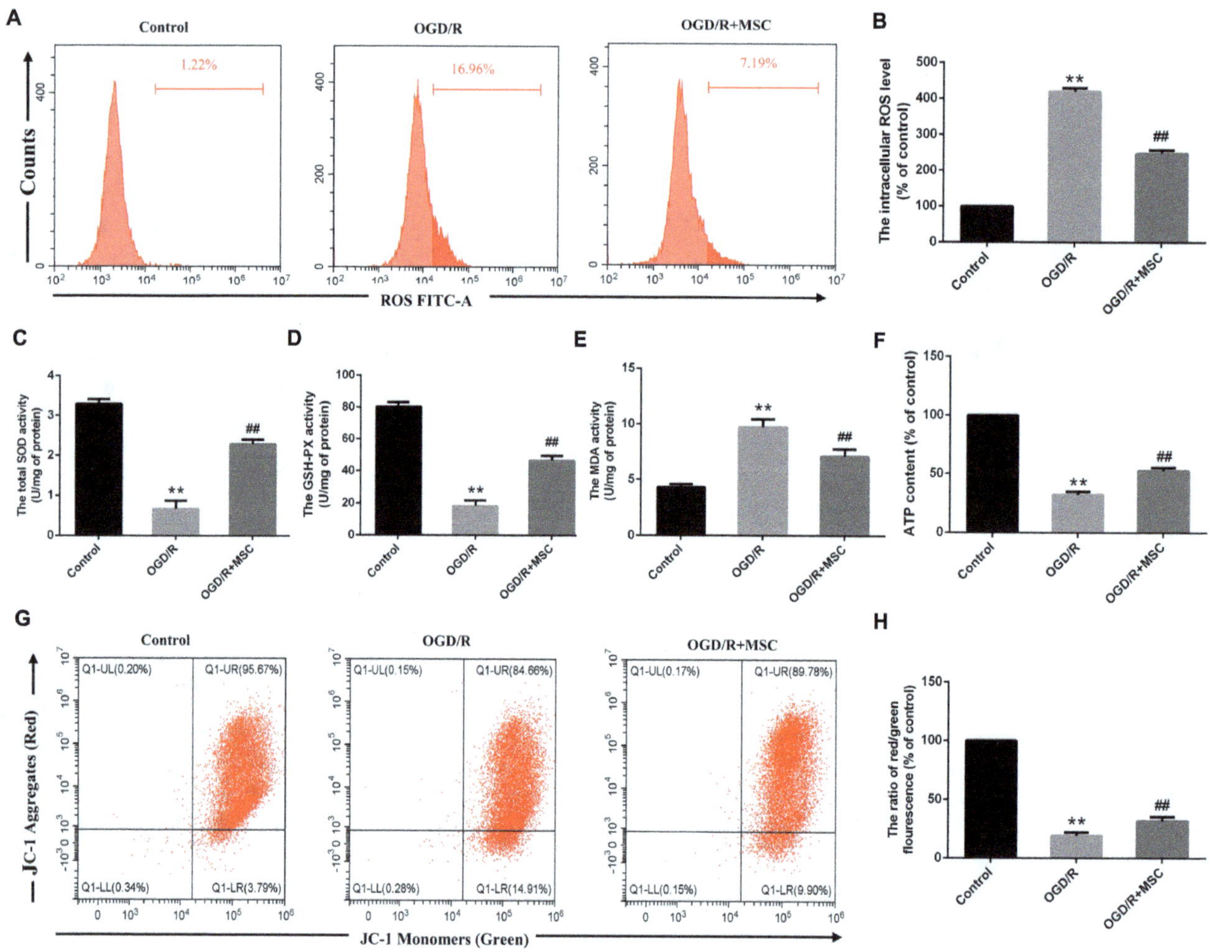

FIGURE 4 | OM-MSCs attenuated OGD/R-induced ROS overproduction and mitochondrial dysfunction in N2a cells. **(A–B)** Intracellular ROS levels in N2a were measured by staining with DCFH-DA, and the determination of ROS levels was carried out using a flow cytometer. **(C–E)** The total SOD, GSH-PX, and MDA activities in N2a were assessed. **(F)** The ATP content in N2a cells was assessed using an ATP assay kit, and data are expressed as a percentage of the control. **(G,H)** N2a was incubated with JC-1, and the determination of mitochondrial membrane potential was carried out using a flow cytometer. The mitochondrial membrane potential in each group was calculated as the ratio of red to green fluorescence. All data are displayed as mean ± SEM (n = 3). (**P < 0.01 vs. control; $^{\#\#}P$ < 0.01 vs. OGD/R).

Damaged Mitochondria Restored by MSC Treatment

Mesenchymal stem cells have been demonstrated to attenuate tissue injury by improving the mitochondrial function with increased mitochondrial membrane potential, enhanced mitochondrial bioenergetics, decreased mitochondrial fragmentation, and increased ATP generation (Garbuzova-Davis et al., 2017; Perico et al., 2017). Accumulating evidence has substantiated that mitochondrial transfer plays a critical role in mediating MSC-based therapy (Liang et al., 2014; Fan et al., 2020). The whole mitochondria can be donated by MSCs and horizontally transferred between different cell types. MSCs derived from various tissues such as bone marrow (Morrison et al., 2017) and adipose tissue (Mahrouf-Yorgov et al., 2017) have shown the capacity to donate mitochondria. One study researched the differential mitochondrial transfer abilities between MSCs derived from various sources; they observed that bone marrow and adipose-derived MSCs have higher mitochondrial transfer compared with dental pulp and Wharton's jelly-derived MSCs (Paliwal et al., 2018). Another study suggested that induced-pluripotent-stem-cell-derived MSCs (iPSC-MSCs) have a higher mitochondrial transfer capacity compared with bone marrow-derived MSCs (Li et al., 2014). The donated mitochondria by iPSC-MSCs are functional, which could rescue anthracycline-induced cardiomyocyte damage (Zhang et al., 2016), oxidative stress-induced corneal epithelial cell injury (Jiang et al., 2016), epithelial cell injury in asthma model (Yao et al., 2018), and retinal ganglion cell degeneration (Jiang et al., 2019). Most of studies demonstrated the functional mitochondrial transfer from MSCs to target cells. However, the mitochondria from somatic cells to MSCs remains unclear. One study suggests that the mitochondrial transfer is unidirectional; they did not detect the transfer of mitochondria from astrocytes or neurons to MSCs (Babenko et al., 2015).

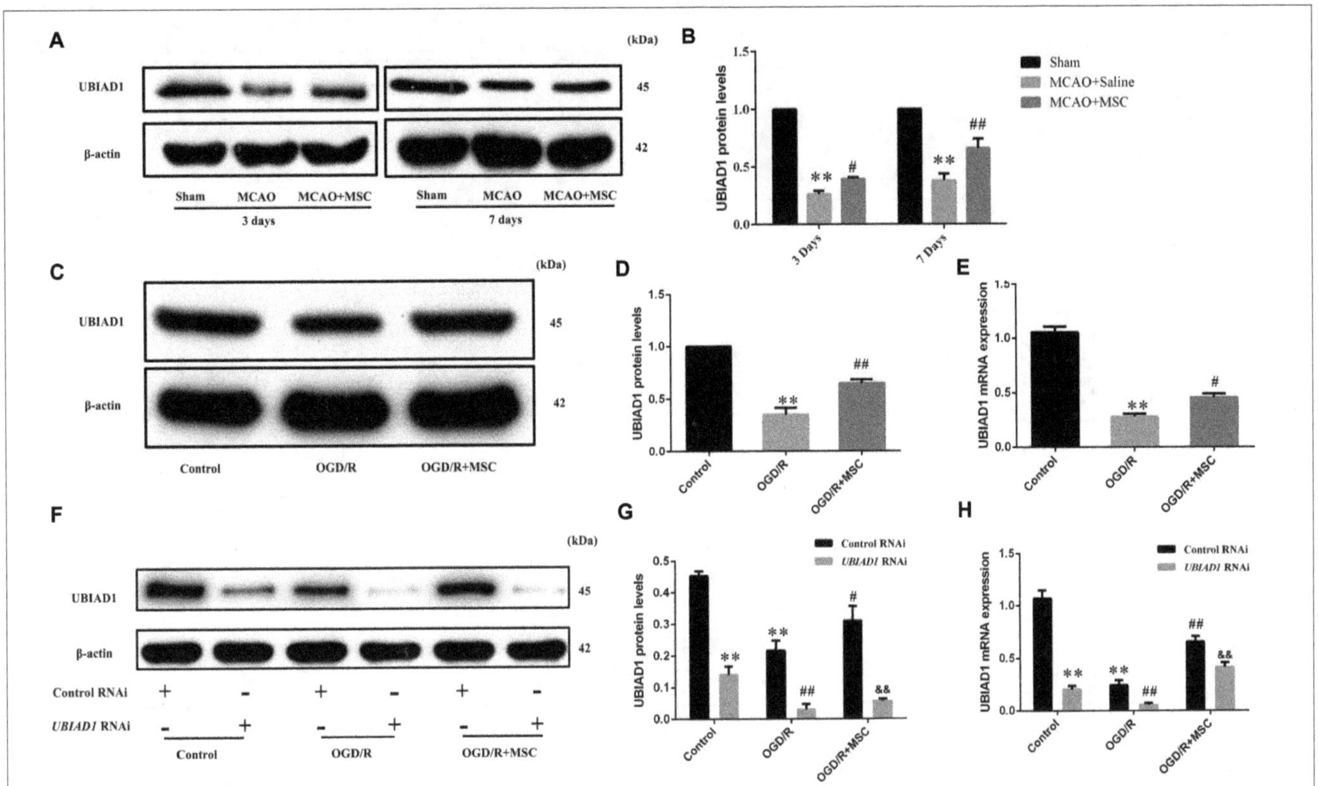

FIGURE 5 | Expression of UBIAD1 upregulated by OM-MSC treatment. **(A,B)** The protein expression of UBIAD1 in the peri-infarct cortex was detected using Western blotting. Data are displayed as mean ± SEM ($n = 3$). ($^{*}P < 0.05$, $^{**}P < 0.01$ vs. sham-operated, $^{\#}P < 0.05$, $^{\#\#}P < 0.01$ vs. MCAO + saline.) **(C,D)** The protein expression of UBIAD1 in the N2a cells was detected using Western blotting. **(E)** The mRNA expression of UBIAD1 in the N2a cells was detected by qPCR. Data are displayed as mean ± SEM ($n = 3$). ($^{*}P < 0.05$, $^{**}P < 0.01$ vs. control, $^{\#}P < 0.05$, $^{\#\#}P < 0.01$ vs. OGD/R.) **(F–H)** Representative expression of UBIAD1 after transfection with siRNA. Data are displayed as mean ± SEM ($n = 3$). ($^{**}P < 0.01$ vs. Con/control RNAi, $^{\#}P < 0.05$, $^{\#\#}P < 0.01$ vs. OGD/R/control RNAi, $^{\&\&}P < 0.01$ vs. OGD/R + MSCs/control RNAi).

Another study demonstrated that the mitochondria could be transferred from injured cardiomyocytes or endothelial cells to MSCs, and the somatic-derived mitochondria could trigger the anti-apoptotic function of MSCs (Mahrouf-Yorgov et al., 2017). Thus, the mitochondrial transfer from somatic cells to MSCs remains in need of further research.

Olfactory Mucosa: A Novel Source of MSCs

Mesenchymal stem cells are found in many organs and tissues in human, containing adipose tissue, bone marrow, umbilical cord, umbilical cord blood, fetal placenta, amniotic membrane, dental pulp, muscle, peripheral blood, cartilage, synovium, tonsil, and thymus (Rui et al., 2018). OM-MSCs, localized in nasal lamina propria, are a novel source of MSCs identified in recent research (Nivet et al., 2011). Compared with MSCs derived from other sources, OM-MSCs have the following advantages: First, OM-MSCs are easily accessible and can be biopsied safely under local anesthetics, which are very suitable for autologous transplantation (Girard et al., 2011). Second, OM-MSCs have specific immunomodulatory abilities. Compared to MSCs derived from other tissue, OM-MSCs have higher anti-apoptotic capacity of non-activation of immune effector cells (Di Trapani et al., 2013) and higher secretion of immunosuppressive cytokines under an inflammatory microenvironment (Jafari et al., 2020). Third, OM-MSCs derived from the neural crest, which maintains in a condition of embryo-like development. OM-MSCs have higher proliferation and capacity of differentiation into dopaminergic neurons (Alizadeh et al., 2019a), which serves as an available candidate for the treatment of neurological diseases. Fourth, the human olfactory mucosa is a kind of permanently self-renewing tissue, which has various cells sustaining its normal function and regeneration (Lindsay et al., 2010). In various disease states, OM-MSCs have been proved to exert protective effects such as supporting hearing regeneration (Pandit et al., 2011; Young et al., 2018), expressing higher immunosuppressive factors in autoimmune arthritis (Rui et al., 2016), restoring memory in amnesic syndrome (Nivet et al., 2011), and differentiating into dopaminergic neuron in Parkinson's disease model (Alizadeh et al., 2019b; Simorgh et al., 2019). In the global cerebral ischemia model, OM-MSCs contribute to the improvement of learning and memory abilities. However, no study investigates the effects of OM-MSCs in mitochondrial and oxidative stress. The present study, conducted in the model of cerebral I/R injury, suggests that the damaged mitochondrial

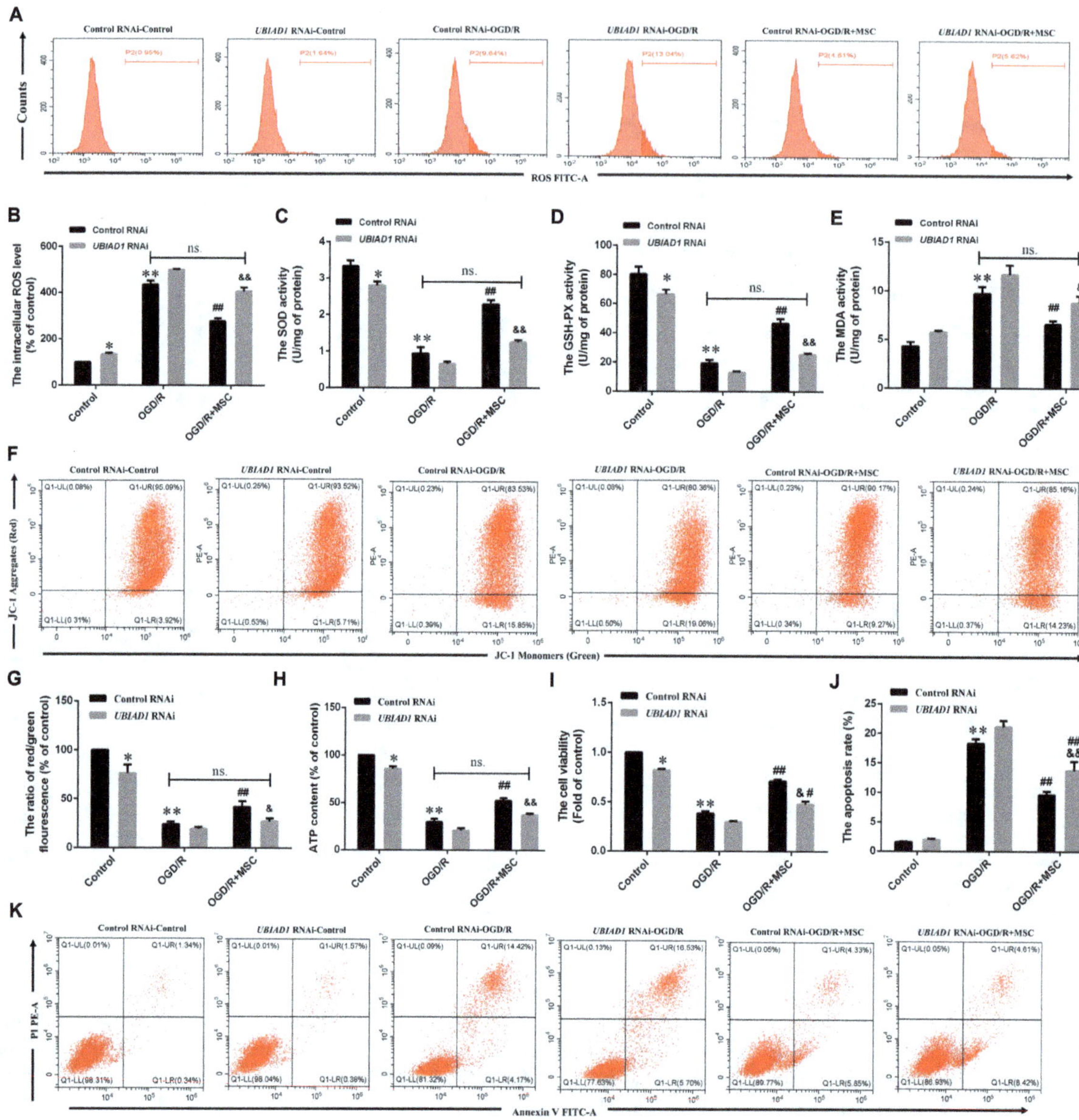

FIGURE 6 | OM-MSCs mitigate OGD/R-induced ROS overproduction and mitochondrial dysfunction through upregulation of UBIAD1. N2a cells were transfected with UBIAD1-specific siRNA (UBIAD1 RNAi) or a non-specific control siRNA. **(A,B)** The intracellular ROS levels were examined by DCFH-DA in N_2a cells knocked down by UBIAD1. **(C–E)** The total SOD, GSH-PX, and MDA activities in N_2a were assessed. **(F–G)** Determination of the mitochondrial membrane potential was carried out using the JC-1 probe. **(H)** The ATP content in N_2a cells. **(I)** The cell viability was examined by CCk-8 assay. **(J,K)** The apoptotic N2a cells knocked down by UBIAD1. All data are displayed as mean ± SEM (n = 3). ($^{*}P < 0.05$, $^{**}P < 0.01$ vs. Con/control RNAi, $^{\#}P < 0.05$, $^{\#\#}P < 0.01$ vs. OGD/R/control RNAi, $^{\&}P < 0.05$, $^{\&\&}P < 0.01$ vs. OGD/R + MSCs/control RNAi, ns. no significance).

and impaired oxidant/antioxidant balance can be substantially mitigated by OM-MSC treatment (**Figure 7C**).

Possible Mechanism of Protective Effect Induced by OM-MSCs in Mitochondrial and Oxidative Stress

UBIAD1 (aka TERE1) is a newly demonstrated human homolog of *Escherichia coli* prenyltransferase menA. In recent studies, the antioxidative stress role of UBIAD1 has been demonstrated in the cardiovascular system (Mugoni et al., 2013), urological cancer (Fredericks et al., 2013b), and pancreatic acinar cells (Nakagawa et al., 2019). We previously found that UBIAD1 also protects against OGD/R-induced excessive oxidative stress through the PI3K/AKT pathway (Huang and Hu, 2018). In the present study, the expression of UBIAD1 was upregulated by OM-MSC transwell coculture, and the protective effects of the OM-MSC *in vitro* model were abolished when UBIAD1 was specifically knocked down. Thus, we proposed that UBIAD1 is essential in OM-MSC-driven suppression of oxidative injury

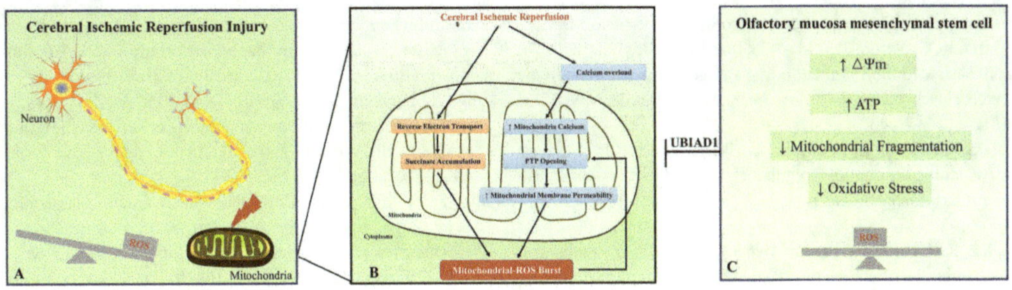

FIGURE 7 | A schematic representation of the proposed mechanism between oxidative stress and mitochondrial dysfunction in cerebral ischemic reperfusion injury. Excessive oxidative stress and mitochondrial dysfunction are fundamental contributors to neuronal injury in cerebral ischemic reperfusion injury **(A)**. Ischemic/reperfusion condition induces calcium overload, which results in calcium influx into mitochondria and mitochondrial permeability transition pore (PTP) opening. Meanwhile, accumulation of intermediate succinate results from reverse electron transport from mitochondrial complex II to complex I. These changes contribute to ROS burst in the neuron. The excessive ROS reversely leads to PTP opening and forms a vicious circle in cerebral ischemic reperfusion injury **(B)**. Our study provides evidence that the impaired oxidant/antioxidant balance and mitochondrial dysfunction could be rescued by OM-MSC treatment **(C)**. Collectively, OM-MSC implantation is a promising approach for the management of cerebral ischemic reperfusion injury.

and improvement of mitochondrial function. However, the mechanism of upregulating UBIAD1 expression by OM-MSC treatment remains in need of further research.

Exosomes derived from stem cells have recently been suggested to have complex functions in cell-to-cell interaction (Chen and Chopp, 2018). Exosomes are comprised of luminal cargo, such as DNA, mRNA, microRNA, long non-coding RNA, proteins, and lipids. These luminal cargoes could be carried by exosomes into the target cells or tissues. Among them, microRNAs are the most investigated in the exosome-mediated intercellular interaction. MicroRNAs are a family of non-coding RNAs of 20–25 nucleotides that modulate the posttranscriptional expression of mRNAs by binding specific seed sequences in the 3′-untranslated region, thereby causing mRNA degeneration, destabilization, and translational inhibition (Lu and Rothenberg, 2018). One microRNA can regulate various different target genes, and one coding gene might be regulated by multiple microRNAs (Miranda et al., 2006). Considering the intercellular interaction role of microRNAs, we further predicted the potential microRNAs targeting UBIAD1 through TargetScan, microRNA, and miRDB software. MicroRNA-4516, microRNA-4478, and microRNA-619-5p are the most potential regulators. Future researches are warranted to explore whether these miRNAs are involved in the neuroprotective effects of OM-MSCs.

AUTHOR CONTRIBUTIONS

JL and YH were responsible for the cellular and animal experiments. JL performed cellular model and transwell coculture and drafted the work. YZ, DD, and WC were responsible for the animal model and preparation of frozen sections of brain tissues. JL and JH performed the Western blot experiments and q-PCR. JL and LG performed the behavioral tests. ML and ZH were responsible for the conception of the work. All authors approved the final version of the manuscript.

CONCLUSION

Our study in an *in vivo* and *in vitro* model of cerebral I/R injury provides evidence that OM-MSCs exert neuroprotective effects by attenuating mitochondrial dysfunction and enhancing antioxidation via upregulation of UBIAD1.

REFERENCES

Alizadeh, R., Bagher, Z., Kamrava, S. K., Falah, M., Ghasemi Hamidabadi, H., Eskandarian Boroujeni, M., et al. (2019a). Differentiation of human mesenchymal stem cells (MSC) to dopaminergic neurons: a comparison between Wharton's Jelly and olfactory mucosa as sources of MSCs. *J. Chem. Neuroanat.* 96, 126–133. doi: 10.1016/j.jchemneu.2019.01.003

Alizadeh, R., Kamrava, S. K., Bagher, Z., Farhadi, M., Falah, M., Moradi, F., et al. (2019b). Human olfactory stem cells: as a promising source of dopaminergic neuron-like cells for treatment of Parkinson's disease. *Neurosci. Lett.* 696, 52–59. doi: 10.1016/j.neulet.2018.12.011

Al-Mufti, F., Amuluru, K., Roth, W., Nuoman, R., El-Ghanem, M., and Meyers, P. M. (2018). Cerebral ischemic reperfusion injury following recanalization of large vessel occlusions. *Neurosurgery* 82, 781–789. doi: 10.1093/neuros/nyx341

Babenko, V. A., Silachev, D. N., Zorova, L. D., Pevzner, I. B., Khutornenko, A. A., Plotnikov, E. Y., et al. (2015). Improving the post-stroke therapeutic potency of mesenchymal multipotent stromal cells by cocultivation with cortical neurons: the role of crosstalk between cells. *Stem Cells Transl. Med.* 4, 1011–1020.

Bederson, J. B., Pitts, L. H., Tsuji, M., Nishimura, M. C., Davis, R. L., and Bartkowski, H. (1986). Rat middle cerebral artery occlusion: evaluation of the model and development of a neurologic examination. *Stroke* 17, 472–476. doi: 10.1161/01.str.17.3.472

Cadenas, E., and Davies, K. J. (2000). Mitochondrial free radical generation, oxidative stress, and aging. *Free Radic. Biol. Med.* 29, 222–230.

Cechetto, D. F., Wilson, J. X., Smith, K. E., Wolski, D., Silver, M. D., and Hachinski, V. C. (1989). Autonomic and myocardial changes in middle cerebral artery occlusion: stroke models in the rat. *Brain Res.* 502, 296–305.

Chance, B., Sies, H., and Boveris, A. (1979). Hydroperoxide metabolism in mammalian organs. *Physiol. Rev.* 59, 527–605. doi: 10.1152/physrev.1979.59.3.527

Chance, B., and Williams, G. R. (1955). A method for the localization of sites for oxidative phosphorylation. *Nature* 176, 250–254. doi: 10.1038/176250a0

Chen, J., and Chopp, M. (2018). Exosome therapy for stroke. *Stroke* 49, 1083–1090. doi: 10.1161/STROKEAHA.117.018292

Chouchani, E. T., Pell, V. R., Gaude, E., Aksentijević, D., Sundier, S. Y., Robb, E. L., et al. (2014). Ischaemic accumulation of succinate controls reperfusion injury through mitochondrial ROS. *Nature* 515, 431–435. doi: 10.1038/nature13909

Christophe, M., and Nicolas, S. (2006). Mitochondria: a target for neuroprotective interventions in cerebral ischemia-reperfusion. *Curr. Pharm. Des.* 12, 739–757. doi: 10.2174/138161206775474242

Di Trapani, M., Bassi, G., Ricciardi, M., Fontana, E., Bifari, F., Pacelli, L., et al. (2013). Comparative study of immune regulatory properties of stem cells derived from different tissues. *Stem Cells Dev.* 22, 2990–3002. doi: 10.1089/scd.2013.0204

Eckert, M. A., Vu, Q., Xie, K., Yu, J., Liao, W., Cramer, S. C., et al. (2013). Evidence for high translational potential of mesenchymal stromal cell therapy to improve recovery from ischemic stroke. *J. Cereb. Blood Flow Metab.* 33, 1322–1334. doi: 10.1038/jcbfm.2013.91

Fan, J. R., Lee, H. T., Lee, W., Lin, C. H., Hsu, C. Y., Hsieh, C. H., et al. (2018). Potential role of CBX7 in regulating pluripotency of adult human pluripotent-like olfactory stem cells in stroke model. *Cell Death Dis.* 9:502.

Fan, X. L., Zhang, Y., Li, X., and Fu, Q. L. (2020). Mechanisms underlying the protective effects of mesenchymal stem cell-based therapy. *Cell. Mol. Life Sci.* 77, 2771–2794.

Fredericks, W. J., McGarvey, T., Wang, H., Zheng, Y., Fredericks, N. J., Yin, H., et al. (2013a). The TERE1 protein interacts with mitochondrial TBL2: regulation of trans-membrane potential, ROS/RNS and SXR target genes. *J. Cell. Biochem.* 114, 2170–2187. doi: 10.1002/jcb.24567

Fredericks, W. J., Yin, H., Lal, P., Puthiyaveettil, R., Malkowicz, S. B., Fredericks, N. J., et al. (2013b). Ectopic expression of the TERE1 (UBIAD1) protein inhibits growth of renal clear cell carcinoma cells: altered metabolic phenotype associated with reactive oxygen species, nitric oxide and SXR target genes involved in cholesterol and lipid metabolism. *Int. J. Oncol.* 43, 638–652. doi: 10.3892/ijo.2013.1985

Friedrich, T., and Böttcher, B. (2004). The gross structure of the respiratory complex I: a Lego system. *Biochim. Biophys. Acta* 1608, 1–9. doi: 10.1016/j.bbabio.2003.10.002

Garbuzova-Davis, S., Haller, E., Lin, R., and Borlongan, C. V. (2017). Intravenously transplanted human bone marrow endothelial progenitor cells engraft within brain capillaries, preserve mitochondrial morphology, and display pinocytotic activity toward blood-brain barrier repair in ischemic stroke rats. *Stem Cells* 35, 1246–1258. doi: 10.1002/stem.2578

GBD 2017 Causes of Death Collaborators (2018). Global, regional, and national age-sex-specific mortality for 282 causes of death in 195 countries and territories, 1980-2017: a systematic analysis for the Global Burden of Disease Study 2017. *Lancet* 392, 1736–1788.

Girard, S. D., Devéze, A., Nivet, E., Gepner, B., Roman, F. S., and Féron, F. (2011). Isolating nasal olfactory stem cells from rodents or humans. *J. Vis. Exp.* 54:2762. doi: 10.3791/2762

Granger, D. N., and Kvietys, P. R. (2015). Reperfusion injury and reactive oxygen species: the evolution of a concept. *Redox Biol.* 6, 524–551. doi: 10.1016/j.redox.2015.08.020

Hacke, W., Kaste, M., Bluhmki, E., Brozman, M., Davalos, A., Guidetti, D., et al. (2008). Thrombolysis with alteplase 3 to 4.5 hours after acute ischemic stroke. *N. Engl. J. Med.* 359, 1317–1329. doi: 10.1056/NEJMoa0804656

Huang, Y., and Hu, Z. (2018). UBIAD1 protects against oxygen-glucose deprivation/reperfusion-induced multiple subcellular organelles injury through PI3K/AKT pathway in N2A cells. *J. Cell. Physiol.* 233, 7480–7496. doi: 10.1002/jcp.26602

Jafari, M., Asghari, A., Delbandi, A. A., Jalessi, M., Jazayeri, M. H., Samarei, R., et al. (2020). Priming TLR3 and TLR4 in human adipose- and olfactory mucosa-derived mesenchymal stromal cells and comparison of their cytokine secretions. *Cytotechnology* 72, 57–68.

Janardhan, V., and Qureshi, A. I. (2004). Mechanisms of ischemic brain injury. *Curr. Cardiol. Rep.* 6, 117–123.

Jang, S., Lewis, T. S., Powers, C., Khuchua, Z., Baines, C. P., Wipf, P., et al. (2017). Elucidating mitochondrial electron transport chain supercomplexes in the heart during ischemia-reperfusion. *Antioxid. Redox Signal.* 27, 57–69. doi: 10.1089/ars.2016.6635

Jiang, D., Gao, F., Zhang, Y., Wong, D. S., Li, Q., Tse, H. F., et al. (2016). Mitochondrial transfer of mesenchymal stem cells effectively protects corneal epithelial cells from mitochondrial damage. *Cell Death Dis.* 7:e2467. doi: 10.1038/cddis.2016.358

Jiang, D., Xiong, G., Feng, H., Zhang, Z., Chen, P., Yan, B., et al. (2019). Donation of mitochondria by iPSC-derived mesenchymal stem cells protects retinal ganglion cells against mitochondrial complex I defect-induced degeneration. *Theranostics* 9, 2395–2410. doi: 10.7150/thno.29422

Jiang, R. H., Wu, C. J., Xu, X. Q., Lu, S. S., Zu, Q. Q., Zhao, L. B., et al. (2018). Hypoxic conditioned medium derived from bone marrow mesenchymal stromal cells protects against ischemic stroke in rats. *J. Cell. Physiol.* 234, 1354–1368. doi: 10.1002/jcp.26931

Jung, J. E., Kim, G. S., Chen, H., Maier, C. M., Narasimhan, P., Song, Y. S., et al. (2010). Reperfusion and neurovascular dysfunction in stroke: from basic mechanisms to potential strategies for neuroprotection. *Mol. Neurobiol.* 41, 172–179. doi: 10.1007/s12035-010-8102-z

Kholodenko, I. V., Yarygin, K. N., Gubsky, L. V., Konieva, A. A., Tairova, R. T., Povarova, O. V., et al. (2012). Intravenous xenotransplantation of human placental mesenchymal stem cells to rats: comparative analysis of homing in rat brain in two models of experimental ischemic stroke. *Bull. Exp. Biol. Med.* 154, 118–123.

Li, X., Zhang, Y., Yeung, S. C., Liang, Y., Liang, X., Ding, Y., et al. (2014). Mitochondrial transfer of induced pluripotent stem cell-derived mesenchymal stem cells to airway epithelial cells attenuates cigarette smoke-induced damage. *Am. J. Respir. Cell Mol. Biol.* 51, 455–465. doi: 10.1165/rcmb.2013-0529OC

Liang, X., Ding, Y., Zhang, Y., Tse, H. F., and Lian, Q. (2014). Paracrine mechanisms of mesenchymal stem cell-based therapy: current status and perspectives. *Cell Transplant.* 23, 1045–1059. doi: 10.3727/096368913x667709

Lindsay, S. L., Riddell, J. S., and Barnett, S. C. (2010). Olfactory mucosa for transplant-mediated repair: a complex tissue for a complex injury? *Glia* 58, 125–134. doi: 10.1002/glia.20917

Liu, H., Honmou, O., Harada, K., Nakamura, K., Houkin, K., Hamada, H., et al. (2006). Neuroprotection by PlGF gene-modified human mesenchymal stem cells after cerebral ischaemia. *Brain* 129(Pt 10), 2734–2745.

Lu, T. X., and Rothenberg, M. E. (2018). MicroRNA. *J. Allergy Clin. Immunol.* 141, 1202–1207. doi: 10.1016/j.jaci.2017.08.034

Mahrouf-Yorgov, M., Augeul, L., Da Silva, C. C., Jourdan, M., Rigolet, M., Manin, S., et al. (2017). Mesenchymal stem cells sense mitochondria released from damaged cells as danger signals to activate their rescue properties. *Cell Death Differ.* 24, 1224–1238. doi: 10.1038/cdd.2017.51

Marchi, S., Giorgi, C., Suski, J. M., Agnoletto, C., Bononi, A., Bonora, M., et al. (2012). Mitochondria-ros crosstalk in the control of cell death and aging. *J. Signal Transduct.* 2012:329635. doi: 10.1155/2012/329635

Marei, H. E., Hasan, A., Rizzi, R., Althani, A., Afifi, N., Cenciarelli, C., et al. (2018). Potential of stem cell-based therapy for ischemic stroke. *Front. Neurol.* 9:34. doi: 10.3389/fneur.2018.00034

Miranda, K. C., Huynh, T., Tay, Y., Ang, Y. S., Tam, W. L., Thomson, A. M., et al. (2006). A pattern-based method for the identification of MicroRNA binding sites and their corresponding heteroduplexes. *Cell* 126, 1203–1217. doi: 10.1016/j.cell.2006.07.031

Morales, C. R., Grigoryeva, L. S., Pan, X., Bruno, L., Hickson, G., Ngo, M. H., et al. (2014). Mitochondrial damage and cholesterol storage in human hepatocellular carcinoma cells with silencing of UBIAD1 gene expression. *Mol. Genet. Metab. Rep.* 1, 407–411. doi: 10.1016/j.ymgmr.2014.09.001

Morrison, T. J., Jackson, M. V., Cunningham, E. K., Kissenpfennig, A., McAuley, D. F., O'Kane, C. M., et al. (2017). Mesenchymal stromal cells modulate macrophages in clinically relevant lung injury models by extracellular vesicle mitochondrial transfer. *Am. J. Respir. Crit. Care Med.* 196, 1275–1286. doi: 10.1164/rccm.201701-0170OC

Mugoni, V., Postel, R., Catanzaro, V., De Luca, E., Turco, E., Digilio, G., et al. (2013). Ubiad1 is an antioxidant enzyme that regulates eNOS activity by CoQ10 synthesis. *Cell* 152, 504–518. doi: 10.1016/j.cell.2013.01.013

Nakagawa, K., Fujiwara, K., Nishimura, A., Murakami, C., Kawamoto, K., Ichinose, C., et al. (2019). UBIAD1 plays an essential role in the survival of pancreatic acinar cells. *Int. J. Mol. Sci.* 20:1971. doi: 10.3390/ijms20081971

Nickerson, M. L., Kostiha, B. N., Brandt, W., Fredericks, W., Xu, K. P., Yu, F. S., et al. (2010). UBIAD1 mutation alters a mitochondrial prenyltransferase to cause Schnyder corneal dystrophy. *PLoS One* 5:e10760. doi: 10.1371/journal.pone.0010760

Nivet, E., Vignes, M., Girard, S. D., Pierrisnard, C., Baril, N., Deveze, A., et al. (2011). Engraftment of human nasal olfactory stem cells restores neuroplasticity in mice with hippocampal lesions. *J. Clin. Invest.* 121, 2808–2820. doi: 10.1172/jci44489

Paliwal, S., Chaudhuri, R., Agrawal, A., and Mohanty, S. (2018). Human tissue-specific MSCs demonstrate differential mitochondria transfer abilities that may determine their regenerative abilities. *Stem Cell Res. Ther.* 9:298.

Pandit, S. R., Sullivan, J. M., Egger, V., Borecki, A. A., and Oleskevich, S. (2011). Functional effects of adult human olfactory stem cells on early-onset sensorineural hearing loss. *Stem Cells* 29, 670–677. doi: 10.1002/stem.609

Park, H. W., Moon, H. E., Kim, H. S., Paek, S. L., Kim, Y., Chang, J. W., et al. (2015). Human umbilical cord blood-derived mesenchymal stem cells improve functional recovery through thrombospondin1, pantraxin3, and vascular endothelial growth factor in the ischemic rat brain. *J. Neurosci. Res.* 93, 1814–1825. doi: 10.1002/jnr.23616

Perico, L., Morigi, M., Rota, C., Breno, M., Mele, C., Noris, M., et al. (2017). Human mesenchymal stromal cells transplanted into mice stimulate renal tubular cells and enhance mitochondrial function. *Nat. Commun.* 8:983.

Reddy, P. V., Rao, K. V., and Norenberg, M. D. (2008). The mitochondrial permeability transition, and oxidative and nitrosative stress in the mechanism of copper toxicity in cultured neurons and astrocytes. *Lab. Invest.* 88, 816–830. doi: 10.1038/labinvest.2008.49

Rodrigo, R., Fernandez-Gajardo, R., Gutierrez, R., Matamala, J. M., Carrasco, R., Miranda-Merchak, A., et al. (2013). Oxidative stress and pathophysiology of ischemic stroke: novel therapeutic opportunities. *CNS Neurol. Disord. Drug Targets* 12, 698–714. doi: 10.2174/1871527311312050015

Rui, K., Lin, X., Tian, J., Wang, X., Sun, L., Hong, X., et al. (2018). Ecto-mesenchymal stem cells: a new player for immune regulation and cell therapy. *Cell. Mol. Immunol.* 15, 82–84. doi: 10.1038/cmi.2017.69

Rui, K., Zhang, Z., Tian, J., Lin, X., Wang, X., Ma, J., et al. (2016). Olfactory ecto-mesenchymal stem cells possess immunoregulatory function and suppress autoimmune arthritis. *Cell. Mol. Immunol.* 13, 401–408. doi: 10.1038/cmi.2015.82

Simorgh, S., Alizadeh, R., Eftekharzadeh, M., Haramshahi, S. M. A., Milan, P. B., Doshmanziari, M., et al. (2019). Olfactory mucosa stem cells: an available candidate for the treatment of the Parkinson's disease. *J. Cell. Physiol.* 234, 23763–23773. doi: 10.1002/jcp.28944

Sugawara, T., and Chan, P. H. (2003). Reactive oxygen radicals and pathogenesis of neuronal death after cerebral ischemia. *Antioxid. Redox Signal.* 5, 597–607. doi: 10.1089/152308603770310266

Tang, J., Hu, Z., Tan, J., Yang, S., and Zeng, L. (2016). Parkin protects against oxygen-glucose deprivation/reperfusion insult by promoting Drp1 degradation. *Oxid. Med. Cell. Longev.* 2016:8474303. doi: 10.1155/2016/8474303

Tsivgoulis, G., Saqqur, M., Sharma, V. K., Brunser, A., Eggers, J., Mikulik, R., et al. (2020). Timing of recanalization and functional recovery in acute ischemic stroke. *J. Stroke* 22, 130–140. doi: 10.5853/jos.2019.01648

Veron, A. D., Bienboire-Frosini, C., Girard, S. D., Sadelli, K., Stamegna, J. C., Khrestchatisky, M., et al. (2018). Syngeneic transplantation of olfactory ectomesenchymal stem cells restores learning and memory abilities in a rat model of global cerebral ischemia. *Stem Cells Int.* 2018:2683969. doi: 10.1155/2018/2683969

Virani, S. S., Alonso, A., Benjamin, E. J., Bittencourt, M. S., Callaway, C. W., Carson, A. P., et al. (2020). Heart disease and stroke statistics-2020 update: a report from the American heart association. *Circulation* 141, e139–e596. doi: 10.1161/cir.0000000000000757

Yao, Y., Fan, X. L., Jiang, D., Zhang, Y., Li, X., Xu, Z. B., et al. (2018). Connexin 43-mediated mitochondrial transfer of iPSC-MSCs alleviates asthma inflammation. *Stem Cell Rep.* 11, 1120–1135. doi: 10.1016/j.stemcr.2018.09.012

Young, E., Westerberg, B., Yanai, A., and Gregory-Evans, K. (2018). The olfactory mucosa: a potential source of stem cells for hearing regeneration. *Regen. Med.* 13, 581–593.

Zhang, Y., Yu, Z., Jiang, D., Liang, X., Liao, S., Zhang, Z., et al. (2016). iPSC-MSCs with high intrinsic MIRO1 and sensitivity to TNF-alpha yield efficacious mitochondrial transfer to rescue anthracycline-induced cardiomyopathy. *Stem Cell Rep.* 7, 749–763. doi: 10.1016/j.stemcr.2016.08.009

Zhao, Y., Yan, F., Yin, J., Pan, R., Shi, W., Qi, Z., et al. (2018). Synergistic interaction between zinc and reactive oxygen species amplifies ischemic brain injury in rats. *Stroke* 49, 2200–2210. doi: 10.1161/strokeaha.118.021179

Zhou, F., Gao, S., Wang, L., Sun, C., Chen, L., Yuan, P., et al. (2015). Human adipose-derived stem cells partially rescue the stroke syndromes by promoting spatial learning and memory in mouse middle cerebral artery occlusion model. *Stem Cell Res. Ther.* 6:92.

Zorov, D. B., Juhaszova, M., and Sollott, S. J. (2006). Mitochondrial ROS-induced ROS release: an update and review. *Biochim. Biophys. Acta* 1757, 509–517. doi: 10.1016/j.bbabio.2006.04.029

Zuo, L., Feng, Q., Han, Y., Chen, M., Guo, M., Liu, Z., et al. (2019). Therapeutic effect on experimental acute cerebral infarction is enhanced after nanoceria labeling of human umbilical cord mesenchymal stem cells. *Ther. Adv. Neurol. Disord.* 12:1756286419859725. doi: 10.1177/1756286419859725

8

Advances of Endothelial Progenitor Cells in the Development of Depression

Nana Yang[1,2†], Shiyu Sun[1†], Guangqing Duan[3], Kaixuan Lv[1], Chen Liang[4], Linlin Zhang[1], Jielun Yu[1,2], Yaohui Tang[5†] and Guohua Lu[6*†]*

[1] School of Bioscience and Technology, Weifang Medical University, Weifang, China, [2] Medical Laboratory Animal Center, Weifang Medical University, Weifang, China, [3] School of Clinical Medicine, Weifang Medical University, Weifang, China, [4] School of Chemical Engineering, Qingdao University of Science & Technology, Qingdao, China, [5] Med-X Research Institute and School of Biomedical Engineering, Shanghai Jiao Tong University, Shanghai, China, [6] School of Psychology, Weifang Medical University, Weifang, China

***Correspondence:**
Yaohui Tang
yaohuitang@sjtu.edu.cn
Guohua Lu
ghluu@126.com

† These authors have contributed equally to this work

Depression is a major psychological disease of human beings. With the severity of depression, it elevates the risk of cardiovascular disease (CVD), especially acute coronary syndrome (ACS), resulting in serious harm to human health. The number of endothelial progenitor cells (EPCs) is closely related to the development of depression. It has been reported that the number of peripheral blood EPCs in patients with depression was reduced. However, effects on the function of EPCs in depression are still unclear. This paper aims to analyze and summarize the research of EPCs in depression, and we envision that EPCs might act as a new target for evaluating the severity of depression and its complications.

Keywords: endothelial progenitor cells, depression, psychological disease, cardiovascular disease, TNF-α

INTRODUCTION

Depression is one of the most common psychological disorders which acts as the primary risk factor for suicide and cardiovascular disease (Shi et al., 2017). As a common disease detrimental to the physical and mental health of human beings, depression is prominently manifested as depressive symptoms, such as reduced activity, memory loss, fatigue, sleep disorders, self-abandonment and even suicide (Malhi and Mann, 2018). Data shows that people aged 55–64 have the highest incidence of depression. In addition, depression is associated with educational level and unhealthy living habits which also serve as an important cause of CVD (Anda et al., 1993; Klakk et al., 2018). Therefore, effective means of evaluating or treating depression have become the main goal of depression research.

EPCs, also known as a precursor of vascular endothelial cells, have the ability to proliferate, migrate, differentiate and form new blood vessels *in vivo* (Kaushik and Das, 2019). Under the stimulation of physiological or pathological factors, EPCs can be migrated from bone marrow to peripheral blood and participate in the repair of damaged blood vessels. Moreover, the proliferation and differentiation of EPCs are vital to tumor angiogenesis. Recently scientists proposed a certain correlation between EPCs and depression (Dome et al., 2009; Di Stefano et al., 2014; Felice et al., 2015). It becomes the most eye-catching whether EPCs can serve as an indicator for the severity of depression.

CHARACTERISTICS OF EPCS

Asahara et al. (1997) discovered precursor cells capable of differentiating into vascular endothelial cells in circulating peripheral blood and named them as vascular EPCs for the first time in 1997. Since then, the characteristics, biological functions and potential therapeutic application of EPC have become hot topics in the field. Functional assays and surface marker-based molecular definition are two essential ways to characterize EPCs. Numerous assays were used to evaluate self-renewal capacity and potency of EPCs, which are crucial to functionally define and classify EPCs (Patel et al., 2016). In addition, it was found that these cells during *in vitro* culture could co-express vascular endothelial growth factor receptor2 (VEGFR-2) and a large number of cell surface markers including CD34, CD133 and von Willebrand factor (VWF), and have the potential to differentiate into mature endothelial cells. Afterward, $CD133^+/CD34^+/VEGFR2^+/CD45^-$ was usually used as the surface marker of EPCs (Pulito-Cueto et al., 2020). The discovery of cell surface markers played an important role in the investigation of human cardiovascular and cerebrovascular diseases and malignant tumors (Chong et al., 2016).

EPCs are usually classified into two types named early EPCs (eEPCs) and late EPCs (lEPCs) (Kou et al., 2020). eEPCs appeared 3–5 days after the onset of *in vitro* colony-forming assays from peripheral blood cells while 2–3 weeks needed for lEPCs (Yuan et al., 2017). Besides the difference on culturing time, cell origin, surface markers, biological functions and further applications are quite distinctive between these two subpopulations. In brief, eEPCs express typical hematopoietic marker CD133 and CD45, monocyte marker CD14 and also endothelial marker like VEGFR2, but negative for CD34, which are considered to have low proliferation and colony-forming abilities but could release numerous cytokines (Ormiston et al., 2015). In contrast, lEPCs are lack of hematopoietic markers CD133 and CD45, but positive for CD34 and VEGFR2 (Yoon et al., 2005). With the high proliferative potential and colony-forming ability, lEPCs could promote vessel formation through angiogenetic pathway (Prasain et al., 2012). EPCs are heterogeneous populations *in vivo*. Thus, it is critical to properly classify EPCs while analyzing them in disease models.

EPCs are rare in peripheral blood but relatively abundant in bone marrow in healthy people. Under a pathological state, EPCs can be migrated from bone marrow to peripheral blood, and promote angiogenesis. For example, in the case of tumorigenesis, tumor tissues can secrete VEGF, SDF-1, G-CSF, GM-CSF and other cytokines and chemokines which mobilize EPCs from bone marrow to tumor site, and participate in tumor angiogenesis (Rana et al., 2018). The aforementioned recruitment of EPCs can also be promoted by some stimulants, such as fibroblast growth factor, epidermal growth factor, and estrogen (Capillo et al., 2003; Vega et al., 2017). Some diseases can be treated or alleviated by changing the number and function of EPCs. For example, EPCs can be employed to repair the heart of patients with myocardial infarction (Bianconi et al., 2018) and improve lung function (Li et al., 2013; Salter and Sehmi, 2017) as well as the angiogenic ability of patients with diabetes (Altabas, 2015). Tagawa et al. (2015) and Morishita et al. (2016) found the number EPCs could be used as an important index to predict the incidence of CVD. EPCs possess a broad application prospect in investigating angiogenesis and cellular treatment of CVD due to their autologous isolation, amplification and transplantation, which is free of rejection reaction.

EFFECT OF DEPRESSION ON THE QUANTITY OF EPCS

The quantity of EPCs is an important index to predict the occurrence and development of various diseases, so what changes will happen to EPCs in patients with depression. Can the quantitative changes in EPCs be a predictor for depression. And how depression affects the quantity of EPCs. Studies of the effect of depression on the number of EPCs are shown in **Table 1**. A total of 10 articles on endothelial progenitor cells and depression were searched in national center for biotechnology information (NCBI). Four papers are associated with depression and the number of EPCs, three articles are related with EPCs numbers after treatment with drugs of depression, and three papers show the correlation between depression, cardiovascular disease and the number of EPCs.

Dome et al. (2009) reported significant lower quantities of mature EPCs ($CD34^+/VEGFR2^+$) and immature EPCs ($CD133^+/VEGFR2^+$) in the peripheral blood of patients with depression than those of control by flow cytometry. Since smoking could lead to significantly reduced quantity of circulating EPCs, the smoking habits of patients and control were matched in the study, demonstrating significant lower levels of mature EPCs in smokers than those in non-smokers from both control and patient groups. Although significantly decreased number of EPCs has been observed in patients with depression, the underlying mechanism still needs further elucidation. That may involve in reduced recruitment as well as the capabilities of survival and differentiation of EPCs. Chen H. et al. (2011) not only investigated the changes of EPCs numbers in peripheral blood but also the vascular function and mental state of patients with depression. Their data showed no significant difference in age, gender, hypercholesterolemia prevalence, smoking, systolic blood pressure, diastolic blood pressure, BMI and serum total cholesterol between the patients with higher and normal depression scale scores. Patients from these two groups received similar medications of antidepressants, antihypertensives and statins. Compared with normal depression score (DS), patients with high DS exhibited significantly lower brachial artery flow-mediated dilation (FMD) and the percentage of $CD34^+/KDR^+$ EPCs, with no significant difference in the percentage of circulating $CD133^+/KDR^+$ EPCs ($P > 0.05$). Yang et al. (2011) also investigated that levels of circulating $CD34^+CD133^+KDR^+$ EPCs and endothelial colony-forming units (CFUs) in patients with depression were lower than that of healthy subjects. EPCs significantly decreased, while the development of depressive symptoms had no significant relationship with changes in EPCs through the year of internship stress (Felice et al., 2015). Altogether, these results indicate

TABLE 1 | Studies of the effect of depression on the number of EPCs.

EPCs markers	Methods of calculating the number of EPCs	Key fingdings	References
$CD34^+/VEGFR2^+$ $CD133^+/VEGFR2^+$	The number of cells per milliliter of blood	The number of mature ($CD34^+/VEGFR2^+$) and immature ($CD133^+/VEGFR2^+$) EPCs were significantly decreased in depression patients, and EPCs levels was significant inverse relationship with the severity of depressive symptoms.	Dome et al., 2009
$CD34^+/VEGFR2^+$ $CD133^+/VEGFR2^+$	The number of EPCs was expressed as absolute EPCs counts divided by the lymphocyte counts.	The percentage of circulating $CD34^+/KDR^+$ EPCs was lower in high depression score than in normal depression score. And the percentage of circulating $CD133^+/KDR^+$ EPCs was no different between two groups.	Chen H. et al., 2011
$CD34^+/KDR^+/CD133^+$	Absolute number of cells per ml of blood	Levels of circulating $CD34^+CD133^+KDR^+$ EPCs and endothelial colony-forming units in patients with depression were lower than that of healthy subjects.	Yang et al., 2011
$CD34^+/VEGFR2^+$	The number of EPCs was quantified as the number of these cells per 10^6 lymphocyte.	There was no significant alteration in CEPCs levels in the course of recovery from major depression.	Dome et al., 2012
$CD34^+/VEGFR2^+$ $CD133^+/VEGFR2^+$	The number of EPCs was expressed as absolute EPCs counts divided by the lymphocyte counts.	In stable angina patients, percentage of circulating $CD34^+/VEGFR2^+$ EPCs and artery flow-mediated dilation in Subjects with high depression or stress score were significantly lower than that in subjects with normal depression or stress score.	Chen et al., 2013
$CD34^+/KDR^+/CD133^+$	Absolute number of cells per ml of blood	The number of EPCs in acute coronary syndrome with major depressive episode showed significant decrease compared with that in acute coronary syndrome without major depressive episode.	Di Stefano et al., 2014
$CD34^+/KDR^+/CD133^+$	Absolute number of cells per ml of blood	Circulating $CD34^+CD133^+KDR^+$ EPCs levels in acute coronary syndromes with affective disorders was significantly lower than that in acute coronary syndromes without affective disorder.	Felice et al., 2015
$CD34^+/CD133^+$	Absolute number of cells per ml of blood	EPCs significantly decreased with the year of intership stress, while the development of depressive symptoms had no significant relationship with changes in EPCs.	Fiedorowicz et al., 2015
$CD45^-/CD146^+/CD31^+$ $CD45^-CD34^+/KDR^+$	Absolute number of cells per ml of blood	CECs ($CD45^-/CD146^+/CD31^+$) counts, soluble VWF and VCAM-1 were statistically increased in diagnosis (MD-0) and gradually decreased during treatment. Conversely, EPCs ($CD45^-/CD34^+/KDR^+$) levels were lower in MD-0, tending to increase throughout treatment.	Lopez-Vilchez et al., 2016
$CD133^+/VEGFR2^+$	the mononuclear cells (lymphocytes, monocytes, and blasts) were set to gates, then the percentage of $CD133^+/VEGFR2^+$ cells in the mononuclear cells was acquired.	CEPCs levels in blood had no significant difference in chronic mild stress (CMS) group, high-fat diet group, high-fat diet with CMS group, and the group of imipramine and pentoxifylline treatment. Chronic pentoxifylline treatment was more effective in increase $CD133^+$ and $VEGFR2^+$ cells in rat thoracic aortae.	Labib et al., 2019

that EPCs numbers in peripheral blood were significantly more in healthy subjects than in subjects with depression, and patients with high DS exhibited significantly lower percentage of $CD34^+/KDR^+$ EPCs compared with normal DS, although the methods of calculating EPCs numbers were different in above studies. Large sample of clinical studies on the correlation between depression and EPCs numbers need to be further studied through consistent method of calculating EPCs.

In order to further clarify the changes of EPCs during the treatment of depression, Dome et al. (2012) evaluated the changes of EPCs in peripheral blood during the recovery of the patients. After receiving antidepressant treatment, the severity of patients with depression was significantly improved through the measurement of Montgomery as berg expression rating scale (MADRS), with improved level of cholesterol but no significant changes in the quantity of EPCs. Recent findings showed that CECs ($CD45^-/CD146^+/CD31^+$) counts, soluble VWF and VCAM-1 were statistically increased in diagnosis (MD-0) and gradually decreased during the selective serotonin reuptake inhibitor escitalopram treatment. Conversely, EPCs ($CD45^-/CD34^+/KDR^+$) levels were lower in MD-0, tending to increase throughout escitalopram treatment. However, the increase level did not reach statistical significance after 24 weeks of antidepressant treatment (Lopez-Vilchez et al., 2016). Sera from patients with depression damage the endothelial cells *in vitro*, such as increased level of intercellular adhesion molecule-1 (ICAM-1), lower level of endothelial nitric oxide synthase (eNOS) and higher reactive oxygen species (ROS) production. There is no damage to endothelial cells in the serum of the depressed patients with escitalopram treatment for 24 weeks. The methods of calculating EPCs numbers and anti-depression treatments were different in above both studies. CEPCs levels in blood had no significant difference in chronic mild stress (CMS) group, high-fat diet group, high-fat diet with CMS group, and the group of imipramine and pentoxifylline treatment. Chronic pentoxifylline treatment was more effective in increase CD133$^+$ and VEGFR2$^+$ cells in rat thoracic aortae (Labib et al., 2019).

The percentage of CD34$^+$/KDR$^+$EPCs (0.029%) in lymphocytes was very low (Vasa et al., 2001), thus the level of EPCs in peripheral blood was difficult to measure and the measurement methods were different. The number of EPCs was quantified as the number of these cells per 10^6 lymphocytes in Dome et al. (2012) research, while calculating EPCs numbers in Lopez-Vilchez et al. (2016) experiment was used absolute number of cells per ml of blood. This may be the reason why the results of the above two clinical experiments are different. In addition, at least 10^6 mononuclear cells in peripherals blood were collected in other study, and EPCs numbers were reported as a percentage of mononuclear cells (Wang et al., 2020). Because of low level of EPCs in peripheral, direct count following with ISHAGE gating would be the promising way.

THE MECHANISM OF DEPRESSION AFFECTING THE QUANTITY AND FUNCTION OF EPCS

Depression May Affect the Quantity and Function of EPCs by Affecting Levels of Inflammatory Mediators

Depression has sustained a series of inflammatory state, and promoted the increase of inflammatory markers, such as tumor necrosis factor-α (TNF-α), C-reactive protein (CRP) and interleukin 6 (IL-6) (Nukina et al., 2001; Pikhart et al., 2009; Liu et al., 2012). Miller et al. (2005) proposed the participation of tumor necrosis factor-α (TNF-α) in the pathogenesis of depression. TNF-α is a multifunctional cytokine that can directly kill tumor cells and can be up-regulated in the case of depression, leading to functional decline in the peripheral immune system (Liu et al., 2012). Afterward, lots of researchers were performed to investigate whether there was a correlation between depression and TNF-α level. Haapakoski et al. (2015) and Ma et al. (2016) found higher TNF-α levels in the peripheral blood of patients with depression than that of control. Similar results were also reported by Fan et al. (2017) from a study with 64 depression patients and 80 healthy controls. Elevated mRNA and protein levels of TNF-α were found in patients with recurrent depressive disorder (RDD), compared with control (Bobińska et al., 2017). Therefore, it could be concluded that the level of TNF-α in the peripheral blood of patients with depression is higher than that of control.

Several studies have reported that the concentration of CRP and IL-6 was higher in depressed patients than in healthy controls. Depression score was positively related to the levels of CRP in a linear manner. After controlling for confounders, the levels of CRP in subjects with depression were higher than that in healthy controls (Pikhart et al., 2009). Another study showed that the levels of fasting CRP were significantly increased in remitted women with major depressive disorder (MDD) versus controls (Kling et al., 2007). High levels of IL-6 in childhood are associated with an increased risk of depression and psychotic experiences (PEs) in a dose-dependent manner in young adulthood (Khandaker et al., 2014). In depression animal models, studies found that restraint stress stimulated the increased levels of IL-6 (Nukina et al., 2001), and administration of LPS or recombinant IL-6 induced depressive-like behaviors (Dantzer et al., 2008; Hayley et al., 2008; Fu et al., 2010; Sukoff Rizzo et al., 2012). Chourbaji et al. (2006) indicated that *IL-6*-deficient mice were resistant to the stress induced by the development of a depressive-like behaviors.

As a cytokine with many biological effects, TNF-α can reduce the number of EPCs in peripheral blood and reduce the function of EPCs. Chen T. G. et al. (2011) investigated the effects of TNF-α (0, 10, 20, 50, and 100 mg/L, respectively) on the proliferation, migration and adhesion of EPCs isolated from human peripheral blood. The results demonstrated that TNF-α significantly reduced the quantity, capabilities of proliferation, migration and adhesion of EPCs, with a negative correlation between the quantity and function of EPCs and concentration of TNF-α. TNF-α could reduce the number of EPCs in peripheral blood of mice in high fat diet mice, and inhibit the proliferation, migration and angiogenic function of EPCs, which could be alleviated by ApoAI analog peptide of Rev-D-4F (Nana et al., 2015). Likewise, CRP significantly inhibited EPCs migration, adhesiveness and proliferation through receptors for advanced glycation end products (RAGE) (Chen et al., 2012), and induced EPCs apoptosis and necrosis (Fujii et al., 2006). Although IL-6 enhances EPCs migration, proliferation, and differentiation (Fan et al., 2008), bone marrow CD34$^+$ cell levels were inversely associated with the inflammatory marker IL-6 in critical limb ischemia patients (Teraa et al., 2013), and the number of EPCs in peripheral blood was negatively correlated with IL-6 levels in rheumatoid arthritis patients (Herbrig et al., 2006).

Despite the decreased number of EPCs and endothelial CFUs in peripheral blood in patients with depression (Yang et al., 2011; Blum et al., 2017), no relevant report on their proliferation, migration and adhesion in patients with depression has been reported to the best of our knowledge. In subjects with depression, the level of TNF-α (Dome et al., 2009; Yang et al., 2011; Labib et al., 2019) and IL-6 (Yang et al., 2011) had a negative correlation with the number of EPC. And there are no relevant studies on whether these inflammatory factors affect endothelial progenitor cell functions in subjects with depression. In consideration of elevated TNF-α in the peripheral blood of patients with depression and its detrimental effects on the proliferation, migration, adhesion of EPCs and angiogenesis, we speculate that depression may affect the function of EPCs by regulating the levels of TNF-α and some inflammatory

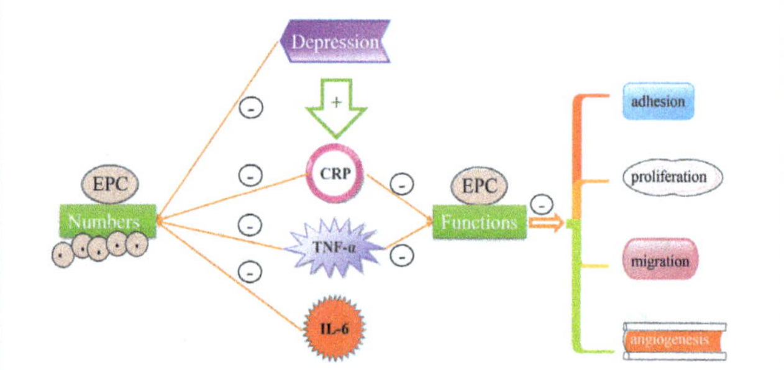

FIGURE 1 | The effect of depression on EPCs numbers and functions. The level of TNF-α, CRP, and IL-6 in peripheral blood of patients with depression are higher than that of healthy subjects, and the level of TNF-α, CRP, and IL-6 have a negative correlation with EPCs numbers. TNF-α and CRP can inhibit the functions of EPCs, such as adhesion, proliferation, migration and angiogenesis *in vitro* study. Whether depression can affect the above functions of EPCs through TNF-α or other inflammatory factors will require further studies.

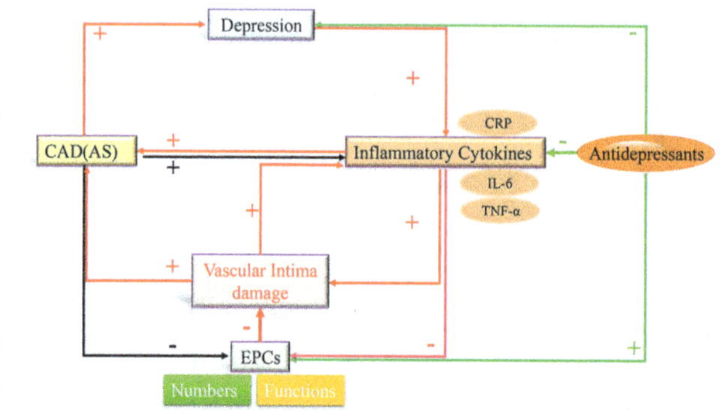

FIGURE 2 | Inflammatory cytokines, depression, CAD, and EPCs dysfunction generate a positive feedback loop. Depression increases the levels of circulating inflammatory cytokines, such as CRP, TNF-α, IL-6 etc., which damage vascular intima, decrease the number of EPCs, and inhibit the functions of EPCs. Dysfunction of EPCs cannot repair the impaired vascular intima, which further stimulates the release of inflammatory factors and induces the development of atherosclerotic cardiovascular. Antidepressants, that inhibit inflammatory cytokines, may reduce the risk of mortality from CAD through improving EPCs functions and repairing intima damage.

mediators in the peripheral blood of patients with depression (see **Figure 1**).

The Occurrence of Depression Complications May Affect the Quantity and Function of EPCs

Depression can lead to many complications such as CVD and diabetes (Ariyo et al., 2000; Bădescu et al., 2016), and McClung et al. (2005) and Aragona et al. (2016) found that these complications can affect the quantity and function of EPCs. Therefore, the direct or indirect effects of depression on the quantity and function of EPCs are worth further investigations. Ariyo et al. (2000) proposed that depression increased the probability of patients to suffer from cardiovascular disease and performed a 6 years' follow up of 4,439 patients with different degrees of depression. It was found that the incidence of coronary heart disease (CAD) was positively correlated with the severity of depression after 6 years, with every 5 scores increase in depression MADRS evaluation corresponding to increased incidence of CAD by 15 % and females exhibiting high incidence than males. In comparison with control, it can be found significantly decreased quantity of EPCs in patients with CAD and the changes in quantity of EPCs serves as an important index to predict the occurrence and development of cardiovascular diseases in the future (McClung et al., 2005; Di Stefano et al., 2014; Felice et al., 2015). Schmidt-Lucke et al. (2019) studied 120 patients consisting of 43 healthy controls, 44 patients with CAD, and 33 patients with acute coronary syndromes (ACS). Their results presented significantly lower level of EPCs in peripheral blood of patients with CAD compared with control, with no significant difference in patients with ACS, however.

Depression increases the probability of cardiovascular complications, and the quantities of EPCs in peripheral blood of patients with CAD are significantly reduced. The changes in quantities of EPCs in patients with depression may be related to its capability of inducing CAD. In stable angina patients, the percentage of circulating $CD34^+/VEGFR2^+$ EPCs and artery flow-mediated dilation in Subjects with high depression or stress score were significantly lower than that in subjects with normal depression or stress scores (Chen et al., 2013). Di Stefano et al. (2014) investigated that the number of EPCs in ACS with major depressive episodes showed significant decrease compared with that in ACS without major depressive episodes. And then Felice et al. (2015) found that circulating $CD34^+CD133^+KDR^+$ EPCs levels in ACS with affective disorders were significantly lower than that in ACS s without the affective disorder.

In the above studies, there was no study on EPCs numbers between depression group and depression subjects with cardiovascular diseases. Whether the number and functions of EPCs predict susceptibility to depression in patients with cardiovascular disease or predicts susceptibility to cardiovascular disease in patients with depression, requires an extensive large sample of clinical research.

We summarized above that EPCs were involved in the development of depression. In addition to depression itself, there are also many pathological conditions such as aging and inflammatory diseases associated with the development of depression. Because of special plasticity of $CD34^+$ cells, $CD34^+$ cells often indicate a distinct subset of cells with progenitor activity (Sidney et al., 2014). Compared with double- or triple-staining in detecting cardiovascular risk, circulating $CD34^+$ cells showed more associated with cardiovascular parameters (Fadini et al., 2006). Mandraffino et al. (2012) observed 100 octogenarians for 7 years, $CD34^+$ cells play an important role in predicting mortality in the elderly. Further studies confirm that the lower levels of circulating $CD34^+$ cells are correlated with increased all causes of deaths, including cardiovascular deaths (Mandraffino et al., 2017). Thus circulating $CD34^+$ cells may be as a marker of health. A total of 4,493 participants without cardiovascular disease were followed for 6 years for the development of CHD and mortality, and results showed that depressive symptoms could be an independent risk factor for CHD development and total mortality (Ariyo et al., 2000). Whether circulating $CD34^+$ cells are associated with depression would be further investigated. A prospective cohort of elderly participants would be needed to be enrolled in mental health study to testify that circulating $CD34^+$ cells may be a marker of mental health and longevity.

There are close correlations among inflammatory factors, depression, CAD, and EPCs, Labib et al. (2019) speculated that pro-inflammatory cytokines-induced dysfunction of circulating EPCs could establish links between depression and atherosclerotic cardiovascular disease. Based on the above conclusions, we propose the hypothesis that depression increases the levels of circulating inflammatory cytokines, such as CRP, TNF-α, IL-6 etc., which damage vascular intima, decrease the number of EPCs, and inhibit the function of EPCs. Dysfunction of EPCs cannot repair the impaired vascular intima, which further stimulates the release of inflammatory factors and induces the development of atherosclerotic cardiovascular and cerebrovascular diseases. Consequently, the development of atherosclerosis and persistent inflammation further aggravate EPCs dysfunction and depression progress. Therefore, inflammatory factors, depression, CAD, and EPCs dysfunction generate a positive feedback loop. Improvement of EPCs numbers and functions may repair the impaired vascular intima, and inhibited the progress of the above positive feedback loop. Antidepressants, that inhibit inflammatory cytokines, may reduce the risk of mortality from CAD through improving EPCs functions and repairing intima damage (see **Figure 2**).

CONCLUSION

Through studying the relationship between EPCs and depression, it was found that the quantity of EPCs was negatively correlated with the severity of depression. And their quantity can be used as an important indicator to predict the occurrence and development of CAD (Ariyo et al., 2000; Werner et al., 2005; Schmidt-Lucke et al., 2019). In the future, EPCs may also serve as an indicator to predict the severity of CAD in depression patients and target for depression patients with CAD. In addition, whether depression affects the quantity and function of EPCs through some specific inflammatory mediators or other diseases such as cardiovascular disease, and then affects the physiology and psychology of human body needs to be further clarified. These unknown factors may become an important research direction for depression targeted prediction and treatment.

AUTHOR CONTRIBUTIONS

YT and GL designed and revised the contents of the manuscript. NY and SS wrote the manuscript and answered the revision. GD and KL designed the figures and searched for references. CL, LZ, and JY made the table and searched for references. All authors listed have made a substantial, direct and intellectual contribution to the work, and approved it for publication.

REFERENCES

Altabas, V. (2015). Diabetes, endothelial dysfunction, and vascular repair: what should a diabetologist keep his eye on. *Int. J. Endocrinol.* 2015:848272. doi: 10.1155/2015/848272

Anda, R., Williamson, D., Jones, D., Macera, C., Eaker, E., Glassman, A., et al. (1993). Depressed affect, hopelessness,and the risk of ischemic heart disease in a cohort of U.S. Adults. *Epidemiology* 4, 285–294. doi: 10.1097/00001648-199307000-00003

Aragona, C. O., Imbalzano, E., Mamone, F., Cairo, V., Lo Gullo, A., D'Ascola, A., et al. (2016). Endothelial progenitor cells for diagnosis and prognosis in cardiovascular disease. *Stem Cells Int.* 2016:8043792. doi: 10.1155/2016/8043792

Ariyo, A. A., Haan, M., Tangen, C. M., Rutledge, J. C., Cushman, M., Dobs, A., et al. (2000). Depressive symptoms and risks of coronary heart disease and mortality in elderly Americans.Cardiovascular Health Study Collaborative Research Group. *Circulation* 102, 1773–1779. doi: 10.1161/01.cir.102.15.177

Asahara, T., Murohara, T., Sullivan, A., Silver, M., van der Zee, R., Li, T., et al. (1997). Isolation of putative progenitor endothelial cells for angiogenesis. *Science* 275, 964–967.

Bădescu, S. V., Tătaru, C., Kobylinska, L., Georgescu, E. L., Zahiu, D. M., Zăgrean, A. M., et al. (2016). The association between diabetes mellitus and depression. *J. Med. Life* 9, 120–125.

Bianconi, V., Sahebkar, A., Kovanen, P., Bagaglia, F., Ricciuti, B., Calabrò, P., et al. (2018). Endothelial and cardiac progenitor cells for cardiovascular repair: a controversial paradigm in cell therapy. *Pharmacol. Ther.* 181, 156–168. doi: 10.1016/j.pharmthera.2017.08.004

Blum, A., Pastukh, N., Zaroura, I., Rotem, J., and Kamal, F. (2017). Impaired ability to grow colonies of endothelial stem cells could be the mechanism explaining the high cardiovascular morbidity and mortality of patients with depression. *QJM* 110, 501–506. doi: 10.1093/qjmed/hcx059

Bobińska, K., Gałecka, E., Szemraj, J., Gałecki, P., and Talarowska, M. (2017). Is there a link between TNF gene expression and cognitive deficits in depression. *Acta Biochim. Pol.* 64, 65–73.

Capillo, M., Mancuso, P., and Gobbi, A. (2003). Continuous infusion of endostatin inhibits differentiation, mobilization, and clonogenic potential of endothelial cell progenitors. *Clin. Cancer Res.* 9, 377–382.

Chen, H., Yiu, K. H., and Tse, H. F. (2011). Relationships between vascular dysfunction, circulating endothelial progenitor cells, and psychological status in healthy subjects. *Depress. Anxiety* 28, 719–727. doi: 10.1002/da.20839

Chen, H., Zhang, L., Zhang, M., Song, X., Zhang, H., Liu, Y., et al. (2013). Relationship of depression, stress and endothelial function in stable angina patients. *Physiol. Behav.* 118, 152–158. doi: 10.1016/j.physbeh.2013.05.024

Chen, J., Jin, J., Song, M., Dong, H., Zhao, G., and Huang, L. (2012). C-reactive protein down-regulates endothelial nitric oxide synthase expression and promotes apoptosis in endothelial progenitor cells through receptor for advanced glycation end-products. *Gene* 496, 128–135. doi: 10.1016/j.gene.2011.12.039

Chen, T. G., Zhong, Z. Y., Sun, G. F., Zhou, Y. X., and Zhao, Y. (2011). Effects of tumour necrosis factor-alpha on activity and nitric oxide synthase of endothelial progenitor cells from peripheral blood. *Cell Prolif.* 44, 352–359. doi: 10.1111/j.1365-2184.2011.00764.x

Chong, M. S., Ng, W. K., and Chan, J. K. (2016). Concise review: endothelial progenitor cells in regenerative medicine: applications and challenges. *Stem Cells Transl. Med.* 5, 530–538. doi: 10.5966/sctm.2015-0227

Chourbaji, S., Urani, A., Inta, I., Sanchis-Segura, C., Brandwein, C., Zink, M., et al. (2006). IL-6 knockout mice exhibit resistance to stress-induced development of depression-like behaviors. *Neurobiol. Dis.* 23, 587–594. doi: 10.1016/j.nbd.2006.05.001

Dantzer, R., O'Connor, J. C., Freund, G. G., Johnson, R. W., and Kelley, K. W. (2008). From inflammation to sickness and depression: when the immune system subjugates the brain. *Nat. Rev. Neurosci.* 9, 46–56. doi: 10.1038/nrn2297

Di Stefano, R., Felice, F., Pini, S., Mazzotta, G., Bovenzi, F. M., Bertoli, D., et al. (2014). Impact of depression on circulating endothelial progenitor cells in patients with acute coronary syndromes: a pilot study. *J. Cardiovasc. Med. (Hagerstown)* 15, 353–359. doi: 10.2459/JCM.0b013e328365c195

Dome, P., Halmai, Z., Dobos, J., Lazary, J., Gonda, X., Kenessey, I., et al. (2012). Investigation of circulating endothelial progenitor cells and angiogenic and inflammatory cytokines during recovery from an episode of major depression. *J. Affect. Disord.* 136, 1159–1163. doi: 10.1016/j.jad.2011.09.027

Dome, P., Teleki, Z., Rihmer, Z., Peter, L., Dobos, J., Kenessey, I., et al. (2009). Circulating endothelial progenitor cells and depression: a possible novel link between heart and soul. *Mol. Psychiatry* 14, 523–531. doi: 10.1038/sj.mp.4002138

Fadini, G. P., de Kreutzenberg, S. V., Coracina, A., Baesso, I., Agostini, C., Tiengo, A., et al. (2006). Circulating CD34+ cells, metabolic syndrome, and cardiovascular risk. *Eur. Heart J.* 27, 2247–2255. doi: 10.1093/eurheartj/ehl198

Fan, N., Luo, Y., Ou, Y., and He, H. (2017). Altered serum levels of TNF-α, IL-6, and IL-18 in depressive disorder patients. *Hum. Psychopharmacol.* 32:e2588. doi: 10.1002/hup.2588

Fan, Y., Ye, J., Shen, F., Zhu, Y., Yeghiazarians, Y., Zhu, W., et al. (2008). Interleukin-6 stimulates circulating blood-derived endothelial progenitor cell angiogenesis in vitro. *J. Cereb. Blood Flow Metab.* 28, 90–98. doi: 10.1038/sj.jcbfm.9600509

Felice, F., Di Stefano, R., Pini, S., Mazzotta, G., Bovenzi, F. M., Bertoli, D., et al. (2015). Influence of depression and anxiety on circulating endothelial progenitor cells in patients with acute coronary syndromes. *Hum. Psychopharmacol.* 30, 183–188. doi: 10.1002/hup.2470

Fiedorowicz, J. G., Ellingrod, V. L., Kaplan, M. J., and Sen, S. (2015). The development of depressive symptoms during medical internship stress predicts worsening vascular function. *J. Psychosom. Res.* 79, 243–245. doi: 10.1016/j.jpsychores.2015.06.004

Fu, X., Zunich, S. M., O'Connor, J. C., Kavelaars, A., Dantzer, R., and Kelley, K. W. (2010). Central administration of lipopolysaccharide induces depressive-like behavior in vivo and activates brain indoleamine 2,3 dioxygenase in murine organotypic hippocampal slice cultures. *J. Neuroinflammation* 7:43. doi: 10.1186/1742-2094-7-43

Fujii, H., Li, S., Szmitko, P. E., Fedak, P. W. M., and Verma, S. (2006). C-reactive protein alters antioxidant defenses and promotes apoptosis in endothelial progenitor cells. *Arterioscler. Thromb. Vasc. Biol.* 26, 2476–2482. doi: 10.1161/01.ATV.0000242794.65541.02

Haapakoski, R., Mathieu, J., Ebmeier, K. P., Alenius, H., and Kivimäki, M. (2015). Cumulative meta-analysis of interleukins 6 and 1β, tumour necrosis factor α and C-reactive protein in patients with major depressive disorder. *Brain Behav. Immun.* 49, 206–215. doi: 10.1016/j.bbi.2015.06.001

Hayley, S., Mangano, E., Strickland, M., and Anisman, H. (2008). Lipopolysaccharide and a social stressor influence behaviour, corticosterone and cytokine levels: divergent actions in cyclooxygenase-2 deficient mice and wild type controls. *J. Neuroimmunol.* 197, 29–36. doi: 10.1016/j.jneuroim.2008.03.015

Herbrig, K., Haensel, S., Oelschlaegel, U., Pistrosch, F., Foerster, S., and Passauer, J. (2006). Endothelial dysfunction in patients with rheumatoid arthritis is associated with a reduced number and impaired function of endothelial progenitor cells. *Ann. Rheum. Dis.* 65, 157–163. doi: 10.1136/ard.2005.035378

Kaushik, K., and Das, A. (2019). Endothelial progenitor cell therapy for chronic wound tissue regeneration. *Cytotherapy* 21, 1137–1150. doi: 10.1016/j.jcyt.2019.09.002

Khandaker, G. M., Pearson, R. M., Zammit, S., Lewis, G., and Jones, P. B. (2014). Association of serum interleukin 6 and C-reactive protein in childhood with depression and psychosis in young adult life: a population-based longitudinal study. *JAMA Psychiatry* 71, 1121–1128. doi: 10.1001/jamapsychiatry.2014.1332

Klakk, H., Kristensen, P. L., Andersen, L. B., Froberg, K., Møller, N. C., and Grøntved, A. (2018). Symptoms of depression in young adulthood is associated with unfavorable clinical- and behavioral cardiovascular disease risk factors. *Prev. Med. Rep.* 11, 209–215. doi: 10.1016/j.pmedr.2018.05.017

Kling, M. A., Alesci, S., Csako, G., Costello, R., Luckenbaugh, D. A., Bonne, O., et al. (2007). Sustained low-grade pro-inflammatory state in unmedicated, remitted women with major depressive disorder as evidenced by elevated serum levels of the acute phase proteins C-reactive protein and serum amyloid A. *Biol. Psychiatry* 62, 309–313. doi: 10.1016/j.biopsych.2006.09.033

Kou, F., Zhu, C., Wan, H., Xue, F., Wang, J., Xiang, L., et al. (2020). Endothelial progenitor cells as the target for cardiovascular disease prediction, personalized prevention, and treatments: progressing beyond the state-of-the-art. *EPMA J.* 11, 629–643. doi: 10.1007/s13167-020-00223-0

Labib, J. M. W., Aboul-Fotouh, S., Habib, M. Z., Mekawy, M. A. E. A., Farrag, K. A., and Abdel-Tawab, A. M. (2019). Pentoxifylline ameliorates chronic stress/high-fat diet-induced vascular wall disease: the role of circulating endothelial progenitor cells. *Naunyn Schmiedebergs Arch. Pharmacol.* 392, 669–683. doi: 10.1007/s00210-019-01627-0

Li, H., Qiang, Y., Wang, G., Wang, L., Yi, J., Jing, H., et al. (2013). Repair of lipopolysaccharide-induced acute lung injury in mice by endothelial progenitor cells,alone and in combination with simvastatin. *Chest* 144, 876–886. doi: 10.1378/chest.12-2429

Liu, Y., Ho, R. C., and Mak, A. (2012). Interleukin (IL)-6, tumour necrosis factor alpha (TNF-alpha) and soluble interleukin-2 receptors (sIL-2R) are elevated in patients with major depressive disorder: a meta-analysis and meta-regression. *J. Affect. Disord.* 139, 230–239. doi: 10.1016/j.jad.2011.08.003

Lopez-Vilchez, I., Diaz-Ricart, M., Navarro, V., Torramade, S., Zamorano-Leon, J., Lopez-Farre, A., et al. (2016). Endothelial damage in major depression patients is modulated by SSRI treatment, as demonstrated by circulating biomarkers and an in vitro cell model. *Transl. Psychiatry* 6:e886. doi: 10.1038/tp.2016.156

Ma, K., Zhang, H., and Baloch, Z. (2016). Pathogenetic and therapeutic applications of tumor necrosis Factor-α (TNF-α) in major depressive disorder: a systematic review. *Int. J. Mol. Sci.* 17:733. doi: 10.3390/ijms17050733

Malhi, G. S., and Mann, J. J. (2018). Depression. *Lancet* 392, 2299–2312. doi: 10.1016/S0140-6736(18)31948-2

Mandraffino, G., Aragona, C. O., Basile, G., Cairo, V., Mamone, F., Morace, C., et al. (2017). CD34+ cell count predicts long lasting life in the oldest old. *Mech. Ageing Dev.* 164, 139–145. doi: 10.1016/j.mad.2017.03.003

Mandraffino, G., Sardo, M. A., Riggio, S., D'Ascola, A., Alibrandi, A., Saitta, C., et al. (2012). Circulating progenitor cells and the elderly: a seven-year observational study. *Exp. Gerontol.* 47, 394–400. doi: 10.1016/j.exger.2012.03.007

McClung, J. A., Naseer, N., Saleem, M., Rossi, G. P., Weiss, M. B., Abraham, N. G., et al. (2005). Circulating endothelial cells are elevated in patients with type 2 diabetes mellitus independently of HbA1C. *Diabetologia* 48, 345–350. doi: 10.1007/s00125-004-1647-5

Miller, G. E., Rohleder, N., Stetler, C., and Kirschbaum, C. (2005). Clinical depression and regulation of the inflammatory response during acute stress. *Psychosom. Med.* 67, 679–687. doi: 10.1097/01.psy.0000174172.82428.ce

Morishita, T., Uzui, H., Ikeda, H., Amaya, N., Kaseno, K., Ishida, K., et al. (2016). Association of CD34/CD133/VEGFR2-positive cell numbers with eicosapentaenoic acid and postprandial hyperglycemia in patients with coronary artery disease. *Int. J. Cardiol.* 221, 1039–1042. doi: 10.1016/j.ijcard.2016.07.079

Nana, Y., Peng, J., Jianlin, Z., Xiangjian, Z., Shutong, Y., Enxin, Z., et al. (2015). Reverse-D-4F increases the number of endothelial progenitor cells and improves endothelial progenitor cell dysfunctions in high fat diet mice. *PLoS One* 10:e0138832. doi: 10.1371/journal.pone.0138832

Nukina, H., Sudo, N., Aiba, Y., Oyama, N., Koga, Y., and Kubo, C. (2001). Restraint stress elevates the plasma interleukin-6 levels in germ-free mice. *J. Neuroimmunol.* 115, 46–52. doi: 10.1016/s0165-5728(01)00260-0

Ormiston, M. L., Toshner, M. R., Kiskin, F. N., Huang, C. J., Groves, E., Morrell, N. W., et al. (2015). Generation and culture of blood outgrowth endothelial cells from human peripheral blood. *J. Vis. Exp.* e53384. doi: 10.3791/53384

Patel, J., Donovan, P., and Khosrotehrani, K. (2016). Concise review: functional definition of endothelial progenitor cells: a molecular perspective. *Stem Cells Transl. Med.* 5, 1302–1306. doi: 10.5966/sctm.2016-0066

Pikhart, H., Hubacek, J. A., Kubinova, R., Nicholson, A., Peasey, A., Capkova, N., et al. (2009). Depressive symptoms and levels of C-reactive protein: a population-based study. *Soc. Psychiatry Psychiatr. Epidemiol.* 44, 217–222. doi: 10.1007/s00127-008-0422-1

Prasain, N., Meador, J. L., and Yoder, M. C. (2012). Phenotypic and functional characterization of endothelial colony forming cells derived from human umbilical cord blood. *J. Vis. Exp.* 62:3872. doi: 10.3791/3872

Pulito-Cueto, V., Remuzgo-Martínez, S., Genre, F., Mora-Cuesta, V. M., Iturbe-Fernández, D., Fernández-Rozas, S., et al. (2020). Endothelial progenitor cells as a potential biomarker in interstitial lung disease associated with rheumatoid arthritis. *J. Clin. Med.* 9:4098. doi: 10.3390/jcm9124098

Rana, D., Kumar, A., and Sharma, S. (2018). Endothelial progenitor cells as molecular targets in vascular senescence and repair. *Curr. Stem Cell Res. Ther.* 13, 438–446. doi: 10.2174/1574888X13666180502100620

Salter, B., and Sehmi, R. (2017). The role of bone marrow-derived endothelial progenitor cells and angiogenic responses in chronic obstructive pulmonary disease. *J. Thorac. Dis.* 9, 2168–2177. doi: 10.21037/jtd.2017.07.56

Schmidt-Lucke, C., Rössig, L., Fichtlscherer, S., Vasa, M., Britten, M., Kämper, U., et al. (2019). Reduced number of circulating endothelial progenitor cells predicts future cardiovascular events proof of concept for the clinical importance of endogenous vascular repair. *Circulation* 111, 2981–2987. doi: 10.1161/CIRCULATIONAHA.104.504340

Shi, M., Sun, H., Xu, Y., Wang, Z., Cui, H., Wang, C., et al. (2017). Methylation status of the serotonin transporter promoter CpG island is associated with major depressive disorder in chinese han population: a Case-Control study. *J. Nerv. Ment. Dis.* 205, 641–646. doi: 10.1097/NMD.0000000000000600

Sidney, L. E., Branch, M. J., Dunphy, S. E., Dua, H. S., and Hopkinson, A. (2014). Concise review: evidence for CD34 as a common marker for diverse progenitors. *Stem Cells* 32, 1380–1389. doi: 10.1002/stem.1661

Sukoff Rizzo, S. J., Neal, S. J., Hughes, Z. A., Beyna, M., Rosenzweig-Lipson, S., Moss, S. J., et al. (2012). Evidence for sustained elevation of IL-6 in the CNS as a key contributor of depressive-like phenotypes. *Transl. Psychiatry* 2:e199. doi: 10.1038/tp.2012.120

Tagawa, S., Nakanishi, C., Mori, M., Yoshimuta, T., Yoshida, S., Shimojima, M., et al. (2015). Determination of early and late endothelial progenitor cells in peripheral circulation and their clinical association with coronary artery disease. *Int. J. Vasc. Med.* 2015:674213. doi: 10.1155/2015/674213

Teraa, M., Sprengers, R. W., Westerweel, P. E., Gremmels, H., Goumans, M. J., Teerlink, T., et al. (2013). Bone marrow alterations and lower endothelial progenitor cell numbers in critical limb ischemia patients. *PLoS One* 8:e55592. doi: 10.1371/journal.pone.0055592

Vasa, M., Fichtlscherer, S., Adler, K., Aicher, A., Martin, H., Zeiher, A. M., et al. (2001). Increase in circulating endothelial progenitor cells by statin therapy in patients with stable coronary artery disease. *Circulation* 103, 2885–2890.

Vega, F. M., Gautier, V., Fernandez Ponce, C. M., Extremera, M. J., Altelaar, A. F. M., Millan, J., et al. (2017). The atheroma plaque secretome stimulates the mobilization of endothelial progenitor cells ex vivo. *J. Mol. Cell Cardiol.* 105, 12–23. doi: 10.1016/j.yjmcc.2017.02.001

Wang, Q. N., Zou, Z. X., Wang, X. P., Zhang, Q., Zhao, Y. Q., Duan, L., et al. (2020). Endothelial progenitor cells induce angiogenesis: a potential mechanism underlying neovascularization in encephaloduroarteriosynangiosis. *Transl. Stroke Res.* 12, 357–365.

Werner, N., Kosiol, S., Schiegl, T., Ahlers, P., Walenta, K., Link, A., et al. (2005). Circulating endothelial progenitor cells and cardiovascular outcomes. *N. Engl. J. Med.* 353, 999–1007. doi: 10.1056/NEJMoa043814

Yang, L., Ruan, L. M., Ye, H. H., Cui, H. B., Mu, Q. T., Lou, Y. R., et al. (2011). Depression is associated with lower circulating endothelial progenitor cells and increased inflammatory markers. *Acta Neuropsychiatr.* 23, 235–240. doi: 10.1111/j.1601-5215.2011.00577.x

Yoon, C. H., Hur, J., Park, K. W., Kim, J. H., Lee, C. S., Oh, I. Y., et al. (2005). Synergistic neovascularization by mixed transplantation of early endothelial progenitor cells and late outgrowth endothelial cells: the role of angiogenic cytokines and matrix metalloproteinases. *Circulation* 112, 1618–1627.

Yuan, J. J., Yang, J., Sun, S. L., Zhang, R., and Xu, Y. M. (2017). Endothelial progenitor cells' classification and application in neurological diseases. *Tissue Eng. Regen. Med.* 14, 327–332. doi: 10.1007/s13770-017-0043-4

9

Combined Transplantation of Olfactory Ensheathing Cells with Rat Neural Stem Cells Enhanced the Therapeutic Effect in the Retina of RCS Rats

Wei Zhai[1,2†], Lixiong Gao[1,2,3†], Linghui Qu[1,2], Yijian Li[1,2], Yuxiao Zeng[1,2], Qiyou Li[1,2], Haiwei Xu[1,2] and Zheng Qin Yin[1,2*]*

[1] Southwest Hospital/Southwest Eye Hospital, Third Military Medical University (Army Medical University), Chongqing, China, [2] Key Lab of Visual Damage and Regeneration & Restoration of Chongqing, Chongqing, China, [3] Department of Ophthalmology, The 6th Medical Center of PLA General Hospital, Beijing, China

***Correspondence:**
Haiwei Xu
haiweixu2001@163.com;
xuhaiwei@tmmu.edu.cn
Zheng Qin Yin
qinzyin@aliyun.com

† These authors have contributed equally to this work

Retinal degenerative diseases (RDDs) are the leading causes of blindness and currently lack effective treatment. Cytotherapy has become a promising strategy for RDDs. The transplantation of olfactory ensheathing cells (OECs) or neural stem cells (NSCs) has recently been applied for the experimental treatment of RDDs. However, the long-term outcomes of single-cell transplantation are poor. The combined transplantation of multiple types of cells might achieve better effects. In the present study, OECs [containing olfactory nerve fibroblasts (ONFs)] and NSCs were cotransplanted into the subretinal space of Royal College of Surgeons (RCS) rats. Using electroretinogram (ERG), immunofluorescence, Western blot, and *in vitro* Transwell system, the differences in the electrophysiological and morphological changes of single and combined transplantation as well as the underlying mechanisms were explored at 4, 8, and 12 weeks postoperation. In addition, using the Transwell system, the influence of OECs on the stemness of NSCs was discovered. Results showed that, compared to the single transplantation of OECs or NSCs, the combined transplantation of OECs and NSCs produced greater improvements in b-wave amplitudes in ERGs and the thickness of the outer nuclear layer at all three time points. More endogenous stem cells were found within the retina after combined transplantation. Glial fibrillary acidic protein (GFAP) expression decreased significantly when NSCs were cotransplanted with OECs. Both the vertical and horizontal migration of grafted cells were enhanced in the combined transplantation group. Meanwhile, the stemness of NSCs was also better maintained after coculture with OECs. Taken together, the results suggested that the combined transplantation of NSCs and OECs enhanced the improvement in retinal protection in RCS rats, providing a new strategy to treat RDDs in the future.

Keywords: retinal degenerative diseases, neural stem cells, olfactory ensheathing cells, combined transplantation, RCS rats

INTRODUCTION

Characterized by the progressive loss or malfunction of retinal cells, retinal degenerative diseases (RDDs) are the leading causes of blindness. One of the most common RDDs is the age-related macular degeneration (AMD) (Veleri et al., 2015). There are two types of AMD, the dry AMD and the wet AMD, whose pathological characteristics are the neovascularization secondary to the stenosis of choroidal vessels and the decrease in phagocytosing function of retina pigment epithelium (RPE) followed by photoreceptor death, respectively (Blasiak, 2020). Data have shown that the general prevalence of all types of AMD is ~8.7%, and the number of individuals affected by AMD will increase to 196 million in 2020 (Wong et al., 2014; Jonas et al., 2017). Currently, there are no specific therapeutic methods for AMD, especially for dry AMD. Treatments for wet AMD such as intravitreous injection do not stop the degeneration of the retina in wet AMD patients (Mavija et al., 2014). As retinal cell loss is the ultimate result of AMD, stem cell therapy is becoming a promising method for treating AMD (Nazari et al., 2015). To better understand the diseases, several RDD models have been developed. Royal College of Surgeons (RCS) rat is one of them. Characterized by the inability of RPE to phagocytose photoreceptor outer segments, RCS rat is the first known animal with inherited retinal degeneration (Strauss et al., 1998). The mutation in this model is found to be the deletion in the receptor tyrosine kinase gene Mertk, which links to the phagocytosing function of RPE (D'Cruz et al., 2000). The RPE dysfunction will further lead to the deposition of photoreceptor outer segments and consequent photoreceptor degeneration. Since its pathology is similar to RDDs, RCS rat is widely used as an animal model to mimic AMD and retinitis pigmentosa (LaVail, 2001). The progress of retinal degeneration is rapid in RCS rats. The response of electroretinograms (ERGs) in RCS rat begin to decrease at postnatal 21 days, reduce by half at postnatal 50 days, and nearly disappear at postnatal 100 days (Pinilla et al., 2005).

As adult stem cells within the central nervous system (CNS), neural stem cells (NSCs) can generate both neurons and glia (Obernier and Alvarez-Buylla, 2019). Several studies have confirmed the therapeutic effect of NSCs after transplantation into the injured CNS (Marsh and Blurton-Jones, 2017). McGill et al. showed that the transplantation of NSCs into the subretinal space (SRS) of RCS rats preserves retinal function and protects photoreceptors from death through the phagocytosis of photoreceptor outer segments (McGill et al., 2012). Lin et al. (2014) found that NSCs have a better proliferative ability than that of retinal progenitor cells (RPCs). Moreover, upon treatment with transforming growth factor beta type III and retinoic acid, NSCs are able to differentiate into opsin-positive retinal cells (Lin et al., 2014). Our previous work also confirmed the preservative effect of the subretinal transplantation of NSCs in rd1 mice (Li et al., 2016). However, there were several problems associated with transplanting NSCs alone into the retina: evidence of functional improvement was only observed during a small treatment window, transplantation triggered gliosis, and the migration of grafted cells was very limited. The combined transplantation of two or more different types of cells might improve the therapeutic effect of stem cell transplantation. Our previous work showed that the combined transplantation of mesenchymal stem cells and RPCs can simultaneously generate synergistic effects after subretinal transplantation (Qu et al., 2017).

As a physiological response to lesion to the CNS, gliosis is a double-edged sword. In the retina, the gliotic response from Müller cells regulates the size of the glial scar, which inhibits transplanted cells from exerting their therapeutic effect. Our previous results showed that olfactory ensheathing cells (OECs), a glial cell type originating from the neocortex of the olfactory bulb, inhibits gliosis in the retinas of RCS rats (Xie et al., 2017). Therefore, transplanting OECs with NSCs might improve the restoration of visual function in RCS rats.

OECs have been reported to support the continuous growth and regeneration of olfactory axons throughout life (Ramon-Cueto and Avila, 1998). Robust studies have confirmed that transplanted OECs exhibit neuroplastic and neuroregenerative effects via interacting with the glial scar and stimulating angiogenesis, axonal outgrowth, and remyelination in the spinal cord injury (Roet and Verhaagen, 2014; Gomez et al., 2018). OECs have also been found to protect visual function. Our previous work showed that OECs restore retinal function and alleviate retinal degeneration in RDD animal models via reducing the gliotic injury response of Müller cells, phagocytosing retinal outer segments, and inhibiting oxidative stress (Huo et al., 2011, 2012; Xie et al., 2017; Xue et al., 2017). We further confirmed that OECs can promote retinal ganglion cell survival and axonal regeneration after optic nerve injury for 3 months (Wang et al., 2017). However, the protective effect of OECs in the eyes is maintained for a relatively short period (Xue et al., 2017). The capacity of OECs to migrate relies on various factors (Gomez et al., 2018). These limitations restrict research on of the treatment of RDDs with OECs.

Because of the potential improvement in the gliotic microenvironment of OECs as well as the therapeutic effect of NSCs, we hypothesized that the cotransplantation of these two cell types might produce a better effect. In the present study, NSCs and OECs were transplanted either singly or in combination into the SRS of RCS rats at early degenerative stage. Using ERGs, immunofluorescence, Western blotting, and an *in vitro* Transwell system, we discovered the efficacy of combined transplantation and explored the possible underlying mechanisms at 4, 8, and 12 weeks postoperation. These three time points covered the moderate to the severe retinal degeneration of RCS rats.

MATERIALS AND METHODS

Animals and Ethics

The RCS (28 days) and Long Evans (LE) rats were obtained from the Animal Research Center of the Third Military Medical University (TMMU). Rats were raised under a 12-h light/dark cycle in the specific pathogen-free room of the Animal Care

Center of the First Affiliated Hospital of TMMU. The breeding of LE rats was performed to harvest embryos as well as the neonatal LE rats. All tissue collection and experimental procedures were performed according to protocols approved by the Institutional Review Board of the TMMU and conformed to the National Institutes of Health (NIH) guidelines on the ethical use of animals.

N Values and Blinding

For *in vivo* study, 18 animals underwent transplantation treatment in each transplantation group at the starting point. On each of the three different posttransplantation time points, six animals in each group were killed after recording ERG. Three animals were used for immunofluorescent test and three animals for Western blot test. In summary, the N value in ERG test was 6; in immunofluorescence, 3; and in Western blot, 3. For *in vitro* study, both immigration and differentiation tests were repeated three times ($N = 3$). Cell harvest was repeated three times in both cells, and identification of cells in each batch was performed to ensure their characteristics ($N = 3$). As for the randomization and blinding, all treatments were randomized, and the persons performing the transplantation surgeries and histological analysis were blinded with respect to the treatment condition.

Isolation, Culture, and Identification of OECs

After LE rats (90 days old) were anesthetized with pentobarbital sodium (10 mg/kg, Sigma-Aldrich), the olfactory bulbs were dissected and removed under a microscope. The glomerular layers of the olfactory bulbs were carefully isolated and cut into small pieces. The tissues were digested in 0.1% trypsin for 15 min at 37°C, and the reaction was stopped by OEC culture medium containing Dulbecco's modified Eagle's medium/F-12 culture medium (DMEM/F-12, 1:1 mixture, HyClone) supplemented with 10% fetal bovine serum (FBS, Gibco) and a mixture of penicillin and streptomycin (PS, 1%, Gibco). Then, the OEC suspension was centrifuged at 1,500 rpm for 5 min and resuspended in OEC culture medium. Then, OECs were plated on 35-mm dishes coated with 10 μg/ml laminin and incubated in a 5% CO_2 saturation-humidity atmosphere at 37°C. The culture medium was changed every 3 days. Subculturing was performed once the cell density was over 80%. OECs were identified at passage 3. After being digested by trypsin, plated on laminin-coated coverslips, and cultured for 3 days, OECs were identified via immunofluorescence. The details are described in section "Immunofluorescence."

Isolation, Culture, and Identification of NSCs

Neural stem cells were harvested from the visual cortex of embryonic LE rats at embryonic day 13.5 and cultured. The maternal LE rats were anesthetized with pentobarbital sodium (10 mg/kg, Sigma-Aldrich), and uteruses containing fetal rats were isolated. The visual cortexes of fetal rats were carefully dissected and cut into small pieces under a microscope. After being digested with Accutase (Innovative Cell Technologies, United States) for 5 min at 37°C and stopped by the NSC culture medium containing DMEM/F-12 (Hyclone) supplemented with B27 (Gibco), glutamine (Gibco), basic fibroblast growth factor (bFGF) (20 ng/ml, Peprotech), epidermal growth factor (EGF) (10 ng/ml, Peprotech), and a mixture of PS (1%, Gibco), NSC suspension was centrifuged at 4,000 rpm for 5 min and resuspended with the NSC culture medium. Then, NSC suspension was transferred to floating culture flasks and incubated in a 5% CO_2 saturation-humidity atmosphere at 37°C. The culture medium was changed every 3 days. For identification, passage 3 floating spheres were directly plated on laminin-coated coverslips or plated after being digested into single cells and cultured for 3 days, after which NSCs were identified via immunofluorescence. The details are described in section "Immunofluorescence."

For flow cytometry, passage 3 floating spheres were digested into single cells. After perforation, washing, and blocking, NSCs were divided into blank and experimental groups. The latter group was incubated with Nestin-FITC, Pax6-FITC, and Sox2-FITC primary antibodies. Then, all groups were tested by flow cytometry (BD, United States).

In vitro Migration and Differentiation Assay of NSCs

Passage 3 NSCs were used to conduct both migration and differentiation assays. For the migration assay, Transwell systems and six-well plates were used. In detail, 2×10^5 rhodamine (Sigma-Aldrich)-labeled NSCs alone or 2×10^5 rhodamine-labeled NSCs together with 2×10^5 Hoechst (Beyotime)-labeled OECs (NSCs + OECs) were added to a single Transwell system. The retinas of neonatal LE rats (postnatal day 0) were removed and isolated on ice after anesthetization with pentobarbital sodium (10 mg/kg, Sigma-Aldrich). The retinas were then placed onto the NSCs or NSCs + OECs in the Transwell system with 1 ml of NSC culture medium. After 12 days of coculture, the retinas were fixed with 4% paraformaldehyde (PFA) for 3 h, followed by dehydration with 30% sucrose for 12 h. Then, the retinas were embedded and sectioned into 10-μm slices. 4′,6-Diamidino-2-phenylindole (DAPI) staining was performed on retinal sections from the NSC group. The sections were observed under an immunofluorescence microscope. For the differentiation assay, Transwell systems and 24-well plates were used. In the control group, 1×10^5 NSCs were plated along the Transwell membrane. In the experimental group, 2×10^5 OECs were first plated in the plates, and 1×10^5 NSCs were then plated on the upper Transwell membrane. All groups were cultured in the NSC culture medium and cultured for 24 h. In the bromodeoxyuridine (BrdU) test, 25 μl of BrdU solution (2 mg/ml) was added to the corresponding wells at 23 h. Then, Transwell membranes containing NSCs were examined via immunofluorescence. The details are described in section "Immunofluorescence." These two assays were repeated three times.

Cell Transplantation

Subretinal transplantation was performed as previously described (Qu et al., 2017). In brief, cell suspensions containing OECs, NSCs, or their combination were injected into the temporal subretinal space of the left eyes of RCS rats 4 weeks postnatally (5 µl/eye). The cell suspension in the combination group was a mixture containing 2.5×10^5 cell NSCs and 2.5×10^5 cells OECs/olfactory nerve fibroblasts (ONFs) per eye. In the single transplantation groups, 2.5×10^5 cells of NSCs per eye were transplanted in the NSC group, and 2.5×10^5 cells OECs/ONFs per eye were transplanted in the OEC group. An identical volume of 0.01 M phosphate-buffered saline (PBS) (5 µl) was injected into the temporal subretinal space of the right eyes of RCS rats. All transplanted cells were labeled with the fluorescent marker CM-DiI (2 mg/ml, Invitrogen). The pupils were dilated with 1% tropicamide (Santen Pharmaceutical Co., Ltd. Osaka, Japan) 30 min before surgery. Once the RCS rats were anesthetized (120 mg/kg ketamine and 20 mg/kg xylazine), a 10-µl Hamilton syringe (30 gauge; Hamilton, NV, United States) containing a cell suspension was tangentially inserted into the subretinal space through the conjunctiva and sclera, which led to a self-sealing wound tunnel. Paracentesis of the anterior chamber was performed to reduce the intraocular pressure and limit the efflux of cells from the injection site. Fundus examinations were performed immediately after the operations. The eyes that did not receive any treatment were labeled as the "blank" group. The eyes in which OECs, NSCs, or OECs + NSCs were transplanted were labeled as the "OEC group," "NSC group," or "OEC + NSC group," respectively. The eyes that were injected with an equal amount of PBS were labeled as the "PBS" group.

ERG Recording

ERG recording was performed 4, 8, and 12 weeks postoperation to evaluate the retinal functional changes, as previously described (Lin et al., 2014). In brief, after being dark adapted for at least 12 h, RCS rats were anesthetized by an intraperitoneal injection of a solution of ketamine (120 mg/kg) and xylazine (20 mg/kg). Pupils were dilated with 1% tropicamide before testing. A heating pad was used to maintain the body temperature at 37°C. Two active gold electrodes were placed on each cornea as recording electrodes. The reference and ground electrodes were subcutaneously inserted into the mid-frontal area of the head and tail, respectively. Light stimulations were delivered with a xenon lamp at 3.0 cd s/m^2. The b-wave amplitudes were recorded and processed by a RETI-Port device (Roland Consult, Brandenburg, Germany). All procedures were performed in a dark room with dim red safety light. When dealing with the results, a and b waves were marked by the typical type of ERG waves as well as the latent period, which was basically achieved by computer and checked by experimenters. If the computer failed to calculate the proper point, experimenters would manually measure the results. Since b wave represents the transduction of extracellular currents and is considered to be the major component of the human ERG recording as used in clinical and experimental analysis of retinal function (Perlman, 1995), we typically focus on the amplitude of b wave in the present study.

Morphological Preparation of Retina

After being anesthetized by 1% pentobarbital (150 mg/kg), RCS rats were perfused with normal saline and 4% PFA via the circulation system on 4, 8, and 12 postoperation weeks as we previously described (Qu et al., 2017). After being enucleated and fixed in 4% PFA for 3 h, eyeballs were incubated in 30% glucose solution overnight. During the embedding of the eyeball, we marked the injection site and placed the embedded eyeball on the slicer. This placement was carried out in two disciplines: (1) The eyeball stood vertically in the embedding medium; (2) the plane formed by 3 points (the injection site, optic disk, and the point opposite to the injection site on the eyeball) kept parallel to the blade (**Supplementary Figure 1**). These disciplines can ensure the consistency in the sections among different groups. Then, 10-µm serially frozen sections were carefully made.

Immunofluorescence

Immunofluorescence of the identification of OECs/ONFs and NSCs as well as NSC differentiation assay were performed as previously described. In detail, after being rinsed in 0.01 M PBS, blocked in 10% of goat serum, and perforated in 0.1% of Triton X-100, coverlids and Transwell membrane were incubated with primary antibodies overnight at 4°C. The coverlids with OECs/ONFs were incubated with anti-P75 (1:500, mouse, Santa) antibodies; the coverlids with NSCs with anti-Nestin (1:500, rabbit, Abcam), anti-GFAP (1:500, mouse, Abcam), and anti-Tuj1 (1:1000, mouse, Beyotime) antibodies; and the Transwell membrane with NSCs with anti-GFAP (1:500, mouse, Abcam), anti-Sox2 (1:500, rabbit, Abcam), anti-Pax6 (1:500, rabbit, Santa), and anti-BrdU antibodies (1:500, Cell Signaling). Cy3- or 488-conjugated secondary antibodies, (Invitrogen) were then implemented (1:400, 37°C, 2 h). Before examination with a confocal laser scanning microscope (Leica, Germany), cells were counterstained with DAPI (Sigma Aldrich). Immunofluorescence of retina sections was also performed as previously described (Gao et al., 2015). In detail, after being washed in 0.01 M PBS, blocked in 10% of goat serum, and perforated in 0.1% of Triton X-100, selected sections were incubated with the primary antibodies, anti-GFAP (1:500, mouse, Abcam), anti-Sox2 (1:500, rabbit, Abcam), and anti-recoverin (1:1000, rabbit, Millipore) antibodies in 1% bovine serum albumin (BSA) at 4°C overnight. Cy3- or 488-conjugated secondary antibodies (Invitrogen) were then implemented (1:400, 37°C, 2 h). Before examination with a confocal laser scanning microscope (Leica, Germany), sections were counterstained with DAPI (Sigma Aldrich).

Western Blot

Animals were euthanized with CO_2 at 4, 8, and 12 weeks postoperation, after which eyeballs were enucleated and retinas were quickly isolated on ice. After being rinsed in 0.01 M PBS and drained, retina tissues were lysed in ice-cold tissue lysis buffer [10% phenylmethylsulfonyl fluoride (PMSF) + 90% radioimmunoprecipitation assay (RIPA)]. The lysates were then centrifuged at 15,000 rpm for 10 min at 4°C. Protein concentration was determined using the BCA Protein Assay (Beyotime). After boiling in loading buffer for 10 min, total proteins (10 μg per slot) were electrophoresed on a 12% sodium dodecyl sulfate polyacrylamide gel and then transferred onto polyvinylidene fluoride membranes. After being blocked in 5% fat-free milk for 2 h at 37°C, membranes were incubated with anti-GFAP antibody (1:500, rabbit, Abcam) and anti-glyceraldehyde 3-phosphate dehydrogenase (anti-GAPDH) (1:1,000, mouse, Proteintech Group) antibody overnight at 4°C. Membranes were then incubated with peroxidase-conjugated immunoglobulin G (1:2,000; Santa Cruz Biotechnology). After being washed in Tris-buffered saline with Tween-20 (TBS-T) and developed in developing solution, membranes were scanned using the Bio-Rad exploding system (Bio-Rad, CA, United States) with corresponding software.

Outer Nuclear Layer Thickness Analysis

Six sections that were cut using the same horizontal angle across the optic disk were chosen to measure the thickness of the outer nuclear layer (ONL). From each section, an average of three areas of the temporal retina area of the optic disk was selected (**Supplementary Figure 2**). The thickness of the ONL was measured by ImageJ (NIH, United States). The average ONL thickness in the three areas represented the ONL thickness of the section.

Quantitative Histological Analysis

To quantitatively analyze the differences in the expression of NSC markers after coculture with OECs, three comparable visual fields from each cell slide were randomly selected. The numbers of BrdU-, Pax6-, Sox2-, and GFAP-positive cells were manually counted and averaged. To quantitatively analyze *in vivo* cell migration after cell transplantation, six sections that were cut using the same horizontal angle across the optic disk were chosen to conduct the migration measurement. In each section, the photos of retina were taken under a 400× microscope. Integrate image of the whole retina was generated by splicing these photos. The distance that transplanted cells migrated within the SRS was measured by ImageJ (NIH, United States) in each integrate image. To quantitatively analyze the status of endogenous stem cell formation after transplantation, at least three sections across the optic disk were selected from each group after Sox2 staining. The number of Sox2-positive cells within a 150 μm × 150 μm square visual field was manually counted and averaged. To semiquantitatively analyze the GFAP expression level, at least three sections across the optic disk were selected from each group after GFAP staining. The density of GFAP-positive cells was recorded by ImageJ (NIH).

Statistical Analysis

Using Statistical Product and Service Solutions software V17.0 (SPSS, Chicago, IL, United States), all quantitative results were analyzed by one-way ANOVA followed by Fisher's protected least-significant difference *post hoc* tests. The data are presented as the mean ± standard error. $P < 0.05$ was considered statistically significant.

RESULTS

Identification of Primarily Isolated OECs and NSCs

The identification of both OECs and NSCs was conducted before transplantation. The OECs and NSC were tested at passage 3. The nerve growth factor receptor P75 and Nestin were used as OECs and NSCs markers, respectively (Xie et al., 2017). The OECs were fusiform shaped with elongated processes (**Supplementary Figures 3A,B**). By immunofluorescence, we found that nearly half of the OEC/ONF mixture expressed P75, and the other half expressed FN (**Supplementary Figures 3C–E**). For NSCs, the results showed that, when floating culture was performed, NSCs exhibited a spherical shape and expressed Nestin (**Supplementary Figures 4A,B**). After being cultured in serum-free media for 2 weeks, NSCs differentiated into neurons and expressed Tuj1 (**Supplementary Figure 4C**). Few GFAP-positive glial cells were observed (**Supplementary Figure 4D**). Flow cytometry showed that a high percentage of NSCs expressed Sox2 (98.86%), Pax6 (98.94%), and Nestin (98.38%) (**Supplementary Figure 4E**). These results confirmed the characteristics of harvested OECs/ONFs and NSCs.

Combined Transplantation Enhanced Electrophysiological Improvement of RCS Rats

After transplantation with a single cell type or a combination of the two cell types, ERGs were recorded, and b-wave amplitude was measured to determine electrophysiological improvements at 4, 8, and 12 weeks postoperation. The results showed that, compared to the blank and the PBS group, the OEC, NSC, and OEC + NSC groups presented significant improvements in b-wave amplitudes at all time points ($P < 0.001$; **Figures 1A–P**). Moreover, there were also differences among three transplantation groups. In particular, at 4 weeks postoperation, the OEC + NSC group showed significantly higher b-wave amplitude than that in either the OEC or the NSC group ($P < 0.001$, $P < 0.001$; **Figures 1C–E,P**). There was no significant difference between the OEC and the NSC group ($P > 0.05$, **Figures 1C–E,P**). At 8 weeks postoperation, the OEC + NSC group still presented significantly higher b-wave amplitude than that in either the OEC or the NSC group ($P < 0.05$, $P < 0.05$; **Figures 1H–J,P**). There was no significant difference between the OEC and the NSC group at 8 weeks postoperation ($P > 0.05$; **Figures 1H–J,P**). However, the retina-function restoration became different at 12 weeks postoperation. Compared to the OEC group, the OEC + NSC group showed a significant increase in b-wave amplitude ($P < 0.001$; **Figures 1M–O,P**), while no statistically significant difference was observed between the

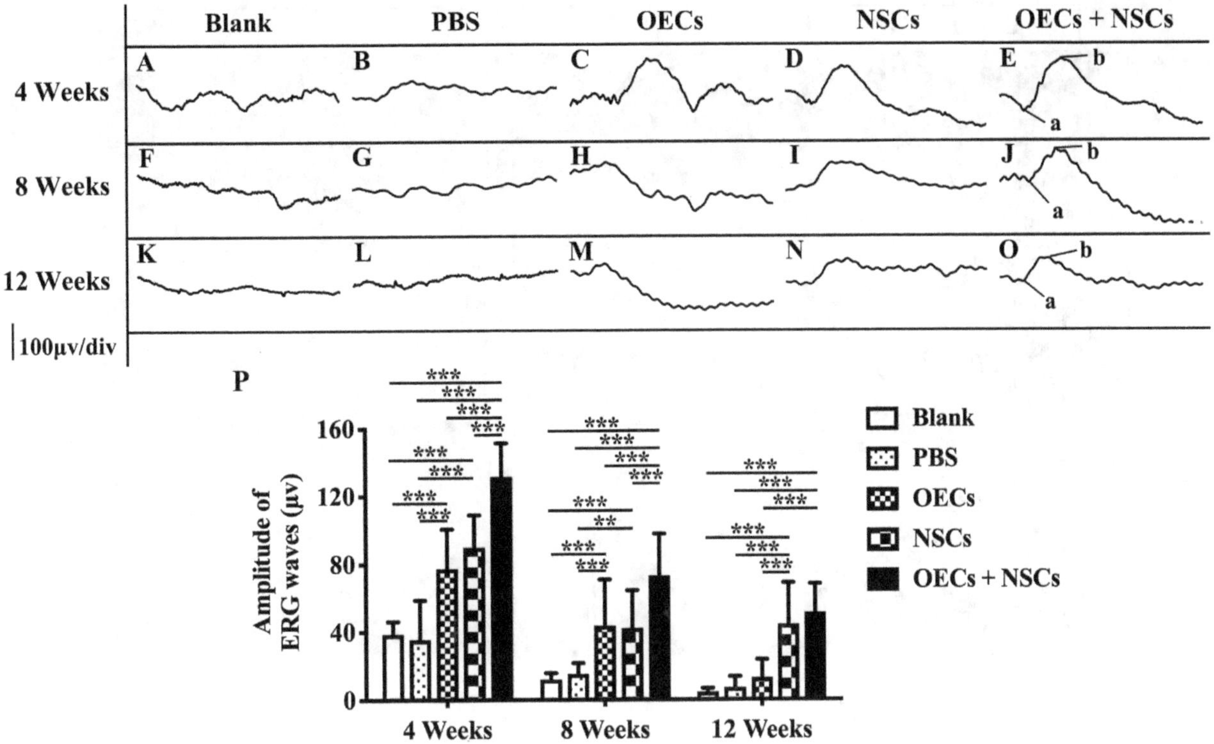

FIGURE 1 | Electrophysiological improvement after combined transplantation of rat olfactory ensheathing cells (OECs) and neural stem cells (NSCs) into the subretinal space of Royal College of Surgeons (RCS) rats. **(A–E)** Representative electroretinogram (ERG) results in the blank, PBS, OEC, NSC, and OEC + NSC groups at 4 weeks postoperation. **(F–J)** Representative ERG results in the blank, PBS, OEC, NSC, and OEC + NSC groups at 8 weeks postoperation. **(K–O)** Representative ERG results in the blank, PBS, OEC, NSC, and OEC + NSC groups at 12 weeks postoperation. **(P)** Relative statistical analysis of all groups. a and b point in **(E,J,O)** indicated the a and b waves, which represented the light absorption of photoreceptors and postsynaptic responses of photoreceptors, respectively. $^{**}P < 0.01$; $^{***}P < 0.001$.

NSC and the OEC + NSC group ($P > 0.05$; **Figures 1N–P**). In addition, the NSC group also presented significantly higher b-wave amplitude than that of the OEC group at 12 weeks postoperation ($P > 0.05$; **Figures 1M,N,P**). These results indicated that, compared with the single transplantation, the combined transplantation of OECs and NSCs enhanced the electrophysiological improvement of RCS.

Combined Transplantation Showed Better Photoreceptor Protection in RCS Rats

To further evaluate the morphological effects of cell transplantation, the ONL thickness was measured. Compared with the blank and the PBS group, three transplantation groups all presented significant protective effects on the ONL ($P < 0.001$; **Figures 2A–P**). Among the transplantation groups, both the OEC + NSC and the OEC groups showed significantly better protective effects on the ONL than that of the NSC group at 4 weeks postoperation ($P < 0.01$, $P < 0.05$; **Figures 2C–E,P**). No significant difference was found between the OEC + NSC and the OEC group at this time point ($P > 0.05$; **Figures 2C,E,P**). At 8 weeks postoperation, the OEC + NSC group displayed a significant increase in ONL thickness compared to that of either the OEC or the NSC group ($P < 0.001$, $P < 0.001$; **Figures 2H–J,P**). There was no significant difference between the OEC and the NSC group at this time point ($P > 0.05$; **Figures 2H,I,P**). A similar change was found at 12 weeks postoperation. The OEC + NSC group still presented a significant higher ONL thickness than that of the two single transplantation groups (OECs, $P < 0.001$; NSCs, $P < 0.05$; **Figures 2M–O,P**). No significant difference was found between the two single transplantation groups ($P > 0.05$; **Figures 2M–O,P**). Moreover, when retinal sections were stained with recoverin to identify photoreceptors, all three cell transplant groups showed evidence of increased numbers of photoreceptors at each time point examined (**Figures 3A–O**). Taken together, these results suggested that combined transplantation showed a better protective effect on the ONL than single transplantation.

Endogenous Stem Cell Activation After Combined Transplantation

To investigate the possibility that endogenous stem cell responses might underlie the apparent therapeutic actions of cell transplantation, we labeled retinas with Sox2 antibodies (Tian et al., 2011). The results showed that Sox2-positive cells

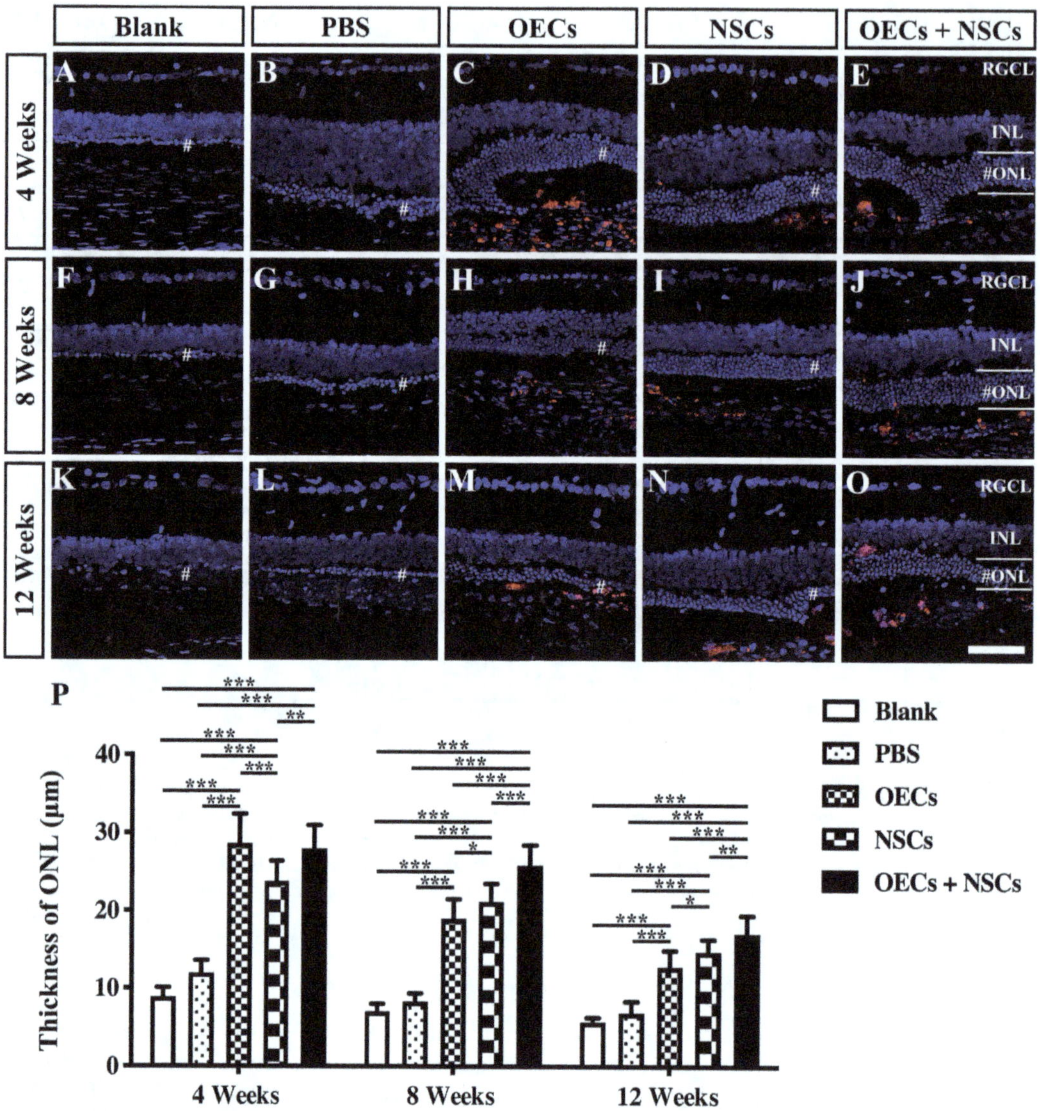

FIGURE 2 | Protection of ONL after combined transplantation of rat olfactory ensheathing cells (OECs) and neural stem cells (NSCs) into the subretinal space of Royal College of Surgeons (RCS) rats. **(A–E)** Retina sections with 4′,6-diamidino-2-phenylindole (DAPI) staining in the blank, PBS, OEC, NSC, and OEC + NSC groups at 4 weeks postoperation. **(F–J)** Retina sections with DAPI staining in the blank, PBS, OEC, NSC, and OEC + NSC groups at 8 weeks postoperation. **(K–O)** Retina sections with DAPI staining in the blank, PBS, OEC, NSC, and OEC + NSC groups at 12 weeks postoperation. The mark "#" represented the out nuclear layer. **(P)** Statistical analysis of the ONL thickness in transplantation groups. Scale bar: **(A–O)** 50 μm. *$P < 0.05$; **$P < 0.01$; ***$P < 0.001$.

within the retina were evenly distributed in the inner nuclear layer. At 4 weeks post-operation, three transplantation groups all presented a significant higher number of Sox2-positive cells than that of either the Blank group or the PBS group ($P < 0.001$; **Figures 4A–E,P**). Among the three transplantation groups, the number of Sox2-positive cells in the OEC + NSC group was significantly higher than that of either the OEC group or the NSC group at 4 weeks post-operation ($P < 0.05$, $P < 0.01$; **Figures 4C–E,P**). This significant difference lasted until 8 weeks post-operation ($P < 0.01$, $P < 0.01$; **Figures 4H–J,P**), while no significant differences in Sox2 expression were observed among the Blank group, the PBS, and the two single transplantation groups at this time point ($P < 0.05$; **Figures 4F–J,P**). However, by 12 weeks post-operation, there was no significant difference among groups ($P < 0.05$; **Figures 4K–O,P**). These results suggested that combined transplantation activated more endogenous stem cells at the early stage of transplantation.

Influence on the Reactive Gliosis of Müller Cells After Combined Transplantation

The reactive gliosis of Müller cells following retina damage become an obstacle to retinal regeneration. To generally analyze the gliosis of Müller cells, we performed the Western blot

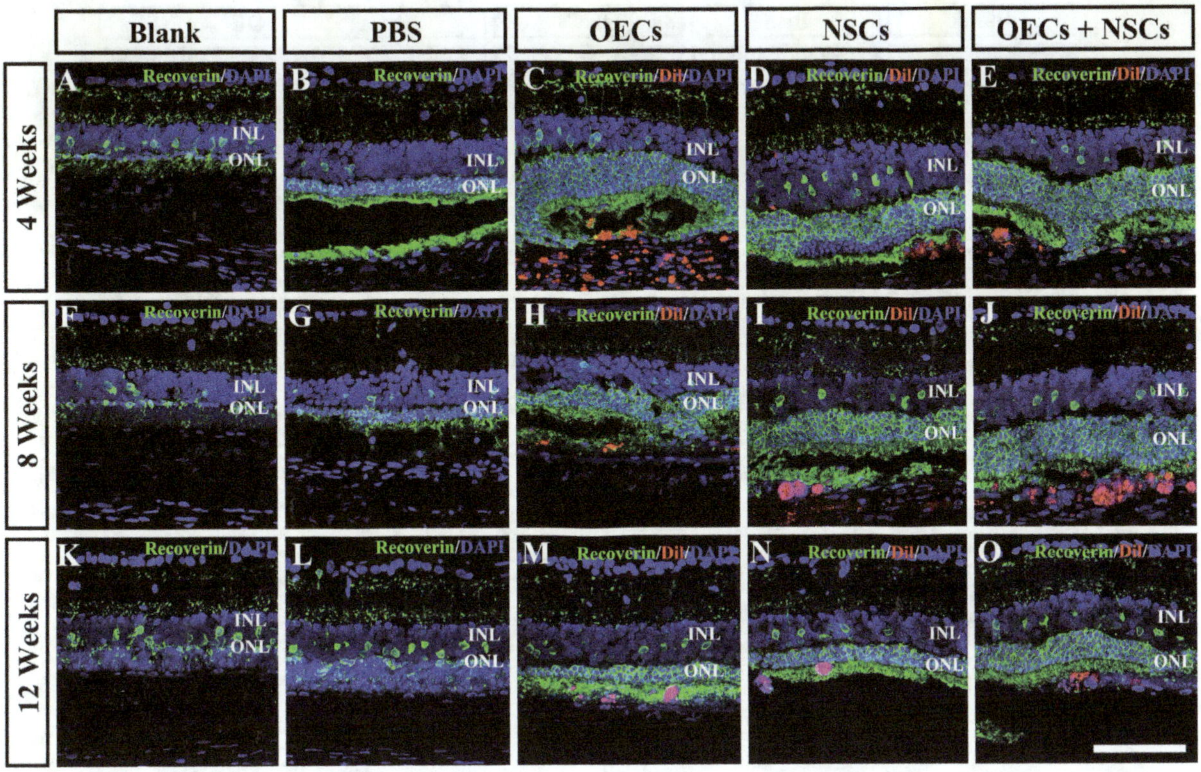

FIGURE 3 | Protection of photoreceptors after combined transplantation of rat olfactory ensheathing cells (OECs) and neural stem cells (NSCs) into the subretinal space of Royal College of Surgeons (RCS) rats. **(A–E)** Immunofluorescence of Recoverin in the blank, PBS, OEC, NSC, and OEC + NSC groups at 4 weeks postoperation. **(F–J)** Immunofluorescence of Recoverin in the blank, PBS, OEC, NSC, and OEC + NSC groups at 8 weeks postoperation. **(K–O)** Immunofluorescence of Recoverin in the blank, PBS, OEC, NSC, and OEC + NSC groups at 12 weeks postoperation. Scale bar: **(A–O)** 75 μm.

analysis and the immunofluorescence of GFAP at 4, 8, and 12 weeks posttransplantation. WB results showed that at 4 weeks postoperation, the blank, the PBS, and the NSC groups presented similar GFAP expression levels, which were significantly higher than those in the OEC group and the OEC + NSC group ($P < 0.05$, $P < 0.01$; **Figures 5A,B**). However, at 8 weeks postoperation, GFAP expression level in three transplantation groups all significantly decreased compared to that in the blank and the PBS groups (OEC, $P < 0.05$; NSC, $P < 0.001$; OEC + NSC, $P < 0.001$; **Figures 5A,B**). The situation remained similar at 12 weeks postoperation; the only difference was that compared with the OEC group, the NSC group presented a significantly lower GFAP expression level ($P < 0.05$, **Figures 5A,B**).

We also divided the retina into upper and lower parts when analyzing the results of immunofluorescence (**Figure 5C**). The boundary was set to be the outer border of ONL (the inner border of SRS) (solid line in **Figure 5C**). The fluorescence intensity (FI) within the inner part and the outer part of the retina was measured and labeled as upper and lower, respectively (**Figure 5C**). Results showed that the trends of GFAP FI in the upper part of the retina were similar to the trends of the GFAP expression level in WB at all three time points (**Figures 5B,S**). Interestingly, the results of GFAP FI in the lower part of the retina showed distinct differences. At 4 weeks postoperation, the OEC and the OEC + NSC group presented significantly lower GFAP FIs than those in the blank, the PBS, and the NSC groups, respectively ($P < 0.05$, $P < 0.01$, $P < 0.01$; **Figures 5D–H,S**). There was no significant difference among the blank, the PBS, and the NSC groups. However, when it came to 8 and 12 weeks postoperation, situation changed dramatically. The NSC and OEC + NSC groups both showed significant increases in GFAP FI compared to those in the blank, the PBS, and the OEC groups at these two time points ($P < 0.001$, $P < 0.001$, $P < 0.001$; **Figures 5I–S**). While the OEC + NSC group displayed a significant decrease in GFAP FI compared to that in the NSC group ($P < 0.001$; **Figures 5L,M,Q,R,S**). Taken together, transplantation of OECs and NSCs can decrease the gliosis following retinal degeneration. Moreover, transplantation of NSCs would bring gliosis to the SRS, which could be inhibited by cotransplantation of OECs.

The Possible Rescuing Mechanism of the Combined Transplantation

To further investigate the mechanism of rescue following cotransplantation, we evaluated the influence of migration and the cell state of NSCs upon coculture with OECs. The migration of transplanted cells within the SRS was first investigated. The results showed that the OEC + NSC group presented

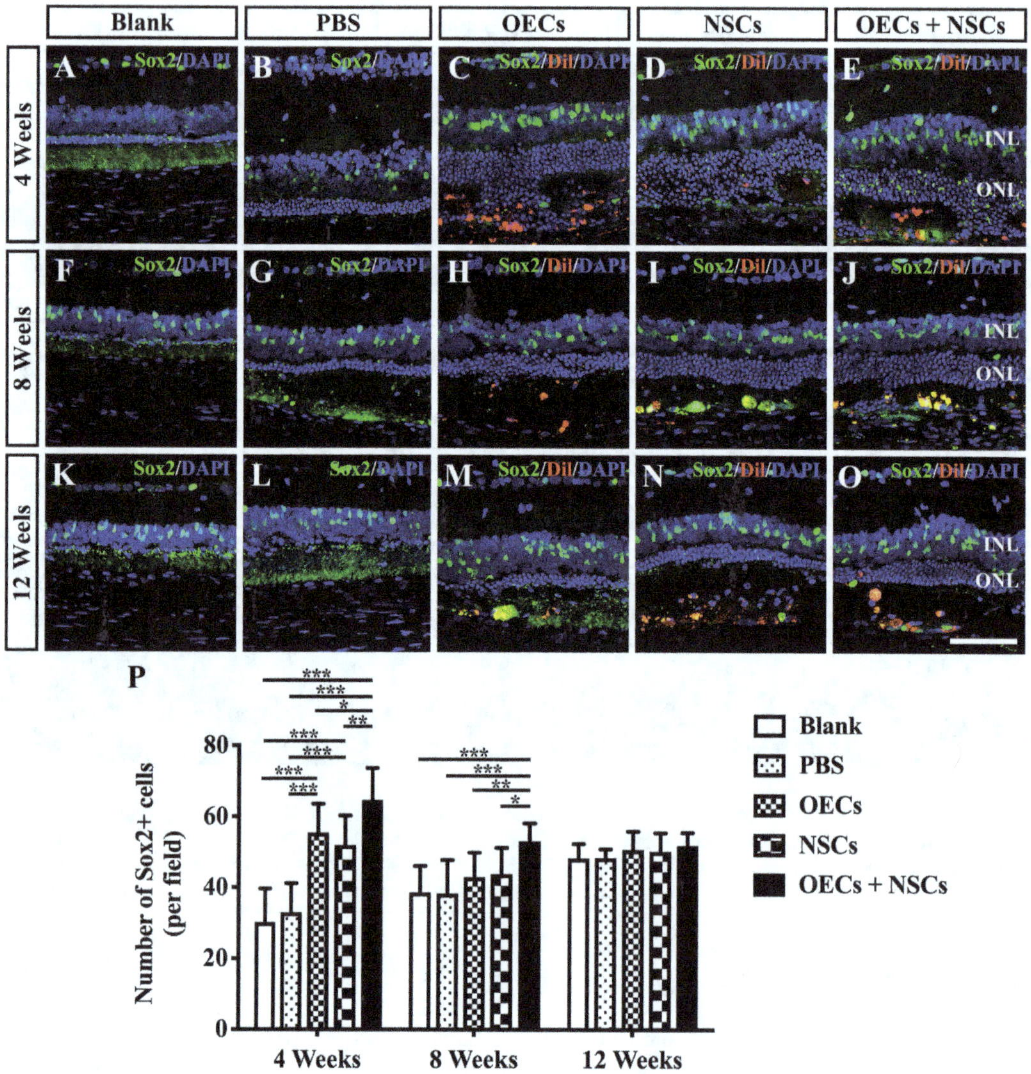

FIGURE 4 | Activation of endogenous retinal stem cells after combined transplantation of rat olfactory ensheathing cells (OECs) and neural stem cells (NSCs) into the subretinal space of Royal College of Surgeons (RCS) rats. **(A–E)** Immunofluorescence of Sox2 in the blank, PBS, OEC, NSC, and OEC + NSC groups at 4 weeks postoperation. **(F–J)** Immunofluorescence of Sox2 in the blank, PBS, OEC, NSC, and OEC + NSC groups at 8 weeks postoperation. **(K–O)** Immunofluorescence of Sox2 in the blank, PBS, OEC, NSC, and OEC + NSC groups at 12 weeks postoperation. **(P)** Statistical analysis of the numbers of Sox2-positive cells in transplantation groups. Scale bar: **(A–O)** 50 μm. $^{*}P < 0.05$; $^{**}P < 0.01$; $^{***}P < 0.001$.

a significantly shorter migration distance compared to that of the NSC group at 4 weeks postoperation ($P < 0.05$; **Figures 6B,C,J**). No significant difference was detected between the two single transplantation groups ($P > 0.05$; **Figures 6A,B,J**). However, by 8 and 12 weeks postoperation, the OEC + NSC group presented a significant increase in migration distance compared to that of the single transplantation groups ($P < 0.05$; **Figures 6D–J**). There was no significant difference between the two single transplantation groups at these two time points ($P > 0.05$; **Figures 6D–H,J**). An *in vitro* Transwell system was used to investigate the cell migration within the retina tissue (**Figure 6K**). After 12 days of culture, we found that there was little vertical migration of NSCs in the retinal tissue from the NSCs alone group (**Figure 6L**), while in the coculture group, there were many more cells that entered the retinal tissue (**Figure 6M**). Both *in vivo* and *in vitro* analyses confirmed the cell-migration-enhancement effect following cotransplantation. To further explore the influence of the cell state of NSCs following cotransplantation, coculture of OECs and NSCs via Transwell system was performed (**Figure 7I**). Results showed no significant difference in proliferation between the NSC single culture group and the NSC + OEC coculture group ($P > 0.05$; **Figures 7A,E,J**). However, the NSC + OEC coculture group presented significant higher Pax6 and Sox2 expressions than those of the NSC single culture group (Pax6, $P < 0.001$; Sox2, $P < 0.001$; **Figures 7B,C,F,G,J**). Meanwhile, a significant decrease in GFAP expression was also found in the NSC + OEC coculture group, compared to the NSC single culture group

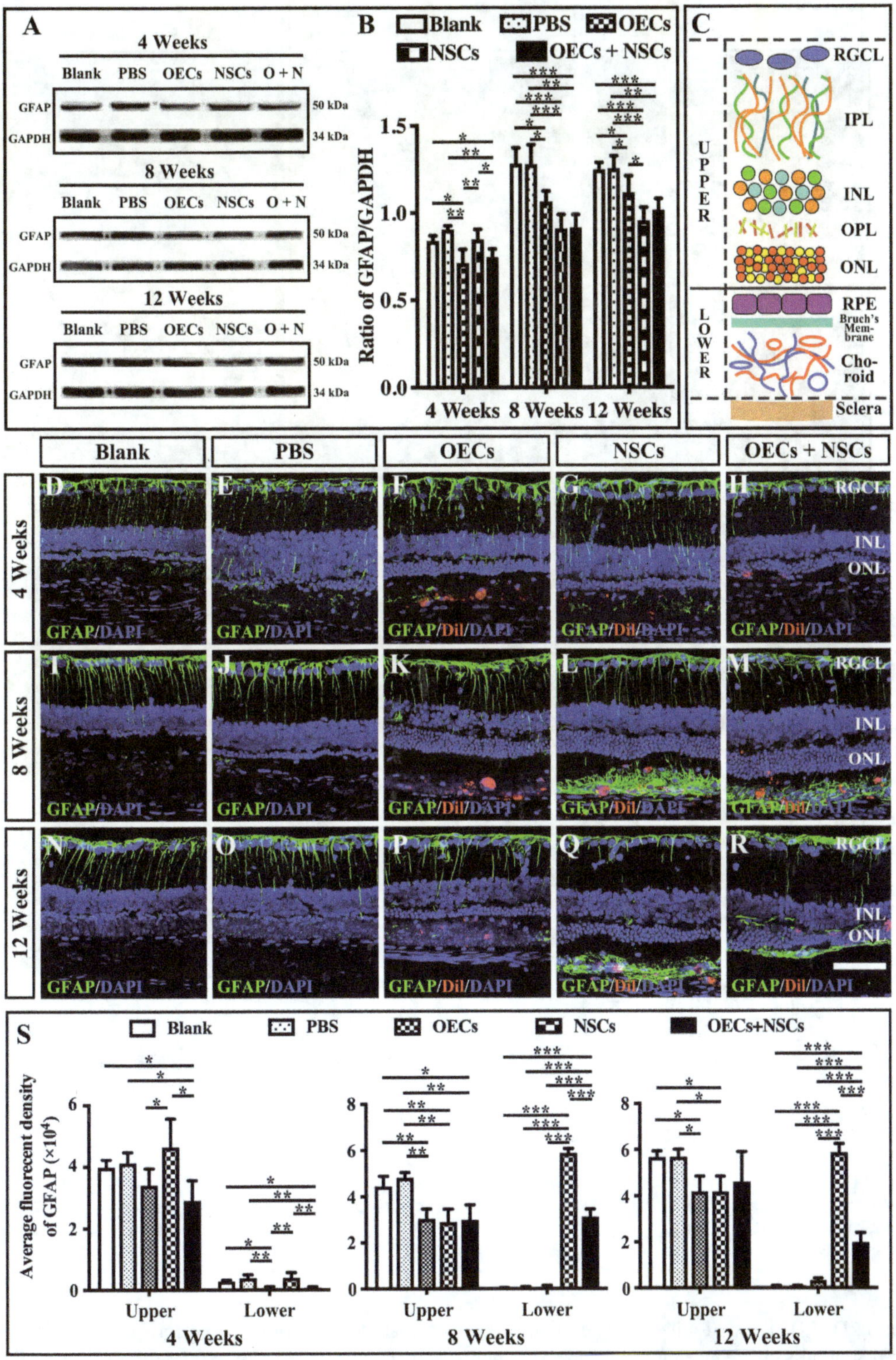

FIGURE 5 | Influence of the reactive gliosis of Muller cells after combined transplantation of rat olfactory ensheathing cells (OECs) and neural stem cells (NSCs) into the subretinal space of Royal College of Surgeons (RCS) rats. **(A,B)** The Western blot analysis of glial fibrillary acidic protein (GFAP) expression after transplantation. **(A)** The band of GFAP and glyceraldehyde 3-phosphate dehydrogenase (GAPDH) at 4, 8, and 12 weeks postoperation. **(B)** Corresponding analysis of the Western blot results. **(C)** Schematic diagram of the division of the retina. Solid line indicated the boundary of the division. **(D–H)** Immunofluorescence of GFAP in the blank, PBS, OEC, NSC, and OEC + NSC groups at 4 weeks postoperation. **(I–M)** Immunofluorescence of GFAP in the blank, PBS, OEC, NSC, and OEC + NSC groups at 8 weeks postoperation. **(N–R)** Immunofluorescence of GFAP in the blank, PBS, OEC, NSC, and OEC + NSC groups at 12 weeks postoperation. **(S)** The average fluorescent intensity in the blank, PBS, OEC, NSC, and OEC + NSC groups of the upper and lower parts of the retina. RGCL, retinal ganglion cell layer; IPL, inner plexiform layer; INL, inner nuclear layer; OPL, outer plexiform layer; ONL, outer nuclear layer. Scale bar: **(A–O)** 50 μm. $*P < 0.05$; $**P < 0.01$; $***P < 0.001$.

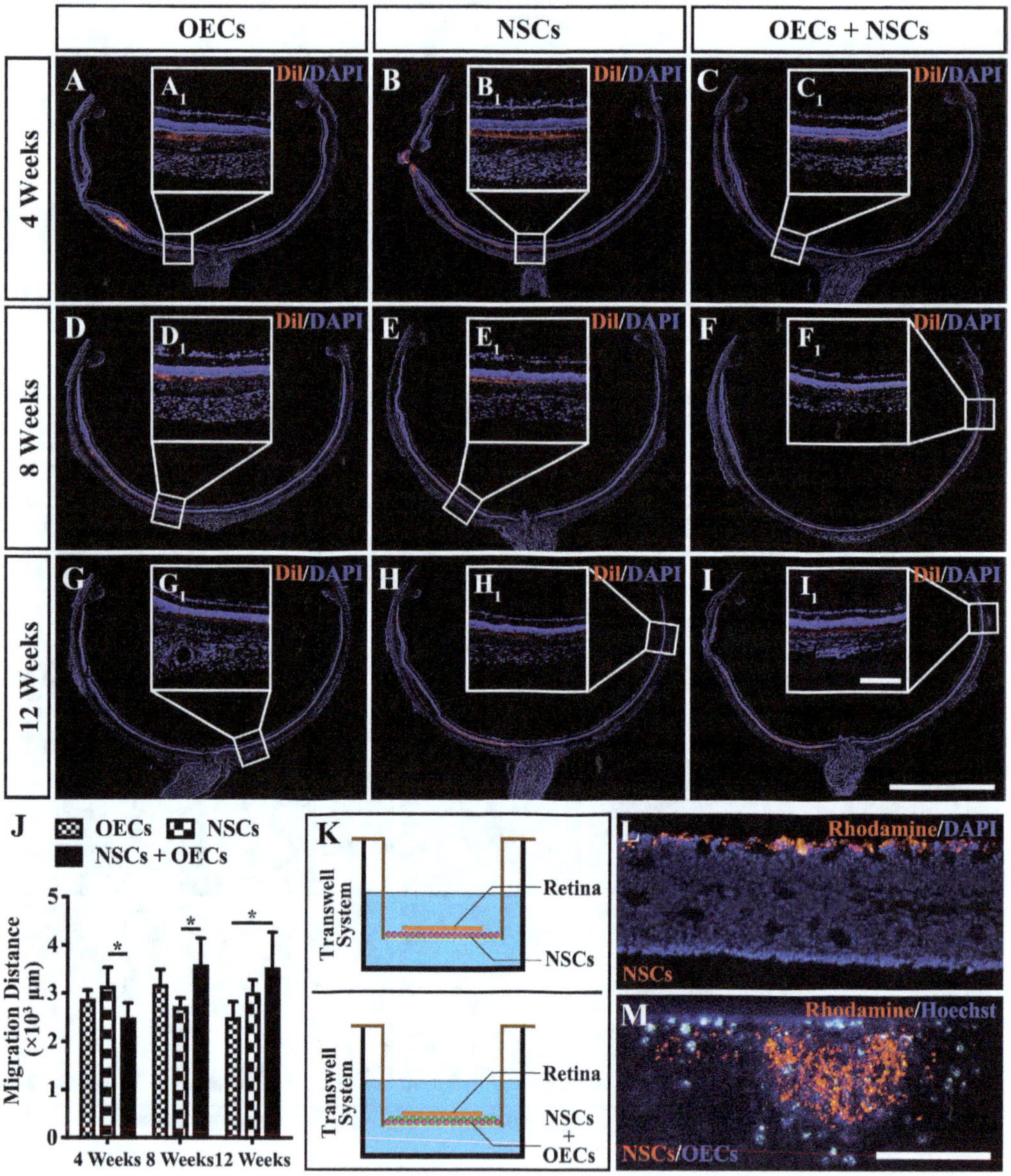

FIGURE 6 | Migration of transplanted cells after combined transplantation of rat olfactory ensheathing cells (OECs) and neural stem cells (NSCs) into the subretinal space of Royal College of Surgeons (RCS) rats. **(A–C)** General microscopy of the retina section in the OEC, NSC, and OEC + NSC groups at 4 weeks postoperation. **(D–F)** General microscopy of the retina section in the OEC, NSC, and OEC + NSC groups at 8 weeks postoperation. **(G–I)** General microscopy of the retina section in OEC, NSC, and OEC + NSC groups at 12 weeks postoperation. **(A_1–I_1)** Relative partial enlargements showed the farthest location of transplanted cells. **(J)** Relative statistical analysis of transplantation groups. **(K)** The schematic diagram of the migration assay performed via Transwell system. **(L,M)** The results of migration analysis. NSCs were stained by Rhodamine, and OECs were stained by Hoechst. Cell migration with NSCs alone **(L)** as well as OECs and NSCs mixture **(M)** were tested. Scale bar: **(B,C)** 200 μm. Scale bar: **(A–I)** 2 mm; **(A_1–I_1)** 200 μm; **(L–M)** 200 μm.$^*P < 0.05$.

($P < 0.001$; **Figures 7D,H,J**). These results suggested that NSCs exhibited enhanced stemness but reduced gliotic tendency when cocultured with OECs.

DISCUSSION

In the current study, the combined transplantation of OECs and NSCs produced a better neuroprotective effect and delayed retinal degeneration in a more effective way than single-cell transplantation in the retinas of RCS rats. Endogenous stem cell activation, enhanced migration of transplanted cells, and stemness maintenance of NSCs following combined transplantation may be possible underlying mechanisms. Comparing to the normal amplitude of ERG b wave in age-corresponding LE rats (about 1,300 ± 200 μV) (Xue et al., 2017), the ratios of the b-wave amplitude between the combined transplantation group and the normal LE rats at 4, 8, and

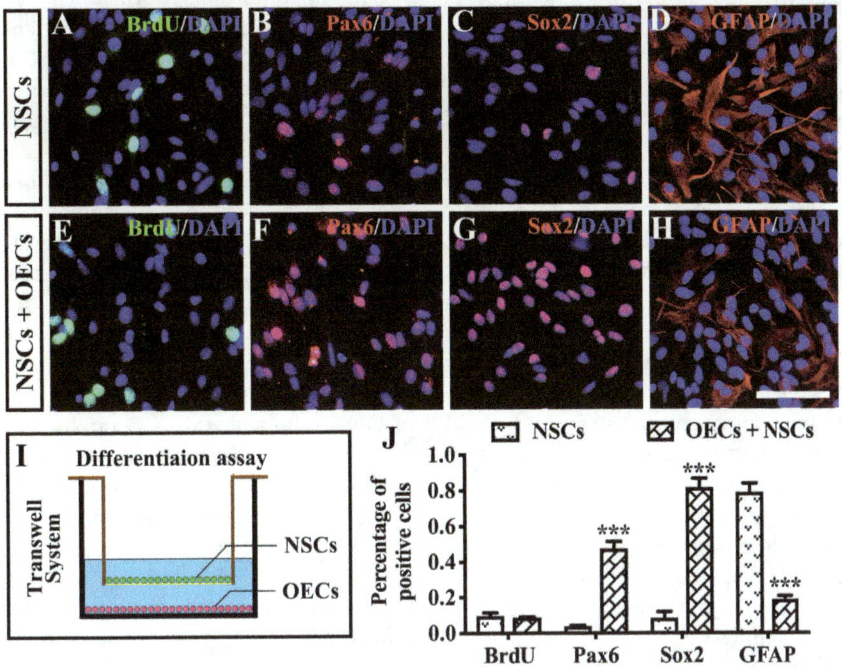

FIGURE 7 | The influence on differentiation of neural stem cells (NSCs) after *in vitro* coculture with olfactory ensheathing cells (OECs) via Transwell system. **(A–H)** Immunofluorescence analysis of BrdU, Pax6, Sox2, and glial fibrillary acidic protein (GFAP) in NSCs cultured along group **(A–D)** and NSCs + OECs coculture group **(E–H)**. **(I)** The schematic diagram of the differentiation assay performed via Transwell system. **(J)** Relative statistical analysis of **(A–H)**. Scale bar: **(A–H)** 25 μm. $^{***}P < 0.001$.

12 weeks postoperation were 10.1, 5.6, and 4%, respectively. Although positive influence was within a limited extent and dropped obviously with time, the better improvement reflected by ERG still lasted for 8 weeks, which is of great significance from bench to bedside. These results indicated that the combined transplantation of OECs and NSCs may be an alternative stem cell therapy for patients with RDDs in the future.

However, in the present study, the observation time lasted for only 12 weeks. The continued functional and morphological improvements were not observed. In addition, no direct measure of visual function was conducted in the present study, which could lead to incomplete assessment and functional bias. More importantly, limited observation period was not sufficient to discover underlying tumor formation possibility and other safety problems. In the future, studies with longer observation period and functional or behavioral tests need to be conducted.

Our previous research confirmed the visual restorative effect of OECs (Huo et al., 2012). The inhibition of gliosis via the downregulation of the Notch signaling pathway in Müller cells might be a possible mechanism (Xie et al., 2017). However, in this study, we found that combined transplantation of OECs and NSCs showed a better preservative effect than that of the OEC single transplantation, and this effect lasted for 8 weeks after transplantation.

Differences in the endogenous stem cell activation may be one of the explanations. In the spinal cord injury, the activation of endogenous stem cells was considered a promising method for spinal cord recovery (Qin et al., 2015). In the retina, Müller cells were reported to be reprogrammed as the progenitor cells and repair the degenerated retina (Jorstad et al., 2017; Yao et al., 2018). Under physiological conditions, Müller cells remain quiescent to avoid depletion of stem-cell pool. On the contrary, they are able to exit from latent state to proliferate and differentiate upon injuries (Madelaine and Mourrain, 2017). Our previous study found that the dedifferentiation of Müller cells was increased after the transplantation of retinal stem cells into the SRS of RCS rats, bringing both morphological and functional improvement to the host (Tian et al., 2011). Besides, Sox2 was identified to be re-expressed in Müller cells after injury, indicating that Müller cells exited the quiescent state and generated new retinal neurons (Gorsuch et al., 2017). As more Sox2-positive cells within the inner nuclear layer were found after combined transplantation in the present study, stronger endogenous repair of retina following activation of more endogenous stem cells might be an underlying mechanism.

Retinal gliosis is a non-neoplastic retinal glial proliferation followed by a complex retinal response participated by Müller cells, microglia cells, and alterations of the vasculature (Bringmann et al., 2006). It acts as a double-edged sword in the pathological process of RCS rats. On the one hand, reactive gliosis is a physical process that can be regarded as a cellular response to protect the retina from further damage. In addition, moderate gliosis can promote retinal repair following pathological insult (Graca et al., 2018). On the other hand, however, gliosis after nervous impairment including spinal cord and retina was widely known as an obstacle for the regeneration of neuron and neural dendrites (Bringmann et al., 2006; Wu et al., 2011).

In the present study, apart from the retinal gliosis, there emerged a certain amount of GFAP-positive neural fibers within the SRS following the NSCs transplantation and the combined transplantation. This NSC-derived gliosis might impede the neural regeneration and the repairing effect resulting from transplanted stem cells. As NSCs were able to differentiate into glial cells, these neural fibers most likely came from the differentiation of NSCs. Basically, although NSCs were widely used to treat various neurodegenerative diseases, the maintaining of stemness had become an imposing barrier. To compensate the drawback, a promising way was to use another kind of stem cell to influence NSCs. Study had shown that coculture of human NSCs with human mesenchymal stem cells could significantly extend the stemness of NSCs via activating Notch-1 signal transduction (Haragopal et al., 2015). In the present study, with *in vitro* Transwell system, the stemness of NSCs was enhanced when OECs and NSCs were cocultured. Combined transplantation of OECs and NSCs also repressed the NSC-derived gliosis within SRS, illustrating that the differentiation of NSCs was inhibited and the stemness of NSCs was maintained by OECs. This gliosis regulating effect of OECs can be attributed to the main function of OECs as supporting cells. However, relatively low retina-protection effect was also observed following OEC single transplantation. As mentioned above, reactive gliosis can also bring about positive effect to retinal repair; inhibition of gliosis might be the cause of the low pro-retina activity. It should be noted that the underlying molecular mechanisms relating to endogenous stem cell activation as well as gliosis inhibition after cotransplantation of OECs and NSCs were not explored in the current study, which should be further explored in the future.

Besides, microglia show close relationship to the gliosis of Müller cells (Gao et al., 2015). As both Müller cells and microglia are responsible for the secretion of neurotrophic factor within the retina, the network formed by microglia–Müller glia–photoreceptors can significantly influence the microenvironment during retinal degeneration (Harada et al., 2002). On the one hand, degenerated photoreceptors induce the activation and migration of microglia from the inner to the outer retina. During this procedure, activated microglia alter the trophic factor production and further cause the gliosis of Müller glia, which can influence the production of neurotrophic factor in Müller glia (Harada et al., 2002). On the other hand, gliosis of Müller glia triggered by photoreceptor degeneration also leads to the apoptosis of neurons resulting in more severe activation of microglia (Telegina et al., 2018). These aspects trap the microglia–Müller glia–photoreceptor network into a vicious cycle and cause the deficiency of trophic factors, which is detrimental to the survival of photoreceptors. However, after combined transplantation of OECs and NSCs, gliosis was found to be reduced in the present study. Besides, our previous work confirmed the inhibition of microglia activation following NSCs transplantation (Li et al., 2016). These situations might improve the neurotrophic factor secretion situation and improve the photoreceptor survival subsequently.

The migration of transplanted cells is essential for the development of their function in the transplanted area. Migration exists in two dimensions: horizontal and vertical. Horizontal migration determines the scope of transplantation (McGill et al., 2012; Peng et al., 2014), while vertical migration determines the function of the inner retinal layer (Santos-Ferreira et al., 2016). Our previous results confirmed that better retinal preservation effects can be derived from combined transplantation due to the enhanced horizontal and vertical migration of transplanted cells (Qu et al., 2017). In the present study, differences in cell migration were also observed between the combined and the single transplantation groups. Grafted cells reached further following combined transplantation. Moreover, NSCs were able to migrate into the inner layer of the retina in the presence of OECs, although this was rarely observed when NSCs were cultured alone. A study showed that, during physiological development, the gene expression products of OECs, including Nelf and Semaphorin 4, are responsible for neuronal migration within the CNS of mice (Geller et al., 2013), indicating the potential migration-enhancing ability of OECs. In addition, OECs are fundamentally characterized by their ability to promote axonal regeneration (Yang et al., 2015). Our previous work also confirmed the stimulating effect on neuronal survival and the outgrowth of OECs, which was due to the phagocytosis of cell debris from OECs (Li et al., 2017). In this way, the processes and integration of NSCs might be extended and promoted following cotransplantation with OECs.

What is more, the fate of transplanted cells should also be taken into consideration. The main functions of NSCs transplanted into the subretinal space in RCS rats were phagocytosis of photoreceptor outer segments, secretion of neurotrophic factors, and inhibition of microglia (McGill et al., 2012; Jung et al., 2013; Li et al., 2016). Although NSCs were observed to be differentiated into photoreceptors and opsin-positive retinal cells, integration of NSC into ONL was hardly found (Nishida et al., 2000; Lin et al., 2014). As for OECs, our previous study showed that the function of OECs transplantation into the subretinal space of RCS rats mainly consisted of two aspects: (1) the microenvironment regulation effects, including secretion of neurotrophic factors and inhibition of the formation of reactive oxygen species; (2) the suppression of the gliotic injury response of the Müller cells (Huo et al., 2011, 2012; Xue et al., 2017). However, the differentiation and integration of OECs following retinal transplantation was not observed. Although there was better retina-protection effect after combined transplantation of OECs and NSCs, current study still presented a decreasing trend in both functional and morphological results. Considering the immune response following exotic cell transplantation, we speculate that the major ending of NSCs and OECs is the apoptosis after a certain time period. Moreover, we could not rule out the possibility that the greater improvements in the cotransplanted condition might be due simply to increased numbers of transplanted cells. The differences in responses to transplants of each cell type separately suggested that the greater improvements were more likely to be due to the combined effects of the two cell types.

Besides, in our previous study, transplantation of OECs mixed with ONFs had been performed. Results showed that OECs (but not ONFs) phagocytose porcine photoreceptor outer segments. The phagocytosis ability was even stronger than RPE (Huo et al., 2012). However, Li et al. also found that OECs and ONFs had synergistic effects in promoting axon regeneration (Li et al., 2005). In the present study, there is roughly 50/50 mixture of OECs and fibroblasts in the transplanted cells, which accords with the cell ratio in our previous study (Huo et al., 2011). OECs are known for their properties like interacting with the glial scar, stimulating angiogenesis, and promoting axonal outgrowth (Roet and Verhaagen, 2014; Gomez et al., 2018). Fibroblasts, both from the olfactory bulb and other parts of the body, can affect the function of other cells and present a lot of different effects. However, as the purpose of this study is to focus on the function of OECs, we referred the mixture of OECs/ONFs as OECs.

Since OEC and NSC transplantation have both entered into the clinical research stage, the advancement of the combined transplantation of OECs and NSCs from bench to bedside in the future is promising. As the main effect of cotransplantation was the activation of endogenous stem cells and improvements in the microenvironment, this therapeutic method could prospectively benefit all types of RDDs as well as other retinal lesions, including glaucoma and ocular trauma. Further research is required to confirm this possibility.

CONCLUSION

In summary, the combined transplantation of NSCs and OECs better preserved retina than transplantation with NSCs or OECs alone, and this effect was verified by improved ERGs and increased ONL thickness in the combined transplantation group. Increased endogenous stem cell activation, better maintenance of NSC stemness, and enhanced migration of transplanted cells following combined transplantation might be potential mechanisms. All of these results illustrated that the combined transplantation of NSCs and OECs might be a possible alternative for the treatment of RDDs.

AUTHOR CONTRIBUTIONS

HX and ZY contributed to the design of the project. WZ and LG contributed to the *in vivo* experiments and discussed the results. WZ, YL, QL, and YZ contributed to the *in vitro* experiments. LG and HX summarized the data and contributed in manuscript preparation. All authors read and approved the manuscript.

REFERENCES

Blasiak, J. (2020). Senescence in the pathogenesis of age-related macular degeneration. *Cell. Mol. Life Sci.* doi: 10.1007/s00018-019-03420-x [Epub ahead of print].

Bringmann, A., Pannicke, T., Grosche, J., Francke, M., Wiedemann, P., Skatchkov, S. N., et al. (2006). Muller cells in the healthy and diseased retina. *Prog. Retin. Eye Res.* 25, 397–424. doi: 10.1016/j.preteyeres.2006.05.003

D'Cruz, P. M., Yasumura, D., Weir, J., Matthes, M. T., Abderrahim, H., LaVail, M. M., et al. (2000). Mutation of the receptor tyrosine kinase gene Mertk in the retinal dystrophic RCS rat. *Hum. Mol. Genet.* 9, 645–651. doi: 10.1093/hmg/9.4.645

Gao, L., Chen, X., Tang, Y., Zhao, J., Li, Q., Fan, X., et al. (2015). Neuroprotective effect of memantine on the retinal ganglion cells of APPswe/PS1DeltaE9 mice and its immunomodulatory mechanisms. *Exp. Eye Res.* 135, 47–58. doi: 10.1016/j.exer.2015.04.013

Geller, S., Kolasa, E., Tillet, Y., Duittoz, A., and Vaudin, P. (2013). Olfactory ensheathing cells form the microenvironment of migrating GnRH-1 neurons during mouse development. *Glia* 61, 550–566. doi: 10.1002/glia.22455

Gomez, R. M., Sanchez, M. Y., Portela-Lomba, M., Ghotme, K., Barreto, G. E., Sierra, J., et al. (2018). Cell therapy for spinal cord injury with olfactory ensheathing glia cells (OECs). *Glia* 66, 1267–1301. doi: 10.1002/glia.23282

Gorsuch, R. A., Lahne, M., Yarka, C. E., Petravick, M. E., Li, J., and Hyde, D. R. (2017). Sox2 regulates Muller glia reprogramming and proliferation in the regenerating zebrafish retina via Lin28 and Ascl1a. *Exp. Eye Res.* 161, 174–192. doi: 10.1016/j.exer.2017.05.012

Graca, A. B., Hippert, C., and Pearson, R. A. (2018). Muller glia reactivity and development of gliosis in response to pathological conditions. *Adv. Exp. Med. Biol.* 1074, 303–308. doi: 10.1007/978-3-319-75402-4_37

Harada, T., Harada, C., Kohsaka, S., Wada, E., Yoshida, K., Ohno, S., et al. (2002). Microglia-Muller glia cell interactions control neurotrophic factor production during light-induced retinal degeneration. *J. Neurosci.* 22, 9228–9236. doi: 10.1523/jneurosci.22-21-09228.2002

Haragopal, H., Yu, D., Zeng, X., Kim, S.-W., Han, I.-B., Ropper, A. E., et al. (2015). Stemness enhancement of human neural stem cells following bone marrow MSC coculture. *Cell Transplant* 24, 645–659. doi: 10.3727/096368915X687561

Huo, S. J., Li, Y., Raisman, G., and Yin, Z. Q. (2011). Transplanted olfactory ensheathing cells reduce the gliotic injury response of Muller cells in a rat model of retinitis pigmentosa. *Brain Res.* 1382, 238–244. doi: 10.1016/j.brainres.2010.12.079

Huo, S. J., Li, Y. C., Xie, J., Li, Y., Raisman, G., Zeng, Y. X., et al. (2012). Transplanted olfactory ensheathing cells reduce retinal degeneration in Royal College of Surgeons rats. *Curr. Eye Res.* 37, 749–758. doi: 10.3109/02713683.2012.697972

Jonas, J. B., Cheung, C. M. G., and Panda-Jonas, S. (2017). Updates on the epidemiology of age-related macular degeneration. *Asia Pac. J. Ophthalmol.* 6, 493–497.

Jorstad, N. L., Wilken, M. S., Grimes, W. N., Wohl, S. G., VandenBosch, L. S., Yoshimatsu, T., et al. (2017). Stimulation of functional neuronal regeneration from Muller glia in adult mice. *Nature* 548, 103–107. doi: 10.1038/nature23283

Jung, G., Sun, J., Petrowitz, B., Riecken, K., Kruszewski, K., Jankowiak, W., et al. (2013). Genetically modified neural stem cells for a local and sustained delivery of neuroprotective factors to the dystrophic mouse retina. *Stem Cells Transl. Med.* 2, 1001–1010. doi: 10.5966/sctm.2013-0013

LaVail, M. M. (2001). Legacy of the RCS rat: impact of a seminal study on retinal cell biology and retinal degenerative diseases. *Prog. Brain Res.* 131, 617–627. doi: 10.1016/s0079-6123(01)31048-8

Li, Y., Field, P. M., and Raisman, G. (2005). Olfactory ensheathing cells and olfactory nerve fibroblasts maintain continuous open channels for regrowth of olfactory nerve fibres. *Glia* 52, 245–251. doi: 10.1002/glia.20241

Li, Y., Zou, T., Xue, L., Yin, Z. Q., Huo, S., and Xu, H. (2017). TGF-beta1 enhances phagocytic removal of neuron debris and neuronal survival by olfactory ensheathing cells via integrin/MFG-E8 signaling pathway. *Mol. Cell. Neurosci.* 85, 45–56. doi: 10.1016/j.mcn.2017.08.006

Li, Z., Zeng, Y., Chen, X., Li, Q., Wu, W., Xue, L., et al. (2016). Neural stem cells transplanted to the subretinal space of rd1 mice delay retinal degeneration by suppressing microglia activation. *Cytotherapy* 18, 771–784. doi: 10.1016/j.jcyt.2016.03.001

Lin, T. C., Hsu, C. C., Chien, K. H., Hung, K. H., Peng, C. H., and Chen, S. J. (2014). Retinal stem cells and potential cell transplantation treatments. *J. Chin. Med. Assoc.* 77, 556–561. doi: 10.1016/j.jcma.2014.08.001

Madelaine, R., and Mourrain, P. (2017). Endogenous retinal neural stem cell reprogramming for neuronal regeneration. *Neural Regen. Res.* 12, 1765–1767. doi: 10.4103/1673-5374.219028

Marsh, S. E., and Blurton-Jones, M. (2017). Neural stem cell therapy for neurodegenerative disorders: The role of neurotrophic support. *Neurochem. Int.* 106, 94–100. doi: 10.1016/j.neuint.2017.02.006

Mavija, M., Alimanovic, E., Jaksic, V., Kasumovic, S. S., Cekic, S., and Stamenkovic, M. (2014). Therapeutic Modalities of Exudative Age-related Macular Degeneration. *Med. Arch.* 68, 204–208.

McGill, T. J., Cottam, B., Lu, B., Wang, S., Girman, S., Tian, C., et al. (2012). Transplantation of human central nervous system stem cells - neuroprotection in retinal degeneration. *Eur. J. Neurosci.* 35, 468–477. doi: 10.1111/j.1460-9568.2011.07970.x

Nazari, H., Zhang, L., Zhu, D., Chader, G. J., Falabella, P., Stefanini, F., et al. (2015). Stem cell based therapies for age-related macular degeneration: the promises and the challenges. *Prog. Retin. Eye Res.* 48, 1–39. doi: 10.1016/j.preteyeres.2015.06.004

Nishida, A., Takahashi, M., Tanihara, H., Nakano, I., Takahashi, J. B., Mizoguchi, A., et al. (2000). Incorporation and differentiation of hippocampus-derived neural stem cells transplanted in injured adult rat retina. *Invest. Ophthalmol. Vis. Sci.* 41, 4268–4274.

Obernier, K., and Alvarez-Buylla, A. (2019). Neural stem cells: origin, heterogeneity and regulation in the adult mammalian brain. *Development* 146:dev156059. doi: 10.1242/dev.156059

Peng, Y., Zhang, Y., Huang, B., Luo, Y., Zhang, M., Li, K., et al. (2014). Survival and migration of pre-induced adult human peripheral blood mononuclear cells in retinal degeneration slow (rds) mice three months after subretinal transplantation. *Curr. Stem Cell Res. Ther.* 9, 124–133. doi: 10.2174/1574888x09666131219115125

Perlman, I. (1995). "The Electroretinogram: ERG," in *Webvision: The Organization of the Retina and Visual System*, eds H. Kolb, E. Fernandez, and R. Nelson (Salt Lake City, UT: University of Utah Health Sciences Center).

Pinilla, I., Lund, R. D., and Sauve, Y. (2005). Cone function studied with flicker electroretinogram during progressive retinal degeneration in RCS rats. *Exp. Eye Res.* 80, 51–59. doi: 10.1016/j.exer.2004.08.012

Qin, Y., Zhang, W., and Yang, P. (2015). Current states of endogenous stem cells in adult spinal cord. *J. Neurosci. Res.* 93, 391–398. doi: 10.1002/jnr.23480

Qu, L., Gao, L., Xu, H., Duan, P., Zeng, Y., Liu, Y., et al. (2017). Combined transplantation of human mesenchymal stem cells and human retinal progenitor cells into the subretinal space of RCS rats. *Sci. Rep.* 7:199. doi: 10.1038/s41598-017-00241-5

Ramon-Cueto, A., and Avila, J. (1998). Olfactory ensheathing glia: properties and function. *Brain Res. Bull.* 46, 175–187. doi: 10.1016/s0361-9230(97)00463-2

Roet, K. C., and Verhaagen, J. (2014). Understanding the neural repair-promoting properties of olfactory ensheathing cells. *Exp. Neurol.* 261, 594–609. doi: 10.1016/j.expneurol.2014.05.007

Santos-Ferreira, T., Llonch, S., Borsch, O., Postel, K., Haas, J., and Ader, M. (2016). Retinal transplantation of photoreceptors results in donor-host cytoplasmic exchange. *Nat. Commun.* 7:13028. doi: 10.1038/ncomms13028

Strauss, O., Stumpff, F., Mergler, S., Wienrich, M., and Wiederholt, M. (1998). The Royal College of Surgeons rat: an animal model for inherited retinal degeneration with a still unknown genetic defect. *Acta Anat.* 162, 101–111. doi: 10.1159/000046474

Telegina, D. V., Kozhevnikova, O. S., and Kolosova, N. G. (2018). Changes in retinal glial cells with age and during development of age-related macular degeneration. *Biochemistry* 83, 1009–1017. doi: 10.1134/S000629791809002X

Tian, C., Zhao, T., Zeng, Y., and Yin, Z. Q. (2011). Increased Muller cell de-differentiation after grafting of retinal stem cell in the sub-retinal space of Royal College of Surgeons rats. *Tissue Eng. Part A* 17, 2523–2532. doi: 10.1089/ten.TEA.2010.0649

Veleri, S., Lazar, C. H., Chang, B., Sieving, P. A., Banin, E., and Swaroop, A. (2015). Biology and therapy of inherited retinal degenerative disease: insights from mouse models. *Dis. Model. Mech.* 8, 109–129. doi: 10.1242/dmm.017913

Wang, Y. H., Yin, Z. Q., and Wang, Y. (2017). Synergistic effect of olfactory ensheathing cells and alpha-crystallin on restoration of adult rat optic nerve injury. *Neurosci. Lett.* 638, 167–174. doi: 10.1016/j.neulet.2016.12.042

Wong, W. L., Su, X., Li, X., Cheung, C. M. G., Klein, R., Cheng, C.-Y., et al. (2014). Global prevalence of age-related macular degeneration and disease burden projection for 2020 and 2040: a systematic review and meta-analysis. *Lancet Glob. Health* 2, e106–e116. doi: 10.1016/S2214-109X(13)70145-1

Wu, J., Stoica, B. A., and Faden, A. I. (2011). Cell cycle activation and spinal cord injury. *Neurotherapeutics* 8, 221–228. doi: 10.1007/s13311-011-0028-2

Xie, J., Huo, S., Li, Y., Dai, J., Xu, H., and Yin, Z. Q. (2017). Olfactory ensheathing cells inhibit Gliosis in retinal degeneration by downregulation of the Muller cell notch signaling pathway. *Cell Transplant* 26, 967–982. doi: 10.3727/096368917X694994

Xue, L., Zeng, Y., Li, Q., Li, Y., Li, Z., Xu, H., et al. (2017). Transplanted olfactory ensheathing cells restore retinal function in a rat model of light-induced retinal damage by inhibiting oxidative stress. *Oncotarget* 8, 93087–93102. doi: 10.18632/oncotarget.21857

Yang, H., He, B. R., and Hao, D. J. (2015). Biological roles of olfactory ensheathing cells in facilitating neural regeneration: a systematic review. *Mol. Neurobiol.* 51, 168–179. doi: 10.1007/s12035-014-8664-2

Yao, K., Qiu, S., Wang, Y. V., Park, S. J. H., Mohns, E. J., Mehta, B., et al. (2018). Restoration of vision after de novo genesis of rod photoreceptors in mammalian retinas. *Nature* 560, 484–488. doi: 10.1038/s41586-018-0425-3

10

A Human iPSC Line Carrying a *de novo* Pathogenic *FUS* Mutation Identified in a Patient with Juvenile ALS Differentiated into Motor Neurons with Pathological Characteristics

Li Chen[1], Yali Wang[2] and Jie Xie[3]*

[1]Department of Neurology, The First Affiliated Hospital of Zhengzhou University, Zhengzhou, China, [2]Department of Neurology, Affiliated Suzhou Hospital of Nanjing Medical University, Suzhou, China, [3]Help Stem Cell Innovations, Nanjing Life Science and Technology Innovation Park, Nanjing, China

***Correspondence:**
Li Chen
chenlisuda@163.com

Human-induced pluripotent stem cells (hiPSCs) are used to establish patient-specific cell lines and are ideal models to mirror the pathological features of diseases and investigate their underlying mechanisms *in vitro*, especially for rare genic diseases. Here, a *de novo* mutation c.1509dupA (p.R503fs) in fused in sarcoma (*FUS*) was detected in a patient with sporadic juvenile amyotrophic lateral sclerosis (JALS). JALS is a rare and severe form of ALS with unclear pathogenesis and no effective cure. An induced pluripotent stem cell (iPSC) line carrying the *de novo* mutation was established, and it represents a good tool to study JALS pathogenesis and gene therapy strategies for the treatment of this condition. The established human iPSC line carrying the *de novo* *FUS* mutation strongly expressed pluripotency markers and could be differentiated into three embryonic germ layers with no gross chromosomal aberrations. Furthermore, the iPSCs could be successfully differentiated into motor neurons exhibiting the pathological characteristics of ALS. Our results indicate that this line may be useful for uncovering the pathogenesis of sporadic JALS and screen for drugs to treat this disorder.

Keywords: juvenile amyotrophic lateral sclerosis, induced pluripotent stem cells, fused in sarcoma, *de novo* mutation, motor neurons

INTRODUCTION

Amyotrophic lateral sclerosis (ALS) is a neurodegenerative disorder characterized by progressive loss of upper and lower motor neurons, resulting in paralysis and death within an average of 3–5 years from disease onset (Fujimori et al., 2018). The mechanisms underlying the pathogenesis of ALS and effective therapeutic strategies targeting this disorder remain unclear and controversial. Juvenile amyotrophic lateral sclerosis (JALS) typically refers to disease onset before 25 years of age, and it is a rare and severe type of classical ALS. Gene mutations are the leading cause of ALS; however, the gene mutation profiles of classical ALS and JALS are different. For instance, mutations in *SOD1, TDP-43, C9ORF72, FUS, ANG, OPTN, UBQLN2,* and *ATXN2* have recently been identified in classical ALS, while mutations in *FUS, SIGMAR1,*

SPG11, ALS2, SOD1, C19ORF12, DDHD1, SETX, and *TARDBP* were detected in patients with JALS (Avemaria et al., 2011; Daoud et al., 2012; Siddiqi et al., 2014; Wu and Fan, 2016; Liu et al., 2017; Ma et al., 2018; Naumann et al., 2019). Whereas 90% of classical ALS cases are sporadic, the proportion falls to approximately 60% for JALS (Zou et al., 2016; Chen et al., 2020). Among the causative genes most commonly mutated in ALS, the variation in the proportion of fused in sarcoma (*FUS*) mutations in classic ALS and JALS is particularly striking. For instance, *FUS* is associated with approximately 3% of familial ALS and 1% of sporadic ALS cases (Guerrero et al., 2016), but with more than 30% of JALS cases (Mackenzie et al., 2011; Zou et al., 2016). We have also previously reported that *FUS* may account for an even greater proportion of cases of sporadic JALS, perhaps as much as 90%.

Juvenile onset and very rapid progression are the main clinical characteristics of sporadic JALS, and the average survival time after onset is only approximately 1.5 years (Zufiria et al., 2016). However, the pathogenesis of sporadic JALS is unclear owing to the difficulty in collecting nerve tissue samples and the very short survival time of patients. The extensive application of induced pluripotent stem cell (iPSC) technology will promote the investigation of sporadic JALS. *FUS*, the most frequently mutated gene related to the pathogenesis of sporadic JALS, consists of 15 exons and encodes a multi-domain, dosage-sensitive protein that is associated with a variety of neurodegenerative diseases, including ALS, frontotemporal lobar degeneration (FTLD), and polyglutamine diseases. Approximately 70 different mutations have been reported in *FUS*, and most of the ALS-causative mutations occur in exons 14 and 15 (Chen et al., 2019). Missense mutations are the most commonly observed mutations, but frameshift or nonsense mutations have also been reported (Kim et al., 2015).

We have previously described a novel heterozygous mutation in exon 14 of the *FUS* gene, c.1509dupA (p.R503fs), chromosome 16, of a 17-year-old girl who had been diagnosed with ALS based on the clinical manifestations, and who had subsequently died of respiratory failure 15 months after onset (Chen et al., 2020). No family history or cognitive impairment was found for the teenager, which hinted that this was a case of sporadic JALS. The *de novo FUS* mutation and the rapid progression of sporadic JALS in this patient caught our attention. Therefore, after obtaining informed consent from the patient and her family members, and after having obtained ethical approval, we decided to reprogram the patient's peripheral blood mononuclear cells (PMBCs) into an iPSC line aiming to provide a cell model for future research on the pathogenesis of JALS and exploration of an effective treatment against this disorder. This was completed under protocol number 2020-KY-098, released by the Ethical Committee of the First Affiliated Hospital of Zhengzhou University.

MATERIALS AND METHODS

Cell Culture and iPSC Reprogramming

iPSC lines were generated from PBMCs of the patient. PBMCs were isolated from the peripheral blood of the patient by centrifugation using a Vacutainer® CPT™ (BD Biosciences, San Jose, CA, USA). After 3–4 days of culture, the PBMCs were transduced with Sendai virus using the CytoTune™-iPS 2.0 Sendai Reprogramming Kit (Thermo Fisher Scientific, Waltham, MA, USA) for 48 h. At day 3 post transduction, the Sendai virus was removed and cells were replated onto 96-well culture plates. Monoclones with dense cell growth formed after continuous culture for approximately 30 days. Healthy clones were selected and replated onto 24-well plates coated with Matrigel (Corning, NY, USA) and cultured in mTeSR1 medium (STEMCELL Technologies, BC, Canada). The mTeSR1 medium was changed every day. When the cell density had reached 70–80% confluence, ACCUTASE (Life Technologies, MD, USA) was used for normal digestion and passaging at a density of 1:4–1:8. During subculture, the ROCK inhibitor y-27632 dihydrochloride (Tocris, Bristol, UK) was added into the medium to promote iPSC adherence and a single-cell layer was formed to cover the surface of the culture vessel. Cells were maintained at 37°C in a humidified atmosphere with 5% CO_2. Pluripotency was assessed by immunofluorescence for OCT4 and flow cytometry for SSEA-4, TRA-1-60, and OCT4.

Flow Cytometry

Cells were detached with ACCUTASE. The fixation/permeabilization procedure was performed using the BD Cytofix/Cytoperm Kit (BD Pharmingen™, CA, USA) according to the manufacturer's instructions. Cells were incubated at room temperature for 1 h. The percentage of PSCs was determined by SSEA-4 and TRA-1-60 antibody staining (diluted in 1% BSA), followed by flow cytometric analysis (Beckman Coulter CytoFLEX, FL, USA; software: CytExpert).

Immunofluorescence Assay

Cells (undifferentiated hiPSCs and trilineage-differentiated cultures) were fixed in 4% PFA for 15 min, permeabilized with Dulbecco's phosphate-buffered saline (DPBS) containing 1% BSA and 0.1% Triton X-100 for 15 min, and blocked in blocking solution (0.1% Triton X-100 and 10% FBS in PBS) for 1 h at room temperature. Samples were then incubated overnight at 4°C with primary antibodies (**Supplementary Table S1**), and then with the appropriate secondary antibodies (**Supplementary Table S1**). Hoechst (Thermo Fisher Scientific, Waltham, MA, USA) was used to stain the nuclei. Images were acquired with an inverted fluorescence microscope (Axio Observer, Zeiss, Jena, Germany).

In vitro Trilineage Differentiation of hiPSCs

At passage 40, hiPSCs were harvested and differentiated into separate lineages. Trilineage differentiation was performed using the STEMdiff™ Trilineage Differentiation Kit (STEMCELL Technologies, BC, Canada) according to the manufacturer's indications. After 5–7 days, cells were fixed in 4% PFA for immunofluorescence staining against specific ectodermal, mesodermal, and endodermal markers.

Motor Neuron Precursor (MNP) Specification and MN Differentiation

To generate MNPs, hiPSCs were first dissociated with dispase (1 mg/ml^{-1}) and split 1:6 on Matrigel-coated plates. The following day, the PSC medium was replaced with a chemically defined neural medium consisting of DMEM/F12, neurobasal medium at 1:1, 0.5× N2, 0.5× B27, 0.1 mM ascorbic acid (Santa Cruz Technologies, Santa Cruz, CA, USA), 1× Glutamax, and 1× penicillin/streptomycin (Invitrogen, Carlsbad, CA, USA). CHIR99021 (3 mM, Torcris, Bristol, UK), 2 mM DMH1 (Torcris, Bristol, UK), and 2 mM SB431542 (Stemgent, Cambridge, MA, USA) were added to the medium. hiPSCs were maintained under this condition for 6 days. The culture medium was changed every other day. Then, as a second step, the cells were first dissociated with dispase (1 mg/ml^{-1}) and split 1:6 with the above-described medium. Retinoic acid (RA; 0.1 mM, Stemgent) and 0.5 mM purmorphamine (Stemgent, MA, USA) were added in combination with 1 mM CHIR99021, 2 mM DMH1, and 2 mM SB431542. The medium was changed every other day. The cells maintained under this condition for 6 days differentiated into *OLIG2*$^{+}$ MNPs.

The *OLIG2*$^{+}$ MNPs were expanded with the medium containing 3 mM CHIR99021, 2 mM DMH1, 2 mM SB431542, 0.1 mM RA, 0.5 mM purmorphamine, and 0.5 mM valproic acid (VPA; Stemgent, Cambridge, MA, USA), and split 1:6 once a week with dispase (1 mg/ml^{-1}) for 2 weeks. Some of the MNPs were frozen in regular freezing medium (DMEM/F12, 10% fetal bovine serum, and 10% DMSO) in liquid nitrogen while the remainder were used for differentiation into MNs.

To induce MN differentiation, *OLIG2*$^{+}$ MNPs were first dissociated with dispase (1 mg/ml^{-1}) and cultured in suspension for 6 days in the above-described neural medium supplemented with 0.5 mM RA and 0.1 mM purmorphamine. The medium was changed every other day. These cells were then dissociated with Accumax (eBioscience Inc., San Diego, CA, USA) into single cells, plated on Matrigel-coated plates, and then cultured with 0.5 mM RA, 0.1 mM purmorphamine, and 0.1 mM compound E (Calbiochem, DA, Germany) for 10 days to mature into MNs.

Karyotyping

Briefly, hiPSCs from passage 10 were treated with 10 μg/ml colcemid (Gibco, NY, USA) for 60 min at 37°C. Subsequently, the cells were dissociated with Accutase, treated with a 0.075 M hypotonic KCl solution, and then fixed in Carnoy's fixative. Five metaphase spreads were prepared and examined by G-banding analysis. Karyotyping was performed at the 450-band level using Genikon software (Nikon, Italia) and described in accordance with ISCN 2016.

Mycoplasma Test

Mycoplasma contamination was checked using the Mycoplasma Detection Kit with UDG PCR Mix and Loading Dye from ExCell Bio^{TM} (Shanghai, China) following the manufacturer's instructions. The results of the mycoplasma test showed that this hiPSC line ZZUNEUi010-A was mycoplasma-free.

Sendai Virus Residue Detection

The presence of Sendai virus residue was detected by PCR and quantitative real-time PCR, as described in section RNA Isolation and Quantitative Real-Time PCR.

Whole-Exome Sequencing

Total DNA was extracted from PBMCs of the patient, the patient's parents, and hiPSCs. Whole-exome sequencing was performed using Illumina paired-end sequencing and Agilent SureSelect Human All Exon V6 for capture and construction, double end (paired-end) sequencing strategy (raw data >12G, Q30 ≥80%). Following the identification of the *FUS* mutation, the result was validated by Sanger sequencing. The primers (5′–3′) for sequencing the targeted mutation were FUS-ex14_F1: TCACATGGGTAAGAAAGGCAGA and FUS-ex14_R1: ACAACCTCAGGTCTTTCCACCT.

Short Tandem Repeat (STR) Analysis

The STR analysis was carried out by Guangzhou Cellcook Biotech Co., Ltd, China. Briefly, gDNA was extracted from PBMCs of the patient and hiPSCs with the DNeasy Blood and Tissue Kit (QIAGEN, Duesseldorf, Germany). Microsatellite DNA locus amplification was carried out by PCR using the STR Multi-Amplification Kit (PowerPlex 18D System) according to the manufacturer's instructions. The PCR products of two multiplexes (STR loci and the sex-linked gene amelogenin) were assayed with 3100 DNA Analyzer (Applied Biosystems®). The following loci were tested: AMEL, D3S1358, D1S1656, D6S1043, D13S317, PentaE, D16S539, D18S51, D2S1338, CSF1PO, Penta D, TH01, vWA, D21S11, D7S820, D5S818, TPOX, D8S1179, D12S391, D19S433, and FGA. The results were analyzed by ABI Genotyper software.

RNA Isolation and Quantitative Real-Time PCR

Total RNA was isolated using RNAiso Plus (Takara, Tokyo, Japan). cDNA was synthesized from 1 μg of total RNA using the Takara PrimeScript^{TM} RT reagent Kit with gDNA Eraser (Takara) according to the manufacturer's instructions. Quantitative real-time PCR was performed in an Applied Biosystems 7500 real-time PCR platform using Fast SYBR Green PCR Master Mix (Applied Biosystems) containing primers for pluripotency genes. The primers (5′–3′) for Sendai virus were F: TCACTAGGTGATATCGAGC and R: ACCAGACAAGAGTTTAAGAGATATGTATC. The primers (5′–3′) for FUS were F: AATAAATTTGGTGGCCCTCGG and R: GTTGCCTCTCCCTTCAGCTT.

Western Blot

Cells were lysized in lysis buffer (1% Triton X-100, 1% SDS, 500 mM NaCl, 1 mM EDTA, and 50 mM Tris-Cl, pH 7.4) with the protease inhibitor cocktail and were boiled in sample buffer. Samples were electrophoresed on 12% Bis-Tris polyacrylamide gels and then were transferred to a PVDF (Millipore) membrane and non-specific bindings were inhibited by 5% milk. Antibodies against FUS (ab124923, Abcam), GAPDH (AF0006, Beyotime, Shanghai, China) were diluted to 1:1,000, respectively and the membranes were incubated overnight at 4°C. After incubation for 1 h at room temperature with second antibodies, proteins were detected using a Chemiluminescence Imaging System (Bio-RAD, ChemiDoc Touch).

Statistical Analysis

Values are expressed as mean ± SD. Statistical significance was calculated with GraphPad Prism 6 (GraphPad Software). A two-tailed non-paired t-test was used to compare differences between two groups. P-values less than 0.05 were considered significant differences.

RESULTS

Resource Utility

We previously reported a *de novo* pathogenic *FUS* mutation (c.1509dupA:p.R503fs) in a patient with sporadic JALS (Chen et al., 2020). In this study, we reprogrammed PBMCs derived from the patient to generate a hiPSC line carrying the novel *FUS* mutation. The specific background information for this hiPSC line (CMF001-A; hPSCreg Name, ZZUNEUi010-A) is shown in hPSCreg[1].

Reprogramming of Cells From the Patient

The hiPSC line (CMF001-A; ZZUNEUi010-A) was established by retrovirus-mediated reprogramming of PBMCs derived from a patient with sporadic JALS. We analyzed the iPSC clones by brightfield microscopy and immunofluorescence staining. After continuous culture of the reprogrammed cells for approximately 30 days, the clones proliferated and spread evenly, as can be seen in **Figure 1A**. G-banded karyotype analysis showed that CMF001-A had a normal 46, XX karyotype (**Figure 1B**). The hiPSCs were cultured in feeder-free conditions after being stably cloned. No Sendai virus residue (**Supplementary Figure S1A**) or mycoplasma contamination was detected (**Supplementary Figure S1B**). To determine whether the mutation present in the hiPSC line was the same as that in the PBMCs obtained from the patient, genomic DNA was extracted for whole-exome sequencing, followed by Sanger-sequencing verification, using the primers FUS-ex14_F1 and FUS-ex14_R1. The results showed that the same heterozygous mutation c.1509dupA (p.R503fs) was present in both the hiPSCs and PBMCs but not in the WT (PBMCs from the patient's mother) sample (**Figure 1C**). The STR results for the hiPSC line also matched those of the PBMCs of the patient, indicating that the hiPSCs were patient-specific (**Supplementary Figure S1C**). These results suggested that the hiPSC line, derived from the patient and harboring the patient's *de novo* mutated gene, was successfully established without contamination or mutation during culture.

Pluripotency and Trilineage Differentiation Potential of the hiPSCs

OCT4 is a key transcription factor for cell reprogramming and functions as a positive regulator of genes required for the maintenance of embryonic stem cell (ESC) pluripotency (Karagiannis et al., 2019). Immunofluorescence staining showed that OCT4 was strongly expressed in the iPSC clones (**Figure 1D**). There was no significant difference between ZZUNEUi010-A and WT in positive staining rate for OCT4 (**Figure 1F**). At passage 10, three pluripotency markers (OCT4, SSEA-4, and TRA-1-60) were utilized to determine the pluripotency potential of the hiPSCs. The positive expression of these markers was measured by flow cytometry using human IgG as the isotype control. The expression rates of OCT4, SSEA-4, and TRA-1-60 were high, showing values of 97.89, 99.78, and 99.57% (**Figure 1E**), respectively, suggesting that this hiPSC line presented pluripotential potential. There was no significant difference between ZZUNEUi010-A and WT in positive rate for those markers (**Figure 1G**). At passage 40, the hiPSCs showed stable proliferative potential. Some of these hiPSCs were harvested and induced to differentiate into the three embryonic germ layers using the STEMdiffTM Trilineage Differentiation Kit after single-cell clone inoculation on culture plates followed by 5–7 days of culture (**Figure 1J**). Three embryonic markers (AFP, an endodermal marker; brachyury, a mesodermal marker; and PAX6, an ectodermal marker) were used to determine the trilineage differentiation potential of the hiPSCs. On day 5, the cells differentiated from the hiPSCs stained strongly for AFP and brachyury respectively, while at day 7 the differentiated cells showed strong PAX6 staining (**Figure 1H**). There was no significant difference between ZZUNEUi010-A and WT in positive staining rate for the three embryonic markers (**Figure 1I**).

MNP Specification and MN Differentiation From the hiPSCs

Twenty-eight days after the induction of differentiation, which involved two stages (MNP specification and MN differentiation) and four steps (Du et al., 2015), the iPSCs differentiated into MNs. In the first stage, hiPSCs differentiated into MNPs over 12 days. The MNPs were allowed to proliferate from day 12 to day 27, and they were identified by OLIG2 staining on day 16 (**Figure 2H**). The representative photos showed that the positive staining rate for OLIG2 expression was 80.79% in wild-type (WT) MNPs and 84.33% in MNPs harboring FUS^{R503fs} (**Figure 2A**). There was no significant difference between FUS^{R503fs}-MNP and WT-MNP in positive staining rate for OLIG2 (**Figure 2B**). On day 28, the MNPs were placed in a special neural medium to induce differentiation into MNs. Subsequently, MNs gradually matured and could be cultured *in vitro* for long periods. After 7 days of differentiation (day 35, **Figure 2H**), differentiating MNs were stained for MAP2 and FUS (**Figure 2C**). We found that both the FUS^{R503fs} group and the WT group developed axons and dendrites, displayed the morphological characteristics of typical neurons, and expressed MAP2 widely. There was no significant difference between FUS^{R503fs}-MNP and WT-MNP in positive staining rate for MAP2 (**Figure 2D**).

The MNs Differentiated From the hiPSCs Showed Pathological Characteristics of JALS

The FUS staining showed interesting results. In the WT group, FUS-positive staining was primarily localized to the nucleus,

[1]https://hpscreg.eu/cell-line/ZZUNEUi010-A

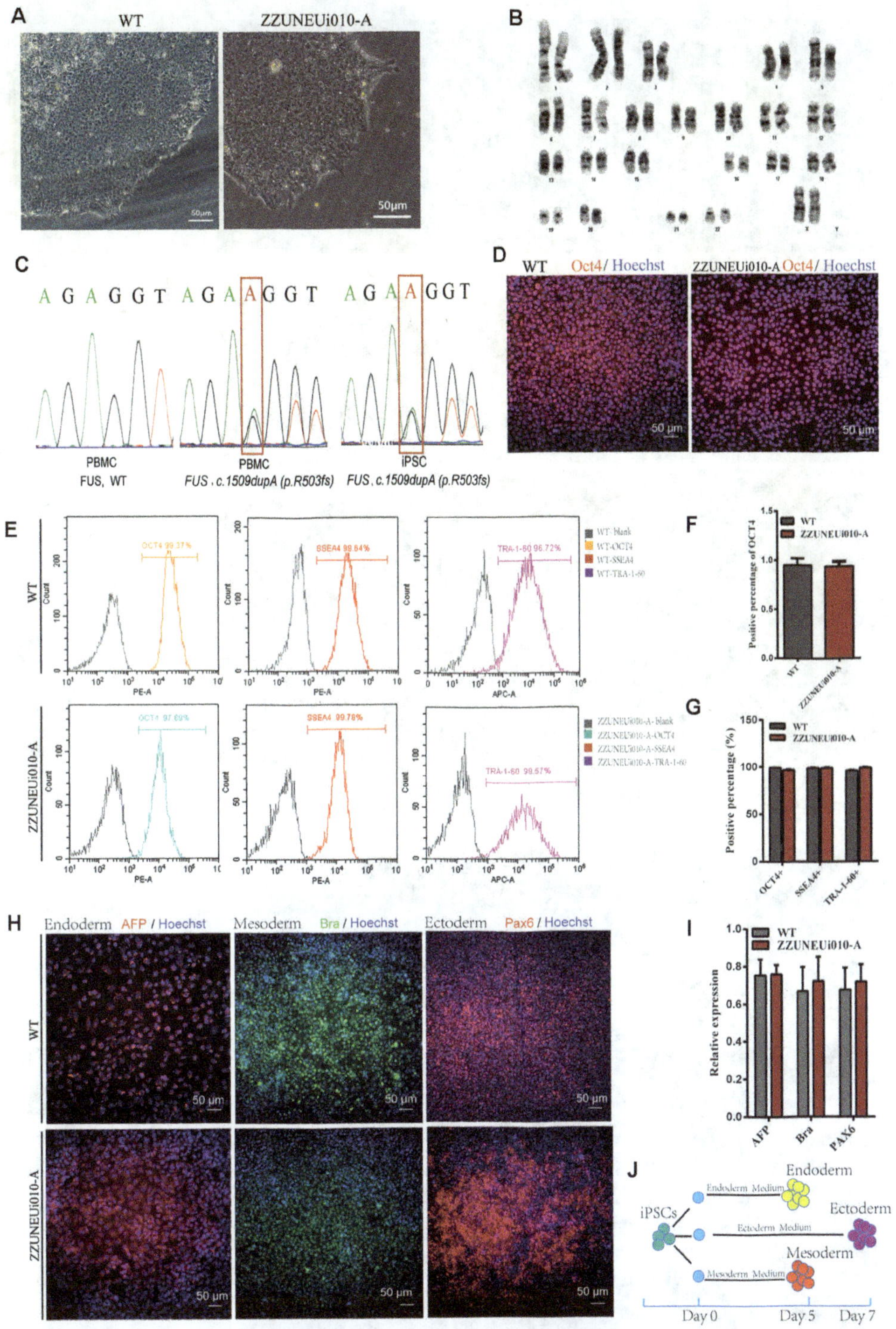

FIGURE 1 | Pluripotency identification and trilineage differentiation of the human-induced pluripotent stem cells (hiPSCs). **(A)** Brightfield images of ZZUNEUi010-A clones (×10). **(B)** Karyotype analysis of ZZUNEUi010-A, 46, XX, without chromosomal aberrations. **(C)** Illumina and Sanger sequencing of PBMCs and ZZUNEUi010-A from this patient detected the same fused in sarcoma (*FUS*) mutation (c.1509dupA:p.R503fs). **(D)** Immunofluorescence images showing that ZZUNEUi010-A colonies stained strongly for the pluripotency marker OCT4. **(E)** Flow cytometry showed that the hiPSCs were strongly positive for the pluripotency markers OCT4 (97.89%), SSEA-4 (99.78%), and TRA-1-60 (99.57%). **(F)**There was no significant difference between ZZUNEUi010-A and WT in positive staining rate for OCT4 ($n = 3$ views, $p > 0.05$). **(G)** There was no significant difference between ZZUNEUi010-A and WT in positive rate for those markers OCT4, SSEA-4, and TRA-1-60 ($n = 3$, $p > 0.05$). **(H)** On day 5, the cells differentiated from the hiPSCs stained strongly for AFP and brachyury respectively, while at day 7 the differentiated cells showed strong PAX6 staining. **(I)** There was no significant difference between ZZUNEUi010-A and WT in positive staining rate for the three embryonic markers ($n = 3$ views, $p > 0.05$). **(J)** Schematic diagram of triploblastic differentiation experiment.

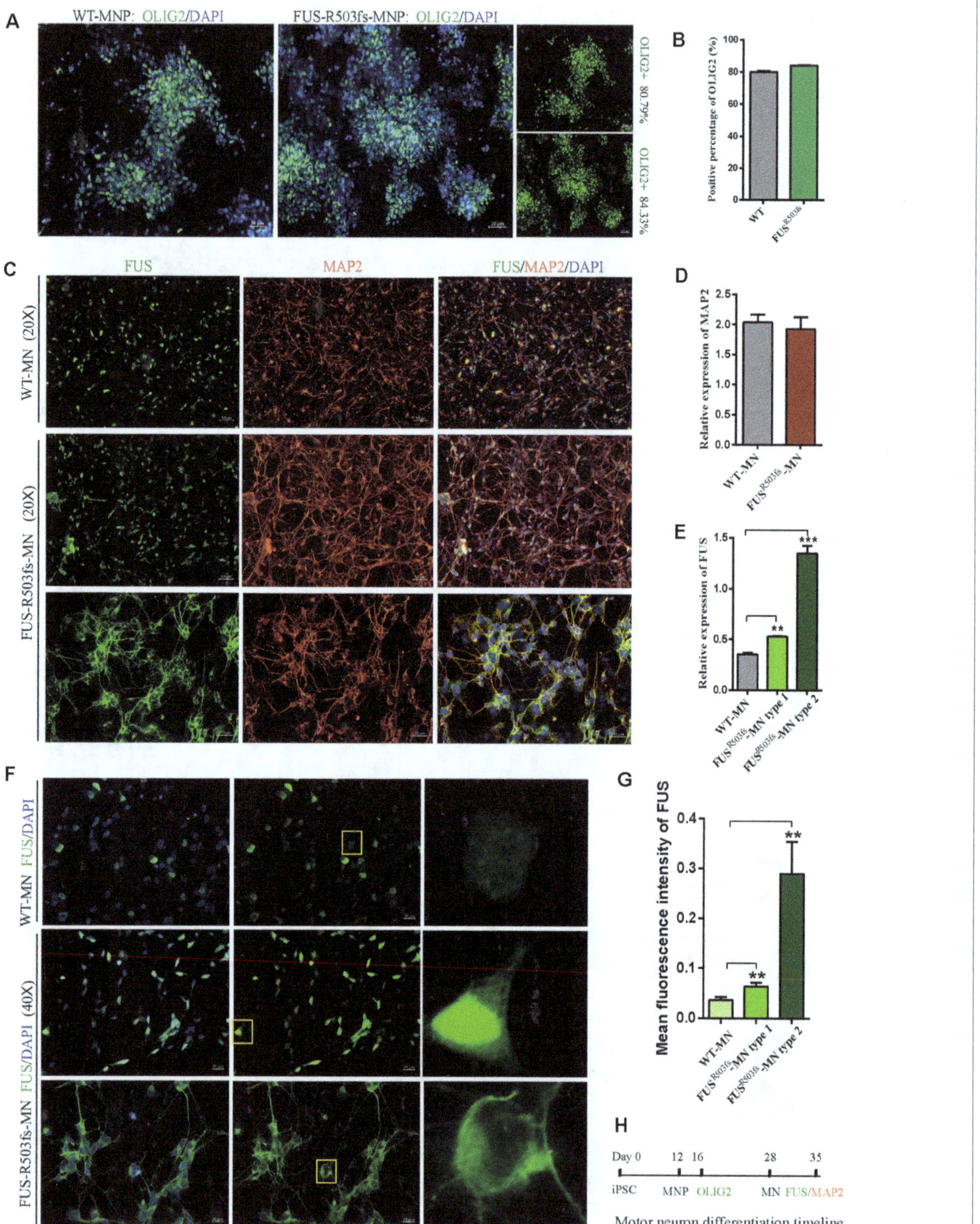

FIGURE 2 | Cells from the hiPSC line (ZZUNEUi010-A) could be differentiated into motor neurons and exhibited pathological characteristics of amyotrophic lateral sclerosis (ALS). **(A)** Representative images of motor neuron precursors (MNPs) on day 16 showing strong OLIG2 (green) staining. **(B)** There was no significant difference between FUSR503fs-MNPs and WT-MNPs in positive staining rate for OLIG2 ($n = 3$, $p > 0.05$). **(C)** Representative images of motor neurons on day 35 showing positive MAP2 (red) and FUS (green) staining. **(D)** There was no significant difference between FUSR503fs-MNP and WT-MNP in positive staining rate for MAP2 ($n = 3$, $p > 0.05$). **(E)** The relative expression of FUS was significantly higher in MNs of the FUSR503fs group than in those of the WT group ($n = 3$, $^{**}p < 0.01$, $^{***}p < 0.001$). **(F)** Representative images of FUS-positive staining MNs in WT, FUS^{R503fs}–MN type 1, and FUS^{R503fs}–MN type 2 groups. **(G)** The mean intracellular FUS fluorescence intensity of single cell was significantly higher in MNs of the FUS^{R503fs} group than in those of the WT group (WT-MNs, $n = 142$; FUS^{R503fs}–MN type 1, $n = 146$; FUS^{R503fs}–MN type 2, $n = 75$; $^{**}p < 0.01$). **(H)** Schematic diagram of motor neuron differentiation process.

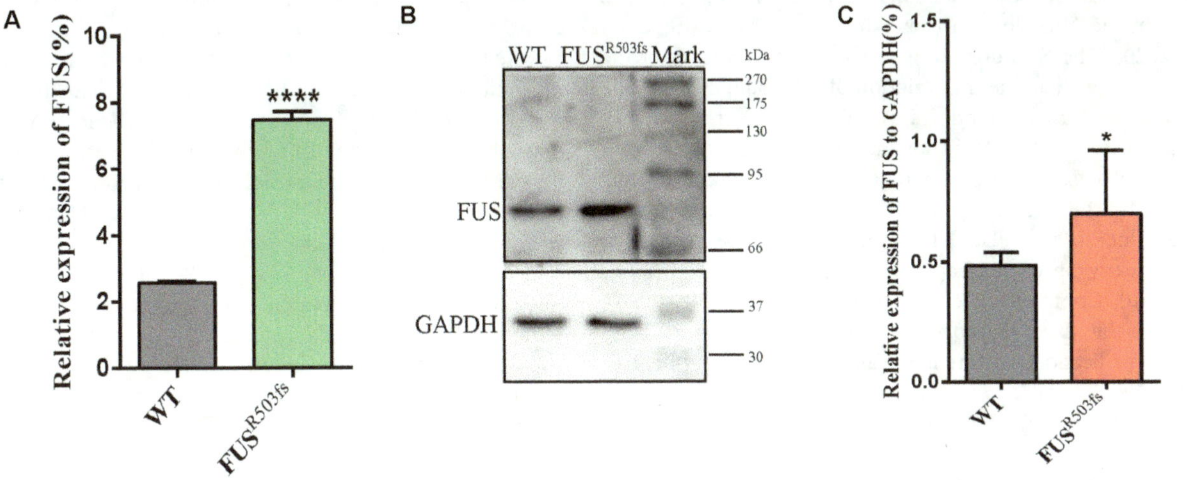

FIGURE 3 | FUS expression was significantly higher in the FUS^{R503fs} than that in WT group. **(A)** The Q-rtPCR results indicated that the mRNA expression level of FUS was significantly increased in FUS^{R503fs} compared with WT group ($n = 6$, $^{****}p < 0.0001$). **(B)** Western blot analysis of FUS protein in neurons of WT and FUS^{R503fs}. Molecular weight markers are shown on the right. **(C)** Quantification of FUS protein levels in FUS^{R503fs} neurons was significantly increased compared with WT group ($n = 8$, $^{*}p > 0.05$).

while in FUS^{R503fs} group, FUS-positive staining appeared to be stronger than that in WT cells and was characterized by two types: FUS^{R503fs}–MN type1, where FUS staining was strong in the nucleus and weak in the cytoplasm; and FUS^{R503fs}–MN type 2, where FUS staining was weak in the nucleus and strong in the cytoplasm, cytoplasmic axons, and dendrites (**Figures 2C,F**). These two types of FUS distribution characteristics did not exist in the WT control group neurons. It was possibly the phenomenon of varying degrees of pathological aggregation and distribution of aberrant FUS in neurons (Scekic-Zahirovic et al., 2017). The relative expression of FUS was significantly higher in MNs of the FUS^{R503fs} group than in those of the WT group (**Figure 2E**). The mean intracellular FUS fluorescence intensity of single cell was significantly higher in MNs of the FUS^{R503fs} group than in those of the WT group (**Figure 2G**). This suggested that the *FUS* mutation-induced pathological characteristics of JALS—aberrant cytoplasmic aggregation and reduced nuclear entry of FUS—could be successfully recapitulated in hiPSC-derived MNs. In order to verify the change of FUS expression caused by this gene mutation in motor neurons, the expression of FUS in FUS^{R503fs} group and WT group was detected by quantitative real-time PCR (Q-rtPCR) and protein immunoimprinting (Western Blot). The expression level of FUS mRNA in FUS^{R503fs} group neurons was significantly higher than that in the WT group (**Figure 3A**). The expression level of FUS protein in FUS^{R503fs} group neurons was significantly higher than that in the WT group (**Figures 3B,C**). Moreover, these results indicated that our hiPSC line, ZZUNEUi010-A, has potential for use as a cell model to study sporadic JALS.

DISCUSSION

With the development of virus-free reprogramming and gene correction technologies, hiPSCs are becoming widely used in classical ALS research. However, JALS cases, especially sporadic JALS, associated with different genetic mutations have increasingly been reported in the past decade due to advances in gene sequencing technology. The hiPSCs derived from our patient with JALS will play an important role in the study of the pathogenesis of this condition and its rapid progress. As previously reported, ALS-causative *FUS* mutations are rare in adults (Guerrero et al., 2016) but more common in JALS (Mackenzie et al., 2011; Zou et al., 2016). However, why the *FUS* mutation appears so disproportionately in different age groups remains unclear. In our study, we successfully established a hiPSC line derived from a patient with sporadic JALS carrying the *de novo FUS* mutation c.1509dupA:p.R503fs. This hiPSC line could be differentiated into motor neurons and showed a typical cytoplasmic *FUS* aggregation, suggesting that it has good potential as a tool for use in JALS research.

FUS is a DNA/RNA-binding protein with functional homology to TDP-43 and belongs to the FET/TET protein family (Guerrero et al., 2016). It is a 526 amino acid, multidomain protein, with an N-terminal transcriptional activation domain, multiple nucleus-binding domains, and a C-terminal nuclear localization signal (NLS; Dormann and Haass, 2013). FUS shuttles continuously between the nucleus and the cytoplasm, regulating gene expression and performing numerous cytoplasmic functions (Zinszner et al., 1997; Birsa et al., 2019) such as DNA damage repair and regulation of mRNA stability, autophagy, RNA granule assembly, and ER-mitochondria associations (An et al., 2019). *FUS* levels in the cytoplasm are extremely low under physiological conditions, while aberrant FUS mislocalizes and accumulates in the cytoplasm in end-stage ALS (Kapeli et al., 2017). In our experiment, this typical pathological change was observed 7 days after the maturation of MNs. The *de novo* mutation

R503fs, like most *FUS* mutations in patients with JALS, is located in the NLS domain (Loughlin and Wilce, 2019; Chen et al., 2020). The NLS domain is essential for FUS to enter the nucleus and regulate the transcription of DNA and cytoplasmic localization (Dormann and Haass, 2013; Casci et al., 2019). FUS mutation in the NLS domain may be associated with cytoplasmic aggregation of aberrant protein. The gain of toxic function in the cytoplasm and the loss of function in the nucleus due to the FUS–NLS mutation may be associated with the early onset and rapid progression of JALS (Baumer et al., 2010; Conte et al., 2012; Scekic-Zahirovic et al., 2016). However, the exact pathogenesis remains unclear, and no effective medication is currently available for the treatment of this disorder.

The difficulty in obtaining nerve tissue from patients and the lack of ideal animal models pose great challenges to the study of sporadic JALS. Models of JALS suffer from the disadvantages of late-onset and the lack of typical pathological characteristics. For instance, Devoy and colleagues created a mouse model of ALS harboring a human-associated frameshift mutation in *FUS* (p.G466VfsX14), and found progressive motor neuron degeneration in 15-month-old mice without aggregation of aberrant proteins (Devoy et al., 2017), which could be explained by the functional redundancy between FUS and other FET proteins that occurs in neurons (Kabashi et al., 2011; Sasayama et al., 2012). Therefore, patient-derived iPSCs may be a better choice as a disease model for sporadic JALS. However, only a few hiPSC lines have been established for use in investigating the pathomechanisms of ALS, including FUS^{P525L} (De Santis et al., 2017; Errichelli et al., 2017; Marrone et al., 2018); FUS^{H517D} (Ichiyanagi et al., 2016); FUS^{R521H}(Guo et al., 2017); $FUS^{R495QfsX527}$ (Naujock et al., 2016); $FUS^{G504Wfs*12}$ and FUS^{Q519E} (Lim et al., 2016); and $FUS^{Asp502Thrfs*}$ and FUS^{R521C} (Higelin et al., 2016). Although all of these iPSC lines can be differentiated into MNs that exhibit typical pathologic characteristics of ALS, most were derived from patients with classical ALS or familial JALS. In our study, c.1509dupA:pR503fs, a *de novo* mutation in *FUS*, was detected in a sporadic JALS case, which will be of great research value. A hiPSC line named ZZUNEUi010-A, and carrying the FUS^{R503fs} mutation, was subsequently generated through the reprogramming of PBMCs derived from the patient with JALS. The hiPSC showed strong expression of pluripotency markers (OCT4, SSEA-4, and TRA-1-60) and could be differentiated into three embryonic germ layers, as evidenced by the detection of AFP (endoderm), brachyury (mesoderm), and Pax6 (ectoderm) marker expression. Moreover, no gross chromosomal aberrations or specific copy number variations were observed. Importantly, we induced the differentiation of the hiPSCs into MNs that exhibited pathological characteristics of ALS. Overexpressed FUS cytoplasmic aggregation were observed in motor neurons differentiated from ZZUNEUi010-A.

In conclusion, a cell model of sporadic JALS (CMF001-A)[1] carrying a *de novo FUS* mutation c.1509dupA:p.R503fs was successfully established. This cell line can be differentiated into motor neurons with pathological features of ALS and represents a good tool as well as a new strategy for exploring the pathogenesis and treatment of sporadic JALS.

AUTHOR CONTRIBUTIONS

LC performed most of the experiments and wrote and submitted the manuscript. YW reviewed and revised the manuscript. JX performed the sequencing analysis.

ACKNOWLEDGMENTS

We sincerely thank all the participants in this study for their cooperation, as well as the patient's parents for supporting our study.

REFERENCES

An, H., Skelt, L., Notaro, A., Highley, J. R., Fox, A. H., La Bella, V., et al. (2019). ALS-linked FUS mutations confer loss and gain of function in the nucleus by promoting excessive formation of dysfunctional paraspeckles. *Acta Neuropathol. Commun.* 7:7. doi: 10.1186/s40478-019-0658-x

Avemaria, F., Lunetta, C., Tarlarini, C., Mosca, L., Maestri, E., Marocchi, A., et al. (2011). Mutation in the senataxin gene found in a patient affected by familial ALS with juvenile onset and slow progression. *Amyotroph. Lateral Scler.* 12, 228–230. doi: 10.3109/17482968.2011.566930

Baumer, D., Hilton, D., Paine, S. M., Turner, M. R., Lowe, J., Talbot, K., et al. (2010). Juvenile ALS with basophilic inclusions is a FUS proteinopathy with FUS mutations. *Neurology* 75, 611–618. doi: 10.1212/WNL.0b013e3181ed9cde

Birsa, N., Bentham, M. P., and Fratta, P. (2019). Cytoplasmic functions of TDP-43 and FUS and their role in ALS. *Semin. Cell Dev. Biol.* 99, 193–201. doi: 10.1016/j.semcdb.2019.05.023

Casci, I., Krishnamurthy, K., Kour, S., Tripathy, V., Ramesh, N., Anderson, E. N., et al. (2019). Muscleblind acts as a modifier of FUS toxicity by modulating stress granule dynamics and SMN localization. *Nat. Commun.* 10:5583. doi: 10.1038/s41467-019-13383-z

Chen, C., Ding, X., Akram, N., Xue, S., and Luo, S. Z. (2019). Fused in sarcoma: properties, self-assembly and correlation with neurodegenerative diseases. *Molecules* 24:1622. doi: 10.3390/molecules24081622

Chen, L., Li, J., Lu, H., and Liu, Y. (2020). A de novo c.1509dupA:p.R503fs mutation of FUS: report of a girl with sporadic juvenile amyotrophic lateral sclerosis. *Amyotroph. Lateral Scler. Frontotemporal Degener.* doi: 10.1080/21678421.2020.1775256. [Epub ahead of print].

Conte, A., Lattante, S., Zollino, M., Marangi, G., Luigetti, M., Del Grande, A., et al. (2012). P525L FUS mutation is consistently associated with a severe form of juvenile amyotrophic lateral sclerosis. *Neuromuscul Disord.* 22, 73–75. doi: 10.1016/j.nmd.2011.08.003

Daoud, H., Zhou, S., Noreau, A., Sabbagh, M., Belzil, V., Dionne-Laporte, A., et al. (2012). Exome sequencing reveals SPG11 mutations causing juvenile ALS. *Neurobiol. Aging* 33, 839.e5–839.e9. doi: 10.1016/j.neurobiolaging.2011.11.012

De Santis, R., Santini, L., Colantoni, A., Peruzzi, G., De Turris, V., Alfano, V., et al. (2017). FUS mutant human motoneurons display altered transcriptome and microRNA Pathways with implications for ALS pathogenesis. *Stem Cell Reports* 9, 1450–1462. doi: 10.1016/j.stemcr.2017.09.004

Devoy, A., Kalmar, B., Stewart, M., Park, H., Burke, B., Noy, S. J., et al. (2017). Humanized mutant FUS drives progressive motor neuron degeneration without aggregation in "FUSDelta14" knockin mice. *Brain* 140, 2797–2805. doi: 10.1093/brain/awx248

Dormann, D., and Haass, C. (2013). Fused in sarcoma (FUS): an oncogene goes awry in neurodegeneration. *Mol. Cell Neurosci.* 56, 475–486. doi: 10.1016/j.mcn.2013.03.006

Du, Z. W., Chen, H., Liu, H., Lu, J., Qian, K., Huang, C. L., et al. (2015). Generation and expansion of highly pure motor neuron progenitors from human pluripotent stem cells. *Nat. Commun.* 6:6626. doi: 10.1038/ncomms7626

Errichelli, L., Dini Modigliani, S., Laneve, P., Colantoni, A., Legnini, I., Capauto, D., et al. (2017). FUS affects circular RNA expression in murine embryonic stem cell-derived motor neurons. *Nat. Commun.* 8:14741. doi: 10.1038/ncomms14741

Fujimori, K., Ishikawa, M., Otomo, A., Atsuta, N., Nakamura, R., Akiyama, T., et al. (2018). Modeling sporadic ALS in iPSC-derived motor neurons identifies a potential therapeutic agent. *Nat. Med.* 24, 1579–1589. doi: 10.1038/s41591-018-0140-5

Guerrero, E. N., Wang, H., Mitra, J., Hegde, P. M., Stowell, S. E., Liachko, N. F., et al. (2016). TDP-43/FUS in motor neuron disease: complexity and challenges. *Prog. Neurobiol.* 145-146, 78–97. doi: 10.1016/j.pneurobio.2016.09.004

Guo, W., Naujock, M., Fumagalli, L., Vandoorne, T., Baatsen, P., Boon, R., et al. (2017). HDAC6 inhibition reverses axonal transport defects in motor neurons derived from FUS-ALS patients. *Nat. Commun.* 8:861. doi: 10.1038/s41467-017-00911-y

Higelin, J., Demestre, M., Putz, S., Delling, J. P., Jacob, C., Lutz, A. K., et al. (2016). FUS mislocalization and vulnerability to DNA damage in ALS patients derived hipscs and aging motoneurons. *Front. Cell. Neurosci.* 10:290. doi: 10.3389/fncel.2016.00290

Ichiyanagi, N., Fujimori, K., Yano, M., Ishihara-Fujisaki, C., Sone, T., Akiyama, T., et al. (2016). Establishment of *in vitro* FUS-associated familial amyotrophic lateral sclerosis model using human induced pluripotent stem cells. *Stem Cell Reports* 6, 496–510. doi: 10.1016/j.stemcr.2016.02.011

Kabashi, E., Bercier, V., Lissouba, A., Liao, M., Brustein, E., Rouleau, G. A., et al. (2011). FUS and TARDBP but not SOD1 interact in genetic models of amyotrophic lateral sclerosis. *PLoS Genet.* 7:e1002214. doi: 10.1371/journal.pgen.1002214

Kapeli, K., Martinez, F. J., and Yeo, G. W. (2017). Genetic mutations in RNA-binding proteins and their roles in ALS. *Hum. Genet.* 136, 1193–1214. doi: 10.1007/s00439-017-1830-7

Karagiannis, P., Takahashi, K., Saito, M., Yoshida, Y., Okita, K., Watanabe, A., et al. (2019). Induced pluripotent stem cells and their use in human models of disease and development. *Physiol. Rev.* 99, 79–114. doi: 10.1152/physrev.00039.2017

Kim, Y.-E., Oh, K.-W., Kwon, M.-J., Choi, W.-J., Oh, S.-I., Ki, C.-S., et al. (2015). De novo FUS mutations in 2 korean patients with sporadic amyotrophic lateral sclerosis. *Neurobiol. Aging* 36, 1604.e17–1604.e9. doi: 10.1016/j.neurobiolaging.2014.10.002

Lim, S. M., Choi, W. J., Oh, K. W., Xue, Y., Choi, J. Y., Kim, S. H., et al. (2016). Directly converted patient-specific induced neurons mirror the neuropathology of FUS with disrupted nuclear localization in amyotrophic lateral sclerosis. *Mol. Neurodegener.* 11:8. doi: 10.1186/s13024-016-0075-6

Liu, Z. J., Lin, H. X., Liu, G. L., Tao, Q. Q., Ni, W., Xiao, B. G., et al. (2017). The investigation of genetic and clinical features in chinese patients with juvenile amyotrophic lateral sclerosis. *Clin. Genet.* 92, 267–273. doi: 10.1111/cge.13015

Loughlin, F. E., and Wilce, J. A. (2019). TDP-43 and FUS-structural insights into RNA recognition and self-association. *Curr. Opin. Struct. Biol.* 59, 134–142. doi: 10.1016/j.sbi.2019.07.012

Ma, L., Shi, Y., Chen, Z., Li, S., and Zhang, J. (2018). A novel SETX gene mutation associated with juvenile amyotrophic lateral sclerosis. *Brain Behav.* 8:e01066. doi: 10.1002/brb3.1066

Mackenzie, I. R., Ansorge, O., Strong, M., Bilbao, J., Zinman, L., Ang, L. C., et al. (2011). Pathological heterogeneity in amyotrophic lateral sclerosis with FUS mutations: two distinct patterns correlating with disease severity and mutation. *Acta Neuropathol.* 122, 87–98. doi: 10.1007/s00401-011-0838-7

Marrone, L., Poser, I., Casci, I., Japtok, J., Reinhardt, P., Janosch, A., et al. (2018). Isogenic FUS-eGFP iPSC reporter lines enable quantification of FUS stress granule pathology that is rescued by drugs inducing autophagy. *Stem Cell Reports* 10, 375–389. doi: 10.1016/j.stemcr.2017.12.018

Naujock, M., Stanslowsky, N., Bufler, S., Naumann, M., Reinhardt, P., Sterneckert, J., et al. (2016). 4-Aminopyridine induced activity rescues hypoexcitable motor neurons from amyotrophic lateral sclerosis patient-derived induced pluripotent stem cells. *Stem Cells* 34, 1563–1575. doi: 10.1002/stem.2354

Naumann, M., Peikert, K., Gunther, R., Van Der Kooi, A. J., Aronica, E., Hubers, A., et al. (2019). Phenotypes and malignancy risk of different FUS mutations in genetic amyotrophic lateral sclerosis. *Ann. Clin. Transl. Neurol.* 6, 2384–2394. doi: 10.1002/acn3.50930

Sasayama, H., Shimamura, M., Tokuda, T., Azuma, Y., Yoshida, T., Mizuno, T., et al. (2012). Knockdown of the drosophila fused in sarcoma (FUS) homologue causes deficient locomotive behavior and shortening of motoneuron terminal branches. *PLoS One* 7:e39483. doi: 10.1371/journal.pone.0039483

Scekic-Zahirovic, J., Oussini, H. E., Mersmann, S., Drenner, K., Wagner, M., Sun, Y., et al. (2017). Motor neuron intrinsic and extrinsic mechanisms contribute to the pathogenesis of FUS-associated amyotrophic lateral sclerosis. *Acta Neuropathol.* 133, 887–906. doi: 10.1007/s00401-017-1687-9

Scekic-Zahirovic, J., Sendscheid, O., El Oussini, H., Jambeau, M., Sun, Y., Mersmann, S., et al. (2016). Toxic gain of function from mutant FUS protein is crucial to trigger cell autonomous motor neuron loss. *EMBO J.* 35, 1077–1097. doi: 10.15252/embj.201592559

Siddiqi, S., Foo, J. N., Vu, A., Azim, S., Silver, D. L., Mansoor, A., et al. (2014). A novel splice-site mutation in ALS2 establishes the diagnosis of juvenile amyotrophic lateral sclerosis in a family with early onset anarthria and generalized dystonias. *PLoS One* 9:e113258. doi: 10.1371/journal.pone.0113258

Wu, C., and Fan, D. (2016). A novel missense mutation of the DDHD1 gene associated with juvenile amyotrophic lateral sclerosis. *Front. Aging Neurosci.* 8:291. doi: 10.3389/fnagi.2016.00291

Zinszner, H., Sok, J., Immanuel, D., Yin, Y., and Ron, D. (1997). TLS (FUS) binds RNA *in vivo* and engages in nucleo-cytoplasmic shuttling. *J. Cell Sci.* 110, 1741–1750.

Zou, Z. Y., Liu, M. S., Li, X. G., and Cui, L. Y. (2016). Mutations in FUS are the most frequent genetic cause in juvenile sporadic ALS patients of chinese origin. *Amyotroph. Lateral Scler. Frontotemporal Degener.* 17, 249–252. doi: 10.3109/21678421.2016.1143012

Zufiria, M., Gil-Bea, F. J., Fernandez-Torron, R., Poza, J. J., Munoz-Blanco, J. L., Rojas-Garcia, R., et al. (2016). ALS: a bucket of genes, environment, metabolism and unknown ingredients. *Prog. Neurobiol.* 142, 104–129. doi: 10.1016/j.pneurobio.2016.05.004

11

Interphotoreceptor Retinoid-Binding Protein (IRBP) in Retinal Health and Disease

Shaoxue Zeng[1,2], *Ting Zhang*[1*], *Michele C. Madigan*[1,3], *Nilisha Fernando*[4], *Riemke Aggio-Bruce*[4,5], *Fanfan Zhou*[6], *Matthew Pierce*[1], *Yingying Chen*[1,2], *Lianlin Huang*[1,3], *Riccardo Natoli*[4,5], *Mark C. Gillies*[1] and *Ling Zhu*[1]

[1] Save Sight Institute, The University of Sydney, Sydney, NSW, Australia, [2] Department of Ophthalmology, West China Hospital, Sichuan University, Chengdu, China, [3] School of Optometry and Vision Sciences, University of New South Wales, Sydney, NSW, Australia, [4] The John Curtin School of Medical Research, The Australian National University, Acton, ACT, Australia, [5] The Australian National University Medical School, The Australian National University, Acton, ACT, Australia, [6] Sydney Pharmacy School, The University of Sydney, Sydney, NSW, Australia

***Correspondence:**
Ting Zhang
ting.zhang@sydney.edu.au

Interphotoreceptor retinoid-binding protein (IRBP), also known as retinol binding protein 3 (RBP3), is a lipophilic glycoprotein specifically secreted by photoreceptors. Enriched in the interphotoreceptor matrix (IPM) and recycled by the retinal pigment epithelium (RPE), IRBP is essential for the vision of all vertebrates as it facilitates the transfer of retinoids in the visual cycle. It also helps to transport lipids between the RPE and photoreceptors. The thiol-dependent antioxidant activity of IRBP maintains the delicate redox balance in the normal retina. Thus, its dysfunction is suspected to play a role in many retinal diseases. We have reviewed here the latest research on IRBP in both retinal health and disease, including the function and regulation of IRBP under retinal stress in both animal models and the human retina. We have also explored the therapeutic potential of targeting IRBP in retinal diseases. Although some technical barriers remain, it is possible that manipulating the expression of IRBP in the retina will rescue or prevent photoreceptor degeneration in many retinal diseases.

Keywords: IRBP gene, photoreceptor degeneration, visual cycle, gene therapy, retinoid

INTRODUCTION

A Brief History of IRBP

Interphotoreceptor retinoid-binding protein (IRBP), also known as retinol binding protein 3 (RBP3), was first discovered in soluble proteins extracted from the bovine interphotoreceptor matrix (IPM), which is located between the neural retina and the retinal pigment epithelium (RPE) (Adler and Severin, 1981). Several unknown proteins were identified, including a 140 kilodalton (kDa) protein (Liou et al., 1982). Liou et al. suggested that this 140 kDa protein might be a transporter protein transferring retinol between the outer segments of rod photoreceptors and the RPE (Liou et al., 1982). They postulated that the molecular weight of IRBP is 260 kDa in its glycosylated form, while its backbone is 140–145 kDa (Liou et al., 1982).

In the 1990's, several studies implicated IRBP in retinal development (Gonzalez-Fernandez et al., 1993; Timmers et al., 1993; Liou et al., 1994; Stenkamp et al., 1998). The messenger RNA (mRNA) expression of IRBP in the retina of embryonic mice was low. mRNA transcripts in mice were first detected at embryonic day 11 and continued to increase to its peak expression on postnatal day 4, after which there was a slow decrease and reached constancy by postnatal day 20 (Liou et al., 1994).

The protein expression of IRBP increased together with its mRNA level during photoreceptor development in bovine and zebrafish embryos (Timmers et al., 1993; Stenkamp et al., 1998). Protein and mRNA expression were also markedly increased in Sprague Dawley rats between postnatal day 1 and 9, a period critical for outer segment formation. Then, IRBP mRNA levels decreased a little and stabilized by postnatal day 20 when the outer segments achieved their adult length (Gonzalez-Fernandez et al., 1993).

IRBP has been shown to be downregulated in animal models of retinal disease, including Abyssinian cats that carry a homozygous IRBP mutation that causes pan-retinal degeneration (Narfstrom et al., 1989). IRBP downregulation has also been reported in mice with induced Müller cell dysfunction and a streptozotocin (STZ)-induced diabetic rat model (Zhu et al., 2015; Malechka et al., 2017). IRBP deficiency also impaired eye growth and compromised retinal health in mice (Wisard et al., 2011). These reports suggest a correlation between IRBP dysregulation and photoreceptor degeneration.

Retinal Location of IRBP

IRBP is secreted by photoreceptors and accumulates in the IPM to facilitate the transport of material for visual pigment regeneration (Ishikawa et al., 2015) (**Figure 1a**). It has only been found in vertebrates (Kusakabe et al., 2009). In the eyes of non-vertebrates, an alternative mechanism of pigment regeneration is confined to photoreceptors (Kusakabe et al., 2009). Vertebrates acquired IRBP in the evolution of the visual cycle to accommodate a complicated visual cycle. IRBP can traffick retinoid, a class of Vitamin A derivatives that includes retinol and retinal, between photoreceptors and RPE cells (Kusakabe et al., 2009). IRBP mRNA was detected in both cones and rods of adult Xenopus retina. However, IRBP in the embryo is synthesized by the central retina and diffuses through the matrix, reaching the peripheral retina (Hessler et al., 1996).

Theo and colleagues found IRBP in bovine pineal gland cells by *in situ* hybridization using IRBP cDNA probes. They discovered that IRBP is highly expressed in a population of pineal cells on mRNA level, but they did not quantify its protein expression (van Veen et al., 1986). Expression of IRBP (Rodrigues et al., 1986), cone arrestin (Craft et al., 1994) and opsins (Blackshaw and Snyder, 1997), all photoreceptors markers, were also found in the pineal gland of Rhesus monkeys, rats and catfish. It remains unclear why there is a morphological and possibly functional analogy between photoreceptors and pinealocytes. The pineal gland, a small neuroendocrine organ that synthesizes and secretes melatonin, is also photosensitive in lower vertebrates (Sapède and Cau, 2013). Some researchers have hypothesized that mammalian pinealocytes might have evolved from photoreceptors (Sapède and Cau, 2013).

Molecular Characteristics and Regulation of IRBP

Mammals have four protein subunits of IRBP (Gonzalez-Fernandez, 2012), each of which consists of ~300 amino acids (Nickerson et al., 2006). Two of the four subunits are similar but have different affinities to all-trans-retinol. Notably, disrupting one subunit does not affect the overall function because other subunits compensate for the dysfunctional subunit (Gonzalez-Fernandez and Ghosh, 2008).

The *IRBP* gene is regulated by cone-rod-homeobox (CRX) and orthodenticle homolog 2 (OTX2), two essential transcriptional factors in photoreceptors (Fei et al., 1999; Nishida et al., 2003) (**Figure 1b**). Studies have suggested that the "cone-rod-homeobox element" is essential for the photoreceptor-specific activity of the *IRBP* promoter *in vivo*. This element is the major binding site of the CRX, which can directly regulate *IRBP* expression (Fei et al., 1999). OTX2 is an upstream regulator of CRX. Both OTX2 and CRX mRNAs have been identified in adult human retinas (Bobola et al., 1999; Nishida et al., 2003; Li et al., 2015). Overexpressing OTX2 increased *irbp* promoter-luciferase activity by 5–7-fold in WERI-Rb1 retinoblastoma cells, suggesting that OTX2 activates the *irbp* promoter (Bobola et al., 1999). Overall, CRX and OTX2 are both specific gene regulators of IRBP that are critical in photoreceptor development.

IRBP IN THE NORMAL RETINA

Retinoids Transport

Retinoid recycling and metabolism within the eye have been extensively studied for decades. George Wald was awarded a Nobel prize in 1967 for discovering the photoreceptive proteins in the eye, the "chromoproteins" (Wald, 1935). Vision is initiated and maintained by their photolysis and regeneration in the visual cycle (Kiser et al., 2014). Photolysis of 11-*cis*-retinal is the only reaction that converts light signals into electrical signals in vertebrate photoreceptors (Molday and Moritz, 2015). Circulation and regeneration of 11-cis-retinal, which relies on IRBP for its transport, is critical for the maintenance of light sensitivity (Liou et al., 1998; Jin et al., 2009). Thus, IRBP is central to developing and maintaining the visual cycle in humans.

The Canonical Visual Cycle

In the canonical visual cycle (**Figure 2A**), IRBP is secreted into the IPM by photoreceptors and rapidly turned over through endocytosis by photoreceptors and the RPE (Gonzalez-Fernandez, 2003). It mediates extracellular diffusion of retinoids during the operation of the retinoid cycle, transporting all-trans-retinol and 11-cis-retinal between the photoreceptors and the RPE (Gonzalez-Fernandez, 2003). A recent study suggested that IRBP may not be the only vehicle and peropsin may also transport all-trans-retinol from the neural retina to RPE (Cook et al., 2017). Human organic anion transporting polypeptide 1A2 (OATP1A2) has been recently found to be expressed at the apical membrane of RPE, where it facilitates the cellular uptake of all-trans-retinol into the RPE cells (Chan et al., 2015).

IRBP is essential for retinoid exchange in the visual cycle. Exogenously applying IRBP protein to the culture medium of rod photoreceptors reduced the level of all-trans-retinal and retinol in outer segments, preventing the formation of light-induced lipofuscin precursor (Chen et al., 2017). Supplementation of IRBP in the culture medium also facilitates the release of 11-cis-retinal from cultured fetal bovine RPE cells (Carlson and Bok, 1992). Addition of the free form IRBP to the RPE apical

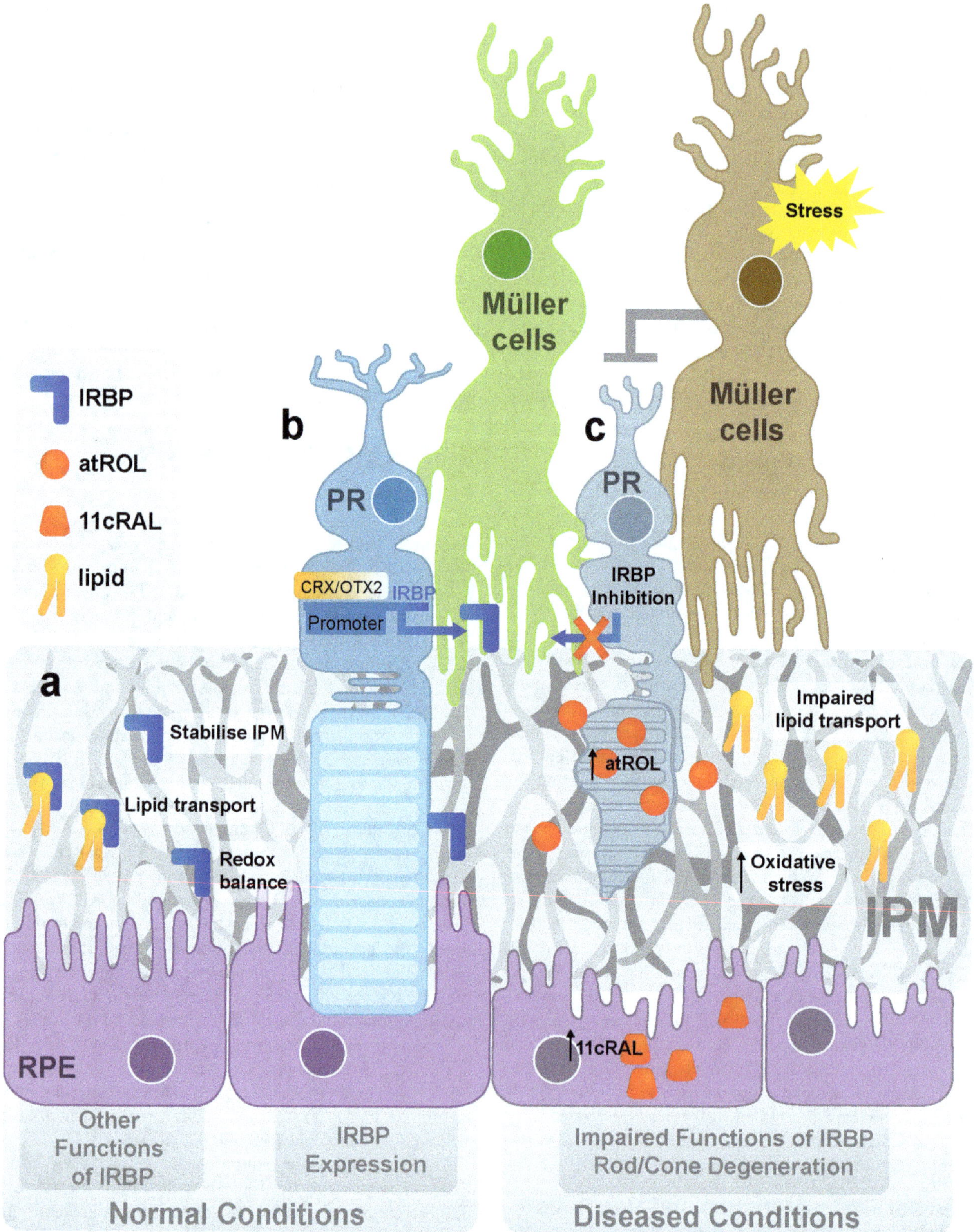

FIGURE 1 | IRBP in normal and diseased retina. **(a)** The expression and functions of IRBP. IRBP is produced and secreted by photoreceptors. The main function of IRBP is to stabilize IPM, transport lipid between photoreceptors and the RPE and maintain the redox balance in the outer retina. **(b)** The regulation of IRBP. OTX2/CRX activates the expression of IRBP by binding to its promoter. **(c)** IRBP dysregulation and rod/cone degeneration under stress. It is proposed that stress

(Continued)

FIGURE 1 | activates Müller cells inhibiting the production of IRBP in photoreceptors, leading to the dysfunction of lipid transport, increased oxidative stress, and accumulation of retinoids (atROL). (IRBP, Interphotoreceptor retinoid-binding protein; CRX, cone-rod-homeobox; OTX2, orthodenticle homolog 2; IPM, interphotoreceptor matrix; RPE, retinal pigment epithelium; PR, photoreceptor; atROL, all-*trans*-retinol; 11cRAL, 11-*cis*-retinal).

membrane resulted in more [3H]-11-cis-retinal release than adding cellular retinaldehyde-binding protein (CRALBP), serum retinol-binding protein (RBP), bovine serum albumin (BSA) or medium devoid of binding proteins (Carlson and Bok, 1992). Light reduced the levels of 11-cis-retinol when IRBP was absent, while it had no effect on 11-cis-retinol levels if IRBP was present. These results indicate IRBP is essential in preserving the isomeric state of retinol (Parker et al., 2011).

The Cone-Specific Visual Cycle

The cone-specific visual cycle (**Figure 2B**) refers to the exclusive ability of cone photoreceptors and Müller cells to convert all-trans-retinol to 11-cis-retinal (Wang and Kefalov, 2009; Sato and Kefalov, 2016). In contrast to the canonical visual cycle, all-trans-retinol is transported to Müller cells, facilitated by IRBP (Mata et al., 2002). IRBP was implicated by the observation that it promoted the uptake of all-trans-retinol and release of 11-cis - retinol in rat Müller cells *in vitro* (Betts-Obregon et al., 2014). Cone electroretinogram (ERG) responses in IRBP knockout mice were reduced compared to that of the control mice, although they had similar cone densities and opsin levels, indicating IRBP is essential for normal cone function or at least for the cone-specific vision cycle (Parker and Crouch, 2010). Müller cells have been found to isomerise all-trans-retinol directly to 11-cis-retinol, which then released for cone photoreceptors (Crouch et al., 1992; Das et al., 1992). Finally, the cone photoreceptor outer segments have been reported to oxidize 11-cis-retinol to 11-cis-retinal (Jones et al., 1989). Recently, further evidence has also implicated IRBP in the retinoid exchange between cones and Müller cells (Tang et al., 2013). IRBP was found to bind to the cone outer segment and Müller cell microvilli pericellular matrices (Garlipp and Gonzalez-Fernandez, 2013). This association can facilitate the delivery and uptake of its retinol ligands.

Lipid Transport

Apart from retinoid transportation, IRBP assists the transport of various essential lipids across the IPM (**Figure 1a**). It has been established that IRBP contains two similar subunits with different affinities for all-trans-retinol. Long-chain fatty acids, such as Docosahexaenoic Acid (DHA), can replace all-trans-retinol from these subunits with lower affinity (Chen et al., 1993). DHA is an essential element in photoreceptor membrane biosynthesis and is thus vital for visual function in photoreceptors (Scott and Bazan, 1989; Jastrzebska et al., 2011). Post-mortem AMD retina and RPE/choroid were reported to have significantly lower levels of polyunsaturated fatty acids, such as DHA, than age-matched normal donors, suggesting the importance of IRBP to photoreceptors by transporting DHA or other fatty acids (Liu et al., 2010) (**Figure 1c**).

Other studies also demonstrated that other fatty acids could also compete with all-trans-retinol for binding IRBP, among which oleic acid had the highest affinity, but it was still significantly lower than that of all-trans-retinol (Semenova and Converse, 2003; Ghosh et al., 2015). A fluorometric titration experiment in which increasing the concentrations of oleic acid gradually reduced the affinity of all-trans-retinol to IRBP provided further evidence that IRBP plays a vital role in lipid transport (Ghosh et al., 2015). Competitive fluorescence and tryptophan-quenching assays also demonstrated that both oleic acid and DHA could displace all-trans-retinol from bovine IRBP, while oleic acid having a relatively higher affinity than DHA. Gas chromatography revealed that oleic acid is the most abundant fatty acid in bovine IPM, suggesting it might play an important role in maintaining the balance and transport of retinoids and fatty acids in the retina (Semenova and Converse, 2003).

Important Roles of IRBP in IPM Integrity

The IPM, a highly organized structure between photoreceptors and the RPE, is essential for maintaining outer retinal homeostasis. Characteristic changes in IPM components occur in retinal degenerations (Ishikawa et al., 2015). For example, photoreceptor degeneration in most retinal diseases begins with the loss of the inner segments/outer segments in the IPM. However, photoreceptors may still survive under these circumstances (Goldberg et al., 2016).

IRBP accounts for ~5% of the total soluble IPM protein (Adler and Evans, 1983) and maybe crucial for the structural integrity of the IPM (**Figure 1a**). Although a large amount of IRBP can be removed by washout with a saline solution, other wash-resistant IRBP that co-localizes in the IPM is concentrated around cone outer segments (cone IPM-associated IRBP) (Garlipp and Gonzalez-Fernandez, 2013). Furthermore, the deduced amino acid sequence of human IRBP contains two receptors for hyaluronan-mediated motility (RHAMM)-like motifs (K321-R329 and K733-R781), suggesting a possible interaction of IRBP with the hyaluronan scaffold (Hollyfield, 1999), a component of non-covalently formed complexes with proteoglycans in the Extracellular Matrix (ECM). Thus, IRBP dysregulation is likely to affect the structural integrity of IPM, which may independently contribute to the development of many retinal diseases.

Antioxidant Activity of IRBP

IRBP has a thiol-dependent antioxidant activity which may protect retinol from oxidation (**Figure 1a**). The thiol-dependent antioxidative activity of IRBP has been evaluated with an assay using myoglobin and 2,2-azinobis (3-ethylbenzothiazoline-6-sulfonic acid) (ABTS). IRBP was found to inhibit the oxidation of ABTS more actively than other free thiol groups or thiol-based reducing enzymes such as thioredoxin, vitamin E analog Trolox, and ovalbumin. Using N-ethylmaleimide (NEM) to inhibit cysteine peptidases by alkylating the active site-thiol on

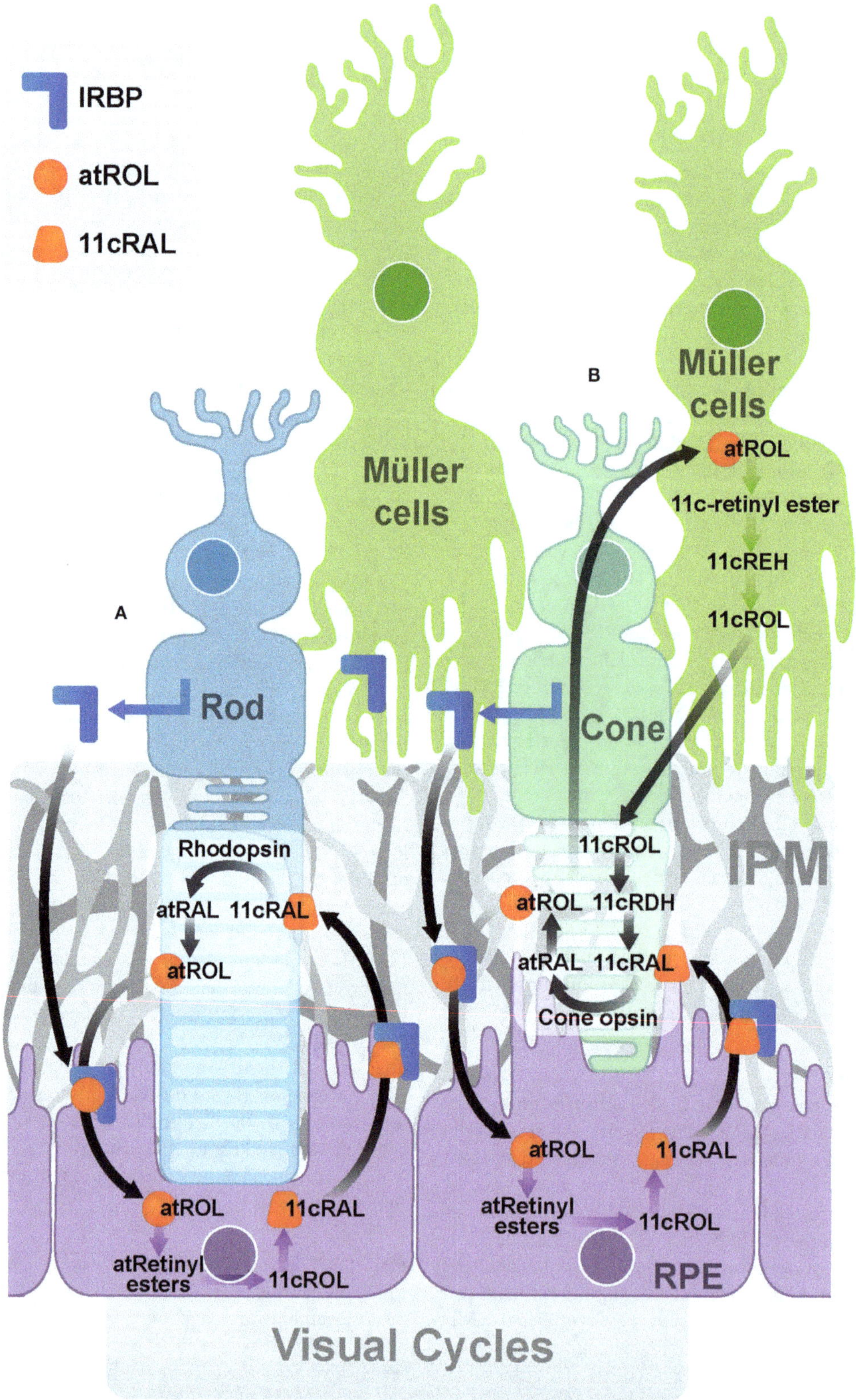

FIGURE 2 | Visual Cycles. **(A)** The canonical visual cycle. IRBP facilitates the transport of retinoids (atROL and 11cRAL) between photoreceptors and the RPE. **(B)** The cone-specific visual cycle between cone photoreceptors and Müller cells. (IRBP, Interphotoreceptor retinoid-binding protein; IPM, interphotoreceptor matrix; RPE, retinal pigment epithelium; atRAL, all-*trans*-retinal; atROL, all-*trans*-retinol; 11cRAL, 11-*cis*-retinal; 11cROL, 11-*cis*-retinol; 11cRDH, 11-*cis*-retinol dehydrogenase; 11cREH, 11-*cis*-Retinyl ester hydrolase).

IRBP also suppressed the antioxidant capacity of IRBP. Thus IRBP contains a thiol-dependent antioxidant site, disruption of the putative ligand-binding site can lead to reduced antioxidant effects (Gonzalez-Fernandez et al., 2014) (**Figure 1c**).

All-trans-retinal causes severe cellular oxidative stress and leads to reactive oxygen species (ROS) accumulation, mitochondrial dysfunction and cell death (Maeda et al., 2009; Chen et al., 2012; Rózanowska et al., 2012). Levels of all-trans-retinal were significantly higher in the retinal explants of IRBP knockout mice than that in wild type mice after 40 min exposure to bright light. This phenomenon was significantly mitigated by adding IRBP in the culture medium (Lee et al., 2016).

IRBP IN RETINAL DISEASES

IRBP in Animal Models of Retinal Disease

Extensive animal studies have shown that IRBP downregulation occurs in the early stages of many various retinal diseases. In fact, it may be a precursor to them (Zhu et al., 2015; Malechka et al., 2017). In 1989, a reduction of IRBP was reported in the early stages of retinal degeneration in a model of Abyssinian cats which carry a homozygous mutation for retinal degeneration (Narfstrom et al., 1989). A mutation in intron 50 of the centrosomal protein 290 (CEP290) gene (IVS50 + 9T > G) induced a stop codon and truncation of the mature protein in this model (Menotti-Raymond et al., 2007). IRBP-immunoreactivity was significantly reduced in the affected retina prior to the development of photoreceptor cell death (Narfstrom et al., 1989). In the same model, mRNA and protein levels of IRBP were significantly reduced as early as 4 weeks of age (Wiggert et al., 1994). Downregulation of IRBP protein was also found in a retinal degeneration 12 (rd12) mouse model with a mutation in the *RPE65* gene (Zheng et al., 2015). We have previously described this in a mouse model in which Müller cell dysfunction can be induced. A 150 kDa protein band was markedly diminished in the transgenic mice retina just 1 week after the Müller cells were disrupted, which was identified as IRBP by mass spectrometry. Photoreceptor degeneration was observed 2 weeks after the Müller cell disruption was induced in this transgenic model, suggesting that photoreceptor degeneration was secondary to Müller cell disruption (Zhu et al., 2015). It was also reported that the IRBP protein level was significantly reduced in the retinas of STZ-induced diabetic rats (Malechka et al., 2017). It has also been reported that IRBP mRNA was reduced in a light-induced retinal degeneration model in rats after animals had been exposed to intense visible light (490–580 nm green light) with an illuminance of ~1,200 lux for 24 h to induce photoreceptor degeneration (Wong et al., 2001). Overall, the early reduction of IRBP in different models of retinal degeneration suggests that IRBP may be a primary defect or an early disease marker in the retina.

IRBP dysregulation has been widely observed in different models of retinal disease. However, the consequences of IRBP downregulation on retinal or ocular health remains poorly understood. Studies on IRBP knockout models have identified an essential role of IRBP in eye growth and retinal health (Wisard et al., 2011; Markand et al., 2016). Eyes of IRBP knockout mice increased in size and weight over the wild type controls even before the mice had opened their eyelids (Wisard et al., 2011). Mice lacking IRBP display severe early and progressive photoreceptor degeneration, characterized by a reduction in both length and numbers of cone sheaths (Sato et al., 2013). IRBP knockout mice also developed profound myopia during the early stages of eye development. These eyes had longer anterior-posterior length, accompanied by a decrease in hyperopic refractive error (Markand et al., 2016). Progressive thinning of the outer nuclear layer was evident, with 20% thinning observed at postnatal day 5, and 38% thinning at day 30. Further studies, using optical coherence tomography (OCT), confirmed the previously reported retinal thinning of the outer nuclear layer in the IRBP knockout mice. Thinning of the outer nuclear layer lasted from postnatal day 15 to at least postnatal day 80 compared to wild type mice. Additionally, the slit-lamp and fundus photographs found no difference between the wild type and knockout mice (Markand et al., 2016). Another study on IRBP knockout mice revealed a loss of photoreceptor nuclei and changes in the structural integrity of retinas at postnatal day 11 and a marked loss in photic sensitivity from Electroretinography (ERG) at postnatal day 30 (Liou et al., 1998).

In summary, IRBP downregulation has been described in photoreceptors in different retinal diseases, but the precise mechanism for IRBP downregulation is still not clear. One possible explanation is that IRBP is a high molecular weight glycoprotein, the synthesis of which requires the production of large amounts of molecular building blocks and consumes a considerable amount of energy. Reduction of photoreceptor metabolism and the visual cycle may be necessary for the retina to survive under stress. Downregulation of IRBP may be beneficial for retina with short-term stress, but it might cause retinal degeneration if stress is persistent.

IRBP and Human Retinal Diseases

It has been established that mutation, dysfunction or downregulation of IRBP can be found in several human eye diseases. For instance, a homozygous missense mutation (p.Asp1080Asn) of *IRBP* was observed in a pedigree of four patients with autosomal recessive retinitis pigmentosa (RP) (den Hollander et al., 2009). Non-sense mutations (c.1530T>A;p.Y510* and c.3454G>T;p.E1152*) in *IRBP* were identified in four children diagnosed with retinal dystrophy and myopia (Arno et al., 2015). OCT images of these patients showed thinning of the central macula and loss of the inner segment ellipsoid band. ERG also revealed a delay and amplitude reduction in cone-specific responses (Arno et al., 2015). Li's group showed that D1080N mutation in IRBP found in patients with RP, abolished the secretion of IRBP and resulted in the formation of insoluble high molecular weight complexes via disulphide bonds. This hindered the transportation of IRBP to the Golgi and caused endoplasmic reticulum (ER) stress, which suggested another mechanism of retinal degeneration caused by IRBP mutation. A heterozygous T-C transition has been identified in autosomal recessive retinitis pigmentosa. IRBP protein was at a significantly lower level in aqueous humor of primary congenital glaucoma patients (Li et al., 2013). Reduced

IRBP mRNA and protein expression were observed in the retinas from diabetic donors when compared with those from non-diabetic donors (Garcia-Ramirez et al., 2009). Analysis of vitreous fluid obtained from clinics revealed that IRBP levels were reduced in the early stages of diabetic retinopathy (Garcia-Ramirez et al., 2009). Vitreous IRBP concentration declined gradually with the increasing severity of diabetic retinopathy in eyes with established retinopathy. IRBP protein levels both in the retina and vitreous of eyes with non-proliferative diabetic retinopathy were higher than those with proliferative diabetic retinopathy (Yokomizo et al., 2019).

Downregulation of IRBP also caused the accumulation of the retinal 'waste product' lipofuscin, which may increase the risk of oxidative damage to the RPE and photoreceptors (Radu et al., 2003) (**Figure 1c**). Lipofuscin is also responsible for retinal autofluorescence in retinal diseases (Birnbach et al., 1994; Marmorstein et al., 2002; Radu et al., 2003). Studies on cryosections of human retinas with AMD have revealed that lipofuscin in the RPE was strongly autofluorescent (Marmorstein et al., 2002). Stargardt's disease is characterized by hyper-autofluorescence and loss of macular photoreceptors which have been correlated with clinical progression of the disease (Birnbach et al., 1994).

We have discussed IRBP dysregulation as an early disease indicator, playing a role in the early stages of retinal pathology. Further studies are required to elucidate all of the mechanisms by which IRBP dysfunction may contribute to retinal disease pathogenesis.

Autoantibodies to IRBP

The retina is an 'immune-privileged' site, which means it may tolerate the introduction of antigens without eliciting an inflammatory immune response (Benhar et al., 2012). This presumably protects the retina from potentially blinding processes such as fibrosis that may result from inflammation. The blood-retinal barrier (BRB) is the interface between systemic circulation and the retina, which is critical for the maintenance of this immune-privilege (Benhar et al., 2012). Breakdown of the BRB, which occurs in many retinal diseases, exposes retinal antigens to the immune system, eliciting an inflammatory response leading to tissue damage and vision loss (Chen et al., 2019).

IRBP peptides can induce experimental autoimmune uveitis, which is a well-estblished model of uveitis (Caspi et al., 1990), as well as the effects of the adaptive immune response in the eye (Agarwal et al., 2012; Kyger et al., 2013). Anti-IRBP autoantibodies have been found in patients with uveitis, RP, Coats disease, AMD and Macular Telangiectasia Type 2 (MacTel) (Solomon et al., 1999; Morohoshi et al., 2012; Kyger et al., 2013; Zhu et al., 2013; Gibbs et al., 2017). A study investigated the autoimmune condition of a girl with a rare triad of RP, Coats disease and uveitis, and found that her peripheral lymphocytes had a specific anamnestic response to IRBP (Solomon et al., 1999), indicating that autoimmunity toward IRBP might play a role in the degeneration of photoreceptors. The IRBP autoantibody was detected in 28% (5 out of 18) of patients with non-infectious uveitis (Gibbs et al., 2017), in 33% (6 out of 18) of patients with AMD and in 24% (11 out of 45) of those with MacTel (Zhu et al., 2013). The detection of IRBP autoantibodies in these patients suggested that such diseases may share some common etiological or pathogenic mechanisms (Zhu et al., 2013). Significant downregulation of IRBP protein expression was also detected in the retina of mice with induced Müller cell dysfunction, which is a model of primary Müller cell loss that has been implicated in the pathogenesis of MacTel (Zhu et al., 2015). Whether these autoantibodies actually cause the reduction of IRBP or whether they are an epiphenomenon is still uncertain, but it appears certain that loss of IRBP is closely related to many retinal diseases.

THE POTENTIAL APPLICATIONS OF IRBP IN TREATMENT FOR RETINAL DISEASES

IRBP could be a potential novel target in treating retinal diseases, considering its essential role in the maintenance of the visual cycle and other physiological functions in the IPM (Gonzalez-Fernandez et al., 2014; Ghosh et al., 2015; Chen et al., 2017). The upregulation of IRBP prevented photoreceptor degeneration in diabetic mice and rats through the regulation of VEGF (Yokomizo et al., 2019). This data indicates that IRBP may also be beneficial in other diseases characterized by photoreceptor degeneration and VEGF dysregulation, such as AMD. Downregulation or dysregulation of IRBP could disrupt the visual cycle which leads to the accumulation of the all-trans-retinal, one component of retinal "waste product" lipofuscin. The effect of IRBP on preventing lipofuscin accumulation could be central for AMD-like diseases (Radu et al., 2003) (**Figure 1c**). Direct upregulation of IRBP through inducing the whole IRBP gene (9.5 k base pairs) warrants further investigation, but the difficulty is that the genetic constructs required would exceed the packing capacity of traditional adenovirus carriers for gene therapy. The insert size could be even longer after adding additional cell-specific promoters such as promoter fragments of cone transducin α (TαC), rhodopsin kinase (GRK1) or cone arrestin (CAR) gene (Kennedy et al., 2007; McDougald et al., 2019) for target-specific expression in photoreceptors. Here, we discuss three alternative ways to upregulate IRBP expression (small molecules, microRNAs, and CRISPRa technique), which we anticipate will protect photoreceptors from degeneration.

Small Molecules

Chemical communication mediated by molecular signaling coordinates cell behavior (Buddingh et al., 2020). Many natural or synthetic chemical compounds which regulate different metabolic or signaling pathways have great therapeutic potential. Small molecules that have the antioxidant, anti-inflammatory and anti-excitotoxic capacity in the central nervous system can be administered to protect photoreceptors (Zhang et al., 2019). For example, simvastatin is a cholesterol-lowering drug and was recently reported to protect Y79 retinoblastoma cells which have many characteristics of photoreceptors, through upregulating IRBP and its transcription factor CRX (Zhang et al., 2019). Some other small molecules acting on the cardiovascular system may also have a neuroprotective effect on photoreceptors (Zhao et al., 2016). Tetramethylpyrazine, which has been widely used in treating cardiovascular diseases for over 40 years, was found to attenuate all-trans-retinal -induced cytotoxicity in the differentiated Y79 cells via suppressing oxidative and nitrosative stress, apoptosis and leukostasis (Zhao et al., 2016). A further study found that the neuroprotective effect of tetramethylpyrazine against all-trans-retinal toxicity is also mediated through upregulating IRBP expression (Wang et al., 2017). Therefore, some small molecules have the potential to protect photoreceptors from stress by increasing IRBP expression.

Small molecules, however, have several shortcomings. Firstly, the downstream signaling cascade reactions and information encoding may vary broadly with different concentrations of signaling molecules (Purves and Fitzpatrick, 2001). Thus, the optimal dosage must be precisely controlled. Also, the enormous quantity of small molecules makes it difficult to find the desired molecule accurately. A luciferase reporter system may overcome this drawback (Xie et al., 2016). The luciferase reporter system can be specifically designed to screen small molecules that upregulate IRBP expression. By inserting the promoter region of human IRBP gene into the luciferase reporter plasmid, small molecules that specifically bind to this promoter region can be identified from the luciferase reporter activity (Miraglia et al., 2011). However, it is difficult to disentangle the downstream signaling through which the candidate small molecules regulate the expression of IRBP. Some molecules acting on Müller cells may also modulate IRBP expression via complex glia-neuron interactions (Vecino et al., 2016). Low specificity, low efficiency and the short-acting time of small molecules also pose a formidable challenge to their successful application (Maxim et al., 2014). A sustained-release formulation is highly desirable in this case. Nanoparticles introduced in the vitreous may sustain the delivery of the encapsulated agents for longer durations (Kompella et al., 2003; Riley and Vermerris, 2017). For example, Anti-VEGF aptamer EYE001 (tested in humans for efficacy) entrapped in Poly lactic-co-glycolic acid (PLGA) microspheres were found to deliver EYE001 in a sustained manner with retained activity *in vitro* and *ex vivo* (Carrasquillo et al., 2003). Lipid nanoparticle approaches with specific and sustained delivery systems are expected to gain more attention in the future. Several other studies have also demonstrated the successful application of nanoparticles in the liver with approval by the US Food and Drug Administration (FDA) (Coelho et al., 2013; Rizk and Tuzmen, 2017).

Therefore, the successful clinical application of synthetic and natural compounds may provide advantages but will present challenges.

MicroRNAs

A microRNA is a small non-coding RNA that can regulate gene expression of complementary mRNAs by binding to the 3′ untranslated region (3′ UTR) (Ambros, 2004; Bartel, 2018). More than 60% of protein-encoding genes are controlled by microRNAs (Friedman et al., 2009). It is recognized that microRNAs play an important role at the post-transcriptional level through degradation and translational repression of their target mRNAs (He and Hannon, 2004). Since the first study in 1993 discovering these post-transcriptional RNA-RNA interactions, microRNAs have attracted lots of attention due to their powerful post-transcriptional role, small size (21 nt), ease of transfection and ability for a single miRNA to regulate whole gene pathways (Lee et al., 1993; Wightman et al., 1993; Filipowicz et al., 2008). 320 and 340 different kinds of microRNAs have been found in mouse retina and RPE/choroid, respectively (Soundara Pandi et al., 2013). Despite miRNA being suggested as being involved in retinal degenerations, no studies have looked at the role of miRNA regulation on IRBP.

Computational analysis using target scan (http://www.targetscan.org/vert_72/) (Agarwal et al., 2015), which used to predict related microRNAs, showed no validated sequences for IRBP gene, but several predicted sequences. Predicted IRBP-related microRNAs have been studied in previous retinal studies, as shown in **Table 1**. Among which, miR-140-3p, miR-210-5p and miR-190b-5p may be of interest in manipulating the expression of IRBP. miR-140-3p participates in RPE cell survival and apoptosis. Loss of circRNA_0084043 depressed high glucose (HG)-induced apoptosis in ARPE-19 cells by upregulating miR-140-3p (Li et al., 2020). Given the essential role inflammation plays in retinal degenerations, miRNA that simultaneously maintains normal retinal function and influence inflammatory processes are of key therapeutic interest.

Once such miRNA miR-146a downregulates various genes involved in normal retinal function and homeostasis, including IL-6, IL-8 (Chen et al., 2010; Li et al., 2010) and is also predicted to regulate IRBP. The exploitation of microRNAs involved in modulating IRBP may be of clinical significance.

CRISPRa Gene Therapy

Since the development of gene-editing techniques such as zinc-finger nucleases (ZFNs) and transcription activator-like effector nucleases (TALENs), the third generation, clustered regularly interspaced short palindromic repeats (CRISPR) technique has recently risen (Doudna and Charpentier, 2014). This technique has made gene therapy easier and more specific for targeting a gene of interest, like IRBP. The recent development of CRISPR interference (CRISPRi) and CRISPR activation (CRISPRa) system that fuses dead nuclease Cas9 (dCas9) to a transcriptional complex, enables inhibiting or activating the transcription of target genes rather than cleaving them (Gilbert et al., 2014). CRISPRa technology also can activate multiple genes simultaneously (McCarty et al., 2020). Multiplex modulation through CRISPRa enables more precise and efficient gene editing, as many human diseases result from mutations in multiple genes (Zlotogora, 2007). Human IRBP is ~9.5 kb in length, which exceeds the standard packing capacity of a virus. This is a significant hurdle to develop gene therapies that target IRBP. We may benefit from boosting the transcription of IRBP through specifically designed guide RNA in the CRISPRa system. Gene therapies using CRISPR technology have already been launched in clinical trials. A patient with a hereditary blindness disorder has become the first to receive a CRISPR/Cas9 gene therapy administered directly into their body recently (Ledford, 2020). Thus, the success of IRBP upregulation using this advanced technology has translational applications.

TABLE 1 | Correlation of predicted microRNA in regulating IRBP and retinal disease.

Predicted miRNA	Reported miRNA	Regulation	Site	Disease or model	References	Others
hsa-miR-152-5p	hsa-miR-152	Down	Human vitreous	AMD vs. Con	Ménard et al., 2016	
hsa-miR-22-3p	hsa-miR-22	Down	Human retina	AMD vs. Con	Lukiw et al., 2012	
hsa-miR-146b-3p	Same	Down	Human vitreous	Diabetic vs. Con	Fulzele et al., 2015	
hsa-miR-3121-5p	hsa-miR-3121	Up	Human serum	dry AMD vs. Con	Szemraj et al., 2015	
hsa-miR-1306-5p	Same	Up	Human plasma	Glaucoma & XFS vs. Con	Hindle et al., 2019	
hsa-miR-3173-5p	hsa-miR-3173	Up	Human aqueous humor	Glaucoma & XFS vs. Con	Hindle et al., 2019	
hsa-miR-4448	Same	Up	Human aqueous humor	Glaucoma & XFS vs. Con	Hindle et al., 2019	
hsa-miR-152-5p	hsa-miR-152	Down	hREC	HG condition vs. Con	Haque et al., 2015	*In vitro*
hsa-miR-185-3p	Same	Down	Rabbit retina	Newborn vs. Adult	Robert et al., 2010	Rabbit
hsa-miR-18a-3p	Same	Down	Human aqueous humor	POAG vs. Cataract	Liu et al., 2018	
hsa-miR-410-3p	Same	Up	Human aqueous humor	POAG vs. Cataract	Liu et al., 2018	
hsa-miR-4433b-3p	Same	Up	Human aqueous humor	POAG vs. cataract	Liu et al., 2018	
hsa-miR-487a-5p	Same	Up	Human aqueous humor	POAG vs. Cataract	Liu et al., 2018	
hsa-miR-501-3p	Same	Up	Human aqueous humor	POAG vs. Cataract	Liu et al., 2018	
hsa-miR-760	Same	Up	Human aqueous humor	POAG vs. Cataract	Liu et al., 2018	
hsa-miR-874-3p	Same	Down	Human aqueous humor	POAG vs. Cataract	Liu et al., 2018	
hsa-miR-3149	Same	Up	Human serum	POAG vs. Con (cataract included)	Liu et al., 2019	
hsa-miR-18a-3p	hsa-miR-18a	Up	Human retina	RB vs. Con	Martin et al., 2013	
hsa-miR-22-3p	hsa-miR-22	Down	Human retina	RB vs. Con	Martin et al., 2013	
hsa-miR-504-3p	hsa-miR-504	Down	Human retina	RB vs. Con	Martin et al., 2013	
hsa-miR-874-3p	hsa-miR-874	Down	Human retina	RB vs. Con	Martin et al., 2013	
hsa-miR-214-3p	Same	Down	Human plasma	ROP vs. Con	Metin et al., 2018	
hsa-miR-223-5p	Same	Up	Human vitreous	severe PVR vs. mild PVR	Toro et al., 2020	
hsa-miR-1909-5p	Same	Up	ARPE-19 cell	TGFβ2 induced EMT vs. Con	Chen et al., 2014	*In vitro*
hsa-miR-223-5p	Same	Up	ARPE-19 cell	TGFβ2 induced EMT vs. Con	Chen et al., 2014	*In vitro*
hsa-miR-146b-3p	hsa-miR-146b-5p	Down	Human plasma	wet AMD vs. Con	Ertekin et al., 2014	
hsa-miR-324-3p	Same	Up	Human plasma	wet AMD vs. Con	Ertekin et al., 2014	Express only in patient group
hsa-miR-410-3p	hsa-miR-410	Down	Human plasma	wet AMD vs. Con	Ertekin et al., 2014	
hsa-miR-574-5p	hsa-miR-574-3p	Down	Human plasma	wet AMD vs. Con	Ertekin et al., 2014	

hREC, human retinal endothelial cells; XFS, exfoliation syndrome; Con, control; AMD, age-related macular degeneration; HG, high glucose; RB, retinoblastoma; POAG, primary open-angle glaucoma; EMT, epithelial-mesenchymal transition; ROP, retinopathy of Prematurity; PVR, proliferative vitreoretinopathy.

CONCLUSION AND FUTURE DIRECTIONS

IRBP is required to maintain the normal functions of the retina, and its downregulation is a common phenomenon at the early stages of photoreceptor degeneration. Although it may be an initial defensive response to retinal stress, the suppression of IRBP is harmful to the health of the photoreceptors in the long term (**Figure 1c**). The close relationship of IRBP downregulation with early symptoms and retinal disease severity forms the basis of its clinical applications as an early diagnostic marker and therapeutic target for many retinal diseases (Garcia-Ramirez et al., 2009; Zhu et al., 2015; Yokomizo et al., 2019). It is expected that restoring the expression of IRBP may slow the degeneration of photoreceptors. Future research involving techniques like CRISPRa-based gene therapy will allow for further exploration of the clinical potential of treating retinal diseases with IRBP-targeted gene therapy.

AUTHOR CONTRIBUTIONS

TZ and LZ conceived the perspective of the work. SZ, TZ, YC, LH, and MP searched and summarized the literature. SZ and TZ drafted the manuscript. TZ, RA-B, LZ, and RN designed the figure. MG, NF, RN, FZ, and MM critically revised the article. All authors commented on and contributed to the final manuscript.

FUNDING

This study was supported by an Australian National Health and Medical Research Council Ideas Grant (APP1188332), grants from the Lowy Medical Research Institute, the Rebecca L. Cooper Medical Research Foundation and the ANU Translational Fellowship. SZ was supported by Merri Borton Scholarship. MG is a Sydney Medical School Fellow and supported by an Australian National Health and Medical Research Council Practitioner Fellowship.

REFERENCES

Adler, A. J., and Evans, C. D. (1983). Rapid isolation of bovine interphotoreceptor retinol-binding protein. *Biochim. Biophys. Acta.* 761, 217–222. doi: 10.1016/0304-4165(83)90068-5

Adler, A. J., and Severin, K. M. (1981). Proteins of the bovine interphotoreceptor matrix: tissues of origin. *Exp. Eye Res.* 32, 755–769. doi: 10.1016/0014-4835(81)90025-7

Agarwal, R. K., Silver, P. B., and Caspi, R. R. (2012). Rodent models of experimental autoimmune uveitis. *Methods Mol. Biol.* 900, 443–469. doi: 10.1007/978-1-60761-720-4_22

Agarwal, V., Bell, G. W., Nam, J. W., and Bartel, D. P. (2015). Predicting effective microRNA target sites in mammalian mRNAs. *eLife* 4:e05005. doi: 10.7554/eLife.05005

Ambros, V. (2004). The functions of animal microRNAs. *Nature* 431, 350–355. doi: 10.1038/nature02871

Arno, G., Hull, S., Robson, A. G., Holder, G. E., Cheetham, M. E., Webster, A. R., et al. (2015). Lack of interphotoreceptor retinoid binding protein caused by homozygous mutation of RBP3 is associated with high myopia and retinal dystrophy. *Invest. Ophthalmol. Vis. Sci.* 56, 2358–2365. doi: 10.1167/iovs.15-16520

Bartel, D. P. (2018). Metazoan microRNAs. *Cell* 173, 20–51. doi: 10.1016/j.cell.2018.03.006

Benhar, I., London, A., and Schwartz, M. (2012). The privileged immunity of immune privileged organs: the case of the eye. *Front. Immunol.* 3:296. doi: 10.3389/fimmu.2012.00296

Betts-Obregon, B. S., Gonzalez-Fernandez, F., and Tsin, A. T. (2014). Interphotoreceptor retinoid-binding protein (IRBP) promotes retinol uptake and release by rat Müller cells (rMC-1) *in vitro*: implications for the cone visual cycle. *Invest. Ophthalmol. Vis. Sci.* 55, 6265–6271. doi: 10.1167/iovs.14-14721

Birnbach, C. D., Jarvelainen, M., Possin, D. E., and Milam, A. H. (1994). Histopathology and immunocytochemistry of the neurosensory retina in fundus flavimaculatus. *Ophthalmology* 101, 1211–1219. doi: 10.1016/S0161-6420(13)31725-4

Blackshaw, S., and Snyder, S. H. (1997). Parapinopsin, a novel catfish opsin localized to the parapineal organ, defines a new gene family. *J. Neurosci.* 17, 8083–8092. doi: 10.1523/JNEUROSCI.17-21-08083.1997

Bobola, N., Briata, P., Ilengo, C., Rosatto, N., Craft, C., Corte, G., et al. (1999). OTX2 homeodomain protein binds a DNA element necessary for interphotoreceptor retinoid binding protein gene expression. *Mech. Dev.* 82, 165–169. doi: 10.1016/S0925-4773(98)00162-2

Buddingh, B. C., Elzinga, J., and van Hest, J. C. M. (2020). Intercellular communication between artificial cells by allosteric amplification of a molecular signal. *Nat. Commun.* 11, 1652. doi: 10.1038/s41467-020-15482-8

Carlson, A., and Bok, D. (1992). Promotion of the release of 11-cis-retinal from cultured retinal pigment epithelium by interphotoreceptor retinoid-binding protein. *Biochemistry* 31, 9056–9062. doi: 10.1021/bi00152a049

Carrasquillo, K. G., Ricker, J. A., Rigas, I. K., Miller, J. W., Gragoudas, E. S., and Adamis, A. P. (2003). Controlled delivery of the anti-VEGF aptamer EYE001 with poly(lactic-co-glycolic)acid microspheres. *Invest. Ophthalmol. Vis. Sci.* 44, 290–299. doi: 10.1167/iovs.01-1156

Caspi, R. R., Chan, C. C., Wiggert, B., and Chader, G. J. (1990). The mouse as a model of experimental autoimmune uveoretinitis (EAU). *Curr. Eye Res.* 9, 169–174. doi: 10.3109/02713689008999438

Chan, T., Zhu, L., Madigan, M. C., Wang, K., Shen, W., Gillies, M. C., et al. (2015). Human organic anion transporting polypeptide 1A2 (OATP1A2) mediates cellular uptake of all-trans-retinol in human retinal pigmented epithelial cells. *Br. J. Pharmacol.* 172, 2343–2353. doi: 10.1111/bph.13060

Chen, C., Adler, L.t, Goletz, P., Gonzalez-Fernandez, F., Thompson, D. A., and Koutalos, Y. (2017). Interphotoreceptor retinoid-binding protein removes all-trans-retinol and retinal from rod outer segments, preventing lipofuscin precursor formation. *J. Biol. Chem.* 292, 19356–19365. doi: 10.1074/jbc.M117.795187

Chen, L. H., Chiou, G. Y., Chen, Y. W., Li, H. Y., and Chiou, S. H. (2010). MicroRNA and aging: a novel modulator in regulating the aging network. *Ageing Res. Rev.* 9(Suppl. 1), S59–S66. doi: 10.1016/j.arr.2010.08.002

Chen, M., Luo, C., Zhao, J., Devarajan, G., and Xu, H. (2019). Immune regulation in the aging retina. *Prog. Retin. Eye Res.* 69, 159–172. doi: 10.1016/j.preteyeres.2018.10.003

Chen, X., Ye, S., Xiao, W., Luo, L., and Liu, Y. (2014). Differentially expressed microRNAs in TGFbeta2-induced epithelial-mesenchymal transition in retinal pigment epithelium cells. *Int. J. Mol. Med.* 33, 1195–1200. doi: 10.3892/ijmm.2014.1688

Chen, Y., Okano, K., Maeda, T., Chauhan, V., Golczak, M., Maeda, A., et al. (2012). Mechanism of all-trans-retinal toxicity with implications for stargardt disease and age-related macular degeneration. *J. Biol. Chem.* 287, 5059–5069. doi: 10.1074/jbc.M111.315432

Chen, Y., Saari, J. C., and Noy, N. (1993). Interactions of all-trans-retinol and long-chain fatty acids with interphotoreceptor retinoid-binding protein. *Biochemistry* 32, 11311–11318. doi: 10.1021/bi00093a007

Coelho, T., Adams, D., Silva, A., Lozeron, P., Hawkins, P. N., Mant, T., et al. (2013). Safety and efficacy of RNAi therapy for transthyretin amyloidosis. *N. Engl. J. Med.* 369, 819–829. doi: 10.1056/NEJMoa1208760

Cook, J. D., Ng, S. Y., Lloyd, M., Eddington, S., Sun, H., Nathans, J., et al. (2017). Peropsin modulates transit of vitamin A from retina to retinal pigment epithelium. *J. Biol. Chem.* 292, 21407–21416. doi: 10.1074/jbc.M117.812701

Craft, C. M., Whitmore, D. H., and Wiechmann, A. F. (1994). Cone arrestin identified by targeting expression of a functional family. *J. Biol. Chem.* 269, 4613–4619.

Crouch, R. K., Hazard, E. S., Lind, T., Wiggert, B., Chader, G., and Corson, D. W. (1992). Interphotoreceptor retinoid-binding protein and alpha-tocopherol preserve the isomeric and oxidation state of retinol. *Photochem. Photobiol.* 56, 251–255. doi: 10.1111/j.1751-1097.1992.tb02154.x

Das, S. R., Bhardwaj, N., Kjeldbye, H., and Gouras, P. (1992). Muller cells of chicken retina synthesize 11-cis-retinol. *Biochem. J.* 285, 907–913. doi: 10.1042/bj2850907

den Hollander, A. I., McGee, T. L., Ziviello, C., Banfi, S., Dryja, T. P., Gonzalez-Fernandez, F., et al. (2009). A homozygous missense mutation in the IRBP gene (RBP3) associated with autosomal recessive retinitis pigmentosa. *Invest. Ophthalmol. Vis. Sci.* 50, 1864–1872. doi: 10.1167/iovs.08-2497

Doudna, J. A., and Charpentier, E. (2014). Genome editing. The new frontier of genome engineering with CRISPR-Cas9. *Science* 346:1258096. doi: 10.1126/science.1258096

Ertekin, S., Yildirim, O., Dinç E., Ayaz, L., Fidanci S. B., and Tamer, L. (2014). Evaluation of circulating miRNAs in wet age-related macular degeneration. *Mol. Vis.* 20, 1057–1066.

Fei, Y., Matragoon, S., Smith, S. B., Overbeek, P. A., Chen, S., Zack, D. J., et al. (1999). Functional dissection of the promoter of the interphotoreceptor retinoid-binding protein gene: the cone-rod-homeobox element is essential for photoreceptor-specific expression *in vivo*. *J. Biochem.* 125, 1189–1199. doi: 10.1093/oxfordjournals.jbchem.a022403

Filipowicz, W., Bhattacharyya, S. N., and Sonenberg, N. (2008). Mechanisms of post-transcriptional regulation by microRNAs: are the answers in sight? *Nat. Rev. Genetics* 9, 102–114. doi: 10.1038/nrg2290

Friedman, R. C., Farh, K. K., Burge, C. B., and Bartel, D. P. (2009). Most mammalian mRNAs are conserved targets of microRNAs. *Genome Res.* 19, 92–105. doi: 10.1101/gr.082701.108

Fulzele, S., El-Sherbini, A., Ahmad, S., Sangani, R., Matragoon, S., El-Remessy, A., et al. (2015). MicroRNA-146b-3p regulates retinal inflammation by suppressing adenosine deaminase-2 in diabetes. *Biomed Res. Int.* 2015:846501. doi: 10.1155/2015/846501

Garcia-Ramirez, M., Hernandez, C., Villarroel, M., Canals, F., Alonso, M. A., Fortuny, R., et al. (2009). Interphotoreceptor retinoid-binding protein (IRBP) is downregulated at early stages of diabetic retinopathy. *Diabetologia* 52, 2633–2641. doi: 10.1007/s00125-009-1548-8

Garlipp, M. A., and Gonzalez-Fernandez, F. (2013). Cone outer segment and Müller microvilli pericellular matrices provide binding domains for interphotoreceptor retinoid-binding protein (IRBP). *Exp. Eye Res.* 113, 192–202. doi: 10.1016/j.exer.2013.02.003

Ghosh, D., Haswell, K. M., Sprada, M., and Gonzalez-Fernandez, F. (2015). Structure of zebrafish IRBP reveals fatty acid binding. *Exp. Eye Res.* 140, 149–158. doi: 10.1016/j.exer.2015.08.026

Gibbs, E., Matsubara, J., Cao, S., Cui, J., and Forooghian, F. (2017). Antigen-specificity of antiretinal antibodies in patients with noninfectious uveitis. *Can. J. Ophthalmol.* 52, 463–467. doi: 10.1016/j.jcjo.2017.03.010

Gilbert, L. A., Horlbeck, M. A., Adamson, B., Villalta, J. E., Chen, Y., Whitehead, E. H., et al. (2014). Genome-scale CRISPR-mediated control of gene repression and activation. *Cell* 159, 647–661. doi: 10.1016/j.cell.2014.09.029

Goldberg, A. F., Moritz, O. L., and Williams, D. S. (2016). Molecular basis for photoreceptor outer segment architecture. *Prog. Retin. Eye Res.* 55, 52–81. doi: 10.1016/j.preteyeres.2016.05.003

Gonzalez-Fernandez, F. (2003). Interphotoreceptor retinoid-binding protein–an old gene for new eyes. *Vis. Res.* 43, 3021–3036. doi: 10.1016/j.visres.2003.09.019

Gonzalez-Fernandez, F. (2012). Interphotoreceptor retinoid binding protein; myths and mysteries. *J. Ophthalmic Vis. Res.* 7, 100–104.

Gonzalez-Fernandez, F., and Ghosh, D. (2008). Focus on molecules: interphotoreceptor retinoid-binding protein (IRBP). *Exp. Eye Res.* 86, 169–170. doi: 10.1016/j.exer.2006.09.003

Gonzalez-Fernandez, F., Sung, D., Haswell, K. M., Tsin, A., and Ghosh, D. (2014). Thiol-dependent antioxidant activity of interphotoreceptor retinoid-binding protein. *Exp. Eye Res.* 120, 167–174. doi: 10.1016/j.exer.2014.01.002

Gonzalez-Fernandez, F., van Niel, E., Edmonds, C., Beaver, H., Nickerson, J. M., Garcia-Fernandez, J. M., et al. (1993). Differential expression of interphotoreceptor retinoid-binding protein, opsin, cellular retinaldehyde-binding protein, and basic fibroblastic growth factor. *Exp. Eye Res.* 56, 411–427. doi: 10.1006/exer.1993.1055

Haque, R., Hur, E. H., Farrell, A. N., Iuvone, P. M., and Howell, J. C. (2015). MicroRNA-152 represses VEGF and TGFbeta1 expressions through post-transcriptional inhibition of (Pro)renin receptor in human retinal endothelial cells. *Mol. Vis.* 21, 224–235.

He, L., and Hannon, G. J. (2004). MicroRNAs: small RNAs with a big role in gene regulation. *Nat. Rev. Genetics* 5, 522–531. doi: 10.1038/nrg1379

Hessler, R. B., Baer, C. A., Bukelman, A., Kittredge, K. L., and Gonzalez-Fernandez, F. (1996). Interphotoreceptor retinoid-binding protein (IRBP): expression in the adult and developing Xenopus retina. *J. Comp. Neurol.* 367, 329–341. doi: 10.1002/(SICI)1096-9861(19960408)367:3<329::AID-CNE1>3.0.CO;2-7

Hindle, A. G., Thoonen, R., Jasien, J. V., Grange, R. M. H., Amin, K., Wise, J., et al. (2019). Identification of candidate miRNA biomarkers for glaucoma. *Invest. Ophthalmol. Vis. Sci.* 60, 134–146. doi: 10.1167/iovs.18-24878

Hollyfield, J. G. (1999). Hyaluronan and the functional organization of the interphotoreceptor matrix. *Invest. Ophthalmol. Vis. Sci.* 40, 2767–2769.

Ishikawa, M., Sawada, Y., and Yoshitomi, T. (2015). Structure and function of the interphotoreceptor matrix surrounding retinal photoreceptor cells. *Exp. Eye Res.* 133, 3–18. doi: 10.1016/j.exer.2015.02.017

Jastrzebska, B., Debinski, A., Filipek, S., and Palczewski, K. (2011). Role of membrane integrity on G protein-coupled receptors: rhodopsin stability and function. *Prog. Lipid Res.* 50, 267–277. doi: 10.1016/j.plipres.2011.03.002

Jin, M., Li, S., Nusinowitz, S., Lloyd, M., Hu, J., Radu, R. A., et al. (2009). The role of interphotoreceptor retinoid-binding protein on the translocation of visual retinoids and function of cone photoreceptors. *J. Neurosci.* 29, 1486–1495. doi: 10.1523/JNEUROSCI.3882-08.2009

Jones, G. J., Crouch, R. K., Wiggert, B., Cornwall, M. C., and Chader, G. J. (1989). Retinoid requirements for recovery of sensitivity after visual-pigment bleaching in isolated photoreceptors. *Proc. Natl. Acad. Sci. U.S.A.* 86, 9606–9610. doi: 10.1073/pnas.86.23.9606

Kennedy, B. N., Alvarez, Y., Brockerhoff, S. E., Stearns, G. W., Sapetto-Rebow, B., Taylor, M. R., et al. (2007). Identification of a zebrafish cone photoreceptor-specific promoter and genetic rescue of achromatopsia in the nof mutant. *Invest. Ophthalmol. Vis. Sci.* 48, 522–529. doi: 10.1167/iovs.06-0975

Kiser, P. D., Golczak, M., and Palczewski, K. (2014). Chemistry of the retinoid (visual) cycle. *Chem. Rev.* 114, 194–232. doi: 10.1021/cr400107q

Kompella, U. B., Bandi, N., and Ayalasomayajula, S. P. (2003). Subconjunctival nano- and microparticles sustain retinal delivery of budesonide, a corticosteroid capable of inhibiting VEGF expression. *Invest. Ophthalmol. Vis. Sci.* 44, 1192–1201. doi: 10.1167/iovs.02-0791

Kusakabe, T. G., Takimoto, N., Jin, M., and Tsuda, M. (2009). Evolution and the origin of the visual retinoid cycle in vertebrates. *Philos. Trans. R Soc. London Series B Biol. Sci.* 364, 2897–2910. doi: 10.1098/rstb.2009.0043

Kyger, M., Worley, A., and Adamus, G. (2013). Autoimmune responses against photoreceptor antigens during retinal degeneration and their role in macrophage recruitment into retinas of RCS rats. *J. Neuroimmunol.* 254, 91–100. doi: 10.1016/j.jneuroim.2012.10.007

Ledford, H. (2020). CRISPR treatment inserted directly into the body for first time. *Nature* 579:185. doi: 10.1038/d41586-020-00655-8

Lee, M., Li, S., Sato, K., and Jin, M. (2016). Interphotoreceptor retinoid-binding protein mitigates cellular oxidative stress and mitochondrial dysfunction induced by all-trans-retinal. *Invest. Ophthalmol. Vis. Sci.* 57, 1553–1562. doi: 10.1167/iovs.15-18551

Lee, R. C., Feinbaum, R. L., and Ambros, V. (1993). The *C. elegans* heterochronic gene lin-4 encodes small RNAs with antisense complementarity to lin-14. *Cell* 75, 843–854. doi: 10.1016/0092-8674(93)90529-Y

Li, G., Luna, C., Qiu, J., Epstein, D. L., and Gonzalez, P. (2010). Modulation of inflammatory markers by miR-146a during replicative senescence in trabecular meshwork cells. *Invest. Ophthalmol. Vis. Sci.* 51, 2976–2985. doi: 10.1167/iovs.09-4874

Li, S., Yang, Z., Hu, J., Gordon, W. C., Bazan, N. G., Haas, A. L., et al. (2013). Secretory defect and cytotoxicity: the potential disease mechanisms for the retinitis pigmentosa (RP)-associated interphotoreceptor retinoid-binding protein (IRBP). *J. Biol. Chem.* 288, 11395–11406. doi: 10.1074/jbc.M112.418251

Li, W. H., Zhou, L., Li, Z., Wang, Y., Shi, J. T., Yang, Y. J., et al. (2015). Zebrafish Lbh-like is required for Otx2-mediated photoreceptor differentiation. *Int. J. Biol. Sci.* 11, 688–700. doi: 10.7150/ijbs.11244

Li, Y., Cheng, T., Wan, C., and Cang, Y. (2020). circRNA_0084043 contributes to the progression of diabetic retinopathy via sponging miR-140-3p and inducing TGFA gene expression in retinal pigment epithelial cells. *Gene* 747:144653. doi: 10.1016/j.gene.2020.144653

Liou, G. I., Bridges, C. D., Fong, S. L., Alvarez, R. A., and Gonzalez-Fernandez, F. (1982). Vitamin A transport between retina and pigment epithelium–an interstitial protein carrying endogenous retinol (interstitial retinol-binding protein). *Vis. Res.* 22, 1457–1467. doi: 10.1016/0042-6989(82)90210-3

Liou, G. I., Fei, Y., Peachey, N. S., Matragoon, S., Wei, S., Blaner, W. S., et al. (1998). Early onset photoreceptor abnormalities induced by targeted disruption of the interphotoreceptor retinoid-binding protein gene. *J. Neurosci.* 18, 4511–4520. doi: 10.1523/JNEUROSCI.18-12-04511.1998

Liou, G. I., Wang, M., and Matragoon, S. (1994). Timing of interphotoreceptor retinoid-binding protein (IRBP) gene expression and hypomethylation in developing mouse retina. *Dev. Biol.* 161, 345–356. doi: 10.1006/dbio.1994.1036

Liu, A., Chang, J., Lin, Y., Shen, Z., and Bernstein, P. S. (2010). Long-chain and very long-chain polyunsaturated fatty acids in ocular aging and age-related macular degeneration. *J. Lipid Res.* 51, 3217–3229. doi: 10.1194/jlr.M007518

Liu, Y. M., Chen, Y., Wang, Y. Y., Zhang, X. Y., Gao, K., Chen, S. D., et al. (2018). microRNA profiling in glaucoma eyes with varying degrees of optic neuropathy by using next-generation sequencing. *Invest. Ophthalmol. Vis. Sci.* 59, 2955–2966. doi: 10.1167/iovs.17-23599

Liu, Y. M., Wang, Y. Y., Chen, Y., Fang, X. L., Wen, T., Xiao, M. L., et al. (2019). Discovery and validation of circulating Hsa-miR-210-3p as a potential biomarker for primary open-angle glaucoma. *Invest. Ophthalmol. Vis. Sci.* 60, 2925–2934. doi: 10.1167/iovs.19-26663

Lukiw, W. J., Surjyadipta, B., Dua, P., and Alexandrov, P. N. (2012). Common micro RNAs (miRNAs) target complement factor H (CFH) regulation in Alzheimer's disease (AD) and in age-related macular degeneration (AMD). *Int. J. Biochem. Mol. Biol.* 3, 105–116.

Maeda, A., Maeda, T., Golczak, M., Chou, S., Desai, A., Hoppel, C. L., et al. (2009). Involvement of all-trans-retinal in acute light-induced retinopathy of mice. *J. Biol. Chem.* 284, 15173–15183. doi: 10.1074/jbc.M900322200

Malechka, V. V., Moiseyev, G., Takahashi, Y., Shin, Y., and Ma, J. X. (2017). Impaired rhodopsin generation in the rat model of diabetic retinopathy. *Am. J. Pathol.* 187, 2222–2231. doi: 10.1016/j.ajpath.2017.06.007

Markand, S., Baskin, N. L., Chakraborty, R., Landis, E., Wetzstein, S. A., Donaldson, K. J., et al. (2016). IRBP deficiency permits precocious ocular development and myopia. *Mol. Vis.* 22, 1291–1308.

Marmorstein, A. D., Marmorstein, L. Y., Sakaguchi, H., and Hollyfield, J. G. (2002). Spectral profiling of autofluorescence associated with lipofuscin, Bruch's membrane, and sub-RPE deposits in normal and AMD eyes. *Invest. Ophthalmol. Vis. Sci.* 43, 2435–2441.

Martin, J., Bryar, P., Mets, M., Weinstein, J., Jones, A., Martin, A., et al. (2013). Differentially expressed miRNAs in retinoblastoma. *Gene* 512, 294–299. doi: 10.1016/j.gene.2012.09.129

Mata, N. L., Radu, R. A., Clemmons, R. C., and Travis, G. H. (2002). Isomerization and oxidation of vitamin a in cone-dominant retinas: a novel pathway for visual-pigment regeneration in daylight. *Neuron* 36, 69–80. doi: 10.1016/S0896-6273(02)00912-1

Maxim, L. D., Niebo, R., and Utell, M. J. (2014). Screening tests: a review with examples. *Inhal. Toxicol.* 26, 811–828. doi: 10.3109/08958378.2014.955932

McCarty, N. S., Graham, A. E., Studen,á L., and Ledesma-Amaro, R. (2020). Multiplexed CRISPR technologies for gene editing and transcriptional regulation. *Nat. Commun.* 11:1281. doi: 10.1038/s41467-020-15053-x

McDougald, D. S., Duong, T. T., Palozola, K. C., Marsh, A., Papp, T. E., Mills, J. A., et al. (2019). CRISPR activation enhances *in vitro* potency of AAV vectors driven by tissue-specific promoters. *Mol. Ther. Methods Clin. Dev.* 13, 380–389. doi: 10.1016/j.omtm.2019.03.004

Ménard, C., Rezende, F. A., Miloudi, K., Wilson, A., Tétreault, N., Hardy, P., et al. (2016). MicroRNA signatures in vitreous humour and plasma of patients with exudative AMD. *Oncotarget* 7, 19171–19184. doi: 10.18632/oncotarget.8280

Menotti-Raymond, M., David, V. A., Schaffer, A. A., Stephens, R., Wells, D., Kumar-Singh, R., et al. (2007). Mutation in CEP290 discovered for cat model of human retinal degeneration. *J. Hered.* 98, 211–220. doi: 10.1093/jhered/esm019

Metin, T., Din,ç E., Görür, A., Erdogan, S., Ertekin, S., Sar,i A. A., et al. (2018). Evaluation of the plasma microRNA levels in stage 3 premature retinopathy with plus disease: preliminary study. *Eye* 32, 415–420. doi: 10.1038/eye.2017.193

Miraglia, L. J., King, F. J., and Damoiseaux, R. (2011). Seeing the light: luminescent reporter gene assays. *Comb. Chem. High Throughput Screen.* 14, 648–657. doi: 10.2174/138620711796504389

Molday, R. S., and Moritz, O. L. (2015). Photoreceptors at a glance. *J. Cell Sci.* 128, 4039–4045. doi: 10.1242/jcs.175687

Morohoshi, K., Ohbayashi, M., Patel, N., Chong, V., Bird, A. C., and Ono, S. J. (2012). Identification of anti-retinal antibodies in patients with age-related macular degeneration. *Exp. Mol. Pathol.* 93, 193–199. doi: 10.1016/j.yexmp.2012.03.007

Narfstrom, K., Nilsson, S. E., Wiggert, B., Lee, L., Chader, G. J., and van Veen, T. (1989). Reduced level of interphotoreceptor retinoid-binding protein (IRBP), a possible cause for retinal degeneration in the Abyssinian cat. *Cell Tissue Res.* 257, 631–639. doi: 10.1007/BF00221474

Nickerson, J. M., Frey, R. A., Ciavatta, V. T., and Stenkamp, D. L. (2006). Interphotoreceptor retinoid-binding protein gene structure in tetrapods and teleost fish. *Mol. Vis.* 12, 1565–1585.

Nishida, A., Furukawa, A., Koike, C., Tano, Y., Aizawa, S., Matsuo, I., et al. (2003). Otx2 homeobox gene controls retinal photoreceptor cell fate and pineal gland development. *Nat. Neurosci.* 6, 1255–1263. doi: 10.1038/nn1155

Parker, R., Wang, J. S., Kefalov, V. J., and Crouch, R. K. (2011). Interphotoreceptor retinoid-binding protein as the physiologically relevant carrier of 11-cis-retinol in the cone visual cycle. *J. Neurosci.* 31, 4714–4719. doi: 10.1523/JNEUROSCI.3722-10.2011

Parker, R. O., and Crouch, R. K. (2010). The interphotoreceptor retinoid binding (IRBP) is essential for normal retinoid processing in cone photoreceptors. *Adv. Exp. Med. Biol.* 664, 141–149. doi: 10.1007/978-1-4419-1399-9_17

Purves, D. A. G., and Fitzpatrick, D. (2001). *Strategies of Molecular Signaling. Neuroscience.* 2nd Edn. Sunderland, MA: Sinauer Associates, Inc.

Radu, R. A., Mata, N. L., Nusinowitz, S., Liu, X. R., Sieving, P. A., and Travis, G. H. (2003). Treatment with isotretinoin inhibits lipofuscin accumulation in a mouse model of recessive Stargardt's macular degeneration. *Proc. Natl. Acad. Sci. U.S.A.* 100, 4742–4747. doi: 10.1073/pnas.0737855100

Riley, M. K., and Vermerris, W. (2017). Recent advances in nanomaterials for gene delivery-a review. *Nanomaterials* 7:94. doi: 10.3390/nano7050094

Rizk, M., and Tuzmen, S. (2017). Update on the clinical utility of an RNA interference-based treatment: focus on Patisiran. *Pharmgenomics. Pers. Med.* 10, 267–278. doi: 10.2147/PGPM.S87945

Robert, E., Anderson, J. G. H., and Matthew, M. (2010). *Retinal Degenerative Disease. Advances in Experimental Medicine and Biology*, Vol. 664 (Springer Science & Bussiness Media), 206–208. Available online at: https://books.google.com.au/books?id=1x_NTfV6xlAC&pg=PA207dq=miR-185+retina&source=bl&ots=7oIO8o0fmV&sig=ACfU3U1aSP6_X_7E5WAetoxtdm-3H8Qsew&hl=en&sa=X&ved=2ahUKEwiftby6loXqAhUkxDgGHQKvAc8Q6AEwA3oECAoQAQ#v=onepage&q=miR-185%20retina&f=false

Rodrigues, M. M., Hackett, J., Gaskins, R., Wiggert, B., Lee, L., Redmond, M., et al. (1986). Interphotoreceptor retinoid-binding protein in retinal rod cells and pineal gland. *Invest. Ophthalmol. Vis. Sci.* 27, 844–850.

Rózanowska, M., Handzel, K., Boulton, M. E., and Rózanowski, B. (2012). Cytotoxicity of all-trans-retinal increases upon photodegradation. *Photochem. Photobiol.* 88, 1362–1372. doi: 10.1111/j.1751-1097.2012.01161.x

Sapède, D., and Cau, E. (2013). The pineal gland from development to function. *Curr. Top. Dev. Biol.* 106, 171–215. doi: 10.1016/B978-0-12-416021-7.00005-5

Sato, K., Li, S., Gordon, W. C., He, J., Liou, G. I., Hill, J. M., et al. (2013). Receptor interacting protein kinase-mediated necrosis contributes to cone and rod photoreceptor degeneration in the retina lacking interphotoreceptor retinoid-binding protein. *J. Neurosci.* 33, 17458–17468. doi: 10.1523/JNEUROSCI.1380-13.2013

Sato, S., and Kefalov, V. J. (2016). cis Retinol oxidation regulates photoreceptor access to the retina visual cycle and cone pigment regeneration. *J. Physiol.* 594, 6753–6765. doi: 10.1113/JP272831

Scott, B. L., and Bazan, N. G. (1989). Membrane docosahexaenoate is supplied to the developing brain and retina by the liver. *Proc. Natl. Acad. Sci. U.S.A.* 86, 2903–2907. doi: 10.1073/pnas.86.8.2903

Semenova, E. M., and Converse, C. A. (2003). Comparison between oleic acid and docosahexaenoic acid binding to interphotoreceptor retinoid-binding protein. *Vis. Res.* 43, 3063–3067. doi: 10.1016/j.visres.2003.09.008

Solomon, A., Banin, E., Anteby, I., and Benezra, D. (1999). Retinitis pigmentosa, coats disease and uveitis. *Eur. J. Ophthalmol.* 9, 202–205. doi: 10.1177/112067219900900307

Soundara Pandi, S. P., Chen, M., Guduric-Fuchs, J., Xu, H., and Simpson, D. A. (2013). Extremely complex populations of small RNAs in the mouse retina and RPE/choroid. *Invest. Ophthalmol. Vis. Sci.* 54, 8140–8151. doi: 10.1167/iovs.13-12631

Stenkamp, D. L., Cunningham, L. L., Raymond, P. A., and Gonzalez-Fernandez, F. (1998). Novel expression pattern of interphotoreceptor retinoid-binding protein (IRBP) in the adult and developing zebrafish retina and RPE. *Mol. Vis.* 4:26.

Szemraj, M., Bielecka-Kowalska, A., Oszajca, K., Krajewska, M., Go,ś R., Jurowski, P., et al. (2015). Serum microRNAs as potential biomarkers of AMD. *Med. Sci. Monit.* 21, 2734–2742. doi: 10.12659/MSM.893697

Tang, P. H., Kono, M., Koutalos, Y., Ablonczy, Z., and Crouch, R. K. (2013). New insights into retinoid metabolism and cycling within the retina. *Prog. Retin. Eye Res.* 32, 48–63. doi: 10.1016/j.preteyeres.2012.09.002

Timmers, A. M., Newton, B. R., and Hauswirth, W. W. (1993). Synthesis and stability of retinal photoreceptor mRNAs are coordinately regulated during bovine fetal development. *Exp. Eye Res.* 56, 257–265. doi: 10.1006/exer.1993.1034

Toro, M. D., Reibaldi, M., Avitabile, T., Bucolo, C., Salomone, S., Rejdak, R., et al. (2020). MicroRNAs in the vitreous humor of patients with retinal detachment and a different grading of proliferative vitreoretinopathy: a pilot study. *Transl. Vis. Sci. Technol.* 9:23 doi: 10.1167/tvst.9.6.23

van Veen, T., Katial, A., Shinohara, T., Barrett, D. J., Wiggert, B., Chader, G. J., et al. (1986). Retinal photoreceptor neurons and pinealocytes accumulate mRNA for interphotoreceptor retinoid-binding protein (IRBP). *FEBS Lett.* 208, 133–137. doi: 10.1016/0014-5793(86)81547-2

Vecino, E., Rodriguez, F. D., Ruzafa, N., Pereiro, X., and Sharma, S. C. (2016). Glia-neuron interactions in the mammalian retina. *Prog. Retin. Eye Res.* 51, 1–40. doi: 10.1016/j.preteyeres.2015.06.003

Wald, G. (1935). Carotenoids and the visual cycle. *J. Gen. Physiol.* 19, 351–371. doi: 10.1085/jgp.19.2.351

Wang, J. S., and Kefalov, V. J. (2009). An alternative pathway mediates the mouse and human cone visual cycle. *Curr. Biol.* 19, 1665–1669. doi: 10.1016/j.cub.2009.07.054

Wang, K., Zhu, X., Zhang, K., Zhou, F., and Zhu, L. (2017). Neuroprotective effect of tetramethylpyrazine against all-trans-retinal toxicity in the differentiated Y-79 cells via upregulation of IRBP expression. *Exp. Cell Res.* 359, 120–128. doi: 10.1016/j.yexcr.2017.08.002

Wiggert, B., van Veen, T., Kutty, G., Lee, L., Nickerson, J., Si, J. S., et al. (1994). An early decrease in interphotoreceptor retinoid-binding protein gene expression in Abyssinian cats homozygous for hereditary rod-cone degeneration. *Cell Tissue Res.* 278, 291–298. doi: 10.1007/s004410050220

Wightman, B., Ha, I., and Ruvkun, G. (1993). Posttranscriptional regulation of the heterochronic gene lin-14 by lin-4 mediates temporal pattern formation in *C. elegans*. *Cell* 75, 855–862. doi: 10.1016/0092-8674(93)90530-4

Wisard, J., Faulkner, A., Chrenek, M. A., Waxweiler, T., Waxweiler, W., Donmoyer, C., et al. (2011). Exaggerated eye growth in IRBP-deficient mice in early development. *Invest. Ophthalmol. Vis. Sci.* 52, 5804–5811. doi: 10.1167/iovs.10-7129

Wong, P., Ulyanova, T., Organisciak, D. T., Bennett, S., Lakins, J., Arnold, J. M., et al. (2001). Expression of multiple forms of clusterin during light-induced retinal degeneration. *Curr. Eye Res.* 23, 157–165. doi: 10.1076/ceyr.23.3.157.5463

Xie, W., Silvers, R., Ouellette, M., Wu, Z., Lu, Q., Li, H., et al. (2016). A luciferase reporter gene system for high-throughput screening of gamma-globin gene activators. *Methods Mol. Biol.* 1439, 207–226. doi: 10.1007/978-1-4939-3673-1_14

Yokomizo, H., Maeda, Y., Park, K., Clermont, A. C., Hernandez, S. L., Fickweiler, W., et al. (2019). Retinol binding protein 3 is increased in the retina of patients with diabetes resistant to diabetic retinopathy. *Sci. Transl. Med.* 11:eaau6627. doi: 10.1126/scitranslmed.aau6627

Zhang, T., Gillies, M., Wang, Y., Shen, W., Bahrami, B., Zeng, S., et al. (2019). Simvastatin protects photoreceptors from oxidative stress induced by all-trans-retinal, through the up-regulation of interphotoreceptor retinoid binding protein. *Br. J. Pharmacol.* 176, 2063–2078. doi: 10.1111/bph.14650

Zhao, Y., Liu, Y., and Chen, K. (2016). Mechanisms and clinical application of tetramethylpyrazine (an interesting natural compound isolated from Ligusticum Wallichii): current status and perspective. *Oxid. Med. Cell. Longev.* 2016:2124638. doi: 10.1155/2016/2124638

Zheng, Q., Ren, Y., Tzekov, R., Hua, S., Li, M., Pang, J., et al. (2015). iTRAQ-based proteomic analysis of visual cycle-associated proteins in RPE of rd12 mice before and after RPE65 gene delivery. *J. Ophthalmol.* 2015:918473. doi: 10.1155/2015/918473

Zhu, L., Shen, W., Lyons, B., Wang, Y., Zhou, F., and Gillies, M. C. (2015). Dysregulation of inter-photoreceptor retinoid-binding protein (IRBP) after induced Muller cell disruption. *J. Neurochem.* 133, 909–918. doi: 10.1111/jnc.13075

Zhu, L., Shen, W., Zhu, M., Coorey, N. J., Nguyen, A. P., Barthelmes, D., et al. (2013). Anti-retinal antibodies in patients with macular telangiectasia type 2. *Invest. Ophthalmol. Vis. Sci.* 54, 5675–5683. doi: 10.1167/iovs.13-12050

Zlotogora, J. (2007). Multiple mutations responsible for frequent genetic diseases in isolated populations. *Eur. J. Hum. Genet.* 15, 272–278. doi: 10.1038/sj.ejhg.5201760

12

Mode-Dependent Effect of Xenon Inhalation on Kainic Acid-Induced Status Epilepticus in Rats

Yurong Zhang[1†], Mengdi Zhang[1†], Jie Yu[1†], Wei Zhu[2], Qiaoyun Wang[1], Xiaohong Pan[1], Xue Gao[1], Jing Yang[1] and Hongliu Sun[1]*

[1] *School of Pharmaceutical Sciences, Binzhou Medical University, Yantai, China,* [2] *Shandong Academy of Medical Sciences, Jinan, China*

***Correspondence:**
Hongliu Sun
sun_china6@163.com

† *These authors have contributed equally to this work*

Previous studies have reported the possible neuroprotective effects of xenon treatment. The purpose of this study was to define the range of effective xenon ratio, most effective xenon ratio, and time-window for intervention in the kainic acid (KA) – induced status epilepticus (SE) rat model. Different ratios of xenon (35% xenon, 21% oxygen, 44% nitrogen, 50% xenon, 21% oxygen, 29% nitrogen, 70% xenon, 21% oxygen, and 9% nitrogen) were used to treat the KA-induced SE. Our results confirmed the anti-seizure role of 50 and 70% xenon mixture, with a stronger effect from the latter. Further, 70% xenon mixture was dispensed at three time points (0 min, 15 min delayed, and 30 min delayed) after KA administration, and the results indicated the anti-seizure effect at all treated time points. The results also established that the neuronal injury in the hippocampus and entorhinal cortex (EC), assessed using Fluoro-Jade B (FJB) staining, were reversed by the xenon inhalation, and within 30 min after KA administration. Our study, therefore, indicates the appropriate effective xenon ratio and time-window for intervention that can depress seizures. The prevention of neuronal injury and further reversal of the loss of effective control of depress network in the hippocampus and EC may be the mechanisms underlying the anti-seizure effect of xenon.

Keywords: status epilepticus, xenon, neuronal injury, kainic acid, seizure

INTRODUCTION

Currently available anti-epileptic interventions such as anti-epileptic drugs, resection, and deep brain stimulation have a limited efficacy (Wang et al., 2015; Falcicchia et al., 2018; Kaur et al., 2019). Most patients with epilepsy depend on anti-epileptic drugs to control seizures. However, these possess a few drawbacks such as drug resistance that influence about one-third of the individuals (Schmidt and Löscher, 2005). The surgical treatment of epilepsy is not suitable for all refractory epilepsy cases (Wiebe et al., 2001; McIntosh et al., 2004). As the recently established epileptic therapy, deep brain stimulation could control a part of refractory epilepsy cases; however, it has a few drawbacks such as selection of stimulation region and parameters for different types of epilepsy (Cohen-Gadol et al., 2003; Theodore and Fisher, 2004) and some other complications (Lesser, 2000; Smyth et al., 2003). Therefore, it is important to develop therapies that can depress seizures and reduce the epileptic brain injury.

Xenon is used as a safe anesthetic for its fewer side effects. In the recent years, reports of neuroprotective effects of xenon have been attracting attention (Metaxa et al., 2014; Yang et al., 2014; Lavaur et al., 2016). Moreover, xenon can significantly inhibit the uptake and efflux of glutamate (Lavaur et al., 2016), and swiftly terminate the synchronous discharge (Uchida et al., 2012). Over excitation induced by elevated concentration of glutamate is closely associated with epileptic development (Sun et al., 2013) and seizures. Neuronal injury and synchronous discharge are the pathological characteristics of epilepsy (During and Spencer, 1993; Chiu et al., 2016; Kim and Kang, 2018). Therefore, we have a reason to speculate that xenon may possess anti-epileptic properties. Considering that xenon has almost no adverse reactions, xenon treatment may be an effective, and safe intervention for epilepsy.

However, several issues, such as the difference in the anti-seizure effect under different xenon ratio, still need to be addressed for xenon to be a potential clinical intervention. Moreover, as the patients generally could not be treated with xenon immediately, there will be a delayed period between the seizure onset and xenon treatment; therefore, it is also necessary to define the effective therapeutic time-window.

Thus, our study aimed to explore the range of xenon ratio for confirming the most effective ratio for seizure therapy. Furthermore, we investigated the difference in the anti-seizure effect upon xenon treatment at different time points from the seizure onset and establish the effective time-window and the best time-window for xenon treatment. Also, our study preliminarily reveals the underlying mechanisms behind the anti-seizure effect of the xenon mixture.

MATERIALS AND METHODS

Animals and Surgery

Male Sprague-Dawley rats (240–260 g, Certificate No. SCXK2014-0006; Jinan Jinfeng Experimental Animal Co., Ltd, Shandong, China) were used in the study. The experiments were conducted in compliance with the National Institutes of Health *Guide for the Care and Use of Laboratory Animal* (NIH Publications No. 80-23, revised 1996) and the ethical principles of Binzhou Medical University Animal Experimentation Committee (approval no. 2016002). Attempts were made to reduce the number of rats used and their suffering. Experimental rats were maintained in individual cages. Water and food were provided *ad libitum*. All investigations and treatments were carried out between 9:00 and 17:00.

Rats were fixed in a stereotactic apparatus (Anhui Zheng Hua Biological Instrument Equipment Co., Ltd., China) after anesthesia (sodium pentobarbital, 50 mg/kg, i.p., CAS, 57-33-0, Xiya Reagent, China). Each rat was implanted a cannula (RSD Life Science, China) into the lateral ventricle (anteroposterior, AP: −1.8 mm, lateral, L: −1.0 mm, and ventral, V: −3.6 mm). The recording stainless steel electrodes (twisted-pair, A.M. Systems, United States) with 0.5 mm uncoated tip were implanted into the right cortex (AP: −3.2 mm, L: −3.0 mm, and V: −1.8 mm). The implanted electrodes were connected with a microsocket. Both the electrodes and cannulas were fixed onto the skull using dental cement as we have reported previously (Sun et al., 2018). The rats were allowed to recover for 7 days before further experiments.

KA Treatment and Xenon Inhalation

Kainic acid (KA) was injected stereotaxically (3.25×10^{-3} mg/kg, 1.25 mg/ml, Sigma–Aldrich) through a cannula into the lateral ventricle to induce the status epilepticus (SE), as described in our previous study (Zhang et al., 2019). All 126 rats showed continuous generalized seizures almost immediately after the KA injection. The seizures were terminated by diazepam injections (1 mg/ml solution at a dose of 0.002 mg/g body weight; intraperitoneally, Sigma–Aldrich) after 60 min.

Rats were randomly selected and treated with different ratio xenon mixture (Dalian Special Gas Co., Ltd., China, 70% xenon, 21% oxygen, 9% nitrogen treatment, 70% xenon group, $n = 9$; 50% xenon, 21% oxygen, 29% nitrogen treatment, 50% xenon group, $n = 10$; or 35% xenon, 21% oxygen, 44% nitrogen treatment, and 35% xenon group, $n = 10$) based on the previous reports of neuroprotective effects by xenon treatment (Dingley et al., 2006; Cattano et al., 2011). Xenon inhalation was performed just after the KA treatment and lasted for 1 h. The rats in control group ($n = 10$) were treated with 21% oxygen, 79% nitrogen (Rulin Gas Co., Ltd., China) instead of xenon mixture (De Deken et al., 2018). The behavior was monitored for 1 h until the injection of diazepam to terminate the seizures, and electroencephalograms (EEGs) were also recorded and digitalized using filters (1 Hz low-pass and 50 Hz high-pass). Frequency spectrum and the power spectrum density (PSD) in EEGs were analyzed using a PowerLab Biological Recording System (1–50 Hz, AD Instruments, Australia) from KA administration to diazepam injection in each group. In different time point xenon-treatment groups, rats were randomly selected and treated with 70% xenon, 21% oxygen, 9% nitrogen delayed 15 min (xenon 15 min group, $n = 9$) or 30 min (xenon 30 min group, $n = 8$) after the KA treatment. The rats in control groups were treated with 21% oxygen, 79% nitrogen at the same time points (15 min or 30 min, $n = 10$/group). The details of the experimental procedure are shown in **Supplementary Figure S1**.

The rats received xenon inhalation in a transparent resin observation box with a bottom air inlet and an upper air gate. The gas mixtures were delivered at a same speed (200 ml/min) regulated by a flow regulator valve (DaTe special gas Ltd., China), which was installed in the gas containers. The rat temperature was stable, and the rats were allowed free movement after xenon inhalation.

Seizure severity was assessed using Racine's criteria (Racine, 1972). Stages 1–3 were considered as focal seizures and stages 4 and 5 were considered as generalized seizures. The placement of cannulas was histologically verified after the experiment. Only the rats with successful implantation in right lateral cerebral ventricle were included in the statistical analysis.

Fluoro-Jade B (FJB) Staining

Fluoro-Jade B (FJB) is a polyanionic fluorescein derivative, which sensitively and specifically binds to the degenerating neurons (Liu et al., 2018). At the designated time points (24 h, 3 or 7 day after KA administration), 4 rats from each group were deeply anesthetized and perfused intracardially with normal saline and 4% paraformaldehyde in PBS sequentially. Using a cryostat (CM3050s, Leica, Germany), 10-μm thick coronal slices were cut from the extracted brains. Firstly, the tissue slides were immersed in 80% alcohol solution containing 1% sodium for 5 min followed by 70% alcohol for 2 min and distilled water for 2 min. Then, in order to ensure the same background between the slides, the slides were immersed in a solution containing 0.06% potassium permanganate for 15 min on a shaking rocker; finally, the slides were rinsed for 2 min in distilled water. The FJB dye powder was used to prepare a 0.01% stock solution. Staining solution was made with 96 ml 0.1% acetic acid and 4 ml FJB stock solution, and was used within 10 min of preparation. The slides were immersed in the staining solution for 20 min and were rinsed for 1 min in distilled water. The slides were then placed in the oven at 50°C for 5 min. Lastly, the slides were immersed in xylene for 1 min and were mounted by neutral balsam and coverslipped with DPX (Sigma, United States). The slides were observed under the epifluorescent microscope (CX41, Olympus, Japan) with blue (495 nm) excitation filter. Positive signals observed in the slides were counted manually.

Statistical Analysis

All data were acquired in a blinded manner and presented as mean ± SEM. Statistical analysis was carried out by SPSS v16.0 (SPSS Inc., Chicago, IL, United States) for Windows. The non-parametric Mann-Whitney *U*-test was used to analyze the cumulative time spent in each seizure stage and the seizure stage at different time points after KA administration. One-way ANOVA with Dunnett's T3 *post hoc* test was used to analyze the effect of different ratio xenon mixtures on cumulative seizure duration and cumulative generalized seizure duration (GSD). Positive signals of FJB staining were also analyzed by one-way ANOVA with Dunnett's T3 *post hoc* test. One-way ANOVA was used to assess the effect of xenon inhalation at different time points (15 or 30 min) on seizure duration and GSD as compared with the controls. For all analyses, a $P < 0.05$ was considered as significant.

RESULTS

The Effect of Different Ratio Xenon Mixture Inhalation on KA-Induced SE

The different ratio xenon mixtures (70% xenon, 21% oxygen, 9% nitrogen, 50% xenon, 21% oxygen, 29% nitrogen, 35% xenon, 21% oxygen, and 44% nitrogen) were inhaled immediately after KA treatment for 1 h. The investigation lasted until the seizures were terminated by diazepam injections after 1 h. The rats in 50 or 70% xenon inhalation groups showed significantly attenuated seizure stage from the second 5 min onward as compared with the control group treated with 21% oxygen, 79% nitrogen ($P = 0.006$ and $P < 0.001$, respectively, **Figure 1A**). The seizure stages in each group were evaluated further. The results showed that the cumulative generalized seizures duration (in stages 4 and 5) was significantly reduced after 50 or 70% xenon treatment (KA, 53.3 min; 50% xenon, 34.6 min; 70% xenon, 3.4 min; $P < 0.001$ and 0.001, **Figure 1B**). Further analysis showed that 70% xenon treatment significantly prolonged the cumulative time in stage 0 (no epileptic seizures, KA group, 0 min; 70% xenon group, 44.1 min; $P < 0.001$, **Figure 1C**), but reduced the time spent in stage 4 and 5 ($P < 0.001$ and 0.001, **Figure 1C**). Moreover, the rats in 50% xenon treated group also spent more time in stage 0 ($P < 0.001$, **Figure 1C**) and focal seizures (stages 1–3, $P < 0.001$, **Figure 1C**), but shorter time in generalized seizures ($P < 0.001$, **Figure 1B**). EEGs were recorded after KA administration for 60 min until the diazepam injection. The cumulative seizure duration was 13.0 min in 70% xenon mixture group and 54.3 min in the control group ($P < 0.001$, **Figure 1D**). The cumulative time in stages 4 and 5, as well as seizure duration in 70% xenon treated group were less than the 50% xenon treated group ($P < 0.001$, **Figure 1B**). The representative EEGs, frequency spectrum, and the PSD changes from each group are shown in **Figure 1E**. The behavioral and EEG results indicate the strong inhibitive effect of 50 and 70% xenon in KA-induced epileptic seizures. 70% xenon mixture led to a stronger anti-seizure effect, while no significant effect was observed in rats treated with 35% xenon.

The Effect of Different Ratio Xenon Mixture Inhalation on KA-Induced Neuronal Degeneration

Fluoro-Jade B, a dye that sensitively and specifically binds to the degenerating neurons, was used to analyze the degenerating neurons in different groups. The positive signals for FJB were found to be increased in the hippocampus and entorhinal cortex (EC) at all the investigated time points (24 h, 3 day, and 7 day) after KA administration. The representative increased FJB signals on day 7 are shown in **Figure 2** (dentate gyrus, $P < 0.001$; CA2, $P < 0.001$; **Figures 2B,G,P**) and entorhinal cortex (EC, $P < 0.001$, **Figures 2L,P**), as compared to the rats treated with saline. However, the increase in the FJB positive staining was attenuated in the 50 and 70% xenon treated groups (**Figures 2D,E,I,J,N,O,Q–S**) and 70% xenon treatment provided a stronger inhibitory effect as compared to the 50% xenon treatment (dentate gyrus, $P < 0.001$, **Figure 2Q**; CA2, $P = 0.019$, **Figure 2R**; EC, $P < 0.001$, **Figure 2S**). With an increase in the ratio of xenon, the FJB positive signal reduced (**Figures 2C–E,H–J,M–O,Q–S**). There was no significant difference between the 35% xenon group and the saline group (**Figures 2C,H,M,Q–S**).

The Time-Window Effect of 70% Xenon Mixture Inhalation Delayed by 15 min or 30 min on the KA-Induced SE

To evaluate the appropriate time window of xenon treatment to inhibit seizures, 70% xenon mixture was inhaled from 15 min or 30 min after KA administration, and the inhalation lasted for 45 min or 30 min, respectively, until the diazepam was injected to terminate the seizures. The behavioral results showed that the rats in xenon 15 min group still significantly prolonged the cumulative time in stages 1–3 ($P < 0.001$, **Figure 3A**) and had no behavioral seizures (stage 0, $P < 0.001$, **Figure 3A**), while the cumulative time in generalized seizures (stages 4

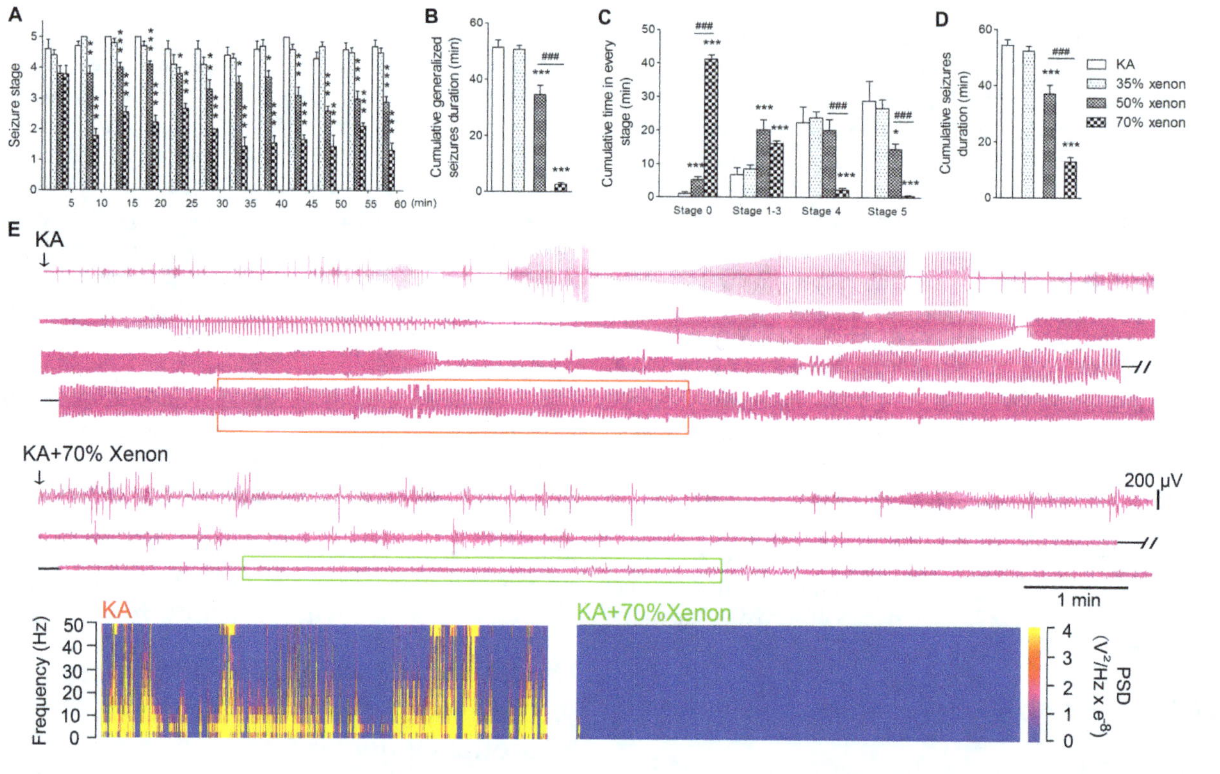

FIGURE 1 | The effect of different ratio xenon mixtures inhalation on the KA-induced SE. **(A)** The change in seizure stage after KA administration. **(B)** Cumulative generalized seizures duration (GSD). **(C)** Cumulative time in each stage. **(D)** Cumulative seizures duration. **(E)** Representative EEGs, frequency spectrum, and the power spectrum density (PSD) changes from each group. KA group, n = 10; 35% xenon group, n = 10; 50% xenon group, n = 10; and 50% xenon group, n = 9. Means ± SEM are shown. The non-parametric Mann-Whitney U test was used to analyze the cumulative time spent in each seizure stage and the seizure stage at different time points after KA administration. One-way ANOVA with Dunnett's T3 *post hoc* test was used to analyze the effect of different ratio xenon mixtures on cumulative seizure duration and cumulative GSD. $^*P < 0.05$, $^{**}P < 0.01$, and $^{***}P < 0.001$ vs. controls. $^{\#\#\#}P < 0.001$ compared with each other.

and 5) was significantly reduced (**Figures 3A,C**). During the xenon inhalation period, the stage of seizures was significantly attenuated from the second 5 min onward in the xenon 15 min group (P = 0.012, **Figure 3D**). The average seizure stage at 60 min was 2.2 in the xenon 15 min group and 4.5 in the control group (P = 0.003, **Figure 3D**). The reduced cumulative seizure duration was also observed in the xenon 15 min group as compared to the control (43.5 and 14.9 min, $P < 0.001$, **Figure 3B**). The similar inhibitive effect was also observed in xenon 30 min group, such as attenuated seizure stage ($P < 0.001$, **Figure 4A**), prolonged cumulative time in stage 0 (P = 0.02) and stages 1–3 ($P < 0.001$, **Figure 4C**), reduced cumulative time in generalized seizure ($P < 0.001$, **Figure 4D**), and cumulative seizure duration ($P < 0.001$, **Figure 4B**). The represented EEGs, frequency spectrum, and the PSD changes are shown in **Figures 3E,F, 4E**. The results indicate the significant inhibitive effect of 70% xenon treatment delayed for 30 min after the epileptic seizures.

The Effect of 70% Xenon Mixture Inhalation Delayed by 15 min or 30 min on KA-Induced Neuronal Degeneration

The FJB staining was performed on day 7 in all the groups: saline group, KA group, and xenon inhaled groups treated at different time points after KA administration (immediately, 15 min, and 30 min). The results showed that 70% xenon mixture treatment noticeably reduced the FJB positive signals, even after delayed for 30 min, compared with the KA group treated with 21% oxygen, 79% nitrogen, in the hippocampus (dentate gyrus, $P < 0.001$, **Figures 5A–E,P**; CA2, $P < 0.001$, **Figures 5F–J,Q**), and EC ($P < 0.001$, **Figures 5K–O,R**). However, with the delay in treatment, the inhibitive effect of xenon mixture was found to be attenuated (**Figures 5C–E,H–J,M–O,P–R**).

DISCUSSION

Our study confirmed the significant therapeutic effect of xenon inhalation on the KA-induced SE. Moreover, we evaluated the therapeutic effect of different proportional xenon gradient at various delayed time-points. We found that the anti-seizure effect is closely associated with the proportion of xenon. The significant therapeutic effect was found in 70% and 50% xenon mixture treated group, but not in 35% xenon treated group. Additionally, both 15 min- and 30 min-delayed treatment displayed a significant therapeutic effect.

Xenon, an inert gas that has been used in the clinics, hardly participates in any chemical reaction and does not undergo

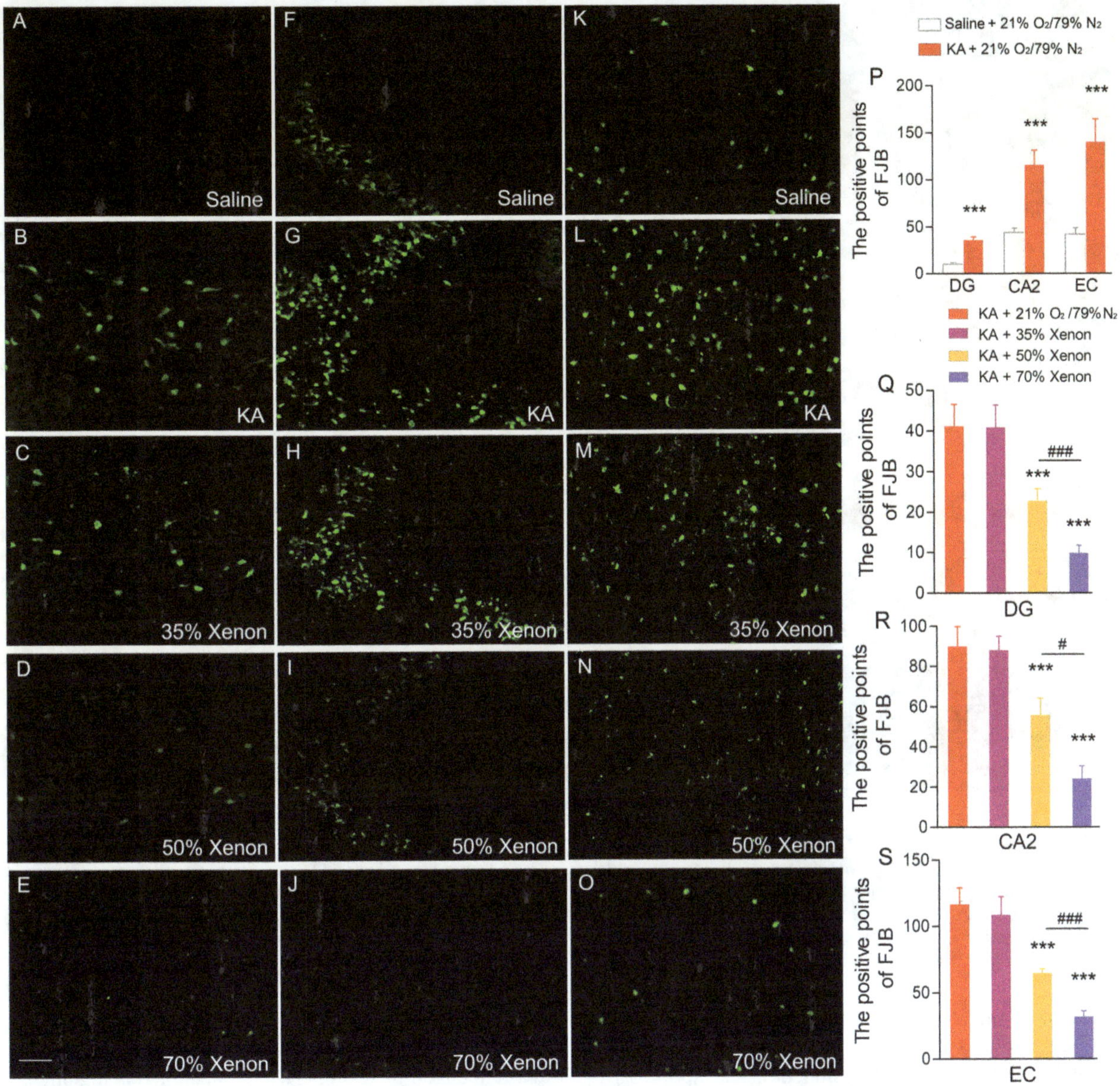

FIGURE 2 | The effect of different ratio xenon mixture inhalation on the KA-induced neuronal degeneration on day 7, 50 and 70%, but not 35% xenon mixture attenuated the neurodegeneration in the hippocampus (**A–E**, dentate gyrus; **F–J**, CA2), and EC **(K–O)** induced by KA administration (n = 4 per group; bar = 50 μm). **(P–S)** Quantified positive signals of FJB staining. Means ± SEM are shown. One-way ANOVA with Dunnett's T3 *post hoc* test was used to analyze the effect of different ratio xenon mixtures on positive signals of FJB staining. $^{***}P < 0.001$ vs. controls. $^{\#}P < 0.05$ and $^{\#\#\#}P < 0.001$ compared with each other.

biological transformation *in vivo*. It is exhaled through the lungs in its original form after inhalation. It has been characterized with non-toxic side effects and high safety. On the other hand, the current anti-epileptic therapies have noticeable problems, such as the serious adverse reactions and drug resistant of anti-epileptic drugs (Schmidt and Löscher, 2005; Chen et al., 2018). Consequently, xenon treatment may possess good and safer prospects as the epilepsy intervention in the clinics.

Previous studies have provided sufficient evidence for the neuroprotective effects of xenon such as attenuation of the neuronal injury, apoptosis, and neuronal loss in several neuronal diseases, such as Alzheimer's disease (Lavaur et al., 2016), unilateral common carotid artery ligation (Metaxa et al., 2014), intrauterine asphyxia (Yang et al., 2012), and neonatal asphyxia (Luo et al., 2008). Furthermore, it has been reported that xenon could effectively terminate the synchronous discharge by suppressing the glutamate intake and release (Uchida et al., 2012). Excessive glutamate-induced excitatory toxicity is an important mechanism in epileptic seizures and propagation (During and Spencer, 1993; Chiu et al., 2016; Goubert et al., 2017; Luna-Munguia et al., 2019). Extracellular glutamate accumulation can over-stimulate the corresponding receptors such as N-methyl-D-aspartate (NMDA) and α-amino-3-hydroxy-5-methyl-4-isoxazolepropionic acid (AMPA) receptors, and

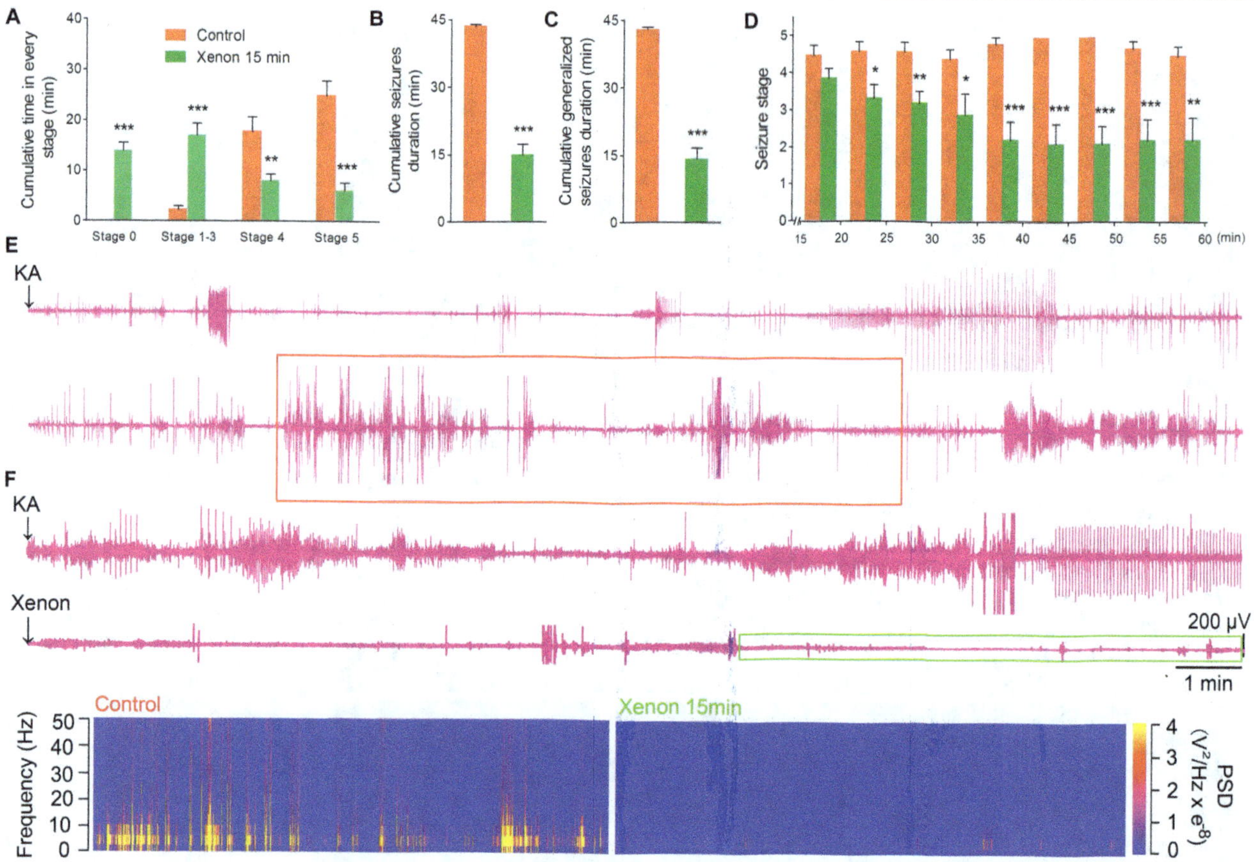

FIGURE 3 | The time-window effect of 70% xenon mixture inhalation delayed by 15 min on the KA-induced SE. Delayed 15 min treatment of 70% xenon mixture (n = 9) significantly prolonged the cumulative time in stages 0–3 and reduced the cumulative time in stages 4 and 5 **(A)**, cumulative seizures duration **(B)**, cumulative GSD **(C)**, and attenuated seizure stage from second 5 min onward after xenon treatment **(D)**, compared with the control group (n = 10). **(E,F)** Representative EEGs, frequency spectrum, and the PSD changes. Means ± SEM are shown. One-way ANOVA was used to assess the role of xenon inhalation at different time points (0 min or delayed 15 min). *P < 0.05, **P < 0.01, and ***P < 0.001 vs. controls.

over-excitation of the NMDA receptors plays an imperative role in the occurrence and development of acute nerve injury. It can eventually lead to neuronal dysfunction and apoptosis by activating the calpain and caspase-3 pathways (Baudry and Bi, 2016; Hoque et al., 2016; Izumida et al., 2017; Deng et al., 2019). On the other hand, synchronous discharge, neuronal injury, and even neuronal loss are the typical electroencephalogram and pathological features of epilepsy (Aracri et al., 2018; Bumanglag and Sloviter, 2018; Chang et al., 2018). For example, apoptosis is known to be involved in the formation of hippocampal sclerosis in the patients with medial temporal lobe epilepsy (Danis et al., 2016). Apoptosis and degeneration of the hippocampal neurons have been observed in the animal model of temporal lobe epilepsy induced by KA administration. In the model of acute temporal lobe epilepsy induced by KA injection into the hippocampal CA3 region, the caspase-3 pathway was activated, which further induced the hippocampal neuron apoptosis (Liang et al., 2016). The slow kindling epilepsy model induced by intraperitoneal injection of pentylenetetrazole (PTZ) was accompanied by the cortical neuron apoptosis (Al-Shorbagy and Nassar, 2017). The anticonvulsant effect by subanesthetic levels of xenon was observed in neonatal asphyxial seizures (Azzopardi et al., 2013). The previous reports, thus, strongly suggest that xenon inhalation may possess a therapeutic role in both epileptic seizures and epilepsy-induced neuronal injury. Additionally, our study further confirms the therapeutic effect of xenon mixture inhalation on the KA-induced SE as well as the neuronal injury.

The therapeutic effects of any drug or treatment are commonly associated with the dose administration. It is well known that too small dose is ineffective and too large dose can lead to toxic effects. Different doses can sometimes even lead to opposite effects. Consequently, the effective range of xenon ratio for the best anti-seizure effect needs to be well-defined. Our study confirms the anti-seizure effect and attenuated neurodegeneration by 50 and 70% xenon mixtures. Moreover, the therapeutic effect of 70% xenon mixture was found to be significantly stronger than the 50% xenon mixture.

Delayed therapeutic period between the seizure onset and the treatment is unavoidable in the clinics. Delayed treatment is often different from immediate drug administration and also from other therapeutic methods such as deep brain stimulation (Wang et al., 2008; Wu et al., 2008). Consequently,

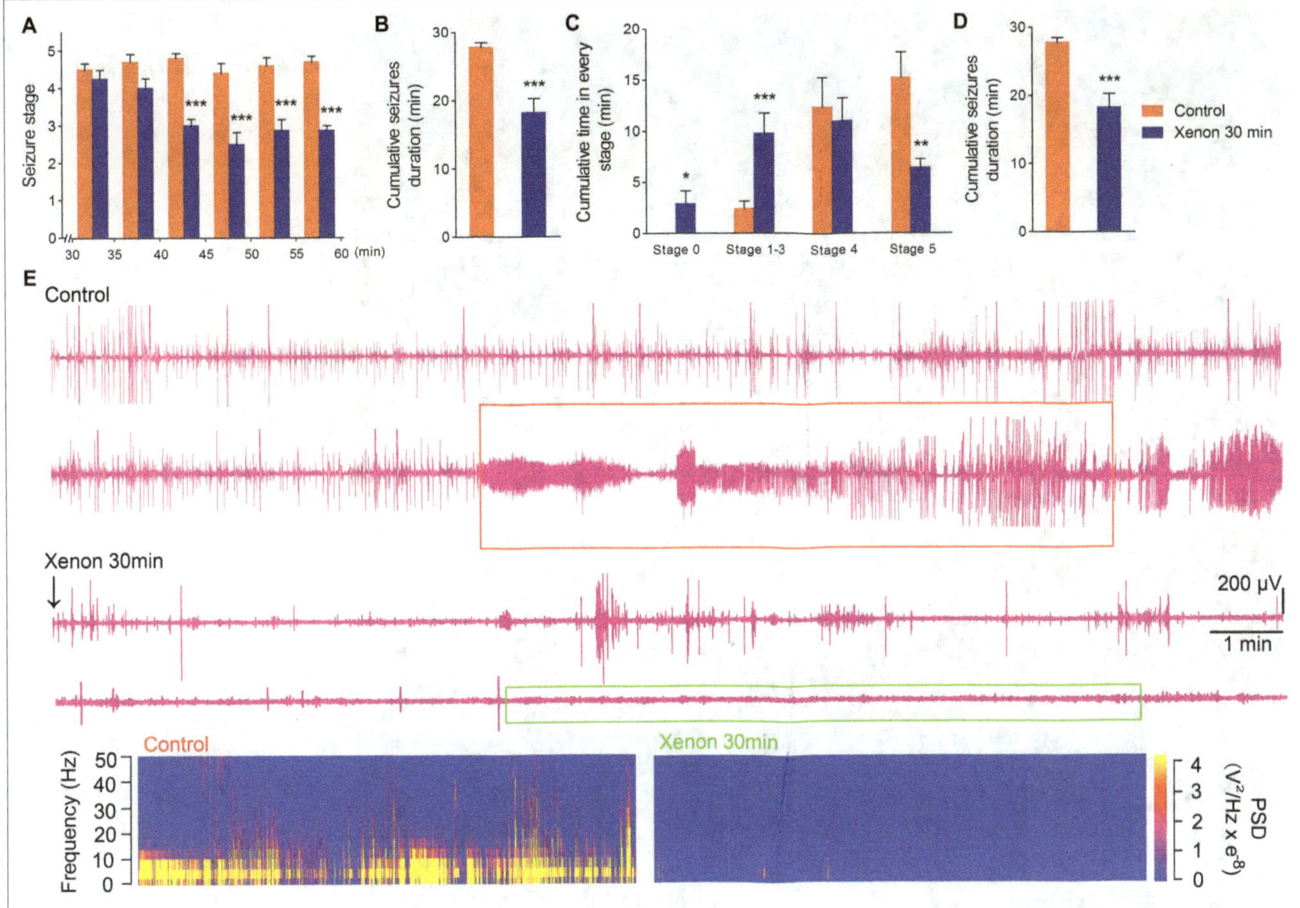

FIGURE 4 | The time-window effect of 70% xenon mixture inhalation delayed by 30 min on the KA-induced SE. Delayed 30 min treatment of 70% xenon mixture (n = 8) provided significant anti-seizure effect from the third 5 min after xenon treatment **(A)**, reduced cumulative seizure duration **(B)**, cumulative time in stages 4 and 5 **(C)**, prolonged the cumulative time in stages 0–3 **(C)**, and cumulative GSD **(D)** compared with the control group (n = 10). Representative EEGs, frequency spectrum, and the PSD changes. **(E)** Means ± SEM are shown. One-way ANOVA was used to assess the role of xenon inhalation at different time points (0 min or delayed 30 min). *P < 0.05, **P < 0.01, and ***P < 0.001 vs. controls.

it is meaningful to define the effective time-window of xenon treatment. Seventy percent xenon mixture, which displayed the strongest effects in our study, was inhaled at different time points after KA administration. Our study indicates that the 70% xenon mixture treatment displayed the anti-seizure effect at all the three time points (0, delayed 15, and delayed 30 min) after KA administration and reduced the extent of neurodegeneration. The results confirm the therapeutic effects of xenon inhalation delayed by less than 30 min after the onset of seizures. These results are meaningful for the possible clinical applications in the future.

Neuronal injury, such as neurodegeneration and apoptosis, are the prominent pathological features of epilepsy in both animal models and human patients (Bumanglag and Sloviter, 2018; Chang et al., 2018). FJB staining was used to evaluate the neuroprotective effects in our study because FJB sensitively and specifically binds to the degenerating neurons (Liu et al., 2018). Our study demonstrated increased FJB positive signals in the hippocampus and EC in KA-induced SE. Moreover, the increased neurodegeneration was reversed after the treatment with appropriate xenon ratio (50 or 70%) and proper delayed time periods (0–30 min), consistent with the effective anti-seizure parameter. The results indicate that the neuroprotective effects of xenon mixture may contribute toward its anti-seizure characteristic.

The FJB staining confirmed the neurodegeneration in the hippocampus and EC, which are the vital brain subregions in the epileptic network (Hsu, 2007; Xu et al., 2010). Hippocampus is considered as the "promoter" or "amplifier" in the epileptic network (Heinemann et al., 1992; Hsu, 2007). The hippocampus was reported to be the primary driving region for the seizures and the pathway for both longitudinal (Derchansky et al., 2006) and contralateral (Blackstad, 1956) epileptiform activity propagation. Several epileptic animal models have shown that a defect in the hippocampus is vital for epileptogenesis (Heinemann et al., 1992; Hsu, 2007). Moreover, the hippocampal CA3 region emits low frequency discharge, which could depress the epileptiform activity generated from the EC, thus, inhibit the activation of CA1-subiculum networks (Barbarosie and Avoli, 1997). The EC, on the other hand, is considered as a gateway to the hippocampus

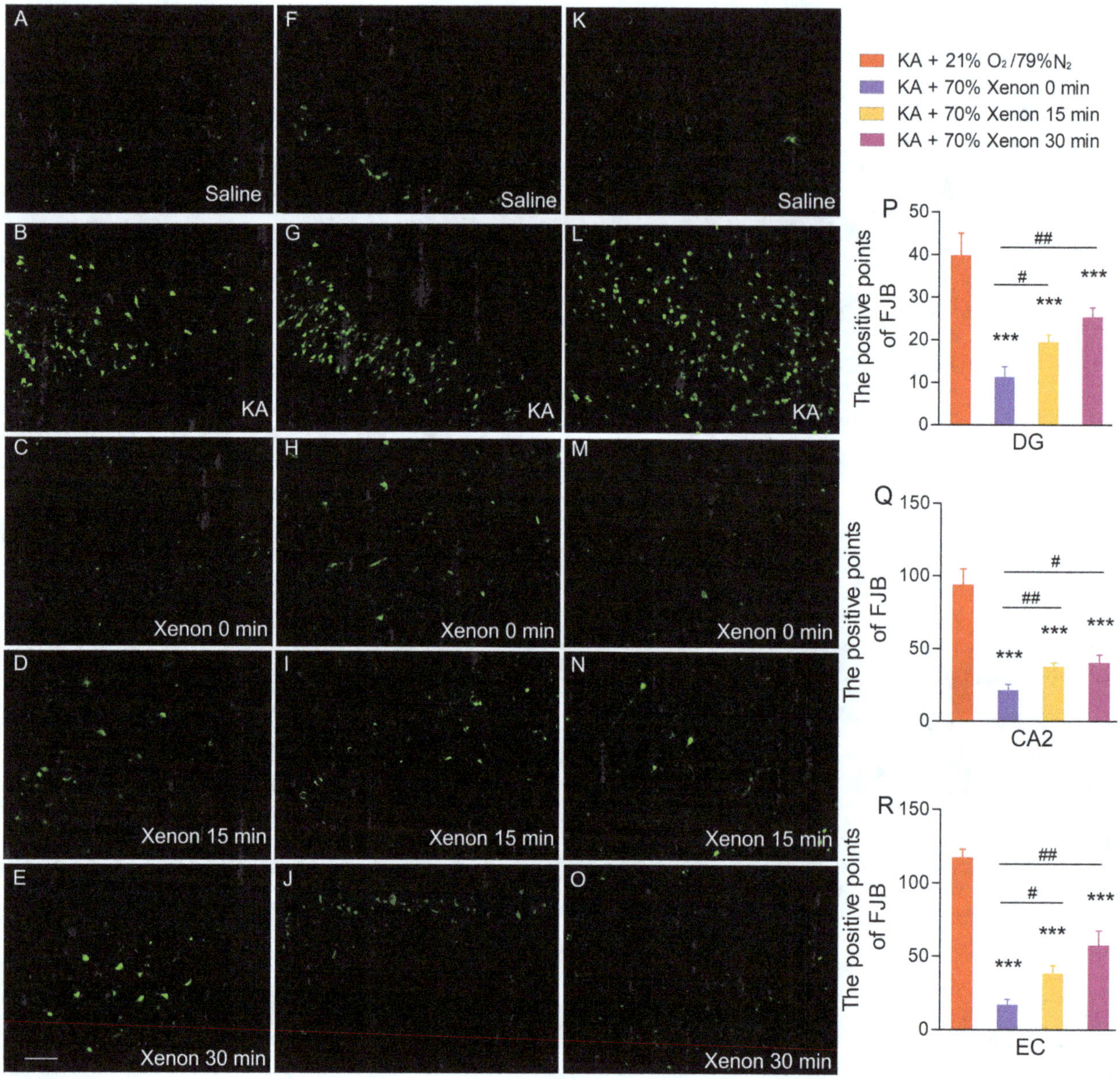

FIGURE 5 | The effect of 70% xenon mixture inhalation delayed by 15 or 30 min on the KA-induced neuronal degeneration. **(A–O)** Different effects of xenon inhalation immediately (delayed 0 min), delayed by 15 min or delayed by 30 min on neurodegeneration induced by KA administration (dentate gyrus, **A–E**; CA2, **F–J**; EC, **K–O**; $n = 4$ per group; bar = 50 μm). **(P–R)** Quantified positive signals of FJB staining. Means ± SEM are shown. One-way ANOVA with Dunnett's T3 *post hoc* test was used to analyze the neuroprotective effect of xenon treatment at different time points. $^{***}P < 0.001$ vs. controls. $^{\#}P < 0.05$, $^{\#\#}P < 0.01$, and $^{\#\#\#}P < 0.001$ compared with each other.

and, similar to the hippocampus, plays a vital role in the development of epilepsy and epileptic seizures (Xu et al., 2010). A remarkable depress network in the EC has been reported during the transition to a seizure (Gnatkovsky et al., 2008). The interictal-like discharges of EC could attenuate the epileptic synchronized activity in the limbic networks (Barbarosie and Avoli, 1997; D'Arcangelo et al., 2005).

These studies indicate that the integrity of the hippocampal and EC neurons may be the vital regulating point in epilepsy. Our research provides the evidence for KA-induced neuronal injury in the hippocampus and EC. The defect in the hippocampal and EC neurons may lead to loss of the effective control over depressed network. Moreover, the results also confirm that an appropriate model of xenon treatment could inhibit seizures and reverse the hippocampal and EC neuronal injury synchronously.

In summary, our study confirms the anti-seizure effect of xenon mixture in the KA-induced SE. Moreover, we optimized the effective xenon ratio and appropriate time-window of therapeutic intervention. The xenon therapeutic effect may be produced by attenuation of the hippocampal and EC neuronal injury and should be further explored as a potential intervention for seizures and epilepsy.

AUTHOR CONTRIBUTIONS

YZ and MZ: study conception and design and data acquisition. JY, WZ, and QW: KA-induced rat model preparation and xenon treatment. XP, XG, and JY: data acquisition and data analysis. HS: study design, data acquisition, and drafting of the manuscript.

ACKNOWLEDGMENTS

We would like to thank Editage for English language editing.

REFERENCES

Al-Shorbagy, M. Y., and Nassar, N. N. (2017). Octreotide ameliorates inflammation and apoptosis in acute and kindled murine PTZ paradigms. *Naunyn Schmiedebergs Arch. Pharmacol.* 390, 61–68. doi: 10.1007/s00210-016-1303-x

Aracri, P., de Curtis, M., Forcaia, G., and Uva, L. (2018). Enhanced thalamo-hippocampal synchronization during focal limbic seizures. *Epilepsia* 59, 1774–1784. doi: 10.1111/epi.14521

Azzopardi, D., Robertson, N. J., Kapetanakis, A., Griffiths, J., Rennie, J. M., Mathieson, S. R., et al. (2013). Anticonvulsant effect of xenon on neonatal asphyxial seizures. *Arch. Dis. Child. Fetal Neonatal Ed.* 98, F437–F439. doi: 10.1136/archdischild-2013-303786

Barbarosie, M., and Avoli, M. (1997). CA3-driven hippocampal-entorhinal loop controls rather than sustains in vitro limbic seizures. *J. Neurosci.* 17, 9308–9314. doi: 10.1523/JNEUROSCI.17-23-09308

Baudry, M., and Bi, X. (2016). Calpain-1 and Calpain-2: the yin and yang of synaptic plasticity and neurodegeneration. *Trends Neurosci.* 39, 235–245. doi: 10.1016/j.tins.2016.01.007

Blackstad, T. W. (1956). Commissural connections of the hippocampal region in the rat, with special reference to their mode of termination. *J. Comp. Neurol.* 105, 417–537. doi: 10.1002/cne.901050305

Bumanglag, A. V., and Sloviter, R. S. (2018). No latency to dentate granule cell epileptogenesis in experimental temporal lobe epilepsy with hippocampal sclerosis. *Epilepsia* 59, 2019–2034. doi: 10.1111/epi.14580

Cattano, D., Valleggi, S., Cavazzana, A. O., Patel, C. B., Ma, D., and Giunta, F. (2011). Xenon exposure in the neonatal rat brain: effects on genes that regulate apoptosis. *Minerva Anestesiol.* 77, 571–578. doi: 10.1016/j.medin.2010.09.005

Chang, W. C., Kudlacek, J., Hlinka, J., Chvojka, J., Hadrava, M., Kumpost, V., et al. (2018). Loss of neuronal network resilience precedes seizures and determines the ictogenic nature of interictal synaptic perturbations. *Nat. Neurosci.* 21, 1742–1752. doi: 10.1038/s41593-018-0278-y

Chen, Z., Brodie, M. J., Liew, D., and Kwan, P. (2018). Treatment outcomes in patients with newly diagnosed epilepsy treated with established and new antiepileptic drugs: a 30- year longitudinal cohort study. *JAMA Neurol.* 75, 279–286. doi: 10.1001/jamaneurol.2017.3949

Chiu, K. M., Lu, C. W., Lee, M. Y., Wang, M. J., Lin, T. Y., and Wang, S. J. (2016). Neuroprotective and anti-inflammatory effects of lidocaine in kainic acid-injected rats. *Neuroreport* 27, 501–507. doi: 10.1097/WNR.0000000000000570

Cohen-Gadol, A. A., Britton, J. W., Wetjen, N. M., Marsh, W. R., Meyer, F. B., and Raffel, C. (2003). Neurostimulation therapy for epilepsy: current modalities and future directions. *Mayo Clin. Proc.* 78, 238–248. doi: 10.4065/78.2.238

Danis, B., van Rikxoort, M., Kretschmann, A., Zhang, J., Godard, P., Andonovic, L., et al. (2016). Differential expression of miR-184 in temporal lobe epilepsy patients with and without hippocampal sclerosis - Influence on microglial function. *Sci. Rep.* 6:33943. doi: 10.1038/srep33943

D'Arcangelo, G., Panuccio, G., Tancredi, V., and Avoli, M. (2005). Repetitive low-frequency stimulation reduces epileptiform synchronization in limbic neuronal networks. *Neurobiol. Dis.* 19, 119–128. doi: 10.1016/j.nbd.2004.11.012

De Deken, J., Rex, S., Lerut, E., Martinet, W., Monbaliu, D., Pirenne, J., et al. (2018). Postconditioning effects of argon or xenon on early graft function in a porcine model of kidney autotransplantation. *Br. J. Surg.* 105, 1051–1060. doi: 10.1002/bjs.10796

Deng, X., Wang, M., Hu, S., Feng, Y., Shao, Y., Xie, Y., et al. (2019). The neuroprotective effect of astaxanthin on pilocarpine-induced status epilepticus in rats. *Front. Cell. Neurosci.* 13:123. doi: 10.3389/fncel.2019.00123

Derchansky, M., Rokni, D., Rick, I. J., Wennberg, R., Bardakjian, B. L., Zhang, L., et al. (2006). Bidirectional multisite seizure propagation in the intact isolated hippocampus: the multifocality of the seizure "focus". *Neurobiol. Dis.* 23, 312–328. doi: 10.1016/j.nbd.2006.03.014

Dingley, J., Tooley, J., Porter, H., and Thoresen, M. (2006). Xenon provides short-term neuroprotection in neonatal rats when administered after hypoxia-ischemia. *Stroke* 37, 501–506. doi: 10.1161/01.STR.0000198867.31134.ac

During, M. J., and Spencer, D. D. (1993). Extracellular hippocampal glutamate and spontaneous seizure in the conscious human brain. *Lancet* 341, 1607–1610. doi: 10.1016/0140-6736(93)90754-5

Falcicchia, C., Simonato, M., and Verlengia, G. (2018). New tools for epilepsy therapy. *Front. Cell. Neurosci.* 12:147. doi: 10.3389/fncel.2018.00147

Gnatkovsky, V., Librizzi, L., Trombin, F., and de Curtis, M. (2008). Fast activity at seizure onset is mediated by inhibitory circuits in the entorhinal cortex in vitro. *Ann. Neurol.* 64, 674–686. doi: 10.1002/ana.21519

Goubert, E., Mircheva, Y., Lasorsa, F. M., Melon, C., Profilo, E., Sutera, J., et al. (2017). Inhibition of the mitochondrial glutamate carrier SLC25A22 in astrocytes leads to intracellular glutamate accumulation. *Front. Cell. Neurosci.* 11:149. doi: 10.3389/fncel.2017.00149

Heinemann, U., Beck, H., Dreier, J. P., Ficker, E., Stabel, J., and Zhang, C. L. (1992). The dentate gyrus as a regulated gate for the propagation of epileptiform activity. *Epilepsy Res. Suppl.* 7, 273–280. doi: 10.1111/j.1528-1157.1992.tb05897.x

Hoque, A., Hossain, M. I., Ameen, S. S., Ang, C. S., Williamson, N., Ng, D. C., et al. (2016). A beacon of hope in stroke therapy- Blockade of pathologically activated cellular events in excitotoxic neuronal death as potential neuroprotective strategies. *Pharmacol. Ther.* 160, 159–179. doi: 10.1016/j.pharmthera.2016.02.009

Hsu, D. (2007). The dentate gyrus as a filter or gate: a look back and a look ahead. *Prog. Brain Res.* 163, 601–613. doi: 10.1016/S0079-6123(07)63032-5

Izumida, H., Takagi, H., Fujisawa, H., Iwata, N., Nakashima, K., Takeuchi, S., et al. (2017). NMDA receptor antagonist prevents cell death in the hippocampal dentate gyrus induced by hyponatremia accompanying adrenal insufficiency in rats. *Exp. Neurol.* 287, 65–74. doi: 10.1016/j.expneurol.2016.08.007

Kaur, U., Chauhan, I., Gambhir, I. S., and Chakrabarti, S. S. (2019). Antiepileptic drug therapy in the elderly: a clinical pharmacological review. *Acta Neurol. Belg.* 19, 163–173. doi: 10.1007/s13760-019-01132-4

Kim, J. E., and Kang, T. C. (2018). Differential roles of mitochondrial translocation of active caspase-3 and HMGB1 in neuronal death induced by status epilepticus. *Front. Cell. Neurosci.* 12:301. doi: 10.3389/fncel.2018.00301

Lavaur, J., Lemaire, M., Pype, J., Le Nogue, D., Hirsch, E. C., and Michel, P. P. (2016). Neuroprotective and neurorestorative potential of xenon. *Cell Death Dis.* 7:e2182. doi: 10.1038/cddis.2016.86

Lesser, R. P. (2000). Ventricular asystole during vagus nerve stimulation for epilepsy in humans. *Neurology* 54:776.

Liang, S., Zhang, L., Yu, X., Zhang, S., Zhang, G., and Ding, P. (2016). Neuroprotective effect of electric conduction treatment on hippocampus cell apoptosis in KA induced acute temporal lobe epileptic rats. *Brain Stimul.* 9, 933–939. doi: 10.1016/j.brs.2016.07.011

Liu, T., Ma, X., Ouyang, T., Chen, H., Xiao, Y., Huang, Y., et al. (2018). Efficacy of 5-aminolevulinic acid-based photodynamic therapy against keloid compromised by downregulation of SIRT1-SIRT3-SOD2-mROS dependent autophagy pathway. *Redox Biol.* 17, 195–203. doi: 10.1016/j.redox.2018.10.011

Luna-Munguia, H., Zestos, A. G., Gliske, S. V., Kennedy, R. T., and Stacey, W. C. (2019). Chemical biomarkers of epileptogenesis and ictogenesis in experimental epilepsy. *Neurobiol. Dis.* 121, 177–186. doi: 10.1016/j.nbd.2018.10.005

Luo, Y., Ma, D., Ieong, E., Sanders, R. D., Yu, B., Hossain, M., et al. (2008). Xenon and sevoflurane protect against brain injury in a neonatal asphyxia model. *Anesthesiology* 109, 782–789. doi: 10.1097/ALN.0b013e3181895f88

McIntosh, A. M., Kalnins, R. M., Mitchell, L. A., Fabinyi, G. C., Briellmann, R. S., and Berkovic, S. F. (2004). Temporal lobectomy: long-term seizure outcome, late recurrence and risks for seizure recurrence. *Brain* 127(Pt 9), 2018–2030. doi: 10.1093/brain/awh221

Metaxa, V., Lagoudaki, R., Meditskou, S., Thomareis, O., Oikonomou, L., and Sakadamis, A. (2014). Delayed post-ischaemic administration of xenon reduces brain damage in a rat model of global ischemia. *Brain Inj.* 28, 349–364. doi: 10.3109/02699052.2013.865273

Racine, R. J. (1972). Modification of seizure activity by electrical stimulation: II motor seizure. *Electroencephalogr. Clin. Neurophysiol.* 32, 281–294. doi: 10.1016/0013-4694(72)90177-0

Schmidt, D., and Löscher, W. (2005). Drug resistance in epilepsy: putative neurobiologic and clinical mechanisms. *Epilepsia* 46, 858–877. doi: 10.1111/j.1528-1167.2005.54904.x

Smyth, M. D., Tubbs, R. S., Bebin, E. M., Grabb, P. A., and Blount, J. P. (2003). Complications of chronic vagus nerve stimulation for epilepsy in children. *J. Neurosurg.* 99, 500–503. doi: 10.3171/jns.2003.99.3.0500

Sun, H. L., Ma, L. Y., Zhang, Y. R., Pan, X. H., Wang, C. Y., Zhang, J. J., et al. (2018). A purinergic P2 receptor family-mediated increase in thrombospondin-1 bolsters synaptic density and epileptic seizure activity in the amygdala-kindling rat model. *Front. Cell. Neurosci.* 12:302. doi: 10.3389/fncel.2018.00302

Sun, H. L., Zhang, S. H., Zhong, K., Xu, Z. H., Feng, B., Yu, J., et al. (2013). A transient upregulation of glutamine synthetase in the dentate gyrus is involved in epileptogenesisinduced by amygdala kindling in the rat. *PLoS One* 8:e66885. doi: 10.1371/journal.pone.0066885

Theodore, W. H., and Fisher, R. S. (2004). Brain stimulation for epilepsy. *Lancet Neurol.* 3, 111–118.

Uchida, T., Suzuki, S., Hirano, Y., Ito, D., Nagayama, M., and Gohara, K. (2012). Xenon-induced inhibition of synchronized bursts in a rat cortical neuronal network. *Neuroscience* 214, 149–158. doi: 10.1016/j.neuroscience.2012.03.063

Wang, S., Wu, D. C., Ding, M. P., Li, Q., Zhuge, Z. B., Zhang, S. H., et al. (2008). Low frequency stimulation of cerebellar fastigial nucleus inhibits amygdaloid kindling acquisition in Sprague–Dawley rats. *Neurobiol. Dis.* 29, 52–58. doi: 10.1016/j.nbd.2007.07.027

Wang, Y., Liang, J., Xu, C. L., Wang, Y., Kuang, Y. F., Xu, Z. H., et al. (2015). Low-frequency stimulation in anterior nucleus of thalamus alleviates kainate-induced chronic epilepsy and modulates the hippocampal EEG rhythm. *Exp. Neurol.* 276, 22–30. doi: 10.1016/j.expneurol.2015.11.014

Wiebe, S., Blume, W. T., Girvin, J. P., Eliasziw, M., and Effectiveness and Efficiency of Surgery for Temporal Lobe Epilepsy Study Group, (2001). A randomized, controlled trial of surgery for temporal lobe epilepsy. *N. Engl. J. Med.* 345, 311–318. doi: 10.1056/nejm200108023450501

Wu, D. C., Xu, Z. H., Wang, S., Fang, Q., Hu, D. Q., Li, Q., et al. (2008). Time-dependent effect of low-frequency stimulation on amygdaloid-kindling seizures in rats. *Neurobiol. Dis.* 31, 74–79. doi: 10.1016/j.nbd.2008.03.007

Xu, Z. H., Wu, D. C., Fang, Q., Zhong, K., Wang, S., Sun, H. L., et al. (2010). Therapeutic time window of low-frequency stimulation at entorhinal cortex for amygdaloid-kindling seizures in rats. *Epilepsia* 51, 1861–1864. doi: 10.1111/j.1528-1167.2010.02663.x

Yang, T., Zhuang, L., Rei Fidalgo, A. M., Petrides, E., Terrando, N., Wu, X., et al. (2012). Xenon and sevofluranne provide analgesia during labor and fetal brain protection in a perinatal rat model of hypoxia-ischemia. *PLoS One* 7:e37020. doi: 10.1371/journal.pone.0037020

Yang, Y. W., Cheng, W. P., Lu, J. K., Dong, X. H., Wang, C. B., Zhang, J., et al. (2014). Timing of xenon-induced delayed postconditioning to protect against spinal cord ischaemia-reperfusion injury in rats. *Br. J. Anaesth.* 117, 168–176. doi: 10.1093/bja/aet352

Zhang, Y. R., Zhu, W., Yu, H. Y., Yu, J., Zhang, M. D., Pan, X. H., et al. (2019). P2Y4/TSP-1/TGF-β1/pSmad2/3 pathway contributes to acute generalized seizures induced by kainic acid. *Brain Res. Bull.* 149, 106–119. doi: 10.1016/j.brainresbull.2019.04.004

13

A Review of Gene, Drug and Cell-Based Therapies for Usher Syndrome

*Lucy S. French[1], Carla B. Mellough[1], Fred K. Chen[1,2,3] and Livia S. Carvalho[1]**

[1]Centre for Ophthalmology and Visual Sciences (incorporating Lions Eye Institute), The University of Western Australia, Nedlands, WA, Australia, [2]Department of Ophthalmology, Royal Perth Hospital, Perth, WA, Australia, [3]Department of Ophthalmology, Perth Children's Hospital, Nedlands, WA, Australia

***Correspondence:**
Livia S. Carvalho
liviacarvalho@lei.org.au

Usher syndrome is a genetic disorder causing neurosensory hearing loss and blindness from retinitis pigmentosa (RP). Adaptive techniques such as braille, digital and optical magnifiers, mobility training, cochlear implants, or other assistive listening devices are indispensable for reducing disability. However, there is currently no treatment to reduce or arrest sensory cell degeneration. There are several classes of treatments for Usher syndrome being investigated. The present article reviews the progress this research has made towards delivering commercial options for patients with Usher syndrome.

Keywords: usher syndrome, gene therapy, cell therapy, ipscs, gene editing, adeno-associated virus, antisense oligonucleotides

INTRODUCTION

Usher syndrome is a group of autosomal recessive disorders characterized by congenital neurosensory hearing loss, progressive night vision impairment, and constriction of the visual field due to retinitis pigmentosa (RP). Some forms of Usher syndrome may also have varying levels of vestibular dysfunction resulting in loss of balance. It is the most common form of inherited deaf-blindness (El-Amraoui and Petit, 2014) affecting an estimated 1 in 6,000 people worldwide (Kimberling et al., 2010). Usher subtypes (1, 2, and 3; **Table 1**) are graded according to the severity of symptoms and age of onset. Type 1 patients are born profoundly deaf and experience pre-pubertal onset of progressive vision loss caused by RP. The majority of type 1 patients also have developmental motor delays caused by vestibular dysfunction. Type 2 patients have mild to moderate congenital hearing loss with RP diagnosed during puberty. Hearing loss in type 3 patients is progressive and post-lingual, while RP onset may be delayed until mid-adulthood (Reiners et al., 2006). The clinical presentation of RP begins with night blindness caused by the degeneration of rod photoreceptor cells. Subsequent constriction of the visual field results in a "tunnel vision" effect caused by the centripetal progression of cone photoreceptor cell loss. In classical RP, the death of cones may be secondary to rod degeneration and this may ultimately lead to complete loss of vision in advanced age (Hartong et al., 2006). Many other inherited retinal diseases are associated with deafness (**Table 2**) such as cone-rod dystrophy and hearing loss-1 (CRDHL1, OMIM #617236), diabetes and deafness, maternally inherited (MIDD, OMIM #520000)

TABLE 1 | Genes and proteins associated with various Usher syndrome subtypes.

Type	Subtype (OMIM REF.)	Gene	Protein	% Cases*	Transcripts	Major transcript	Exons
1	1B (#276900)	*MYO7A*	MYOSIN 7A	21	14	7,483 bp; NM_000260.4	56
	1C (#276904)	*USH1C*	HARMONIN	2	11	2,232 bp; NM_005709.4	29
	1D (#601067)	*CDH23*	CADHERIN 23	6	19	11,138 bp; NM_022124.6	71
	1F (#602083)	*PCDH15*	PROTOCADHERIN 15	3	36	6,983 bp; NM_001142763.2	48
	1G (#606943)	*USH1G*	SANS	1	2	3,558 bp; NM_173477.5	4
2	2A (#276901)	*USH2A*	USHERIN	50	5	6,372 bp; NM_007123.6	72
	2C (#605472)	*ADGRV1*	ADHESION G-PROTEIN COUPLED RECEPTOR-V1	5	37	19,557 bp; NM_032119.4	91
	2D (#611383)	*WHRN*	WHIRLIN	0.4	9	3,989 bp; XM_011518485.1	21
3	3A (#276902)	*CLRN1*	CLARIN-1	2	8	2,087 bp; NM_174878.3	6
	3B (#614504)	*HARS*	HISTIDYL-TRNA SYNTHETASE	-	16	1,948 bp; NM_002109.6	13
Modifier	-	*PDZD7***					

**Frequencies were calculated in a 2019 study (Jouret et al., 2019) of 684 patients with dual vision and hearing loss. A proportion of these patients did not have mutations in any of the Usher genes tested. Candidate genes are not shown. **Denotes the modifier gene PDZD7, which contributes to the phenotype of Usher 2A patients through interactions with USHERIN, but is not independently causal of Usher syndrome (Ebermann et al., 2010).*

TABLE 2 | List of non-Usher syndromes that cause hearing loss and inherited retinal disease.

Disease	Omim reference	Gene	Retinal phenotype	Systemic phenotype/s
Cone-rod dystrophy and hearing loss 1 (CRDHL1)	#617236	*CEP78*	Cone-rod dystrophy	Hearing loss
Cone-rod dystrophy and hearing loss 2 (CRDHL2)	#618358	*CEP250*	Cone-rod dystrophy	Early-onset sensorineural hearing loss
Leber congenital amaurosis with early-onset deafness (LCAEOD)	#617879	*TUBB4B*	Leber congenital amaurosis	Early-onset deafness
Polyneuropathy, hearing loss, ataxia, retinitis pigmentosa, and cataract	#612674	*ABHD12*	Retinitis pigmentosa	Hearing loss, polyneuropathy, ataxia
Diabetes and deafness, maternally inherited (MIDD)	#520000	*MTTL1*	Macular dystrophy	Adult-onset sensorineural hearing loss and diabetes, ptosis, cardiomyopathy, myopathy, renal failure, neuropsychiatric symptoms

See www.omim.org for further information.

and Leber congenital amaurosis (LCA) with early-onset deafness (LCAEOD, OMIM #617879). This review is limited to the combination of RP and deafness, the classical presentation of Usher syndrome.

The Genetics and Biology of Usher Syndrome

Usher syndrome is caused by autosomal recessive inheritance of mutations in Usher genes known to encode proteins involved in transmembrane adhesion, scaffolding and motor transport. Ten causative genes have so far been identified. Inheritance of hypomorphic alleles with missense mutations often causes non-syndromic deafness, while nonsense and cryptic splice-site mutations result in Usher syndrome (Ahmed et al., 2008; Bademci et al., 2016). Certain types of mutations in Usher genes may also cause non-syndromic RP (Seyedahmadi et al., 2004). Digenic inheritance of mutations in separate Usher genes has also been proposed to be causative of Usher syndrome (Zheng et al., 2005; Bonnet et al., 2011), but remains controversial (Jouret et al., 2019). Additionally, the disparity in phenotypes and progression rates between monozygotic twins is suggestive of environmental influence (Liu et al., 1999). A summary of identified genes and their proteins can be seen in **Table 1**.

Usher proteins are localized both to the inner ear and retina. In the inner ear, Usher proteins are present in the cochlea and vestibular organs, accounting for the balance and deafness phenotypes. Usher genes instruct the differentiation of mechano-sensitive hair cells and affect hair bundle organization during development. Functional Usher proteins are also essential for the transduction of electrical signals in mature stereocilia (Grati and Kachar, 2011; Cosgrove and Zallocchi, 2014; El-Amraoui and Petit, 2014; Pepermans et al., 2014; Mathur and Yang, 2015, 2019; Han et al., 2018). In the retina, Usher proteins are found in the light-sensitive photoreceptor neurons. They have been implicated in intracellular trafficking at the connecting cilium, which links the photoreceptor's inner and outer segments, and facilitates the movement of phototransduction proteins and lipids to the outer segment (Liu et al., 1997; Mathur and Yang, 2015). Knowledge of the function of Usher proteins in the retina is limited by a lack of effective animal models (El-Amraoui and Petit, 2014), perhaps due to their association with calyceal processes, the microvilli which protrude from the apical region of the inner segment and surround the connecting cilium of human, but not murine, photoreceptors (Sahly et al., 2012; Schietroma et al., 2017).

Current Trials and Pre-clinical Studies for Usher Syndrome Treatment

Cochlear implants are often provided to patients with type 1 Usher syndrome due to the profound deafness at birth. Those with type 2 or 3 Usher syndrome may benefit from hearing aids or cochlear implant later in life if the mild congenital deafness progresses. However, there is currently no treatment available to prevent or reverse the inevitable retinal degeneration associated with RP. There is some evidence for use of dietary supplements to delay the progression of RP, including vitamin A (Berson et al., 1993, 2018), omega-3 fish oils (Berson et al., 2004, 2012), N-acetylcysteine (Campochiaro et al., 2020)[1] and the antioxidant taurine (Trouillet et al., 2018). Although some of these have been disputed (Rayapudi et al., 2013; Hoffman et al., 2014), the lack of strong evidence for efficacy may be related to the heterogeneity of study cohort, small sample size, and short follow-up study design. Additionally, vector-induced expression of ciliary neurotrophic factor (CNTF) was shown to be neuroprotective in a mouse model of RP (Lipinski et al., 2015), yet long-term follow up of three RP patients treated with sustained-release CNTF delivered intravitreally *via* encapsulated cell technology implant (NT-501, Neurotech Pharmaceuticals Inc.), one of which had Usher-associated RP, showed no significant difference in visual acuity (Talcott et al., 2011). In this review article, we will summarise the major classes of ongoing investigations for gene and mutation-specific treatment of hearing and visual loss and tissue regeneration and the progress and challenges in delivering therapeutic outcomes to patients with Usher syndrome. The main approaches discussed can be divided into DNA intervention, RNA intervention and cell replacement and an overview of the different approaches is shown in **Figure 1**.

DNA INTERVENTIONS

Non-viral Systems

DNA interventions are possible due to engineered or recombinant vectors (broadly classified as viral and non-viral) that are synthesized to carry genetic material (Sengillo et al., 2016). Non-viral vectors (nanoparticles) are non-immunogenic, readily customizable, and may package up to 20 kb of material, allowing the delivery of large therapeutic genes (Moore et al., 2018).

Lipid-based nanoparticles, which form a protective bilayer around transported genetic components, have been explored as potential treatments for hearing loss. Zou et al. (2017) investigated the functional, inflammatory, and apoptotic response to liposome nanoparticle delivery to the murine inner ear and found no adverse reactions. Additionally, Gao et al. (2018) reported that cationic lipid-mediated delivery of Cas9:guideRNA complexes to the Beethoven (*Bth*) mouse model of deafness selectively disrupted the dominant mutant *Tmc1* allele, reducing hearing loss. Inner hair cells (IHCs) and outer hair cells (OHCs) had significantly improved survival rates and auditory brainstem response (ABR) thresholds compared to uninjected controls (Gao et al., 2018). Lipid-based nanoparticles have also been used to deliver base editing machinery in a proof-of-concept study to ameliorate hearing loss (discussed further in section Base Editing; Yeh et al., 2018).

Supraparticles are colloidal nanoparticle aggregates with a larger drug loading capacity relative to individual nanoparticles (Sperling and Gradzielski, 2017). Supraparticles have already been used to deliver the developmental neurotrophin, brain-derived neurotrophic factor (BDNF), to the inner ear of a hearing loss guinea pig model (Wang et al., 2014). BDNF is required for the maintenance of spiral ganglion neurons (SGNs; Ylikoski et al., 1993) and may serve to protect or regenerate SGNs as well as promote synaptic regeneration at the ribbon synapse (Suzuki et al., 2016). Supraparticles offer sustained longer-term release of BDNF and maintained near-wild-type numbers of SGNs in the guinea pig cochlea (Wang et al., 2014). However, delivery of supraparticles may mediate the unintentional movement of small nanoparticles from the inner ear to the cerebral spinal fuid *via* the cochlear aqueduct (Zhang et al., 2013).

Another significant advancement in the non-viral delivery platform was the development of the polyethylene glycol-substituted 30-mer lysine peptide (CK30-PEG) nanoparticles, which have been used successfully in a cystic fibrosis clinical trial (Konstan et al., 2004). CK30-PEG nanoparticles may also be effective for the treatment of ocular diseases, as they show a retinal targeting efficiency comparable to viral vectors up to 2 weeks post-injection (Farjo et al., 2006; Han et al., 2012a). Appreciable transgene expression mediated by CK30-PEG nanoparticle gene delivery has been reported in retinal degeneration models, including an autosomal dominant model of RP (Cai et al., 2010; Han et al., 2012b). In the autosomal RP model study, nanoparticles containing the mouse opsin promoter and the *Prph2* gene were subretinally delivered to *Rds* mice carrying a haploinsufficiency mutation, resulting in wild-type-level recovery of cone function but a modest restoration of rod function (Cai et al., 2010). Modified CK30-PEG nanoparticles with a photoreceptor-specific promoter also led to the structural and functional rescue of Stargardt-associated pathology in the *Abca4*$^{-/-}$ mouse model of vision loss (Han et al., 2012b). Recent studies have used solid lipid nanoparticles (SLNs) as a more efficient non-viral delivery system and a study on a mouse model of x-linked juvenile retinoschisis showed transduction of retinal pigment epithelium (RPE) and photoreceptors and improved retinal phenotype (Apaolaza et al., 2016).

Though nanoparticle delivery is promising and continues to be developed, there are several roadblocks to its clinical application, including biodegradability, biocompatibility, and non-specific transfection (Yin et al., 2014; Chen et al., 2016). Topical delivery is the simplest and most patient-friendly form of administration. However, ocular barriers and the long diffusion pathway prevent therapeutic levels from reaching the retina (Bisht et al., 2018). The intraocular injection route would, therefore, be preferable, though the transient expression associated with non-viral gene delivery (Bisht et al., 2018; Huang and Chau, 2019) would necessitate repeated injections and

[1] https://tinyurl.com/r93ejqs

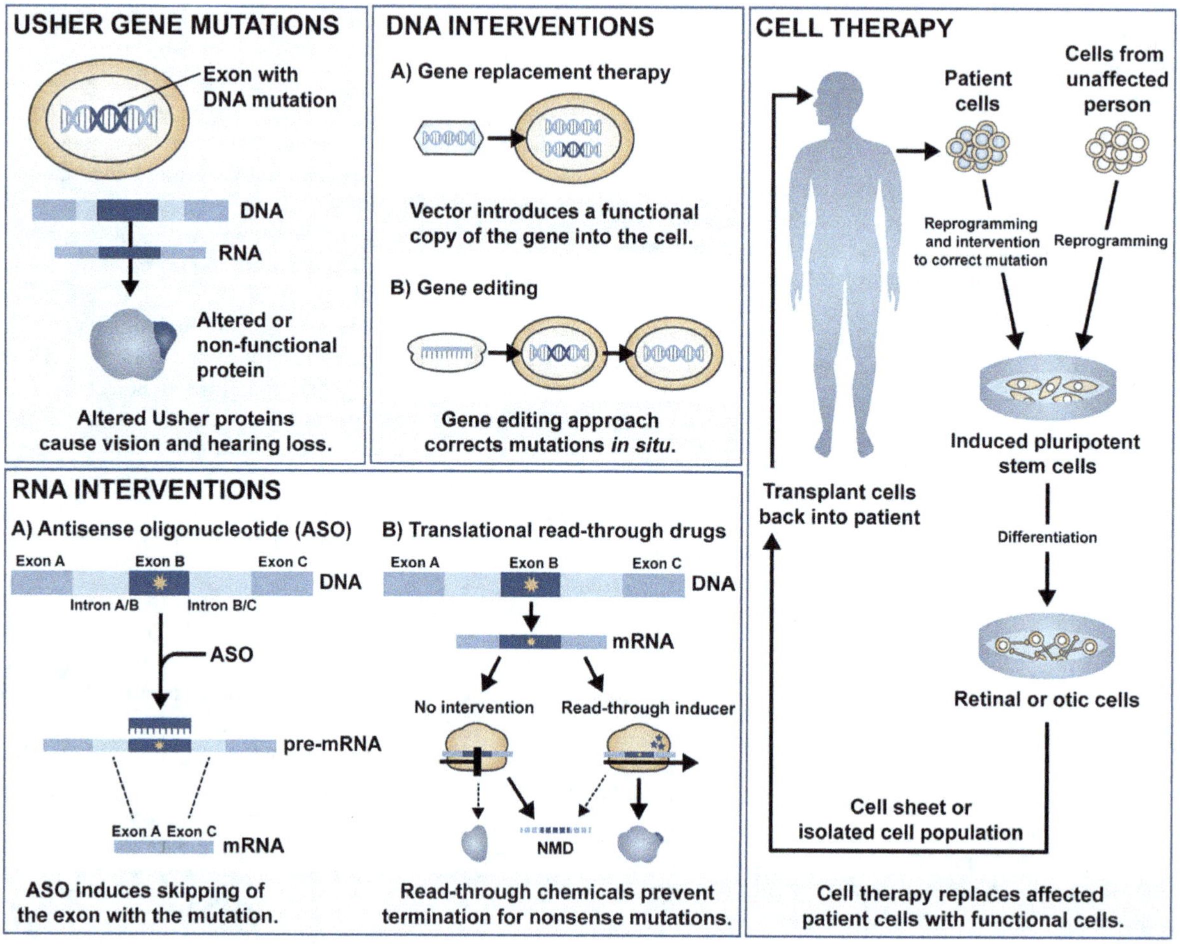

FIGURE 1 | Overview of the different therapeutic approaches for Usher syndrome.

consequently, an increase in their associated risks. Thus far, no pre-clinical studies have investigated nanoparticle delivery of Usher genes but the large packaging capacity of non-viral vectors does increase the potential of this type of platform for treatment development for Usher-causing mutations in the large genes such as *CDH23, PCDH15, ADGRV1, MYO7A*, and *USH2A* (**Table 1**).

Viral-Based Gene Replacement Therapy

With the advantage of increased efficiency over non-viral vectors (Nayerossadat et al., 2012), viral vectors currently represent the most promising approach to therapeutic genetic interventions. Recombinant viral vectors utilize the inherent capability of viruses for cellular transduction to deliver genetic material to donor cells *in vivo*. However, viral vectors come with their limitations, including limited packaging capacities. Potential immunogenicity is also an obstacle to clinical application, though advances in molecular biology have allowed for the separation of wild-type viral coding genes and cis-acting sequences. Consequently, segregation can now produce viral vectors that do not reconstitute by recombination into productive viral particles but still maintain viral infectivity capacity (Kay et al., 2001).

Gene replacement therapy involves the introduction of a non-mutant copy of the affected gene to restore expression to the host cell or tissue (Sengillo et al., 2016). One of the first described, and most frequently used vectors, adenovirus, has a large (~40 kb) packaging capacity and has been used in over 400 clinical trials (Wold and Toth, 2013). In a study by Bennett et al. (1996), an adenovirus-based vector was used to deliver the Phosphodiesterase β subunit gene to *Rd1* mice, delaying photoreceptor degeneration. However, many people have circulating antibodies against adenoviruses, so their use in gene therapies is limited due to a high immunogenic response (DiCarlo et al., 2018). Furthermore, adenovirus has shown to have low tropism for photoreceptor cells *in vivo* (Li et al., 1994), an additional limitation that instigated the search for other types of viral vectors with improved targeting in the eye. The sections below will discuss the two most promising vectors: lentivirus and adeno-associated virus (AAV).

Lentiviral Vectors

Lentiviral vectors, derived from the human immunodeficiency virus, were also initially favored by researchers for clinical application due to their relatively large packaging capacity (reviewed in Zheng et al., 2018). However, whilst their immunogenicity is low, their tendency to integrate into the genome calls into question their clinical applicability (Zheng et al., 2018). Pioneering work by Hashimoto et al. (2007) led to the development of a lentiviral-mediated *MYO7A* gene delivery vector that produced wild-type protein levels in RPE cells in culture, as well as rescuing phenotypic RPE defects *in vitro*. *In vivo*, melanosome mislocalization and opsin accumulation at the photoreceptor connecting cilium was corrected in the *Myo7a*-deficient *shaker1* mouse model of Usher 1B (Hashimoto et al., 2007). Subsequently, UshStat, a therapeutic recombinant vector expressing the human *MYO7A*, was developed using the equine infectious anaemia virus (Zallocchi et al., 2014), a non-pathogenic, non-primate-derived lentivirus capable of transducing human cells (Mitrophanous et al., 1999; Mazarakis et al., 2001). Subretinal delivery of UshStat to *shaker1* mice was shown to protect photoreceptors from light-induced degeneration and demonstrated tolerability in non-human primates, which led to this approach progressing to a clinical trial for the prevention of RP in Usher 1B patients (NCT02065011; results pending).

Adeno-Associated Viral Vectors (AAV)

Safety is of principal concern in all viral vector technologies and as AAVs are not known to cause any human pathogenesis, they have emerged as the vector of choice for gene therapy applications worldwide. AAVs have low immunogenicity, high cellular specificity, and long-lasting gene expression. AAVs typically persist as episomes [Zheng et al., 2018; circular genomes which replicate independently of their host (Watanabe and Fukasawa, 1961)], reducing the risk of insertional mutagenesis. Vandendriessche et al. (2007) directly compared lentiviral and AAV vectors and found that AAV serotypes 8 and 9 induced higher expression of transgenic factor IX protein than lentiviruses in mouse models of hemophilia B and severe combined immunodeficiency, without interacting with proinflammatory cytokines. AAVs also hold potential for the treatment of Usher retinal degeneration as they can efficiently transduce photoreceptors and RPE (Rodrigues et al., 2018). Successful treatment of vision loss in LCA type 2 (LCA2) and choroideremia *via* AAV-mediated delivery of the *RPE65* and *CHM* genes, respectively, have been demonstrated in human clinical trials (reviewed in Russell et al., 2017). The LCA2 trials have now resulted in the first-ever Food and Drug Administration (FDA)-approved and European Medicines Agency (EMA)-approved AAV-based gene therapy drug for LCA2 due to recessive *RPE65* mutations (Luxturna® or Luxturna™/voretigene neparvovec-ryzl or voretigene neparvovec, Roche and Novartis).

A major disadvantage of AAV vectors is their relatively small (4.7 kb) packaging capacity (Zheng et al., 2018); however, *in vivo* delivery of the smaller Usher genes has been investigated using AAV vectors. Delivery of rAAV2/8 containing *USH1G* cDNA to the inner ear improved hearing and hair cell disorganization in the Usher 1G mice (Emptoz et al., 2017). Durable inner ear expression of Usher genes has also resulted in *in vivo* protein restoration of WHIRLIN (Zou et al., 2011) and CLARIN-1 in several studies (Dinculescu et al., 2016; Geng et al., 2017; Dulon et al., 2018; György et al., 2019).

Transduction efficiency is a crucial factor in the success of gene therapy and is highly cell and serotype-dependent. Several synthetically-produced AAV variants have been investigated for Usher syndrome treatments. One example is the synthetic rAAV2/Anc80L65, which may transduce close to 100% of IHCs and 90% of OHCs in mice (Landegger et al., 2017). Using rAAV2/Anc80L65, Pan et al. (2017) demonstrated gene and protein recovery of Harmonin in an Usher 1C mouse model. Deverman et al. (2016) reported a highly efficient synthetic vector, AAV9-PHP.B, which showed robust transduction efficiency in the retina of a dominant RP mouse model (Giannelli et al., 2017). Further, AAV9-PHP.B carrying a green fluorescent protein (GFP) reporter transduced the inner ear and retina of wild-type mice at a rate of 60–80% for IHCs, 30–40% for OHCs and 70–80% for photoreceptors (György et al., 2019). When used to package the *Clrn-1* gene, AAV9-PHP.B mediated rescue of low-frequency hearing (4–8 kHz) in a mouse model of Usher 3A deafness (György et al., 2019). However, the tropism and potential toxicity of the AAV9-PHP.B vector in the central nervous system (CNS) of non-human primates is still under study (Hordeaux et al., 2018; reviewed in Deverman et al., 2018). The potential, therefore, for treatment of Usher syndrome using certain AAV-based vectors is high, but the large size of some Usher genes, including some of the most common forms of Usher such as USH1B (*MYO7A*), does present unique challenges for the field.

Oversized Adeno-Associated Viral Vectors

As mentioned above, the packaging limitation of AAVs precludes delivery of several of the larger Usher genes, including *MYO7A* (USH1B; Jaijo et al., 2006) and *PCDH15* (USH1F; Alagramam et al., 2001). Some studies, however, have pushed the limits of AAV packaging capacity by using oversized AAV transgene constructs (fAAV). A study by Allocca et al. (2008) identified rAAV2/5 as being an efficient packager of up to 8.9 kb of genetic material, allowing, in theory, the large retinal disease genes *MYO7A*, *ABCA4* and *CEP290* to be packaged and delivered subretinally. They showed that fAAV2/5 *ABCA4* delivery led to a stable improvement of morphological abnormalities and retinal dysfunction associated with a mouse model of Stargardt disease. Additionally, Colella et al. (2013) identified defects in light/dark adaptation in *shaker1* mice, which was improved by fAAV2/5 delivery of *MYO7A*. However, other studies have shown that these vectors do not contain full-length genes, but instead tend to contain heterogeneous mixtures of gene fragments, most of which are truncated and typically less than 5 kb (Dong et al., 2010). Furthermore, Grieger and Samulski (2005) demonstrated that AAV vectors carrying genomes larger than 5 kb had less efficient transduction capacity due to a post entry preferential degradation of AAVs carrying larger genomes. Though the truncated genomes may reassemble before transcription and form full-length proteins inside the cell (Lopes et al., 2013), the lack of characterization and heterogeneity of this approach limits its clinical use.

Multi-Adeno-Associated Viral Vector Systems

As an alternative to the oversized AAV approach, different research groups have tested the use of dual and triple AAV vector systems. Multi or dual AAV systems use the inherent concatemerization of AAV genomes *in vivo* to form full-length cDNA which can then be transcribed into functional mRNA within the transduced cell (Trapani et al., 2014). Different methods are used to deliver dual/multivectors and are usually divided into trans-splicing, overlap, and hybrid approaches. Trans-splicing vectors, first described by Yan et al. (2000), separates the gene of interest into 5′ (left) and 3′ (right) halves, with the 3′ end of the left construct containing a splice donor site, which concatemerises with the splice acceptor site located on the 5′ end of the right construct. In the overlap approach, the 3′ end of the right half and the 5′ end of the left half contain a highly recombinogenic overlapping region which mediates homologous recombination between gene segments (Duan et al., 2001). This sequence can be from an external gene, like alkaline phosphatase, or directly from the gene of interest. Finally, hybrid vectors contain both the splice donor/acceptor sites of trans-splicing vectors and the recombinogenic properties of overlapping vectors, allowing reconstitution to occur *via* either method (Ghosh et al., 2008).

Several research groups have published proof-of-concept studies using multi or dual AAV systems, including for inherited retinal diseases (Colella et al., 2014; Dyka et al., 2014, 2019; McClements et al., 2019). Multi-AAV systems have been used to express *MYO7A in vivo* and *in vitro* with equal or higher efficiency than fAAV delivery (Dyka et al., 2014), and to produce full-length mRNA with 100% fidelity to the target cDNA (Dyka et al., 2014). Additionally, Maddalena et al. (2018) expanded the transfer capacity further by using a triple therapeutic vector system to incorporate large genes such as *ALMS1* and the Usher 1D gene, *CDH23*. Truncated protein products were detected in eyes treated with *CDH23* but not *ALMS1* triple AAV vectors. A functional response was not recorded in *CDH23*-treated mice but treated *Alms1*$^{-/-}$ mice showed a non-significant increase in outer nuclear layer thickness and transient (2–6 months) improvement of electroretinogram a- and b-wave measurements (Maddalena et al., 2018).

Despite the promising potential of dual AAV approaches, most studies show very low levels of protein expression using this system. Dyka et al. (2014) quantified the relative expression of full-length cDNA mediated by dual AAV reconstitution and found hybridizing vectors to be 2–3 fold more efficient than overlap and trans-splicing counterparts in *MYO7A*-transfected HEK293T cells. Data published by Colella et al. (2014) found that hybrid and trans-splice dual AAV reconstitution achieved approximately one-quarter of the photoreceptor expression levels induced by single AAV vectors in the large white pig retina. In their 2018 article, Maddalena et al. (2018) reported that the co-transduction rate for triple AAV vectors was <6% of single vector expression in HEK293T cells. Interestingly, this efficiency increased when vectors were delivered subretinally to mice and pigs (photoreceptor expression = 27 ± 6% and 39 ± 17% of that induced by single AAV, respectively). In a recent study, Carvalho et al. (2017) tested the *in vitro* and *in vivo* expression of the different dual AAV approaches and found hybrid vectors to have superior rates of reconstitution. They showed that reconstitution efficiency in HEK293T cells for trans-splicing, hybrid, and overlapping vectors was 10.3%, 15.3%, and 17.4%, respectively. The efficiency of overlapping vectors was found to be gene-specific as it was dependent on the length of the recombinogenic region and showed no detectable levels *in vivo*. *In vivo* subretinal delivery to the mouse retina resulted in full-length protein expression in 9.07% and 1.78% of cells transduced with hybrid and trans-splicing vectors, respectively (Carvalho et al., 2017). Interestingly, they were the first to show a discrepancy in reconstituted mRNA and protein levels after dual AAV delivery which may be explained by transcript instability of longer mRNA sequences (Feng and Niu, 2007).

An alternative to dual AAV systems may be dual-intein splicing, which is based on protein rather than mRNA reconstitution, but is still delivered by AAV vectors. Inteins are segments of proteins that have been described as "protein introns," as they can excise themselves from the sequence of peptides and join the flanking portions (the exteins) together. Recently, Tornabene et al. (2019) demonstrated reporter protein levels in C57BL/6J mice retinae induced by dual-intein splicing to be comparable to single AAV transduction and significantly higher than dual AAV transduction. This approach was also shown to alleviate retinal degeneration in animal models of Leber Congenital Amaurosis (*Rd16* mice) and Stargardt disease (*Abca4*$^{-/-}$ mice). Physiological symptoms of retinal degeneration, including RPE lipofuscin accumulation in *Abca4*$^{-/-}$ mice, and outer nuclear layer thinning in *Rd16* mice, were significantly reduced in the dual-intein treated mice. Additionally, pupillary light responses increased by ~20% in treated *Rd16* mice compared to untreated *Rd16* controls (Tornabene et al., 2019). However, the viability of this approach for the delivery of large genes has not yet been tested in models of Usher syndrome.

DNA Editing

While gene delivery aims to replace a defective gene with a functional copy, DNA editing attempts to directly correct the mutation *in vivo*, which also allows the repaired gene to be expressed under endogenous regulators. Furthermore, the size of the gene is not a limiting factor. Repair of single base transitions or transversions at the DNA level can be achieved through the induction of a DNA break to facilitate incorporation of the correct DNA base (Kim et al., 1996; Christian et al., 2010; Jinek et al., 2012; Cong et al., 2013; Mali et al., 2013) while targeted conversion of a single DNA base can reverse a transitional mutation (Komor et al., 2016; Gaudelli et al., 2017).

Gene Editing

The first attempt using gene editing for Usher syndrome was a study by Overlack et al. (2012) that used zinc finger nucleases (ZFNs) to target the p.R31X mutation in the human *Ush1c* gene. Their *in vitro* results show partial repair of the *Ush1c* gene and recovery of harmonin protein in a p.R31X cell line after transfection with ZFNs. In recent years, however, the discovery of the clustered, regularly interspaced, palindromic repeats (CRISPR) and CRISPR-associated protein 9 (Cas9) system have changed the field of gene editing profoundly. Their higher efficiency, simplicity, and targeting capacity have made them the preferred technology of choice for gene editing (Khan, 2019).

The CRISPR/Cas9 system uses an adapted microbial immune technique for precise editing of the genome (Jinek et al., 2012; Cong et al., 2013; Mali et al., 2013). A guide RNA is engineered to target a specific locus, which is then digested by the Cas9 endonuclease (Jinek et al., 2012; Cong et al., 2013; Mali et al., 2013). Either of two endogenous mechanisms repairs double-stranded breaks in the genome. The clinically-favorable homology-directed repair relies on an intact template strand whilst error-prone non-homologous end-joining is independent of a template sequence, and often results in insertions or deletions (Adli, 2018).

CRISPR/Cas9 has been used to study potential treatments for inherited retinal diseases. Moreno et al. (2018) used the viral delivery of CRISPR components to downregulate *Nrl* in mouse models of non-Usher RP. *Nrl* is a transcriptional regulator that indirectly determines whether photoreceptors develop into rods or cones (Cheng et al., 2004; Moreno et al., 2018). The knockdown of *Nrl* expression transformed rods into cone-like cells that did not experience rod-specific degeneration. This leads to a decrease in rods and therefore night blindness, but an increase in daylight vision (Moreno et al., 2018). However, mouse models of Usher 1C and 1G have cones that are sensitive to oxidative stress, indicating that increasing the number of cone-like cells may not be a viable option for Usher RP treatment (Trouillet et al., 2018).

Viral delivery of CRISPR/Cas9 to fibroblasts derived from an Usher 2A patient showed some rescue of Usherin expression, however, the efficiency of restoration was not significant enough to proceed to clinical trials (Fuster-García et al., 2017). Additionally, induced pluripotent stem cells (iPSCs) derived from a patient with a mutation in *MYO7A* were investigated by Tang et al. (2016). Stereocilia-like protrusions from patient iPSCs developed disorganized morphology which was rescued by CRISPR/Cas9 editing (Tang et al., 2016), indicating a potential treatment avenue for Usher 1B patients that deserves further investigation.

Recently, the CRISPR/Cas9-based tool EDIT-101 received FDA approval for a clinical trial in LCA10 patients (NCT03872479), leading to the first-ever direct human administration of CRISPR/Cas9[2]. EDIT-101 is a gene editing therapeutic which utilizes a Staphylococcus aureus Cas9 guide RNA which has high specificity for the c.2991 + 1655A >G transversion in intron 26 (IVS26) of the CEP290 gene, limiting off-target effects (Maeder et al., 2019). Recently, the company behind EDIT-101, Editas[3], has publically announced the utilization of a new therapeutic, EDIT-102, to target Usher type 2A[4]. Another recent update is the introduction of a new gene editing tool, "prime editing," which utilizes a synthetic fusion of an altered Cas9 endonuclease and reverse transcriptase, directed by a gene editing guide RNA (Anzalone et al., 2019). This technology shows promising specificity and broad applicability to a large number of human pathogenic mutations, although further studies will be needed to confirm its viability for treating Usher syndrome.

Though the potential of CRISPR/Cas9 to make precise and patient-specific gene corrections with high targeting capacity is attractive, the efficacy of *in vivo* delivery is still in doubt due to potential off-target effects (Fu et al., 2013; Kuscu et al., 2014). Advancements in the field since its original application in gene editing, including the development of anti-CRISPR proteins to limit non-specific activity (Pawluk et al., 2016; Rauch et al., 2017; Nakamura et al., 2019), have potential to address some limitations of this technology and further studies could help validate CRISPR/Cas9-based approaches for the treatment of Usher syndrome.

Base Editing

The advent of CRISPR/Cas9 gene editing systems has also allowed for the development of base editing, which utilizes inactivated Cas nucleases and single-stranded editing enzymes to replace one nucleobase with another without creating double-stranded breaks (Rees and Liu, 2018). Two classes of base editing technology have so far been reported; cytosine base editors (CBEs; Komor et al., 2016) which convert C-G base pairs to T-A, and adenine base editors (ABEs), which convert A-T to G-C (Gaudelli et al., 2017). Both CBEs and ABEs introduce transition mutations between chemically similar base pairs, which can theoretically correct over 60% of human pathogenic point mutations (Rees and Liu, 2018). However, transversion base editing (e.g., G-C to C-G) remains an elusive target.

Base editing has already been used in proof-of-concept studies in mouse models of Duchenne muscular dystrophy (Ryu et al., 2018), phenylketonuria (Villiger et al., 2018), hereditary tyrosinemia type I (Song et al., 2020) and hypercholesterolemia through targeting the *PCSK9* gene (Chadwick et al., 2017). Importantly, base editing has also been used to investigate enhanced cellular reprogramming of supporting cells to cochlea hair cells to mediate hearing loss (Yeh et al., 2018). C-T conversion of a single base in the *CTNNB1* gene prevented phosphorylation and degradation of β-catenin protein, leading to a 7-fold increase of β-catenin levels in HEK293T cultures. Consequently, there was an increase in signaling to the Wnt pathway, which is crucial to the development of sensory hair cells. When delivered to mice *via* intracochlear injection, base editing resulted in the differentiation of cochlea supporting cells to MYO7A-expressing hair cells (Yeh et al., 2018).

RNA INTERVENTION

RNA splicing can be altered to either restore exons lost due to aberrant splicing induced by the mutation, or induce exon skipping to remove nonsense or missense mutations in coding regions. RNA intervention resulting in altered splicing

[2]https://www.nature.com/articles/d41586-020-00655-8
[3]https://www.editasmedicine.com
[4]https://ir.editasmedicine.com/news-releases/news-release-details/editas-medicine-reports-recent-progress-jp-morgan-healthcare-0

is achieved through the binding of an antisense oligonucleotide to RNA strands (reviewed by Havens et al., 2013). Base editing of RNA is also possible (Bass, 2002). Finally, translational readthrough techniques aim to suppress protein truncation by overriding mutations that cause premature termination of the translation machinery. The reader is referred to a review outlining the therapeutic progress of translational readthrough-inducing compounds in the treatment of inherited diseases (Nagel-Wolfrum et al., 2016).

RNA-Based Drug Interventions

RNA molecules designed to silence or interfere with toxic gain-of-function mutations have been investigated in the treatment of several non-Usher models of genetic hearing loss. Notably, Maeda et al. (2005) identified a short interfering RNA (siRNA) that targets the R75W allele variant in GJB2, a common cause of autosomal recessive hearing loss. The synthetic siRNA suppressed GJB2 in HEK293T cells by >80% for more than 120 h. When administered to the GJB2 mouse model, siRNA treatment suppressed GJB2 expression by >70% and prevented hearing loss (Maeda et al., 2005, 2007). Another study by Shibata et al. (2016) developed a microRNA targeting non-syndromic deafness caused by gain-of-function mutations in *TMC1*. The microRNA was packaged *via* AAV2/9 and delivered to *Bth* mice, resulting in significant preservation of hearing for up to 21 weeks relative to untreated controls. The animals who responded best to treatment maintained ABR thresholds 40 dB greater than untreated counterparts (Shibata et al., 2016).

Antisense Oligonucleotides

ASOs are short nucleic acid sequences that modulate gene expression *via* complementary binding to mRNA. ASOs are often synthesized to activate ribonuclease H (RNAse-H, which degrades mRNA) or target splicing defects (Goyal and Narayanaswami, 2018). ASOs can then be designed to target a pathogenic mutation; hence, the size of the gene is not an obstacle as with gene delivery techniques. However, the half-life of the drug means re-occurring invasive administration to the eye rather than one-off treatments, as is the case with a gene therapy approach.

ASO treatment has been widely investigated for retinal disease therapies (Huang et al., 2017). Recently, up to nine antisense-oligonucleotide variants were identified for the treatment of Stargardt disease caused by the intronic c.4539 + 2001G >A mutation in the large *ABCA4* gene (Garanto et al., 2019). In a separate study, screening of ASOs led to the discovery of QR-110 (Dulla et al., 2018), a therapeutic for LCA type 10 (LCA10), caused by the c.2991 + 1655A >G mutation, in the *CEP290* gene (Den Hollander et al., 2006). Like Usher syndrome, the pathogenesis of LCA10 is caused by defects in the photoreceptor connecting cilium. QR-110 was able to restore wild-type *CEP290* transcript and protein expression in mutant fibroblasts and decrease ciliopathy in 3-dimensional (3D) retinal organoids (explained further in section Cell Therapy for Retinal Degeneration). The drug was also well-tolerated in mice, rabbits, and non-human primates (Dulla et al., 2018), leading to the approval of phase I/IIa clinical trial (NCT03140969) by ProQR Therapeutics[5].

ASOs have also been developed for autosomal dominant RP (Naessens et al., 2019), as well as Usher 2A-associated RP (Slijkerman et al., 2016). An engineered ASO targeting a pseudo-exon-causing mutation in *USH2A* (c.7595–2144A >G) displayed splice-correcting properties in patient-derived fibroblasts (Slijkerman et al., 2016). The same company that funded this study, ProQR Therapeutics, has also investigated an ASO, QR-421a, for the treatment of Usher 2A-associated RP caused by a common exon 13 mutation (c.2299delG; Van Diepen et al., 2019). The efficacy of this treatment was initially demonstrated in patient-derived retinal organoids and animal models (**Table 3**), with exon-skipping capability maintained in cynomolgus monkeys for more than 100 days post-treatment. They also showed that Usherin protein was present in wild-type zebrafish larvae at the photoreceptor connecting cilium but absent in untreated c.2299delG zebrafish. Partial restoration of correctly localized usherin expression was observed in treated larvae as well as improved ERG recordings compared to untreated zebrafish (Van Diepen et al., 2019; conference abstract). The results of this study have allowed for advancement to a clinical trial (NCT03780257), which has already led to the first treatment to the eyes of an Usher 2A patient using QR-421a[6] (preliminary findings reported on the 20th of April 2020 here: https://ir.proqr.com/news-releases/news-release-details/proqr-announces-positive-findings-interim-analysis-phase-12).

Several studies have investigated ASO-29 treatment of the Usher 1C, c.216G >A mouse model, which exhibits hearing loss as well as disrupted exploratory movements, indicating vestibular dysfunction. ASO-29 administration to the inner ear of c.216G >A mouse pups has been shown to improve both the auditory and vestibular response (Lentz et al., 2013; Vijayakumar et al., 2017; Donaldson et al., 2018). However, mounting evidence has demonstrated an age threshold for effective delivery, with early-treated mice responding better to treatment. Ponnath et al. (2018) demonstrated that both outer and inner hair cell function was effectively restored by ASO-29 treatment, but the threshold for effective outer hair cell treatment (post-natal day 5) was lower than that of inner hair cell treatment (post-natal day 7). Considering these promising results in the auditory and vestibular response of USH1C mice, it would be interesting to assess the efficacy of ASO-29 towards the vision loss phenotype. However, the limited visual phenotype of the USH1C mice model means alternative models will need to be used for testing ASO-29.

Translational Readthrough Inducing Drugs

Nonsense mutations cause premature termination codons (PTCs), leading to the production of a truncated polypeptide or targeting of the transcript by nonsense-mediated decay (NMD). When a translation termination codon (UAA, UGA, UAG) enters the ribosomal A site, it is recognized by eukaryotic release factor 1 (eRF1), which changes conformation when bound

[5]https://www.proqr.com

[6]https://www.globenewswire.com/news-release/2019/03/11/1751124/0/en/ProQR-Doses-First-Patient-in-Phase-1-2-STELLAR-Trial-of-QR-421a-for-Usher-Syndrome-Type-2.html

TABLE 3 | Summary of transgenic animal disease models discussed in this review.

Human disease (equivalent)	Model	Animal	Gene	Mutation	Therapy tested	Reference
Usher 1B	*Shaker1* ($Sh1^{-/-}$)	Mouse	Myo7a	G720X induced by ENU	AAV2/2-mediated gene delivery	Colella et al. (2013)
Usher 1C caused by c.216G >A mutation	*Ush1c c.216G >A*	Mouse	Ush1C	c.216G >A	AAV2/Anc80L65-mediated gene delivery	Pan et al. (2017)
					ASO-mediated suppression of exon 3 cryptic splice site (ASO-29)	Lentz et al. (2013), Vijayakumar et al. (2017), Donaldson et al. (2018) and Ponnath et al. (2018)
Usher 1G	$Ush1g^{-/-}$	Mouse	Ush1g	$Ush1g^{fl/fl}$ mice targeting exon 2, crossed with *PGK-cre* mice	rAAV2/8-mediated gene delivery	Emptoz et al. (2017)
Usher 2A	$Ush2a^{rmcm1}$	Zebrafish	Ush2a	Homozygous premature stop mutations in exon 13	ASO-mediated exon 13 skipping (QR-421a)	Van Diepen et al. (2019)
Usher 2D caused by compound heterozygous Q103 ×/c.837 + 1G >A mutation	$Whirlin^{-/-}$	Mouse	Whrn	Targeted deletion of exon 1	AAV2/8-mediated gene delivery	Zou et al. (2011)
Usher 3A	*Clrn1* KO	Mouse	Clrn1	Targeted deletion of *Clrn1* upstream promoter, first exon, 269 bp of the first intron	AAV-mediated gene delivery	Dinculescu et al. (2016), Geng et al. (2017) and György et al. (2019)
	$Clrn1^{ex4-/-}$	Mouse	Clrn1	Early ubiquitous deletion of Clrn1 exon 4	AAV-mediated gene delivery	Dulon et al. (2018)
	$Clrn1^{ex4fl/fl}$Myo15-$Cre^{+/-}$	Mouse	Clrn1	Hair cell-specific postnatal deletion of Clrn1 exon 4	AAV-mediated gene delivery	Dulon et al. (2018)
Non-syndromic hearing loss (DFNA36)	*Beethoven (Bth)*	Mouse	Tmc1	c.1235T>A	Lipid-mediated delivery of cas9:gRNA complex	Gao et al. (2018)
Non-syndromic hearing loss (SLC26A4)	*Slc26a4-null*	Mouse	Slc26a4	Mutations not reported	iPSC otic progenitors	Chen et al. (2018)
Non-syndromic hearing loss (GJB2)	*GJB2*_R75W-eGFP	Mouse	Gjb2	R75W	siRNA	Maeda et al. (2005)

Abbreviations: ENU, N-ethyl-N-nitrosourea; AAV, adeno-associated virus; ASO, Antisense oligonucleotides; iPSC, induced pluripotent stem cells; siRNA, small interfering RNA.

to eRF3, initiating a signal cascade that results in hydrolytic cleavage of the polypeptide by eRF1. Translation termination is a competition between recognition by eRF1 leading to termination and continuation of the translation process, leading to readthrough or natural suppression. Natural suppression occurs >0.1% of the time in non-mutant transcripts, but PTCs increase the rate of natural suppression to up to 1% (Keeling et al., 2012). Nonsense suppression therapies aim to increase translational readthrough, avoiding the production of a truncated protein. The two main obstacles with this type of therapy are avoiding mRNA degradation by NMD, and avoiding the promotion of non-specific readthrough (Frischmeyer and Dietz, 1999; Keeling and Bedwell, 2011; Wang and Gregory-Evans, 2015; Dabrowski et al., 2018; Campofelice et al., 2019).

Nonsense suppression has clear advantages over gene-based therapies; they are not gene- or mutation-specific, the size of the gene is not an issue, and they allow the cell to maintain expression regulation. However, they are associated with nephrotoxic (Mingeot-Leclercq and Tulkens, 1999) and ototoxic (Selimoglu, 2007) effects and often result in the incorporation of a non-cognate amino acid, which may alter protein function. A recent advancement in the field was the development of anticodon engineered transfer RNAs (ACE-tRNAs), which recognize and suppress stop codons while encoding the cognate amino acid of the non-mutant polypeptide (Lueck et al., 2019). This method induced a specific readthrough of *CFTR* mutations *in vitro*. Low levels of off-target suppression were detected, depending on the genetic environment of the mutation, but the authors suggest that endogenous cellular pathways to degrade incorrectly transcribed proteins would be sufficient to offset these effects (Lueck et al., 2019).

Several studies have investigated nonsense-mediated therapies for the treatment of Usher syndrome. Initial studies focussed on nb30, a synthetic derivative developed from the commercial aminoglycoside, paromomycin (Nudelman et al., 2006). Nb30 was shown to induce significant nonsense suppression of a common *USH1C* mutation (p.R31X) with reduced toxicity and increased biocompatibility compared to commercial aminoglycosides (Goldmann et al., 2010). Similarly, favorable toxicity profiles were observed in nb30-mediated suppression of *PCDH15* nonsense mutations relative to traditional antibiotics (Rebibo-Sabbah et al., 2007). Subsequently, a second paromomycin derivative, nb54, was identified as a promising drug candidate for nonsense suppression which demonstrated readthrough ability in several prominent disease genes, including Usher 1F syndrome (Nudelman et al., 2009). In this study, Nudelman et al. (2009) show that nb54 is capable of inducing *in vitro* stop codon suppression 3–5-fold times more efficient for *PCDH15* (Usher 1F) mutations p.R3X and p.R245X compared to three other aminoglycoside compounds.

PTC-124 (Ataluren) is another translational readthrough inducer which has been studied for application in cystic fibrosis and Duchenne muscular dystrophy cases (Welch et al., 2007), though success has thus far been limited in clinical trials (Wilschanski et al., 2011; Kerem et al., 2014; McDonald et al., 2017). PTC-124-induced readthrough *in vitro* has been demonstrated in the common p.Trp3955* mutation, which accounts for 13% of *USH2A* mutations (Neuhaus et al., 2017). Additionally, PTC-124 treatment induced *in vitro* and *in vivo* readthrough of *USH1C* mutations, leading to the recovery of protein function with superior biocompatibility to gentamicin and nb30 (Goldmann et al., 2011). Finally, Goldmann et al. (2012) compared nb30, nb54, and PTC-124 for translational readthrough of *USH1C* mutations and identified both PTC-124 and nb54 as ideal drug candidates. Though promising, functional rescue of Usher phenotypes *in vivo* will nonetheless be necessary to evaluate the efficacy of both PTC-124 and nb54.

CELL REPLACEMENT THERAPY

Cellular therapies are therapeutic approaches that aim to regenerate damaged tissue by introducing replacement donor cells (Zakrzewski et al., 2019). This approach relies on the survival of connecting interneurons when the stimulus-receiving neuron has already degenerated in the cochlea or retina. Typically, progenitor cells derived from stem cells are used as an unlimited source of donor cells. Stem cells are pluripotent progenitors that can differentiate into cell lineages from each of the three germ layers. Early stem cell studies relied on the controversial use of human embryonic stem cells (hESCs; Omole and Fakoya, 2018) until the discovery in 2007 by Shinya Yamanaka and colleagues that adult human somatic cells, such as adult fibroblasts, could be reprogrammed back into a pluripotent state (Takahashi et al., 2007). They achieved this through the expression of characteristic pluripotent markers *SOX2*, *OCT3/4*, *KLF4*, and *c-MYC*, delivered to the cell *via* retroviral vectors (Takahashi et al., 2007). As the stem cells carry the same genetic information as the somatic cells from which they were made, this breakthrough allowed human diseases to be replicated and studied non-invasively in the laboratory for the first time (Omole and Fakoya, 2018).

Cell Therapy for Retinal Degeneration

iPSCs can aggregate to form "embryoid bodies" (aggregates of cells thought to mimic the early embryo), which can be prompted to differentiate into specific lineages (Omole and Fakoya, 2018), including 3D retinal organoids (Nakano et al., 2012; Phillips et al., 2012; Zhong et al., 2014; Mellough et al., 2015, 2019). These organoids closely resemble *in vivo* human eye development, forming eye field-like domains that form 3D retinal neuroepithelium and RPE. The neural retina and RPE domains transition into a pseudo optic cup-like structure and retinal progenitor cells differentiate into neurons including horizontal cells, amacrine cells, and retinal ganglion cells (Zhong et al., 2014). Organoids develop a defined outer nuclear layer containing radially-aligned photoreceptors with detectable inner and outer segments connected by a connecting cilium (Zhong et al., 2014; Mellough et al., 2015, 2019; Lowe et al., 2016; Parfitt et al., 2016). These photoreceptors express phototransduction proteins, including opsins, and can respond to a light stimulus (Zhong et al., 2014; Hallam et al., 2018; Mellough et al., 2019).

Retinal organoids are a good resource of photoreceptor progenitors for transplantation. These progenitors can be enriched before transplantation for a homogenous cell population *via* cell sorting for dissociated cell transplants (Lakowski et al., 2015, 2018). MacLaren et al. (2006) transplanted developing mouse photoreceptors into the retinae of degenerative mouse models with successful integration and differentiation, noting an optimal therapeutic window corresponding to the specification of rod cell fate. Subsequently, photoreceptor transplantation was reported by several research groups (Pearson et al., 2012; Barber et al., 2013; Gonzalez-Cordero et al., 2013; Singh et al., 2013; Santos-Ferreira et al., 2016), including Barber et al. (2013) who looked into six mouse models of inherited retinal degenerations (IRDs) with environmental, disease course, and genetic factors influencing the integration outcome (Barber et al., 2013). Though these initial photoreceptor transplantation experiments seemed promising, the transfer of cytoplasmic material between labeled donors and host photoreceptors has been attributed to the seemingly high rates of integration (Santos-Ferreira et al., 2016; Singh et al., 2016). Alternatively, a stem cell-derived retinal sheet may be transplanted into the recipient's eye (Assawachananont et al., 2014). Though clinical trials have focussed on RPE transfer in patients with age-related macular degeneration (AMD) and Stargardt Disease (NCT03102138, NCT02941991, NCT01469832), none thus far have reported results on photoreceptor transplants. However, a first-in-human phase I/II dose-escalation study is currently examining the safety and tolerability of human retinal progenitor cells in patients with RP (NCT02464436, estimated completion date: July 2021). Furthermore, proof-of-principle studies in mouse and non-human primates showed survival of retinal sheets [containing hESC-derived retinal cells (Shirai et al., 2016), hiPSC-derived RPE cells (Mandai et al., 2017) or hiPSC-derived retinal progenitors (Tu et al., 2019)] post-transplantation, and improvement of light sensitivity (Shirai et al., 2016; Mandai et al., 2017; Tu et al., 2019).

Cell Therapy for Neurosensory Deafness

Stem cell therapy also has the potential to treat hearing loss by replacing cochlear hair cells, although this is a difficult environment for this approach. The hostile high potassium luminal fluid (endolymph) environment of the cochlear duct poses significant challenges for cell survival. Further, the robust tight junctions in the auditory epithelium in the organ of Corti make donor cellular integration difficult. A recent article reported the ability of iPSC-derived otic progenitors to form connections with co-cultured spiral ganglion-like cells (Chen et al., 2018). Subsequently, these cells were detached and administered to the inner ear *via* round window injection in hearing loss *Slc26a4* mice. Cells positive for the hair cell marker, *MYO7A* (also causative of Usher 1B), were detected in the organ of Corti, indicating the potential of this method to deliver progenitor cells capable of migrating, differentiating and forming appropriate connections in the inner ear (Chen et al., 2018). However, the progenitors did *not* form an organized stereocilial arrangement, which is critical to hearing, and disrupted in Usher cases (Mathur and Yang, 2015).

Limitations of Cell Therapies

The development of cellular therapies presents a promising and broad approach to treat a host of sensory diseases regardless of genetic etiology. However, several safety concerns remain to be solved. Of principal concern is the potential for donor cells to cause neoplastic changes in the tissue, particularly those that are derived from stem cells. Mouse embryonic stem cells injected into the subretinal space have been shown to induce tumor formation (Arnhold et al., 2004), though human iPSC-derived photoreceptor progenitor delivery to *rd1* mice found no evidence of abnormal cell growth (Barnea-Cramer et al., 2016). Additionally, Chen et al. (2018) injected mouse models with hiPSC-derived otic progenitors and hiPSC controls to examine the tumorigenicity of hiPSC-derived cells. They reported tumor formation in hiPSC-injected tissue, but not in tissue injected with hiPSC-derived progenitors (Chen et al., 2018). Exclusion of undifferentiated stem cells before transplantation could further decrease the risk of neoplastic changes in the host. In a recent study, Gagliardi et al. (2018) demonstrated the potential of this strategy by safely transplanting CD73+ photoreceptor precursors separated from a cell population using magnetic-activated cell sorting into the eyes of P23H-transgenic rats.

Despite the multitude of challenges in delivering stem cell therapeutics safely and effectively, regenerative medicine remains attractive to patients, industry, and commercial clinics. Several stem cell-based clinical trials are listed on clinicaltrials.gov for AMD (NCT01736059, NCT01920867) and other ocular diseases (NCT01920867), including RP (NCT02320812). Recently, the stem cell ophthalmology treatment study (SCOTS) published results from a phase I/IIa study of five Usher syndrome patients with bilaterally-treated eyes (Weiss and Levy, 2019). Autologous bone marrow-derived stem cells were separated from bone marrow aspirate and administered into patients'

TABLE 4 | Active clinical trials for Usher patients.

Identifier	Status	Intervention	Usher subtypes	Country
NCT02065011	NR	UshStat—lentiviral delivery of MYO7A	Usher 1B	USA, France
NCT01530659	NR	NT-501—ciliary neurotrophic factor intraocular implant	Usher types 2 and 3	USA
NCT03780257	R	QR-421a—antisense oligonucleotide to induce skipping of exon 13	Usher 2A due to mutations in exon 13	USA, Belgium, Canada, France
NCT03011541	R	Autologous bone marrow-derived stem cells	All	USA

All trials listed are for the treatment or prevention of Usher-associated retinal degeneration. R, currently recruiting; NR, not recruiting. See www.clinicaltrials.gov for further information.

eyes *via* a combination of retrobulbar, sub-Tenon, intravitreal, subretinal, and intra-optic injections, as chosen by the patient and depending on the extent of visual loss and relative risks. Increases in visual acuity were noted in 80% of treated eyes with a statistical significance of $p < 0.01$. There was no reported visual loss nor any complications up to a year post-treatment (Weiss and Levy, 2019). However, a continual follow up is needed to confirm the long-term visual acuity effects and determine whether the patients remain free of adverse events.

CONCLUSIONS

Significant advancements have been made in the last several years towards the treatment and prevention of sensory loss in Usher patients. Particularly, gene and cell therapies pose attractive, potentially one-off solutions that would reduce the burden of invasive re-administration of medications to the eye. Though the majority of studies are currently proof-of-principle treatment strategies using animal models of disease (summarized in **Table 3**), it is highly possible that the results from current ongoing clinical trials may translate into effective new treatments (see **Table 4** for a summary of ongoing clinical trials). However, more temporary therapeutics such as ASOs and translational readthrough inhibitors may also offer a significant reduction in disability. Continuing to improve the safety and efficacy of these treatment approaches is of utmost importance to provide commercially available options for Usher syndrome patients.

AUTHOR CONTRIBUTIONS

LF wrote the first draft of the manuscript. FC, CM, and LC contributed to manuscript revision, reading, and approval of the submitted version.

REFERENCES

Adli, M. (2018). The CRISPR tool kit for genome editing and beyond. *Nat. Commun.* 9:1911. doi: 10.1038/s41467-018-04252-2

Ahmed, Z. M., Riazuddin, S., Aye, S., Ali, R. A., Venselaar, H., Anwar, S., et al. (2008). Gene structure and mutant alleles of PCDH15: nonsyndromic deafness DFNB23 and type 1 Usher syndrome. *Hum. Genet.* 124, 215–223. doi: 10.1007/s00439-008-0543-3

Alagramam, K. N., Murcia, C. L., Kwon, H. Y., Pawlowski, K. S., Wright, C. G., and Woychik, R. P. (2001). The mouse Ames waltzer hearing-loss mutant is caused by mutation of Pcdh15, a novel protocadherin gene. *Nat. Genet.* 27, 99–102. doi: 10.1038/83837

Allocca, M., Doria, M., Petrillo, M., Colella, P., Garcia-Hoyos, M., Gibbs, D., et al. (2008). Serotype-dependent packaging of large genes in adeno-associated viral vectors results in effective gene delivery in mice. *J. Clin. Invest.* 118, 1955–1964. doi: 10.1172/JCI34316

Anzalone, A. V., Randolph, P. B., Davis, J. R., Sousa, A. A., Koblan, L. W., Levy, J. M., et al. (2019). Search-and-replace genome editing without double-strand breaks or donor DNA. *Nature* 576, 149–157. doi: 10.1038/s41586-019-1711-4

Apaolaza, P. S., del Pozo-Rodríguez, A., Solinís, M. A., Rodríguez, J. M., Friedrich, U., Torrecilla, J., et al. (2016). Structural recovery of the retina in a retinoschisin-deficient mouse after gene replacement therapy by solid lipid nanoparticles. *Biomaterials* 90, 40–49. doi: 10.1016/j.biomaterials.2016.03.004

Arnhold, S., Klein, H., Semkova, I., Addicks, K., and Schraermeyer, U. (2004). Neurally selected embryonic stem cells induce tumor formation after long-term survival following engraftment into the subretinal space. *Invest. Ophthalmol. Vis. Sci.* 45, 4251–4255. doi: 10.1167/iovs.03-1108

Assawachananont, J., Mandai, M., Okamoto, S., Yamada, C., Eiraku, M., Yonemura, S., et al. (2014). Transplantation of embryonic and induced pluripotent stem cell-derived 3D retinal sheets into retinal degenerative mice. *Stem Cell Reports* 2, 662–674. doi: 10.1016/j.stemcr.2014.03.011

Bademci, G., Foster, J. II., Mahdieh, N., Bonyadi, M., Duman, D., Cengiz, F. B., et al. (2016). Comprehensive analysis *via* exome sequencing uncovers genetic etiology in autosomal recessive nonsyndromic deafness in a large multiethnic cohort. *Genet. Med.* 18, 364–371. doi: 10.1038/gim.2015.89

Barber, A. C., Hippert, C., Duran, Y., West, E. L., Bainbridge, J. W. B., Warre-Cornish, K., et al. (2013). Repair of the degenerate retina by photoreceptor transplantation. *Proc. Natl. Acad. Sci. U S A* 110, 354–359. doi: 10.1073/pnas.1212677110

Barnea-Cramer, A. O., Wang, W., Lu, S.-J., Singh, M. S., Luo, C., Huo, H., et al. (2016). Function of human pluripotent stem cell-derived photoreceptor progenitors in blind mice. *Sci. Rep.* 6:29784. doi: 10.1038/srep29784

Bass, B. L. (2002). RNA editing by adenosine deaminases that act on RNA. *Annu. Rev. Biochem.* 71, 817–846. doi: 10.1146/annurev.biochem.71.110601.135501

Bennett, J., Tanabe, T., Sun, D., Zeng, Y., Kjeldbye, H., Gouras, P., et al. (1996). Photoreceptor cell rescue in retinal degeneration (Rd) mice by *in vivo* gene therapy. *Nat. Med.* 2, 649–654. doi: 10.1038/nm0696-649

Berson, E. L., Rosner, B., Sandberg, M. A., Hayes, K. C., Nicholson, B. W., Weigel-DiFranco, C., et al. (1993). A randomized trial of vitamin A and vitamin E supplementation for retinitis pigmentosa. *Arch. Ophthalmol.* 111, 761–772. doi: 10.1001/archopht.1993.01090060049022

Berson, E. L., Rosner, B., Sandberg, M. A., Weigel-DiFranco, C., and Willett, W. C. (2012). Ω-3 intake and visual acuity in patients with retinitis pigmentosa receiving vitamin A. *Arch. Ophthalmol.* 130, 707–711. doi: 10.1001/archophthalmol.2011.2580

Berson, E. L., Rosner, B., Sandberg, M. A., Weigel-DiFranco, C., Moser, A., Brockhurst, R. J., et al. (2004). Further evaluation of docosahexaenoic acid in patients with retinitis pigmentosa receiving vitamin A treatment: subgroup analyses. *Arch. Ophthalmol.* 122, 1306–1314. doi: 10.1001/archopht.122.9.1306

Berson, E. L., Weigel-DiFranco, C., Rosner, B., Gaudio, A. R., and Sandberg, M. A. (2018). Association of vitamin a supplementation with disease course in children with retinitis pigmentosa. *JAMA Ophthalmol.* 136, 490–495. doi: 10.1001/jamaophthalmol.2018.0590

Bisht, R., Mandal, A., Jaiswal, J. K., and Rupenthal, I. D. (2018). Nanocarrier mediated retinal drug delivery: overcoming ocular barriers to treat posterior eye diseases. *Wiley Interdiscip. Rev. Nanomed. Nanobiotechnol.* 10:e1473. doi: 10.1002/wnan.1473

Bonnet, C., Grati, M. H., Marlin, S., Levilliers, J., Hardelin, J.-P., Parodi, M., et al. (2011). Complete exon sequencing of all known Usher syndrome genes greatly improves molecular diagnosis. *Orphanet J. Rare Dis.* 6:21. doi: 10.1186/1750-1172-6-21

Cai, X., Conley, S. M., Nash, Z., Fliesler, S. J., Cooper, M. J., and Naash, M. I. (2010). Gene delivery to mitotic and postmitotic photoreceptors *via* compacted DNA nanoparticles results in improved phenotype in a mouse model of retinitis pigmentosa. *FASEB J.* 24, 1178–1191. doi: 10.1096/fj.09-139147

Campochiaro, P. A., Iftikhar, M., Hafiz, G., Akhlaq, A., Tsai, G., Wehling, D., et al. (2020). Oral N-acetylcysteine improves cone function in retinitis pigmentosa patients in phase I trial. *J. Clin. Invest.* 130, 1527–1541. doi: 10.1172/JCI132990

Campofelice, A., Lentini, L., Di Leonardo, A., Melfi, R., Tutone, M., Pace, A., et al. (2019). Strategies against nonsense: oxadiazoles as translational readthrough-inducing drugs (TRIDs). *Int. J. Mol. Sci.* 20:3329.doi: 10.3390/ijms20133329

Carvalho, L. S., Turunen, H. T., Wassmer, S. J., Luna-Velez, M. V., Xiao, R., Bennett, J., et al. (2017). Evaluating efficiencies of dual AAV approaches for retinal targeting. *Front. Neurosci.* 11:503. doi: 10.3389/fnins.2017.00503

Chadwick, A. C., Wang, X., and Musunuru, K. (2017). *In vivo* base editing of pcsk9 (proprotein convertase subtilisin/kexin type 9) as a therapeutic alternative to genome editing. *Arterioscler. Thromb. Vasc. Biol.* 37, 1741–1747. doi: 10.1161/atvbaha.117.309881

Chen, J., Guo, Z., Tian, H., and Chen, X. (2016). Production and clinical development of nanoparticles for gene delivery. *Mol. Ther. Methods Clin. Dev.* 3:16023. doi: 10.1038/mtm.2016.23

Chen, J., Hong, F., Zhang, C., Li, L., Wang, C., Shi, H., et al. (2018). Differentiation and transplantation of human induced pluripotent stem cell-derived otic epithelial progenitors in mouse cochlea. *Stem Cell Res. Ther.* 9:230. doi: 10.1186/s13287-018-0967-1

Cheng, H., Khanna, H., Oh, E. C., Hicks, D., Mitton, K. P., and Swaroop, A. (2004). Photoreceptor-specific nuclear receptor NR2E3 functions as a transcriptional activator in rod photoreceptors. *Hum. Mol. Genet.* 13, 1563–1575. doi: 10.1093/hmg/ddh173

Christian, M., Cermak, T., Doyle, E. L., Schmidt, C., Zhang, F., Hummel, A., et al. (2010). Targeting DNA double-strand breaks with TAL effector nucleases. *Genetics* 186, 757–761. doi: 10.1534/genetics.110.120717

Colella, P., Sommella, A., Marrocco, E., Di Vicino, U., Polishchuk, E., Garcia Garrido, M., et al. (2013). Myosin7a deficiency results in reduced retinal activity which is improved by gene therapy. *PLoS One* 8:e72027. doi: 10.1371/journal.pone.0072027

Colella, P., Trapani, I., Cesi, G., Sommella, A., Manfredi, A., Puppo, A., et al. (2014). Efficient gene delivery to the cone-enriched pig retina by dual AAV vectors. *Gene Ther.* 21, 450–456. doi: 10.1038/gt.2014.8

Cong, L., Ran, F. A., Cox, D., Lin, S., Barretto, R., Habib, N., et al. (2013). Multiplex genome engineering using CRISPR/Cas systems. *Science* 339, 819–823. doi: 10.1126/science.1231143

Cosgrove, D., and Zallocchi, M. (2014). Usher protein functions in hair cells and photoreceptors. *Int. J. Biochem. Cell Biol.* 46, 80–89. doi: 10.1016/j.biocel.2013.11.001

Dabrowski, M., Bukowy-Bieryllo, Z., and Zietkiewicz, E. (2018). Advances in therapeutic use of a drug-stimulated translational readthrough of premature termination codons. *Mol. Med.* 24, 25–25. doi: 10.1186/s10020-018-0024-7

Den Hollander, A. I., Koenekoop, R. K., Yzer, S., Lopez, I., Arends, M. L., Voesenek, K. E. J., et al. (2006). Mutations in the CEP290 (NPHP6) gene are a frequent cause of Leber congenital amaurosis. *Am. J. Hum. Genet.* 79, 556–561. doi: 10.1086/507318

Deverman, B. E., Pravdo, P. L., Simpson, B. P., Kumar, S. R., Chan, K. Y., Banerjee, A., et al. (2016). Cre-dependent selection yields AAV variants for widespread gene transfer to the adult brain. *Nat. Biotechnol.* 34, 204–209. doi: 10.1038/nbt.3440

Deverman, B. E., Ravina, B. M., Bankiewicz, K. S., Paul, S. M., and Sah, D. W. (2018). Gene therapy for neurological disorders: progress and prospects. *Nat. Rev. Drug Discov.* 17, 641–659. doi: 10.1038/nrd.2018.110

DiCarlo, J. E., Mahajan, V. B., and Tsang, S. H. (2018). Gene therapy and genome surgery in the retina. *J. Clin. Invest.* 128, 2177–2188. doi: 10.1172/jci120429

Dinculescu, A., Stupay, R. M., Deng, W. T., Dyka, F. M., Min, S. H., Boye, S. L., et al. (2016). AAV-mediated clarin-1 expression in the mouse retina: implications for USH3A gene therapy. *PLoS One* 11:e0148874. doi: 10.1371/journal.pone.0148874

Donaldson, T. N., Jennings, K. T., Cherep, L. A., McNeela, A. M., Depreux, F. F., Jodelka, F. M., et al. (2018). Antisense oligonucleotide therapy rescues disruptions in organization of exploratory movements associated with Usher syndrome type 1C in mice. *Behav. Brain Res.* 338, 76–87. doi: 10.1016/j.bbr.2017.10.012

Dong, B., Nakai, H., and Xiao, W. (2010). Characterization of genome integrity for oversized recombinant AAV vector. *Mol. Ther.* 18, 87–92. doi: 10.1038/mt.2009.258

Duan, D., Yue, Y., and Engelhardt, J. F. (2001). Expanding AAV packaging capacity with trans-splicing or overlapping vectors: a quantitative comparison. *Mol. Ther.* 4, 383–391. doi: 10.1006/mthe.2001.0456

Dulla, K., Aguila, M., Lane, A., Jovanovic, K., Parfitt, D. A., Schulkens, I., et al. (2018). Splice-modulating oligonucleotide QR-110 restores CEP290 mRNA and function in human c.2991+1655A>G LCA10 models. *Mol. Ther. Nucleic Acids* 12, 730–740. doi: 10.1016/j.omtn.2018.07.010

Dulon, D., Papal, S., Patni, P., Cortese, M., Vincent, P. F., Tertrais, M., et al. (2018). Clarin-1 gene transfer rescues auditory synaptopathy in model of Usher syndrome. *J. Clin. Invest.* 128, 3382–3401. doi: 10.1172/jci94351

Dyka, F. M., Boye, S. L., Chiodo, V. A., Hauswirth, W. W., and Boye, S. E. (2014). Dual adeno-associated virus vectors result in efficient *in vitro* and *in vivo* expression of an oversized gene, MYO7A. *Hum. Gene Ther. Methods* 25, 166–177. doi: 10.1089/hgtb.2013.212

Dyka, F. M., Molday, L. L., Chiodo, V. A., Molday, R. S., and Hauswirth, W. W. (2019). Dual *ABCA4*-AAV vector treatment reduces pathogenic retinal A2E accumulation in a mouse model of autosomal recessive stargardt disease. *Hum. Gene Ther.* 30, 1361–1370. doi: 10.1089/hum.2019.132

Ebermann, I., Phillips, J. B., Liebau, M. C., Koenekoop, R. K., Schermer, B., Lopez, I., et al. (2010). PDZD7 is a modifier of retinal disease and a contributor to digenic Usher syndrome. *J. Clin. Invest.* 120, 1812–1823. doi: 10.1172/JCI39715

El-Amraoui, A., and Petit, C. (2014). The retinal phenotype of Usher syndrome: pathophysiological insights from animal models. *C. R. Biol.* 337, 167–177. doi: 10.1016/j.crvi.2013.12.004

Emptoz, A., Michel, V., Lelli, A., Akil, O., Boutet de Monvel, J., Lahlou, G., et al. (2017). Local gene therapy durably restores vestibular function in a mouse model of Usher syndrome type 1G. *Proc. Natl. Acad. Sci. U S A* 114, 9695–9700. doi: 10.1073/pnas.1708894114

Farjo, R., Skaggs, J., Quiambao, A. B., Cooper, M. J., and Naash, M. I. (2006). Efficient non-viral ocular gene transfer with compacted DNA nanoparticles. *PLoS One* 1:e38. doi: 10.1371/journal.pone.0000038

Feng, L., and Niu, D.-K. (2007). Relationship between mRNA stability and length: an old question with a new twist. *Biochem. Genet.* 45, 131–137. doi: 10.1007/s10528-006-9059-5

Frischmeyer, P. A., and Dietz, H. C. (1999). Nonsense-mediated mRNA decay in health and disease. *Hum. Mol. Genet.* 8, 1893–1900. doi: 10.1093/hmg/8.10.1893

Fu, Y., Foden, J. A., Khayter, C., Maeder, M. L., Reyon, D., Joung, J. K., et al. (2013). High-frequency off-target mutagenesis induced by CRISPR-Cas nucleases in human cells. *Nat. Biotechnol.* 31, 822–826. doi: 10.1038/nbt.2623

Fuster-García, C., García-García, G., González-Romero, E., Jaijo, T., Sequedo, M. D., Ayuso, C., et al. (2017). USH2A gene editing using the CRISPR system. *Mol. Ther. Nucleic Acids* 8, 529–541. doi: 10.1016/j.omtn.2017.08.003

Gagliardi, G., Ben M'Barek, K., Chaffiol, A., Slembrouck-Brec, A., Conart, J.-B., Nanteau, C., et al. (2018). Characterization and transplantation of CD73-positive photoreceptors isolated from human iPSC-derived retinal organoids. *Stem Cell Reports* 11, 665–680. doi: 10.1016/j.stemcr.2018.07.005

Gao, X., Tao, Y., Lamas, V., Huang, M., Yeh, W.-H., Pan, B., et al. (2018). Treatment of autosomal dominant hearing loss by *in vivo* delivery of genome editing agents. *Nature* 553, 217–221. doi: 10.1038/nature25164

Garanto, A., Duijkers, L., Tomkiewicz, T. Z., and Collin, R. W. J. (2019). Antisense oligonucleotide screening to optimize the rescue of the splicing defect caused by the recurrent deep-intronic *ABCA4* variant c.4539+2001G >A in stargardt disease. *Genes* 10:452. doi: 10.3390/genes10060452

Gaudelli, N. M., Komor, A. C., Rees, H. A., Packer, M. S., Badran, A. H., Bryson, D. I., et al. (2017). Programmable base editing of A•T to G•C in genomic DNA without DNA cleavage. *Nature* 551, 464–471. doi: 10.1038/nature24644

Geng, R., Omar, A., Gopal, S. R., Chen, D. H., Stepanyan, R., Basch, M. L., et al. (2017). Modeling and preventing progressive hearing loss in usher syndrome III. *Sci. Rep.* 7:13480. doi: 10.1038/s41598-017-13620-9

Ghosh, A., Yue, Y., Lai, Y., and Duan, D. (2008). A hybrid vector system expands adeno-associated viral vector packaging capacity in a transgene-independent manner. *Mol. Ther.* 16, 124–130. doi: 10.1038/sj.mt.6300322

Giannelli, S. G., Luoni, M., Castoldi, V., Massimino, L., Cabassi, T., Angeloni, D., et al. (2017). Cas9/sgRNA selective targeting of the P23H Rhodopsin mutant allele for treating retinitis pigmentosa by intravitreal AAV9.PHP.B-based delivery. *Hum. Mol. Genet.* 27, 761–779. doi: 10.1093/hmg/ddx438

Goldmann, T., Overlack, N., Möller, F., Belakhov, V., Van Wyk, M., Baasov, T., et al. (2012). A comparative evaluation of NB30, NB54 and PTC124 in translational read-through efficacy for treatment of an USH1C nonsense mutation. *EMBO Mol. Med.* 4, 1186–1199. doi: 10.1002/emmm.201201438

Goldmann, T., Overlack, N., Wolfrum, U., and Nagel-Wolfrum, K. (2011). PTC124-mediated translational readthrough of a nonsense mutation causing Usher syndrome type 1C. *Hum. Gene Ther.* 22, 537–547. doi: 10.1089/hum.2010.067

Goldmann, T., Rebibo-Sabbah, A., Overlack, N., Nudelman, I., Belakhov, V., Baasov, T., et al. (2010). Beneficial read-through of a USH1C nonsense mutation by designed aminoglycoside NB30 in the retina. *Invest. Ophthalmol. Vis. Sci.* 51, 6671–6680. doi: 10.1167/iovs.10-5741

Gonzalez-Cordero, A., West, E. L., Pearson, R. A., Duran, Y., Carvalho, L. S., Chu, C. J., et al. (2013). Photoreceptor precursors derived from three-dimensional embryonic stem cell cultures integrate and mature within adult degenerate retina. *Nat. Biotechnol.* 31, 741–747. doi: 10.1038/nbt.2643

Goyal, N., and Narayanaswami, P. (2018). Making sense of antisense oligonucleotides: a narrative review. *Muscle Nerve* 57, 356–370. doi: 10.1002/mus.26001

Grati, M. H., and Kachar, B. (2011). Myosin VIIa and sans localization at stereocilia upper tip-link density implicates these Usher syndrome proteins in mechanotransduction. *Proc. Natl. Acad. Sci. U S A* 108, 11476–11481. doi: 10.1073/pnas.1104161108

Grieger, J. C., and Samulski, R. J. (2005). Packaging capacity of adeno-associated virus serotypes: impact of larger genomes on infectivity and postentry steps. *J. Virol.* 15, 9933–9944. doi: 10.1128/jvi.79.15.9933-9944.2005

György, B., Meijer, E. J., Ivanchenko, M. V., Tenneson, K., Emond, F., Hanlon, K. S., et al. (2019). Gene transfer with AAV9-PHP.B rescues hearing in a mouse model of usher syndrome 3A and transduces hair cells in a non-human primate. *Mol. Ther. Methods Clin. Dev.* 13, 1–13. doi: 10.1016/j.omtm. 2018.11.003

Hallam, D., Hilgen, G., Dorgau, B., Zhu, L., Yu, M., Bojic, S., et al. (2018). Human-induced pluripotent stem cells generate light responsive retinal organoids with variable and nutrient-dependent efficiency. *Stem Cells* 36, 1535–1551. doi: 10.1002/stem.2883

Han, Z., Conley, S. M., Makkia, R. S., Cooper, M. J., and Naash, M. I. (2012a). DNA nanoparticle-mediated ABCA4 delivery rescues Stargardt dystrophy in mice. *J. Clin. Invest.* 122, 3221–3226. doi: 10.1172/jci64833

Han, Z., Conley, S. M., Makkia, R., Guo, J., Cooper, M. J., and Naash, M. I. (2012b). Comparative analysis of DNA nanoparticles and AAVs for ocular gene delivery. *PLoS One* 7:e52189. doi: 10.1371/journal.pone.0052189

Han, S., Liu, X., Xie, S., Gao, M., Liu, F., Yu, S., et al. (2018). Knockout of ush2a gene in zebrafish causes hearing impairment and late onset rod-cone dystrophy. *Hum. Genet.* 137, 779–794. doi: 10.1007/s00439-018-1936-6

Hartong, D. T., Berson, E. L., and Dryja, T. P. (2006). Retinitis pigmentosa. *Lancet* 368, 1795–1809. doi: 10.1016/S0140-6736(06)69740-7

Hashimoto, T., Gibbs, D., Lillo, C., Azarian, S. M., Legacki, E., Zhang, X. M., et al. (2007). Lentiviral gene replacement therapy of retinas in a mouse model for Usher syndrome type 1B. *Gene Ther.* 14, 584–594. doi: 10.1038/sj.gt.3302897

Havens, M. A., Duelli, D. M., and Hastings, M. L. (2013). Targeting RNA splicing for disease therapy. *Wiley Interdiscip. Rev. RNA* 4, 247–266. doi: 10.1002/wrna.1158

Hoffman, D. R., Hughbanks-Wheaton, D. K., Pearson, N. S., Fish, G. E., Spencer, R., Takacs, A., et al. (2014). Four-year placebo-controlled trial of docosahexaenoic acid in X-linked retinitis pigmentosa (DHAX trial): a randomized clinical trial. *JAMA Ophthalmol.* 132, 866–873. doi: 10.1001/jamaophthalmol.2014.1634

Hordeaux, J., Wang, Q., Katz, N., Buza, E. L., Bell, P., and Wilson, J. M. (2018). The neurotropic properties of AAV-PHP.B are limited to C57BL/6J mice. *Mol. Ther.* 26, 664–668. doi: 10.1016/j.ymthe.2018.01.018

Huang, X., and Chau, Y. (2019). Intravitreal nanoparticles for retinal delivery. *Drug Discov. Today* 24, 1510–1523. doi: 10.1016/j.drudis.2019.05.005

Huang, D., Fletcher, S., Wilton, S. D., Palmer, N., McLenachan, S., Mackey, D. A., et al. (2017). Inherited retinal disease therapies targeting precursor messenger ribonucleic acid. *Vision* 1:22. doi: 10.3390/vision1030022

Jaijo, T., Aller, E., Oltra, S., Beneyto, M., Najera, C., Ayuso, C., et al. (2006). Mutation profile of the MYO7A gene in Spanish patients with Usher syndrome type I. *Hum. Mutat.* 27, 290–291. doi: 10.1002/humu.9404

Jinek, M., Chylinski, K., Fonfara, I., Hauer, M., Doudna, J. A., and Charpentier, E. (2012). A programmable dual-RNA-guided DNA endonuclease in adaptive bacterial immunity. *Science* 337, 816–821. doi: 10.1126/science.1225829

Jouret, G., Poirsier, C., Spodenkiewicz, M., Jaquin, C., Gouy, E., Arndt, C., et al. (2019). Genetics of Usher syndrome: new insights from a meta-analysis. *Ontol. Neurotol.* 40, 121–129. doi: 10.1097/mao.0000000000002054

Kay, M. A., Glorioso, J. C., and Naldini, L. (2001). Viral vectors for gene therapy: the art of turning infectious agents into vehicles of therapeutics. *Nat. Med.* 7, 33–40. doi: 10.1038/83324

Keeling, K. M., and Bedwell, D. M. (2011). Suppression of nonsense mutations as a therapeutic approach to treat genetic diseases. *Wiley Interdiscip. Rev. RNA* 2, 837–852. doi: 10.1002/wrna.95

Keeling, K. M., Wang, D., Conard, S. E., and Bedwell, D. M. (2012). Suppression of premature termination codons as a therapeutic approach. *Crit. Rev. Biochem. Mol. Biol.* 47, 444–463. doi: 10.3109/10409238.2012.694846

Kerem, E., Konstan, M. W., De Boeck, K., Accurso, F. J., Sermet-Gaudelus, I., Wilschanski, M., et al. (2014). Ataluren for the treatment of nonsense-mutation cystic fibrosis: a randomised, double-blind, placebo-controlled phase 3 trial. *Lancet Respir. Med.* 2, 539–547. doi: 10.1016/S2213-2600(14)70100-6

Khan, S. H. (2019). Genome-editing technologies: concept, pros and cons of various genome-editing techniques and bioethical concerns for clinical application. *Mol Ther. Nucleic Acids* 16, 326–334. doi: 10.1016/j.omtn.2019.02.027

Kim, Y. G., Cha, J., and Chandrasegaran, S. (1996). Hybrid restriction enzymes: zinc finger fusions to Fok I cleavage domain. *Proc. Natl. Acad. Sci. U S A* 93, 1156–1160. doi: 10.1073/pnas.93.3.1156

Kimberling, W. J., Hildebrand, M. S., Shearer, A. E., Jensen, M. L., Halder, J. A., Trzupek, K., et al. (2010). Frequency of Usher syndrome in two pediatric populations: implications for genetic screening of deaf and hard of hearing children. *Genet Med.* 12, 512–516. doi: 10.1097/gim.0b013e3181e5afb8

Komor, A. C., Kim, Y. B., Packer, M. S., Zuris, J. A., and Liu, D. R. (2016). Programmable editing of a target base in genomic DNA without double-stranded DNA cleavage. *Nature* 533, 420–424. doi: 10.1038/nature17946

Konstan, M. W., Davis, P. B., Wagener, J. S., Hilliard, K. A., Stern, R. C., Milgram, L. J. H., et al. (2004). Compacted DNA nanoparticles administered to the nasal mucosa of cystic fibrosis subjects are safe and demonstrate partial to complete cystic fibrosis transmembrane regulator reconstitution. *Hum. Gene Ther.* 15, 1255–1269. doi: 10.1089/hum.2004.15.1255

Kuscu, C., Arslan, S., Singh, R., Thorpe, J., and Adli, M. (2014). Genome-wide analysis reveals characteristics of off-target sites bound by the Cas9 endonuclease. *Nat. Biotechnol.* 32, 677–683. doi: 10.1038/nbt.2916

Lakowski, J., Gonzalez-Cordero, A., West, E. L., Han, Y.-T., Welby, E., Naeem, A., et al. (2015). Transplantation of photoreceptor precursors isolated *via* a cell surface biomarker panel from embryonic stem cell-derived self-forming retina. *Stem Cells* 33, 2469–2482. doi: 10.1002/stem.2051

Lakowski, J., Welby, E., Budinger, D., Di Marco, F., Di Foggia, V., Bainbridge, J. W. B., et al. (2018). Isolation of human photoreceptor precursors *via* a cell surface marker panel from stem cell-derived retinal organoids and fetal retinae. *Stem Cells* 36, 709–722. doi: 10.1002/stem.2775

Landegger, L. D., Pan, B., Askew, C., Wassmer, S. J., Gluck, S. D., Galvin, A., et al. (2017). A synthetic AAV vector enables safe and efficient gene transfer to the mammalian inner ear. *Nat. Biotechnol.* 35, 280–284. doi: 10.1038/nbt.3781

Lentz, J. J., Jodelka, F. M., Hinrich, A. J., McCaffrey, K. E., Farris, H. E., Spalitta, M. J., et al. (2013). Rescue of hearing and vestibular function by antisense oligonucleotides in a mouse model of human deafness. *Nat. Med.* 19, 345–350. doi: 10.1038/nm.3106

Li, T., Adamian, M., Roof, D. J., Berson, E. L., Dryja, T. P., Roessler, B. J., et al. (1994). *In vivo* transfer of a reporter gene to the retina mediated by an adenoviral vector. *Invest. Ophthalmol. Vis. Sci.* 35, 2543–2549.

Lipinski, D. M., Barnard, A. R., Singh, M. S., Martin, C., Lee, E. J., Davies, W. I. L., et al. (2015). CNTF gene therapy confers lifelong neuroprotection in a mouse model of human retinitis pigmentosa. *Mol. Ther.* 23, 1308–1319. doi: 10.1038/mt.2015.68

Liu, X. Z., Hope, C., Liang, C. Y., Zou, J. M., Xu, L. R., Cole, T., et al. (1999). A mutation (2314delG) in the Usher syndrome type IIA gene: high prevalence and phenotypic variation. *Am. J. Hum. Genet.* 64, 1221–1225. doi: 10.1086/302332

Liu, X., Vansant, G., Udovichenko, I. P., Wolfrum, U., and Williams, D. S. (1997). Myosin VIIa, the product of the Usher 1B syndrome gene, is concentrated in the connecting cilia of photoreceptor cells. *Cell Motil. Cytoskeleton* 37, 240–252. doi: 10.1002/(sici)1097-0169(1997)37:3<240::aid-cm6>3.0.co;2-a

Lopes, V. S., Boye, S. E., Louie, C. M., Boye, S., Dyka, F., Chiodo, V., et al. (2013). Retinal gene therapy with a large MYO7A cDNA using adeno-associated virus. *Gene Ther.* 20, 824–833. doi: 10.1038/gt.2013.3

Lowe, A., Harris, R., Bhansali, P., Cvekl, A., and Liu, W. (2016). Intercellular adhesion-dependent cell survival and ROCK-regulated actomyosin-driven forces mediate self-formation of a retinal organoid. *Stem Cell Reports* 6, 743–756. doi: 10.1016/j.stemcr.2016.03.011

Lueck, J. D., Yoon, J. S., Perales-Puchalt, A., Mackey, A. L., Infield, D. T., Behlke, M. A., et al. (2019). Engineered transfer RNAs for suppression of premature termination codons. *Nat. Commun.* 10:822. doi: 10.1038/s41467-019-08329-4

MacLaren, R. E., Pearson, R. A., MacNeil, A., Douglas, R. H., Salt, T. E., Akimoto, M., et al. (2006). Retinal repair by transplantation of photoreceptor precursors. *Nature* 444, 203–207. doi: 10.1038/nature05161

Maddalena, A., Tornabene, P., Tiberi, P., Minopoli, R., Manfredi, A., Mutarelli, M., et al. (2018). Triple vectors expand AAV transfer capacity in the retina. *Mol. Ther.* 26, 524–541. doi: 10.1016/j.ymthe.2017.11.019

Maeda, Y., Fukushima, K., Kawasaki, A., Nishizaki, K., and Smith, R. J. (2007). Cochlear expression of a dominant-negative GJB2 R75W construct delivered through the round window membrane in mice. *Neurosci. Res.* 58, 250–254. doi: 10.1016/j.neures.2007.03.006

Maeda, Y., Fukushima, K., Nishizaki, K., and Smith, R. J. H. (2005). *In vitro* and *in vivo* suppression of GJB2 expression by RNA interference. *Hum. Mol. Genet.* 14, 1641–1650. doi: 10.1093/hmg/ddi172

Maeder, M. L., Stefanidakis, M., Wilson, C. J., Baral, R., Barrera, L. A., Bounoutas, G. S., et al. (2019). Development of a gene-editing approach to restore vision loss in Leber congenital amaurosis type 10. *Nat. Med.* 25, 229–233. doi: 10.1038/s41591-018-0327-9

Mali, P., Yang, L., Esvelt, K. M., Aach, J., Guell, M., Dicarlo, J. E., et al. (2013). RNA-guided human genome engineering *via* Cas9. *Science* 339, 823–826. doi: 10.1126/science.1232033

Mandai, M., Fujii, M., Hashiguchi, T., Sunagawa, G. A., Ito, S.-I., Sun, J., et al. (2017). iPSC-derived retina transplants improve vision in rd1 end-stage retinal-degeneration mice. *Stem Cell Reports* 8, 69–83. doi: 10.1016/j.stemcr.2016.12.008

Mathur, P., and Yang, J. (2015). Usher syndrome: hearing loss, retinal degeneration and associated abnormalities. *Biochim. Biophys. Acta* 1852, 406–420. doi: 10.1016/j.bbadis.2014.11.020

Mathur, P. D., and Yang, J. (2019). Usher syndrome and non-syndromic deafness: functions of different whirlin isoforms in the cochlea, vestibular organs and retina. *Hear. Res.* 375, 14–24. doi: 10.1016/j.heares.2019.02.007

Mazarakis, N. D., Azzouz, M., Rohll, J. B., Ellard, F. M., Wilkes, F. J., Olsen, A. L., et al. (2001). Rabies virus glycoprotein pseudotyping of lentiviral vectors enables retrograde axonal transport and access to the nervous system after peripheral delivery. *Hum. Mol. Genet.* 10, 2109–2121. doi: 10.1093/hmg/10.19.2109

McClements, M. E., Barnard, A. R., Singh, M. S., Issa, P. C., Jiang, Z., Radu, R. A., et al. (2019). An AAV dual vector strategy ameliorates the stargardt phenotype in adult *Abca4-/-* mice. *Hum. Gene Ther.* 30, 590–600. doi: 10.1089/hum.2018.156

McDonald, C. M., Campbell, C., Torricelli, R. E., Finkel, R. S., Flanigan, K. M., Goemans, N., et al. (2017). Ataluren in patients with nonsense mutation Duchenne muscular dystrophy (ACT DMD): a multicentre, randomised, double-blind, placebo-controlled, phase 3 trial. *Lancet* 390, 1489–1498. doi: 10.1016/S0140-6736(17)31611-2

Mellough, C. B., Collin, J., Khazim, M., White, K., Sernagor, E., Steel, D. H. W., et al. (2015). IGF-1 signaling plays an important role in the formation of three-dimensional laminated neural retina and other ocular structures from human embryonic stem cells. *Stem Cells* 33, 2416–2430. doi: 10.1002/stem.2023

Mellough, C. B., Collin, J., Queen, R., Hilgen, G., Dorgau, B., Zerti, D., et al. (2019). Systematic comparison of retinal organoid differentiation from human pluripotent stem cells reveals stage specific, cell line and methodological differences. *Stem Cells Transl. Med.* 8, 694–706. doi: 10.1002/sctm.18-0267

Mingeot-Leclercq, M.-P., and Tulkens, P. M. (1999). Aminoglycosides: nephrotoxicity. *Antimicrob. Agents Chemother.* 43, 1003–1012. doi: 10.1128/AAC.43.5.1003

Mitrophanous, K. A., Yoon, S., Rohll, J. B., Patil, D., Wilkes, F. J., Kim, V. N., et al. (1999). Stable gene transfer to the nervous system using a non-primate lentiviral vector. *Gene Ther.* 6, 1808–1818. doi: 10.1038/sj.gt.3301023

Moore, N. A., Morral, N., Ciulla, T. A., and Bracha, P. (2018). Gene therapy for inherited retinal and optic nerve degenerations. *Expert Opin. Biol. Ther.* 18, 37–49. doi: 10.1080/14712598.2018.1389886

Moreno, A. M., Fu, X., Zhu, J., Katrekar, D., Shih, Y. V., Marlett, J., et al. (2018). in situ gene therapy *via* AAV-CRISPR-Cas9-mediated targeted gene regulation. *Mol. Ther.* 26, 1818–1827. doi: 10.1016/j.ymthe.2018.04.017

Naessens, S., Ruysschaert, L., Lefever, S., Coppieters, F., and De Baere, E. (2019). Antisense oligonucleotide-based downregulation of the g56r pathogenic variant causing NR2E3-associated autosomal dominant retinitis pigmentosa. *Genes* 10:363. doi: 10.3390/genes10050363

Nagel-Wolfrum, K., Moller, F., Penner, I., Baasov, T., and Wolfrum, U. (2016). Targeting nonsense mutations in diseases with translational read-through-inducing drugs (TRIDs). *BioDrugs* 30, 49–74.doi: 10.1007/s40259-016-0157-6

Nakamura, M., Srinivasan, P., Chavez, M., Carter, M. A., Dominguez, A. A., La Russa, M., et al. (2019). Anti-CRISPR-mediated control of gene editing and synthetic circuits in eukaryotic cells. *Nat. Commun.* 10:194. doi: 10.1038/s41467-018-08158-x

Nakano, T., Ando, S., Takata, N., Kawada, M., Muguruma, K., Sekiguchi, K., et al. (2012). Self-formation of optic cups and storable stratified neural retina from human ESCs. *Cell Stem Cell* 10, 771–785. doi: 10.1016/j.stem.2012.05.009

Nayerossadat, N., Maedeh, T., and Ali, P. A. (2012). Viral and nonviral delivery systems for gene delivery. *Adv. Biomed. Res.* 1:27. doi: 10.4103/2277-9175.98152

Neuhaus, C., Eisenberger, T., Decker, C., Nagl, S., Blank, C., Pfister, M., et al. (2017). Next-generation sequencing reveals the mutational landscape of clinically diagnosed Usher syndrome: copy number variations, phenocopies, a predominant target for translational read-through and PEX26 mutated in Heimler syndrome. *Mol. Genet. Genomic Med.* 5, 531–552. doi: 10.1002/mgg3.312

Nudelman, I., Rebibo-Sabbah, A., Cherniavsky, M., Belakhov, V., Hainrichson, M., Chen, F., et al. (2009). Development of novel aminoglycoside (NB54) with reduced toxicity and enhanced suppression of disease-causing premature stop mutations. *J. Med. Chem.* 52, 2836–2845. doi: 10.1021/jm801640k

Nudelman, I., Rebibo-Sabbah, A., Shallom-Shezifi, D., Hainrichson, M., Stahl, I., Ben-Yosef, T., et al. (2006). Redesign of aminoglycosides for treatment of human genetic diseases caused by premature stop mutations. *Bioorg. Med. Chem. Lett.* 16, 6310–6315. doi: 10.1016/j.bmcl.2006.09.013

Omole, A. E., and Fakoya, A. O. J. (2018). Ten years of progress and promise of induced pluripotent stem cells: historical origins, characteristics, mechanisms, limitations and potential applications. *PeerJ* 6:e4370. doi: 10.7717/peerj.4370

Overlack, N., Goldmann, T., Wolfrum, U., and Nagel-Wolfrum, K. (2012). Gene repair of an Usher syndrome causing mutation by zinc-finger nuclease mediated homologous recombination. *Invest. Ophthalmol. Vis. Sci.* 53, 4140–4146. doi: 10.1167/iovs.12-9812

Pan, B., Askew, C., Galvin, A., Heman-Ackah, S., Asai, Y., Indzhykulian, A. A., et al. (2017). Gene therapy restores auditory and vestibular function in a mouse model of Usher syndrome type 1c. *Nat. Biotechnol.* 35, 264–272. doi: 10.1038/nbt.3801

Parfitt, D. A., Lane, A., Ramsden, C., Jovanovic, K., Coffey, P. J., Hardcastle, A. J., et al. (2016). Using induced pluripotent stem cells to understand retinal ciliopathy disease mechanisms and develop therapies. *Biochem. Soc. Trans.* 44, 1245–1251. doi: 10.1042/bst20160156

Pawluk, A., Amrani, N., Zhang, Y., Garcia, B., Hidalgo-Reyes, Y., Lee, J., et al. (2016). Naturally occurring off-switches for CRISPR-Cas9. *Cell* 167, 1829.e9–1838.e9. doi: 10.1016/j.cell.2016.11.017

Pearson, R. A., Barber, A. C., Rizzi, M., Hippert, C., Xue, T., West, E. L., et al. (2012). Restoration of vision after transplantation of photoreceptors. *Nature* 485, 99–103. doi: 10.1038/nature10997

Pepermans, E., Michel, V., Goodyear, R., Bonnet, C., Abdi, S., Dupont, T., et al. (2014). The CD2 isoform of protocadherin-15 is an essential component of the tip-link complex in mature auditory hair cells. *EMBO Mol. Med.* 6, 984–992. doi: 10.15252/emmm.201403976

Phillips, M. J., Wallace, K. A., Dickerson, S. J., Miller, M. J., Verhoeven, A. D., Martin, J. M., et al. (2012). Blood-derived human iPS cells generate optic vesicle-like structures with the capacity to form retinal laminae and develop synapses. *Invest. Ophthalmol. Vis. Sci.* 53, 2007–2019. doi: 10.1167/iovs.11-9313

Ponnath, A., Depreux, F. F., Jodelka, F. M., Rigo, F., Farris, H. E., Hastings, M. L., et al. (2018). Rescue of outer hair cells with antisense oligonucleotides in usher mice is dependent on age of treatment. *J. Assoc. Res. Otolaryngol.* 19, 1–16. doi: 10.1007/s10162-017-0640-x

Rauch, B. J., Silvis, M. R., Hultquist, J. F., Waters, C. S., Mcgregor, M. J., Krogan, N. J., et al. (2017). Inhibition of CRISPR-Cas9 with bacteriophage proteins. *Cell* 168, 150.e10–158.e10. doi: 10.1016/j.cell.2016.12.009

Rayapudi, S., Schwartz, S. G., Wang, X., and Chavis, P. (2013). Vitamin A and fish oils for retinitis pigmentosa. *Cochrane Database Syst. Rev.* 12:CD008428. doi: 10.1002/14651858.cd008428

Rebibo-Sabbah, A., Nudelman, I., Ahmed, Z. M., Baasov, T., and Ben-Yosef, T. (2007). *In vitro* and *ex vivo* suppression by aminoglycosides of PCDH15 nonsense mutations underlying type 1 Usher syndrome. *Hum. Genet.* 122, 373–381. doi: 10.1007/s00439-007-0410-7

Rees, H. A., and Liu, D. R. (2018). Base editing: precision chemistry on the genome and transcriptome of living cells. *Nat. Rev. Genet.* 19, 770–788. doi: 10.1038/s41576-018-0059-1

Reiners, J., Nagel-Wolfrum, K., Jurgens, K., Marker, T., and Wolfrum, U. (2006). Molecular basis of human Usher syndrome: deciphering the meshes of the Usher protein network provides insights into the pathomechanisms of the Usher disease. *Exp. Eye Res.* 83, 97–119. doi: 10.1016/j.exer.2005.11.010

Rodrigues, G. A., Shalaev, E., Karami, T. K., Cunningham, J., Slater, N. K. H., and Rivers, H. M. (2018). Pharmaceutical development of AAV-based gene therapy products for the eye. *Pharm. Res.* 36:29. doi: 10.1007/s11095-018-2554-7

Russell, S., Bennett, J., Wellman, J. A., Chung, D. C., Yu, Z.-F., Tillman, A., et al. (2017). Efficacy and safety of voretigene neparvovec (AAV2-hRPE65v2) in patients with RPE65-mediated inherited retinal dystrophy: a randomised, controlled, open-label, phase 3 trial. *Lancet* 390, 849–860. doi: 10.1016/S0140-6736(17)31868-8

Ryu, S.-M., Koo, T., Kim, K., Lim, K., Baek, G., Kim, S.-T., et al. (2018). Adenine base editing in mouse embryos and an adult mouse model of Duchenne muscular dystrophy. *Nat. Biotechnol.* 36, 536–539. doi: 10.1038/nbt.4148

Sahly, I., Dufour, E., Schietroma, C., Michel, V., Bahloul, A., Perfettini, I., et al. (2012). Localization of Usher 1 proteins to the photoreceptor calyceal processes, which are absent from mice. *J. Cell Biol.* 199, 381–399. doi: 10.1083/jcb.201202012

Santos-Ferreira, T., Llonch, S., Borsch, O., Postel, K., Haas, J., and Ader, M. (2016). Retinal transplantation of photoreceptors results in donor-host cytoplasmic exchange. *Nat. Commun.* 7:13028. doi: 10.1038/ncomms13028

Schietroma, C., Parain, K., Estivalet, A., Aghaie, A., Boutet De Monvel, J., Picaud, S., et al. (2017). Usher syndrome type 1-associated cadherins shape the photoreceptor outer segment. *J. Cell Biol.* 216, 1849–1864. doi: 10.1083/jcb.201612030

Selimoglu, E. (2007). Aminoglycoside-induced ototoxicity. *Curr. Pharm. Des.* 13, 119–126. doi: 10.2174/138161207779313731

Sengillo, J. D., Justus, S., Tsai, Y. T., Cabral, T., and Tsang, S. H. (2016). Gene and cell-based therapies for inherited retinal disorders: an update. *Am. J. Med. Genet. C Semin. Med. Genet.* 172, 349–366. doi: 10.1002/ajmg.c.31534

Seyedahmadi, B. J., Rivolta, C., Keene, J. A., Berson, E. L., and Dryja, T. P. (2004). Comprehensive screening of the USH2A gene in Usher syndrome type II and non-syndromic recessive retinitis pigmentosa. *Exp. Eye Res.* 79, 167–173. doi: 10.1016/j.exer.2004.03.005

Shibata, S. B., Ranum, P. T., Moteki, H., Pan, B., Goodwin, A. T., Goodman, S. S., et al. (2016). RNA interference prevents autosomal-dominant hearing loss. *Am. J. Hum. Genet.* 98, 1101–1113. doi: 10.1016/j.ajhg.2016.03.028

Shirai, H., Mandai, M., Matsushita, K., Kuwahara, A., Yonemura, S., Nakano, T., et al. (2016). Transplantation of human embryonic stem cell-derived retinal tissue in two primate models of retinal degeneration. *Proc. Natl. Acad. Sci. U S A* 113, E81–E90. doi: 10.1073/pnas.1512590113

Singh, M. S., Balmer, J., Barnard, A. R., Aslam, S. A., Moralli, D., Green, C. M., et al. (2016). Transplanted photoreceptor precursors transfer proteins to host photoreceptors by a mechanism of cytoplasmic fusion. *Nat. Commun.* 7:13537. doi: 10.1038/ncomms13537

Singh, M. S., Charbel Issa, P., Butler, R., Martin, C., Lipinski, D. M., Sekaran, S., et al. (2013). Reversal of end-stage retinal degeneration and restoration of visual function by photoreceptor transplantation. *Proc. Natl. Acad. Sci. U S A* 110, 1101–1106. doi: 10.1073/pnas.1119416110

Slijkerman, R. W., Vaché, C., Dona, M., García-García, G., Claustres, M., Hetterschijt, L., et al. (2016). Antisense oligonucleotide-based splice correction for USH2A-associated retinal degeneration caused by a frequent deep-intronic mutation. *Mol. Ther. Nucleic Acids* 5:e381. doi: 10.1038/mtna.2016.89

Song, C.-Q., Jiang, T., Richter, M., Rhym, L. H., Koblan, L. W., Zafra, M. P., et al. (2020). Adenine base editing in an adult mouse model of tyrosinaemia. *Nat. Biomed. Eng.* 4, 125–130. doi: 10.1038/s41551-019-0357-8

Sperling, M., and Gradzielski, M. (2017). Droplets, evaporation and a superhydrophobic surface: simple tools for guiding colloidal particles into complex materials. *Gels* 3:15. doi: 10.3390/gels3020015

Suzuki, J., Corfas, G., and Liberman, M. (2016). Round-window delivery of neurotrophin 3 regenerates cochlear synapses after acoustic overexposure. *Sci. Rep.* 6:24907. doi: 10.1038/srep24907

Takahashi, K., Tanabe, K., Ohnuki, M., Narita, M., Ichisaka, T., Tomoda, K., et al. (2007). Induction of pluripotent stem cells from adult human fibroblasts by defined factors. *Cell* 131, 861–872. doi: 10.1016/j.cell.2007.11.019

Talcott, K. E., Ratnam, K., Sundquist, S. M., Lucero, A. S., Lujan, B. J., Tao, W., et al. (2011). Longitudinal study of cone photoreceptors during retinal degeneration and in response to ciliary neurotrophic factor treatment. *Invest. Ophthalmol. Vis. Sci.* 52, 2219–2226. doi: 10.1167/iovs.10-6479

Tang, Z.-H., Chen, J.-R., Zheng, J., Shi, H.-S., Ding, J., Qian, X.-D., et al. (2016). Genetic correction of induced pluripotent stem cells from a deaf patient with MYO7A mutation results in morphologic and functional recovery of the derived hair cell-like cells. *Stem Cells Transl. Med.* 5, 561–571. doi: 10.5966/sctm.2015-0252

Tornabene, P., Trapani, I., Minopoli, R., Centrulo, M., Lupo, M., De Simone, S., et al. (2019). Intein-mediated protein trans-splicing expands adeno-associated virus transfer capacity in the retina. *Sci. Transl. Med.* 11:eaav4523. doi: 10.1126/scitranslmed.aav4523

Trapani, I., Colella, P., Sommella, A., Iodice, C., Cesi, G., De Simone, S., et al. (2014). Effective delivery of large genes to the retina by dual AAV vectors. *EMBO Mol. Med.* 6, 194–211. doi: 10.1002/emmm.201302948

Trouillet, A., Dubus, E., Degardin, J., Estivalet, A., Ivkovic, I., Godefroy, D., et al. (2018). Cone degeneration is triggered by the absence of USH1 proteins but prevented by antioxidant treatments. *Sci. Rep.* 8:1968. doi: 10.1038/s41598-018-20171-0

Tu, H.-Y., Watanabe, T., Shirai, H., Yamasaki, S., Kinoshita, M., Matsushita, K., et al. (2019). Medium- to long-term survival and functional examination of human iPSC-derived retinas in rat and primate models of retinal degeneration. *EBioMedicine* 39, 562–574. doi: 10.1016/j.ebiom.2018.11.028

Van Diepen, H., Dulla, K., Chan, H., Schulkens, I., Beumer, W., Vorthoren, L., et al. (2019). QR-421a, an antisense oligonucleotide, for the treatment of retinitis pigmentosa due to USH2A exon 13 mutations. *Invest. Ophthalmol. Vis. Sci.* 60:3250.

Vandendriessche, T., Thorrez, L., Acosta-Sanchez, A., Petrus, I., Wang, L., Ma, L., et al. (2007). Efficacy and safety of adeno-associated viral vectors based on serotype 8 and 9 vs. lentiviral vectors for hemophilia B gene therapy. *J. Thromb. Haemost.* 5, 16–24. doi: 10.1111/j.1538-7836.2006.02220.x

Vijayakumar, S., Depreux, F. F., Jodelka, F. M., Lentz, J. J., Rigo, F., Jones, T. A., et al. (2017). Rescue of peripheral vestibular function in Usher syndrome mice using a splice-switching antisense oligonucleotide. *Hum. Mol. Genet.* 26, 3482–3494. doi: 10.1093/hmg/ddx234

Villiger, L., Grisch-Chan, H. M., Lindsay, H., Ringnalda, F., Pogliano, C. B., Allegri, G., et al. (2018). Treatment of a metabolic liver disease by *in vivo* genome base editing in adult mice. *Nat. Med.* 24, 1519–1525. doi: 10.1038/s41591-018-0209-1

Wang, X., and Gregory-Evans, C. Y. (2015). Nonsense suppression therapies in ocular genetic diseases. *Cell. Mol. Life Sci.* 72, 1931–1938. doi: 10.1007/s00018-015-1843-0

Wang, Y., Wise, A. K., Tan, J., Maina, J. W., Shepherd, R. K., and Caruso, F. (2014). Mesoporous silica supraparticles for sustained inner-ear drug delivery. *Small* 10, 4244–4248. doi: 10.1002/smll.201401767

Watanabe, T., and Fukasawa, T. (1961). Episome-mediated transfer of drug resistance in Enterobacteriaceae. I. Transfer of resistance factors by conjugation. *J. Bacteriol.* 81, 669–678.

Weiss, J. N., and Levy, S. (2019). Stem Cell Ophthalmology Treatment Study (SCOTS): bone marrow derived stem cells in the treatment of Usher syndrome. *Stem Cell Investig.* 6:31. doi: 10.21037/sci.2019.08.07

Welch, E. M., Barton, E. R., Zhuo, J., Tomizawa, Y., Friesen, W. J., Trifillis, P., et al. (2007). PTC124 targets genetic disorders caused by nonsense mutations. *Nature* 447, 87–91. doi: 10.1038/nature05756

Wilschanski, M., Miller, L. L., Shoseyov, D., Blau, H., Rivlin, J., Aviram, M., et al. (2011). Chronic ataluren (PTC124) treatment of nonsense mutation cystic fibrosis. *Eur. Respir. J.* 38, 59–69. doi: 10.1183/09031936.00120910

Wold, W. S. M., and Toth, K. (2013). Adenovirus vectors for gene therapy, vaccination and cancer gene therapy. *Curr. Gene Ther.* 13, 421–433. doi: 10.2174/1566523213666131125095046

Yan, Z., Zhang, Y., and Duan, D. (2000). Trans-splicing vectors expand the utility of adeno-associated virus for gene therapy. *Proc. Natl. Acad. Sci. U S A* 97, 6716–6721. doi: 10.1073/pnas.97.12.6716

Yeh, W.-H., Chiang, H., Rees, H. A., Edge, A. S. B., and Liu, D. R. (2018). *In vivo* base editing of post-mitotic sensory cells. *Nat. Commun.* 9:2184. doi: 10.1038/s41467-018-04580-3

Yin, H., Kanasty, R. L., Eltoukhy, A. A., Vegas, A. J., Dorkin, J. R., and Anderson, D. G. (2014). Non-viral vectors for gene-based therapy. *Nat. Rev. Genet.* 15, 541–555. doi: 10.1038/nrg3763

Ylikoski, J., Pirvola, U., Moshnyakov, M., Palgi, J., Arumäe, U., and Saarma, M. (1993). Expression patterns of neurotrophin and their receptor mRNAs in the rat inner ear. *Hear. Res.* 65, 69–78. doi: 10.1016/0378-5955(93)90202-c

Zakrzewski, W., Dobrzyński, M., Szymonowicz, M., and Rybak, Z. (2019). Stem cells: past, present, and future. *Stem Cell Res. Ther.* 10:68. doi: 10.1186/s13287-019-1165-5

Zallocchi, M., Binley, K., Lad, Y., Ellis, S., Widdowson, P., Iqball, S., et al. (2014). EIAV-based retinal gene therapy in the shaker1 mouse model for usher syndrome type 1B: development of UshStat. *PLoS One* 9:e94272. doi: 10.1371/journal.pone.0094272

Zhang, X., Chen, G., Wen, L., Yang, F., Shao, A.-L., Li, X., et al. (2013). Novel multiple agents loaded PLGA nanoparticles for brain delivery *via* inner ear administration: *in vitro* and *in vivo* evaluation. *Eur. J. Pharm. Sci.* 48, 595–603. doi: 10.1016/j.ejps.2013.01.007

Zheng, C. X., Wang, S. M., Bai, Y. H., Luo, T. T., Wang, J. Q., Dai, C. Q., et al. (2018). Lentiviral vectors and adeno-associatedvirus vectors: useful tools for gene transfer in pain research. *Anat. Rec.* 301, 825–836. doi: 10.1002/ar.23723

Zheng, Q. Y., Yan, D., Ouyang, X. M., Du, L. L., Yu, H., Chang, B., et al. (2005). Digenic inheritance of deafness caused by mutations in genes encoding cadherin 23 and protocadherin 15 in mice and humans. *Hum. Mol. Genet.* 14, 103–111. doi: 10.1093/hmg/ddi010

Zhong, X., Gutierrez, C., Xue, T., Hampton, C., Vergara, M. N., Cao, L. H., et al. (2014). Generation of three-dimensional retinal tissue with functional photoreceptors from human iPSCs. *Nat. Commun.* 5:4047. doi: 10.1038/ncomms5047

Zou, J., Feng, H., Sood, R., Kinnunen, P. K. J., and Pyykko, I. (2017). Biocompatibility of liposome nanocarriers in the rat inner ear after intratympanic administration. *Nanoscale Res. Lett.* 12:372. doi: 10.1186/s11671-017-2142-5

Zou, J., Luo, L., Shen, Z., Chiodo, V. A., Ambati, B. K., Hauswirth, W. W., et al. (2011). Whirlin replacement restores the formation of the USH2 protein complex in whirlin knockout photoreceptors. *Invest. Ophthalmol. Vis. Sci.* 52, 2343–2351. doi: 10.1167/iovs.10-6141

14

Comparison of CRISPR/Cas Endonucleases for *in vivo* Retinal Gene Editing

Fan Li[1,2], Kristof Wing[1], Jiang-Hui Wang[3], Chi D. Luu[3,4], James A. Bender[5], Jinying Chen[1,6], Qi Wang[1], Qinyi Lu[1], Minh Thuan Nguyen Tran[1], Kaylene M. Young[1], Raymond C. B. Wong[3,4], Alice Pébay[7,8], Anthony L. Cook[5], Sandy S. C. Hung[3,4], Guei-Sheung Liu[1,4†] and Alex W. Hewitt[1,3*†]*

[1] Menzies Institute for Medical Research, University of Tasmania, Hobart, TAS, Australia, [2] State Key Laboratory of Ophthalmology, Zhongshan Ophthalmic Centre, Sun Yat-sen University, Guangzhou, China, [3] Centre for Eye Research Australia, Royal Victorian Eye and Ear Hospital, East Melbourne, VIC, Australia, [4] Ophthalmology, Department of Surgery, The University of Melbourne, Parkville, VIC, Australia, [5] Wicking Dementia Research and Education Centre, University of Tasmania, Hobart, TAS, Australia, [6] Department of Ophthalmology, The First Affiliated Hospital of Jinan University, Guangzhou, China, [7] Department of Surgery, Royal Melbourne Hospital, The University of Melbourne, Parkville, VIC, Australia, [8] Department of Anatomy and Neuroscience, The University of Melbourne, Parkville, VIC, Australia

***Correspondence:**
Guei-Sheung Liu
rickliu0817@gmail.com
Alex W. Hewitt
hewitt.alex@gmail.com

† These authors have contributed equally to this work

CRISPR/Cas has opened the prospect of direct gene correction therapy for some inherited retinal diseases. Previous work has demonstrated the utility of adeno-associated virus (AAV) mediated delivery to retinal cells *in vivo*; however, with the expanding repertoire of CRISPR/Cas endonucleases, it is not clear which of these are most efficacious for retinal editing *in vivo*. We sought to compare CRISPR/Cas endonuclease activity using both single and dual AAV delivery strategies for gene editing in retinal cells. Plasmids of a dual vector system with SpCas9, SaCas9, Cas12a, CjCas9 and a sgRNA targeting *YFP*, as well as a single vector system with SaCas9/YFP sgRNA were generated and validated in YFP-expressing HEK293A cell by flow cytometry and the T7E1 assay. Paired CRISPR/Cas endonuclease and its best performing sgRNA was then packaged into an AAV2 capsid derivative, AAV7m8, and injected intravitreally into CMV-Cre:Rosa26-YFP mice. SpCas9 and Cas12a achieved better knockout efficiency than SaCas9 and CjCas9. Moreover, no significant difference in *YFP* gene editing was found between single and dual CRISPR/SaCas9 vector systems. With a marked reduction of YFP-positive retinal cells, AAV7m8 delivered SpCas9 was found to have the highest knockout efficacy among all investigated endonucleases. We demonstrate that the AAV7m8-mediated delivery of CRISPR/SpCas9 construct achieves the most efficient gene modification in neurosensory retinal cells *in vivo*.

Keywords: CRISPR (clustered regularly interspaced short palindromic repeats), retina, retinal dystrophy, gene editing, AAV (adeno-associated virus)

INTRODUCTION

Being discovered as a critical component of some bacterial and archaea, acting to counter viral intrusion (Jinek et al., 2012), the Clustered Regularly Interspaced Short Palindromic Repeats (CRISPR)/CRISPR-associated protein (Cas) system has been successfully repurposed for efficient genome editing in mammalian cells (Cong et al., 2013; Mali et al., 2013). This has opened the door to direct gene correction therapy for many inherited retinal diseases. Nevertheless, one of the greatest challenges is the efficient delivery of the CRISPR/Cas genome-editing system to the target tissues or cells in living organisms. Due to the large size of the commonly used SpCas9 (*Streptococcus pyogenes*, ~4.2 kb) and the loading capacity of some currently available viral vectors for ocular gene therapy such as adeno-associated virus (AAV), recent studies have demonstrated that a dual AAV2 system can be used to deliver CRISPR/Cas9 to effectively perform DNA editing in retinal cells in adult mice (Bakondi et al., 2016; Hung et al., 2016; Latella et al., 2016; Ruan et al., 2017; Yu et al., 2017; Li et al., 2019). Despite the success of this dual-vector strategy, it is challenging to transduce two AAVs into one cell and clearly activity of the CRISPR/Cas machinery requires the receipt of both the endonuclease and sgRNA expression cassettes.

With the expanding repertoire of CRISPR/Cas endonucleases, various CRISPR/Cas systems have been developed that utilize smaller Cas endonuclease from different bacterial species, such as Cas12a (*Acidaminococcus*,~3.9 kb or *Lachnospiraceae*, ~3.7 kb), SaCas9 (*Staphylococcus aureus*, 3.2 kb), CjCas9 (*Campylobacter jejuni*, 2.9 kb), NmCas9 (*Neisseria meningitidis*, ~3.2 kb), making it possible to use a single vector to package both the Cas endonuclease and its sgRNA. A handful of studies have reported the successful *in vivo* genome editing of SaCas9 (Maeder et al., 2019), CjCas9 (Kim et al., 2017; Jo et al., 2019), Cas12a (Koo et al., 2018), and NmeCas9 (Xia et al., 2018) in retinal cells. These various CRISPR/Cas systems differ in their editing efficacy, packageability and protospacer-adjacent motif (PAM) requirement (listed in **Supplementary Table S2**), which largely expands the *in vivo* application of CRISPR/Cas based genome editing in various tissues or cells. There have been a small number of studies, which have applied all-in-one AAV vector-mediated CRISPR/Cas genome editing in different cells including retinal pigment epithelium cells. Eunji and colleagues reported the successful disruption of the *Vegfa* or *Hif1a* genes in mouse RPE cells using single AAV-CjCas9 (Kim et al., 2017). Other groups have utilized a single AAV vector to deliver SaCas9, or NmeCas9 to a variety of somatic tissue in mice (Ran et al., 2015; Ibraheim et al., 2018; Jarrett et al., 2018; Pan et al., 2018; Xu et al., 2019). Despite the encouraging *in vivo* application of these CRISPR/Cas systems, delivered via dual or all-in-one vectors, it is not clear which are the most efficacious for retinal editing *in vivo*.

The aim of this study was to directly compare the CRISPR/Cas endonuclease activity of single/dual AAV strategies for retinal gene editing in the transgenic mice expressing a yellow fluorescent protein (YFP) reporter. To achieve this, we designed YFP-targeting sgRNAs for each Cas endonuclease and quantified the editing efficiency, indicated by the disruption of YFP *in vitro* and *in vivo*.

MATERIALS AND METHODS

Ethics Approval and Animal Maintenance

All experimental studies were performed in accordance with the Association for Research in Vision and Ophthalmology Statement for the Use of Animals in Ophthalmic and Vision Research and the requirements of the National Health and Medical Research Council of Australia (Australian Code of Practice for the Care and Use of Animals for Scientific Purposes). This study was approved by the Animal Ethics Committees of the University of Tasmania (Reference Number A0014827). CMV-Cre and Rosa26-YFP transgenic mouse lines were maintained on a C57BL/6 background and intercrossed to generate experimental offspring that were heterozygous for each transgene. Adult (8–12 weeks old) CMV-Cre:Rosa26-YFP transgenic mice (YFP mice), which express YFP throughout the retina, were maintained and bred at the University of Tasmania (Hobart, TAS, Australia). Animals were group housed with same-sex littermates in Optimice micro-isolator cages (Animal Care Systems, Centennial, CO, United States) with uninhibited access to food and water. They were maintained on a 12 h light (50 lux illumination) and 12 h dark (<10 lux illumination) cycle, at 20°C to minimize possible light-induced damage to the eye.

Design and Construction of Cas Endonucleases and sgRNAs Vectors

Single guide RNAs targeting the same 5′ region of the *YFP* gene were designed using a CRISPR design tool[1] with different relevant PAM sites (**Figure 1A**). Briefly, three sgRNAs for SpCas9 (referred as SpCas9-YFP sgRNA1, 2, and 3), two sgRNAs with different lengths for Cas12a (referred as Cas12a-YFP sgRNA 20 and 23 nt), two sgRNAs for CjCas9 (referred CjCas9-YFP sgRNA1 and 2) and one sgRNA for SaCas9 (referred as SaCas9-YFP sgRNA, as only one possible PAM site was found in that region) were designed. These sgRNAs were then cloned into the AAV-U6-sgRNA-hSyn-mCherry vector (Addgene #87916). A control sgRNA, targeting the *LacZ* gene (5′-TGCGAATACGCCCACGCGAT-3′), was designed based on a previous study by Swiech et al. (2015) and LacZ sgRNA plasmids were generated and used for *in vitro* validation.

Cas endonuclease plasmids were generated following different cloning approaches. The AAV-miniCMV-SpCas9 (SpCas9) construct was generated by replacing the CMV promoter with a miniCMV promoter in the AAV-CMV-SpCas9 plasmid (Addgene #107024) via *Age*I and *Xba*I restriction enzyme sites. Other CRISPR/Cas endonucleases (SaCas9, Cas12a, and CjCas9) were subcloned from AAV-CMV:NLS-SaCas9-NLS-3xHA-bGHpA;U6: *Bsa*I-sgRNA (kindly provided by Feng Zhang; Addgene #61591), pcDNA3.1-hAsCpf1 (kindly provided by Feng Zhang; Addgene #69982) and CjCas9 (kindly provided by Feng Zhang; Addgene #68338) into AAV-CMV-SpCas9 plasmid by replacing SpCas9.

All-in-one single vector, AAV-miniCMV-SaCas9/YFP sgRNA or AAV-miniCMV-SaCas9/LacZ sgRNA were generated based on the AAV-CMV:NLS-SaCas9-NLS-3xHA-bGHpA;U6: *Bsa*I-sgRNA (kindly provided by Feng Zhang; Addgene #61591) by replacing the CMV promoter with miniCMV promoter, adding SpA terminator and inserting YFP sgRNA or LacZ sgRNA.

Cell Culture and Transfection

HEK293A cells that stably express YFP (HEK293A-YFP) were generated as previously described (Hung et al., 2016, 2018). Cells

[1] http://crispr.mit.edu

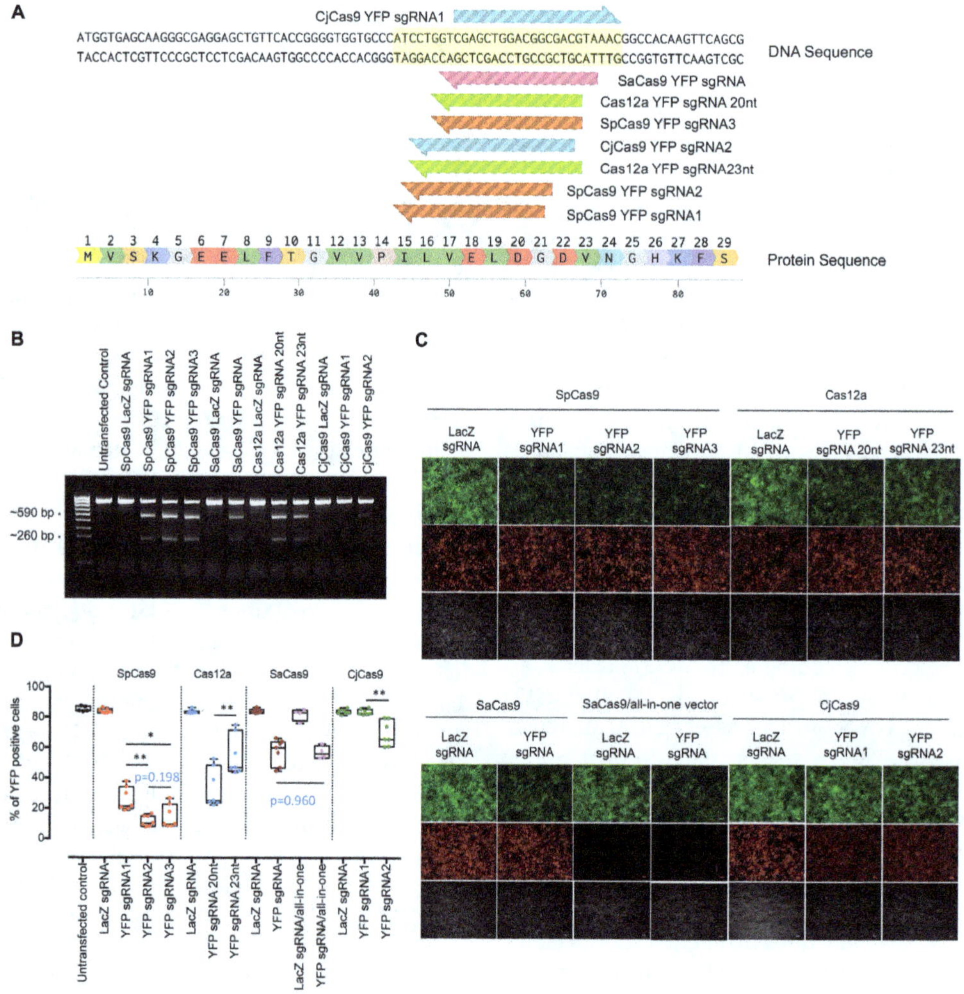

FIGURE 1 | *In vitro* YFP sgRNA validation and selection. **(A)** *YFP*-targeting sequence for sgRNA design. *YFP*-targeting sgRNAs were designed (3 sgRNAs for SpCas9, 1 sgRNA for SaCas9, 2 sgRNAs for Cas12a, and 2 for CjCas9). **(B)** T7E1 assay to detect cleavage efficiency for *YFP*. Expected cleavage products by T7E1 were detected in 2% TAE gel. * Cleavage products around 590 and 260 bp. **(C)** Representative fluorescence microscopy images showing YFP expression in cells transfected with different CRISPR/Cas constructs. Scale bar: 100 μm. **(D)** Flow cytometry analysis for sgRNA selection. Data are represented as mean ± SEM for 4–7 independent replicates. Intergroup comparisons were performed using a one-way ANOVA and corrected for multiple comparisons. HEK293A cells without YFP expression were also included as negative control. No significant difference in *YFP* editing was observed between single and dual CRISPR/SaCas9 vector systems (p = 0.9608). Selected sgRNAs for *in vivo* testing were SpCas9 YFPsgRNA2, Cas12a YFP sgRNA20nt, and CjCas9 YFPsgRNA2. *$p < 0.05$, **$p < 0.01$.

were maintained in Dulbecco's modified Eagle's media (DMEM) (catalog no. 11965118; Life Technologies Australia, Mulgrave, VIC, Australia) supplemented with 10% fetal bovine serum (Sigma-Aldrich, St. Louis, MO, United States), 2 mM glutamine (catalog no. 2503008; Life Technologies Australia), antibiotic-antimycotic (catalog no. 15240062; Life Technologies Australia) in a humidified 5% CO_2 atmosphere at 37°C. HEK293A-YFP cells were transfected with 750 ng of Cas endonuclease plasmid (under CMV promoter) and 750 ng of related sgRNA plasmid, or 750 ng of single SaCas9 plasmid, using lipofectamine 2000 (catalog no. 11668019; Life Technologies Australia), according to manufacturer's instructions. YFP expression was evaluated 10 days later by collecting images of the cell cultures using a fluorescent microscope and by performing a flow cytometric analysis. Genomic DNA was extracted from cells after each treatment and used to carry out a T7 endonuclease 1 (T7E1) assay. The detailed information of reagents is provided in **Supplementary Table S1**.

For Cas endonuclease detection, HEK293A cells were transfected with 1000 ng of the Cas endonuclease plasmid (under the miniCMV promoter) or the CjCas9 plasmid (under CMV promoter), and protein lysates were generated 2 days later to perform a Western blot analysis.

Genomic DNA Extraction and T7E1 Mismatch Detection Assay

Genomic DNA was extracted with QuickExtract DNA Extraction Solution (catalog no. QE09050; Lucigen, Biosearch technologies, Middleton, WI, United States) and used as the DNA template for PCR reactions performed using KAPA HiFi HotStart DNA Polymerase (catalog no. KR0369; Roche Diagnostics Australia, North Ryde, NSW, Australia) with primers listed in **Supplementary Table S3** (CMV SeqFWD forward and EYFP SURVEYOR reverse primers). PCR products were then denatured at 95°C for 10 min and gradually lowered to room temperature to allow for DNA heteroduplex formation, which were then digested by T7 Endonuclease I (catalog no. M0302S; New England Biolabs, Ipswich, MA, United States) following the manufacturer's instructions. The digested products were visualized on 2% (w/v) agarose gels.

Western Blot Analysis

To validate that Cas protein expression was being driven effectively by the Cas endonuclease plasmids, HEK293A cells were transfected with AAV-miniCMV-SpCas9, AAV-miniCMV-SaCas9, AAV-miniCMV-Cas12a, AAV-miniCMV-CjCas9, and AAV-CMV-CjCas9 (under CMV promoter) plasmids. Cells were collected at day 2 post-transfection, and protein was extracted as described previously (Li et al., 2019). Protein samples were separated by using NuPAGE Electrophoresis system (Life Technologies Australia), after which proteins were transferred to polyvinylidene fluoride (PVDF) membranes (catalog no. 162-0177; Bio-Rad Laboratories; Hercules, CA, United States). Membranes were blocked with 5% (w/v) skim milk in TBS-T (10 mM Tris, 150 mM NaCl, and 0.05% Tween-20) at room temperature for 1 h and then incubated with a mouse monoclonal HA antibody (F-7) (1:500 dilution; catalog no. sc-7392; Santa Cruz Biotechnology, Dallas, TX, United States) or mouse monoclonal β-actin antibody (1:1000 dilution; catalog no. catalog no. MAB 1501; Merck Millipore, Burlington, MA, United States) at room temperature for 1 h. Membranes were washed, further incubated with a horseradish peroxidase-conjugated goat anti-mouse secondary antibody (1:5000 dilution; catalog no. A-11045; Life Technologies Australia) at room temperature for 1 h, and developed using the Amersham ECL Prime Western Blotting Detection Kit (catalog no. RPN2232; GE Healthcare Australia, Parramatta, NSW, Australia).

Viral Production

The AAV7m8 vectors were prepared by transfecting HEK293D cells (kindly provided by Ian Alexander, Children's Medical Research Institute, Australia) with the AAV-miniCMV-Cas (SpCas9, SaCas9, Cas12a, and CjCas9), AAV-CMV-CjCas9 or AAV-CMV-mCherry, selected YFP targeting sgRNAs or AAV-miniCMV-SaCas9/YFP sgRNA (all-in-one single vector) plasmids, helper plasmid (pXX6; kindly provided by Richard Samulski, The University of North Carolina School of Medicine, United States) and AAV7m8 capsid plasmid (Addgene #64839) using the calcium phosphate method (Hung et al., 2016). Viral vectors were purified using an AAVpro® Purification Kit (All Serotypes) (catalog no. 6666; Clontech Laboratories, Mountain View, CA, United States) 48 h after viral transduction. Viral titrations were determined by real-time quantitative PCR using a Fast SYBR Green Master Mix (catalog no. 4385612; Life Technologies Australia) with AAV-ITR primers (**Supplementary Table S3**). The titrations of AAV7m8 were provided in **Supplementary Table S4**.

Intravitreal Injection

Mice were anesthetized with an intraperitoneal injection of ketamine (60 mg/kg) and xylazine (10 mg/kg). Bioccular, intravitreal injections were performed under a surgical microscope, using a hand-pulled glass needle connected to a 10 μL Hamilton syringe (Bio-Strategy, Broadmeadows, VIC, Australia), as described previously (Hung et al., 2016, 2018). Eyes with severe surgical or post-operative complications such as ocular hemorrhage or inflammation were excluded from the study. A scleral incision was made on the nasal region with a 30G needle before the glass needle was inserted into the center of vitreous cavity to inject 1 μL of the dual vector system (2.5×10^9vg AAV7m8-Cas endonuclease and 2.5×10^9vg AAV7m8-YFP sgRNA), the SaCas9 single vector system (2.5×10^9vg AAV7m8-miniCMV-SaCas9/YFP sgRNA and 2.5×10^9vg AAV7m8-mCherry) or the control vector (2.5×10^9vg AAV7m8-mCherry). A total of 150 YFP transgenic mice were randomly allocated to the following groups: mCherry control ($n = 11$), AAV7m8-miniCMV-SpCas9 ($n = 12$), AAV7m8-miniCMV-SaCas9 ($n = 20$), AAV7m8-miniCMV-Cas12a ($n = 15$), AAV7m8-CMV-CjCas9 ($n = 11$), AAV7m8-miniCMV-CjCas9 ($n = 9$) and AAV7m8-miniCMV-SaCas9/YFP sgRNA ($n = 20$), receiving the same viral injection regimen in each eye.

Retinal Flat Mounts and Histology

Enucleated eyes were immersion fixed in ice-cold 4% (w/v) paraformaldehyde in PBS for 1 h before the retina was removed using a dissecting microscope as described previously (Hung et al., 2018). Processed retinal flat mounts were stained with NucBlue™ Live ReadyProbes™ Reagent (catalog no. R37605; Life Technologies Australia) for 20 min at room temperature before mounting with Dako Fluorescent mounting medium (catalog no. s3020; DAKO, Carpinteria, CA, United States). For histological assessment, eyes were fixed in 4% paraformaldehyde (w/v) in PBS for 1 h and embedded in optimal cutting temperature compound (Leica Biosystems, Germany) and stored at −80°C until cryosectioning. Serial 10 to 20-μm-thick cryosections were collected directly onto FLEX glass slides, followed by staining and mounting. Images of the retina were collected using an Olympus VS120 Slide Scanner or Perkin Elmer Spinning Disk Confocal Microscope (Zeiss spinning disk, Germany).

Retinal Dissociation and Flow Cytometry Analysis

Retinas were rapidly dissected and digested using a papain dissociation kit (catalog no. LK003176; Worthington Biochemical Corporation, Lakewood, NJ, United States) following the manufacturer's instructions to obtain a homogenous cell suspension. After dissociation, retinal cells were resuspended in FACS buffer (1% Bovine Serum Albumin in Phosphate Buffered Saline) and stained with DAPI (5 μg/mL;

catalog no. D1306; Life Technologies Australia) to exclude dead cells. Dissociated retinal cells from C57BL/6 mice were used as a negative control for YFP expression. Live retinal cells with mCherry (532 nm, 622/22 nm) and/or YFP (488 nm, 513/26 nm) expression were detected by flow cytometry (MoFlo ASTRIOS; Beckman Coulter, Brea, CA, United States). We quantified the proportion of mCherry-labeled cells that co-labeled for YFP in each retina using FlowJo analysis software (FlowJo®; FlowJo LLC, Ashland, OR, United States). Eyes with severe surgical complications such as cataract or retinal detachment or those with negligible mCherry expression were excluded from the final FACS analysis.

Statistical Analysis

GraphPad Prism7 software (GraphPad Software, Inc., La Jolla, CA, United States) was used for statistical analyses. The D'Agostino-Pearson test for normality was performed. Data are represented as mean ± SEM, and were analyzed using unpaired one-way analyses of variance (ANOVA). A value of $p < 0.05$ was considered statistically significant.

RESULTS

In vitro YFP sgRNA Selection and Cas Endonuclease Validation

To select the most effective sgRNA for each Cas endonuclease, we first validated the on-target editing efficacy of different Cas endonucleases together with their respective sgRNAs using a T7E1 assay in HEK293A-YFP cells. Robust cleavage activity was evident in the groups transfected with the Cas endonuclease and their respective *YFP*-targeting sgRNAs, except for those treated with either CjCas9-*YFP* targeting constructs or *LacZ*-targeting controls (**Figure 1B**). Here, SpCas9-*YFP* targeting constructs were the most efficacious at knocking out *YFP* transgene expression, followed by Cas12a-*YFP* and SaCas9-*YFP* targeting constructs (**Figure 1C**).

The *YFP* disruption efficacy for each CRISPR/Cas construct was further quantified through flow cytometric analysis (**Figure 1D**). Compared to LacZ sgRNA counterparts, the percentage of *YFP*-expressing cells was significantly reduced by those transfected with SpCas9 and a *YFP*-targeting sgRNA (*YFP* sgRNA1: 26.0 ± 2.9%, $n = 7$, $p < 0.0001$; *YFP* sgRNA2: 11.5 ± 1.3%, $n = 7$, $p < 0.0001$; and *YFP* sgRNA3: 14.7 ± 2.9%, $n = 7$, $p < 0.0001$). Similarly, Cas12a-targeting conditions resulted in appreciable *YFP* transgene knockout with a preference for a 20 nt-protospacer (*YFP* sgRNA 20 nt: 33.6 ± 4.9%, $n = 7$, $p < 0.0001$; and 23 nt sgRNA: 55.0 ± 5.0%, $n = 7$, $p < 0.0001$). Comparatively, CjCas9 was less effective at abrogating *YFP* transgene expression (*YFP* sgRNA2: 69.5 ± 3.1%, $n = 7$, $p = 0.0011$) and failed to induce significant gene knockout in one of the conditions (CjCas9-*YFP* sgRNA1: 83.7 ± 0.7%, $n = 7$, $p = 0.999$); while there was no significant difference in editing efficiency ($p = 0.9608$) between the use of the SaCas9 single CRISPR construct (SaCas9/*YFP*-targeting sgRNA: 57.3 ± 3.2%, $n = 4$) and the dual CRISPR/Cas construct system (SaCas9 and its *YFP*-targeting sgRNA: 57.0 ± 2.0%, $n = 7$). The most effective *YFP*-targeting sgRNA for each Cas endonuclease (*YFP* sgRNA2 for SpCas9, 20 nt *YFP* sgRNA for Cas12a and *YFP* sgRNA2 for CjCas9) were selected for subsequent *in vivo* testing.

To validate the protein expression of HA-tagged Cas endonuclease in a recombinant AAV vector (driven by minimal promoter, miniCMV or full-length CMV promoter), HEK293A cells were transfected with AAV-miniCMV-SpCas9, AAV-miniCMV-SaCas9, AAV-miniCMV-Cas12a, AAV-miniCMV-CjCas9, and AAV-CMV-CjCas9. Cas endonuclease protein expression was evident with the use of the minimal promoter, except for AAV-miniCMV-CjCas9 (**Supplementary Figure S1**), which required the full-length CMV promoter to drive transgene expression. Therefore, four AAV-miniCMV-Cas endonucleases (SpCas9, SaCas9, and Cas12a) and the AAV-CMV-CjCas9 plasmid were used along with their selected sgRNAs for further *in vivo* CRISPR/Cas editing comparison.

In vivo AAV7m8 Delivery of CRISPR/Cas in the Mouse Retina

AAV7m8-mediated gene expression (mCherry) and distribution were assessed on retinal sectioning/flatmounts of the CMV-Cre:*Rosa26-YFP* mouse eye 5 months after intravitreal injection (**Figure 2A**). Retinal flatmount images from AAV7m8-CRISPR/Cas-injected retina showed robust expression of mCherry, although there was variation in fluorescence intensity across quadrants (**Figure 2B**). Fluorescence images revealed AAV7m8 transduction (as indicated by mCherry expression) was visible throughout the retina, including the ganglion cell layer (GCL), inner nuclear layer (INL) and even some parts of the retinal outer nuclear layer (ONL), with major expression within INL (**Figure 2C** and **Supplementary Figure S3**). Moreover, *YFP* expression could be found in all the layers of the retina with no observable difference between AAV7m8-CRISPR/Cas-treated mice and control mice.

Comparison of YFP Knockout in the Mouse Retina With Different Cas Endonucleases Constructs

Five dual AAV7m8-CRISPR/Cas constructs (miniCMV-SpCas9, miniCMV-SaCas9, miniCMV-Cas12a, miniCMV-CjCas9, and CMV-CjCas9) with their selected *YFP*-targeting sgRNA and a single all-in-one AAV7m8-SaCas9 CRISPR construct (miniCMV-SaCas9/YFP-targeting sgRNA) were used to compare the editing efficacy in the retinal cell *in vivo* (**Figure 3A**). To evaluate and compare the *YFP* knockout *in vivo* delivered by AAV7m8-mediated different CRISPR/Cas system, the percentage of *YFP* disruption among mCherry positive retinal cells was quantified by flow cytometry (**Figure 3B**). The flow cytometric gating strategy is shown in Supplementary data (**Supplementary Figure S2**). Representative dot plots in **Figure 3B** illustrate the difference in *YFP* disruption in retinal cells receiving AAV7m8-SpCas9 CRISPR vector or control vector. Differences in AAV7m8 transduction efficiency were observed between CRISPR/Cas treatment groups (**Figure 3C**) with a lower percentage of mCherry positive cells observed in the retinas transfected with AAV7m8-Cas12a (35.1 ± 2.8%, $n = 15$) and AAV7m8-CjCas9

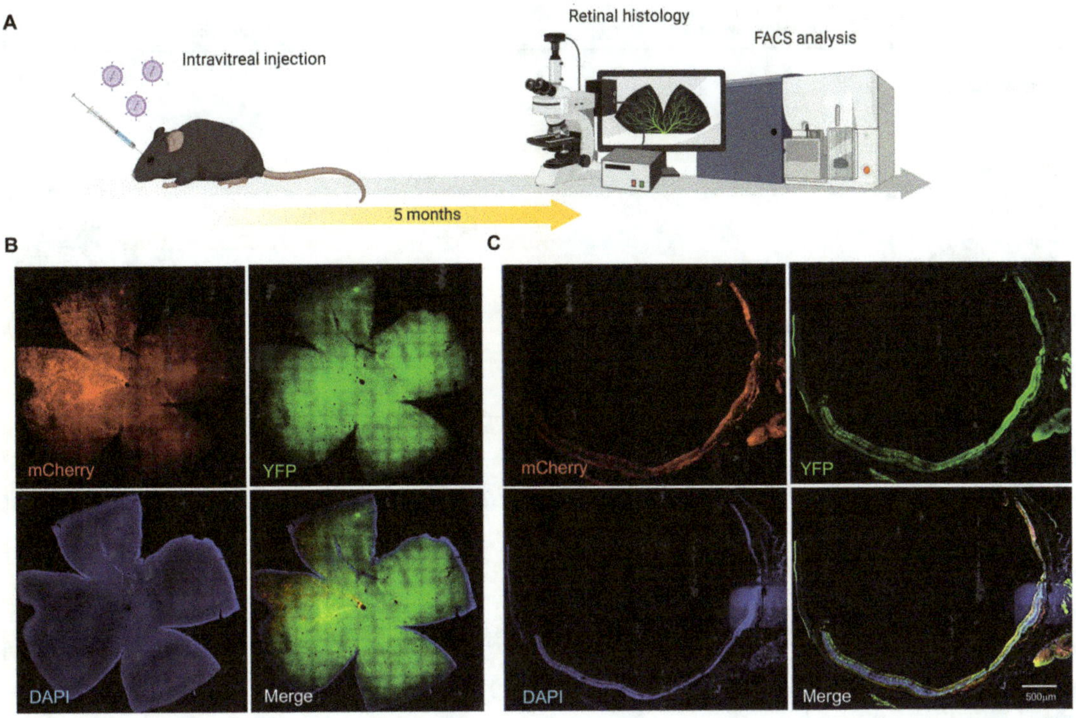

FIGURE 2 | AAV7m8 mediated delivery of CRISPR/Cas to the mouse retina *in vivo*. **(A)** Schematic diagram of in vivo experiment. Mice were sacrificed 5 months after intravitreal injection. **(B)** Representative cross section image from retina co-transduced with AAV7m8-CRISPR/Cas and its selected YFP sgRNA. Mouse ID 29, right eye Robust AAV7m8 transductions in the retina were found. Scale bar: 200 μm. Images were taken by a Zeiss spinning disk confocal microscope. **(C)** Representative retinal whole-mount images from a mouse eye receiving AAV7m8-CRISPR/Cas and its selected YFP sgRNA. Mouse ID 76, right eye. Scale bar: 500 μm. Images were taken using an Olympus Slide Scanner. Please see **Supplementary Figure S3** for representative cross-sectional images with higher magnification.

(28.6 ± 3.4%, $n = 9$) vectors. Retinas receiving AAV7m8-SpCas9 and AAV7m8-SaCas9 (both single and dual vector system) vectors had a relatively high proportion of mCherry expression (50.0 ± 4.6%, $n = 12$; 52.0 ± 4.3%, $n = 20$; 57.7 ± 3.3%, $n = 19$ respectively). For *YFP* disruption, AAV7m8-SpCas9 vector (18.9 ± 2.9%, $n = 12$) had the highest knockout efficiency of YFP among all the CRISPR/Cas systems, followed by AAV7m8-SaCas9 (single vector system: 8.4 ± 3.4%, $n = 20$; dual vector system: 9.8 ± 2.6%, $n = 20$) and Cas12a (5.4 ± 2.0%, $n = 15$), while AAV7m8-CjCas9 showed no disruption of *YFP* expression (**Figure 3D**). Moreover, there was no significant difference in the *YFP* disruption in the retinas receiving single and dual AAV7m8-SaCas9 vectors (single vector system: 8.4 ± 3.4% vs. dual vector system: 9.8 ± 2.6%, $n = 20$, $p = 0.9994$) (**Figure 3D**). Despite efficiency *in vivo* YFP knockout in animals administered AAV7m8-SpCas9, AAV7m8-SaCas9 and AAV7m8-Cas12a vectors, there was a high degree of variability between individual animals within identical treatment groups.

DISCUSSION

In this study, we provide a direct comparison of the efficacy for retinal editing *in vivo* with four different currently available CRISPR/Cas systems. Here, we showed that SpCas9 and Cas12a achieved better knockout efficiency than SaCas9 and CjCas9 *in vitro*. AAV7m8-packaged CRISPR/Cas construct with SpCas9 was found to have the highest editing efficacy among all Cas endonucleases *in vivo*. No significant difference in *YFP* gene editing was found between single and dual CRISPR/SaCas9 vector systems *in vitro* and *in vivo*.

This study was based on our previous work, which used AAV2-mediated delivery of CRISPR/Cas9 to achieve efficient gene editing in the inner layer of retina in Thy1-*YFP* mice (Hung et al., 2016). To assess and compare the genome efficiency in the whole retina, we applied a different murine model CMV-Cre:*Rosa26-YFP* transgenic mice (*YFP* mouse) which express YFP throughout the retina. To this end, we used the AAV7m8-pseudotype, an AAV2-based variant with enhanced retinal transduction when delivered through intravitreal injection (Dalkara et al., 2013; Khabou et al., 2016). As the degeneration of RPE and photoreceptors are involved in the majority of inherited retinal diseases, efficient gene delivery of CRISPR constructs to the outer layer of retina is imperative for therapeutic retinal gene editing. Subretinal injection of conventional AAVs (e.g., AAV2) has high photoreceptor transduction rate, but it is surgically challenging with more complications. In addition, the cellular transduction is confined within the injection bubble of the retina. mCherry expression was generally observed in at least two quadrants on retinal flat mount (**Figure 2B**), and any variation in distribution of transfection is likely to be

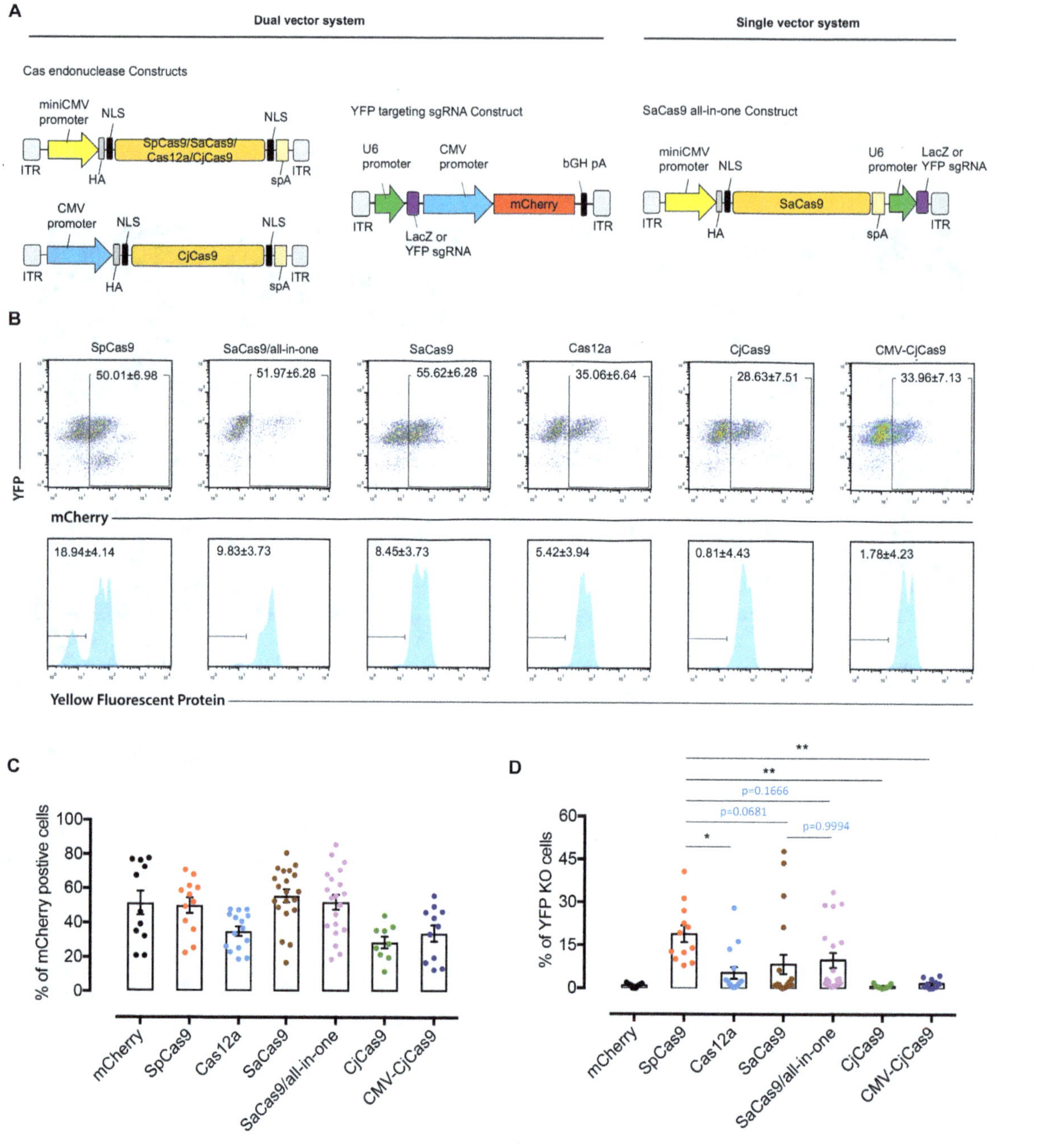

FIGURE 3 | Comparison of YFP disruption in retinal cells with different CRISPR/Cas systems delivered by AAV7m8. **(A)** Schematic of the dual and single vector systems. For dual vector plasmids, the Cas endonuclease was driven by miniCMV or CMV promoter, whilst the sgRNA was driven by U6 promoter and mCherry under the control of CMV promoter to confirm vector transfection. For the single vector systems, an all-in-one plasmid with SaCas9 was designed with the Cas endonuclease being driven by a miniCMV promoter and sgRNA by U6 promoter. For Cas12a, we used the Cas endonuclease from *Acidaminococcus* (originally designated AsCpf1). A hemagglutinin (HA) tag was fused to the C-terminus of Cas endonuclease in the vector. **(B)** Representative FACS plots of dissociated retinal cells receiving different AAV7m8-CRISPR/Cas/AAV7m8-YFP sgRNA. The histograms in the lower panels (lower panel) were based on mCherry gating. Dissociated cells from one retina were used in each group. **(C)** Comparison of AAV7m8 transduction in the retina indicated by mCherry expression by FACS. Data are presented as mean ± SEM for 9–20 independent samples in each group. The D'Agostino-Pearson normality test was performed, and all groups were found to have a Gaussian distribution. Statistical analysis between groups was performed using one-way ANOVA followed by multiple comparisons test. **(D)** Comparison of *YFP* disruption in mCherry positive cells by FACS. Data are presented as mean ± SEM for 9–20 independent samples in each group. Data in two groups (Cas12a and Dual SaCas9) were found to not pass the D'Agostino-Pearson normality test, and as such the Kruskal–Wallis test was used. *$p < 0.05$, **$p < 0.01$, ***$p < 0.001$.

due to stochastic or technical issues. Nonetheless, our study shows that AAV7m8-mediated CRISPR/Cas has reasonable pan-retinal transduction.

The stringent design of this study ensured a fair comparison of editing efficiency between different CRISPR/Cas systems. First, we analyzed the *YFP* coding sequence for all potential PAM sites for each Cas endonuclease and then designed sgRNA targeting *YFP* within a similar region. Previous work has shown that both guide RNA sequence and target gene-chromatin accessibility can directly influence CRISPR/Cas editing efficiencies. An important limitation of this work is the fact that we could not directly compare identical sequences across all endonucleases. Although we sought to target the same region within YFP, given their differing PAM requirements, each endonuclease had a different guide sequence, which may have directly biased editing efficiencies. Further, these PAM restrictions also limited the number of guide RNAs which could be directly tested. Naturally, there is a tradeoff between targeting the same genic region and ensuring similar *'a priori'* guide RNA efficacy, and it is important to note that both factors should be considered in direct head-to-head comparisons. To further mitigate biases, we employed the same ubiquitous promoter (CMV for *in vitro* sgRNA selection, miniCMV for *in vivo*) for each endonuclease, and employed for the same virus. The only exception to this design was the use of the more potent CMV promoter for *in vivo* CjCas9 constructs, due to its poor expression on western blots of *in vitro* HEK293A cells. Despite this modification, CjCas9 barely demonstrated *YFP* knockout on flow cytometric analysis of *in vivo* specimens. We hypothesize that variation in CjCas9 codon-optimization may account for the differences observed in study compared to that reported by other groups (Kim et al., 2017).

We additionally found differences in gene knockout efficiency between *in vitro* and *in vivo* modes. For the *in vitro* study, SpCas9 outperformed Cas12a, followed by SaCas9 and CjCas9. For *in vivo* samples, SpCas9 remained the best-performing Cas endonuclease among all, without a clear trend among the other Cas orthologs. Initially, we hypothesized that the single all-in-one SaCas9 vector expressing both the SaCas endonuclease and its respective sgRNA may have a competitive or even higher editing efficiency compared to dual-vector mediated-editing with SpCas9, but we did not observe this result in our *in vivo* test.

In this proof-of-concept study, we engineered CMV-Cre:Rosa26-YFP mice. The principal advantage of screening a reporter gene at the Rosa26 locus, is that genomic edits could be readily quantified (using flow cytometry), at a single site. That is, only a single integration event would have occurred for our chosen reporter. Nevertheless, there are important limitations of this model, and it must be noted that Rosa26 locus is naturally permissive for DNA targeting. As such, our results may not be directly transferrable or representative for other endogenous genes, and likely represent the upper bounds for editing efficiencies.

In summary, we demonstrate that AAV7m8-mediated delivery of a SpCas9 construct appeard to achieve the most efficient gene modification in retinal cells *in vivo* among four currently available CRISPR/Cas systems. Ongoing research investigating different guide sequences at different loci is required before firm conclusions regarding retinal cell gene editing of different endonucleases can be made.

AUTHOR CONTRIBUTIONS

FL, AH, G-SL, and SH conceptualized and designed the study. FL, KW, J-HW, CL, JB, JC, QW, QL, and MN performed all laboratory-based experiments. KY, RW, AP, and AC provided reagents and resources for this work. FL wrote the original draft, with all authors providing reviewing and editing. G-SL and AH jointly supervised this work. All authors contributed to the article and approved the submitted version.

FUNDING

This work was supported by funding from a Bayer Global Ophthalmology Award, The Ophthalmic Research Institute of Australia, the Royal Hobart Hospital Research Foundation, an Australian National Health and Medical Research Council (NHMRC) grant (GNT1123329), an NHMRC Practitioner Fellowship (AH, GNT1103329), a NHMRC Career Development Award (KY, GNT1045240), and an NHMRC Research Fellowship NHMRC Senior Research Fellowship (AP, GNT1154389). Centre for Eye Research Australia receives Operational Infrastructure Support from the Victorian Government.

SUPPLEMENTARY MATERIAL

FIGURE S1 | The *in vitro* validation of Cas endonuclease expression.

FIGURE S2 | Representative FACS plot showing gating strategy.

FIGURE S3 | Representative retinal cross section with high magnification.

TABLE S1 | Summary of reagents and resources used.

TABLE S2 | Comparison of Cas orthologs for *in vivo* retinal gene editing application.

TABLE S3 | Sequence of primers for sgRNA cloning, vector construction, sequencing and qPCR analysis.

TABLE S4 | AAV7m8 titrations.

REFERENCES

Bakondi, B., Lv, W., Lu, B., Jones, M. K., Tsai, Y., Kim, K. J., et al. (2016). In Vivo CRISPR/Cas9 gene editing corrects retinal dystrophy in the S334ter-3 rat model of autosomal dominant retinitis pigmentosa. *Mol. Ther.* 24, 556–563. doi: 10.1038/mt.2015.220

Cong, L., Ran, F. A., Cox, D., Lin, S., Barretto, R., Habib, N., et al. (2013). Multiplex genome engineering using CRISPR/Cas systems. *Science* 339, 819–823.

Dalkara, D., Byrne, L. C., Klimczak, R. R., Visel, M., Yin, L., Merigan, W. H., et al. (2013). In vivo-directed evolution of a new adeno-associated virus for therapeutic outer retinal gene delivery from the vitreous. *Sci. Transl. Med.* 5:189ra76. doi: 10.1126/scitranslmed.3005708

Huang, X., Zhou, G., Wu, W., Duan, Y., Ma, G., Song, J., et al. (2017). Genome editing abrogates angiogenesis in vivo. *Nat. Commun.* 8:112.

Hung, S. S., Li, F., Wang, J.-H., King, A. E., Bui, B. V., Liu, G.-S., et al. (2018). Methods for in vivo CRISPR/Cas editing of the adult murine retina. *Methods*

Mol. Biol. 1715, 113–133. doi: 10.1007/978-1-4939-7522-8_9

Hung, S. S. C., Chrysostomou, V., Li, F., Lim, J. K. H., Wang, J.-H., Powell, J. E., et al. (2016). AAV-mediated CRISPR/Cas gene editing of retinal cells in vivo. *Invest. Ophthalmol. Vis. Sci.* 57, 3470–3476. doi: 10.1167/iovs.16-19316

Ibraheim, R., Song, C.-Q., Mir, A., Amrani, N., Xue, W., and Sontheimer, E. J. (2018). All-in-one adeno-associated virus delivery and genome editing by *Neisseria meningitidis* Cas9 in vivo. *Genome Biol.* 19:137.

Jain, A., Zode, G., Kasetti, R. B., Ran, F. A., Yan, W., Sharma, T. P., et al. (2017). CRISPR-Cas9-based treatment of myocilin-associated glaucoma. *Proc. Natl. Acad. Sci. U.S.A.* 114, 11199–11204. doi: 10.1073/pnas.1706193114

Jarrett, K. E., Lee, C., De Giorgi, M., Hurley, A., Gillard, B. K., Doerfler, A. M., et al. (2018). Somatic editing of Ldlr with adeno-associated viral-CRISPR is an efficient tool for atherosclerosis research. *Arterioscler. Thromb. Vasc. Biol.* 38, 1997–2006. doi: 10.1161/atvbaha.118.311221

Jinek, M., Chylinski, K., Fonfara, I., Hauer, M., Doudna, J. A., and Charpentier, E. (2012). A programmable dual-RNA-guided DNA endonuclease in adaptive bacterial immunity. *Science* 337, 816–821. doi: 10.1126/science.1225829

Jo, D. H., Koo, T., Cho, C. S., Kim, J. H., Kim, J.-S., and Kim, J. H. (2019). Long-Term effects of in vivo genome editing in the mouse retina using *Campylobacter jejuni* Cas9 expressed via adeno-associated virus. *Mol. Ther.* 27, 130–136. doi: 10.1016/j.ymthe.2018.10.009

Khabou, H., Desrosiers, M., Winckler, C., Fouquet, S., Auregan, G., Bemelmans, A.-P., et al. (2016). Insight into the mechanisms of enhanced retinal transduction by the engineered AAV2 capsid variant -7m8. *Biotechnol. Bioeng.* 113, 2712–2724. doi: 10.1002/bit.26031

Kim, E., Koo, T., Park, S. W., Kim, D., Kim, K., Cho, H.-Y., et al. (2017). In vivo genome editing with a small Cas9 orthologue derived from *Campylobacter jejuni. Nat. Commun.* 8:14500.

Koo, T., Park, S. W., Jo, D. H., Kim, D., Kim, J. H., Cho, H.-Y., et al. (2018). CRISPR-LbCpf1 prevents choroidal neovascularization in a mouse model of age-related macular degeneration. *Nat. Commun.* 9:1855.

Latella, M. C., Di Salvo, M. T., Cocchiarella, F., Benati, D., Grisendi, G., Comitato, A., et al. (2016). In vivo editing of the human mutant rhodopsin gene by electroporation of plasmid-based CRISPR/Cas9 in the Mouse Retina. *Mol. Ther. Nucleic Acids* 5:e389. doi: 10.1038/mtna.2016.92

Li, F., Hung, S. S. C., Mohd Khalid, M. K. N., Wang, J.-H., Chrysostomou, V., Wong, V. H. Y., et al. (2019). Utility of self-destructing CRISPR/Cas constructs for targeted gene editing in the retina. *Hum. Gene Ther.* 30, 1349–1360. doi: 10.1089/hum.2019.021

Maeder, M. L., Stefanidakis, M., Wilson, C. J., Baral, R., Barrera, L. A., Bounoutas, G. S., et al. (2019). Development of a gene-editing approach to restore vision loss in Leber congenital amaurosis type 10. *Nat. Med.* 25, 229–233.

Mali, P., Yang, L., Esvelt, K. M., Aach, J., Guell, M., DiCarlo, J. E., et al. (2013). RNA-guided human genome engineering via Cas9. *Science* 339, 823–826. doi: 10.1126/science.1232033

Pan, X., Philippen, L., Lahiri, S. K., Lee, C., Park, S. H., Word, T. A., et al. (2018). In Vivo Ryr2 editing corrects *Catecholaminergic* polymorphic ventricular tachycardia. *Circ. Res.* 123, 953–963.

Ran, F. A., Cong, L., Yan, W. X., Scott, D. A., Gootenberg, J. S., Kriz, A. J., et al. (2015). In vivo genome editing using *Staphylococcus aureus* Cas9. *Nature* 520, 186–191. doi: 10.1038/nature14299

Ruan, G.-X., Barry, E., Yu, D., Lukason, M., Cheng, S. H., and Scaria, A. (2017). CRISPR/Cas9-mediated genome editing as a therapeutic approach for leber congenital amaurosis 10. *Mol. Ther.* 25, 331–341. doi: 10.1016/j.ymthe.2016.12.006

Swiech, L., Heidenreich, M., Banerjee, A., Habib, N., Li, Y., Trombetta, J., et al. (2015). In vivo interrogation of gene function in the mammalian brain using CRISPR-Cas9. *Nat. Biotechnol.* 33, 102–106. doi: 10.1038/nbt.3055

Tsai, Y.-T., Wu, W.-H., Lee, T.-T., Wu, W.-P., Xu, C. L., Park, K. S., et al. (2018). Clustered regularly interspaced short palindromic repeats-based genome surgery for the treatment of autosomal dominant retinitis pigmentosa. *Ophthalmology* 125, 1421–1430. doi: 10.1016/j.ophtha.2018.04.001

Xia, C.-H., Ferguson, I., Li, M., Kim, A., Onishi, A., Li, L., et al. (2018). Essential function of NHE8 in mouse retina demonstrated by AAV-mediated CRISPR/Cas9 knockdown. *Exp. Eye Res.* 176, 29–39. doi: 10.1016/j.exer.2018.06.026

Xu, L., Lau, Y. S., Gao, Y., Li, H., and Han, R. (2019). Life-long AAV-mediated CRISPR genome editing in dystrophic heart improves cardiomyopathy without causing serious lesions in mdx Mice. *Mol. Ther.* 27, 1407–1414. doi: 10.1016/j.ymthe.2019.05.001

Yu, W., Mookherjee, S., Chaitankar, V., Hiriyanna, S., Kim, J.-W., Brooks, M., et al. (2017). Nrl knockdown by AAV-delivered CRISPR/Cas9 prevents retinal degeneration in mice. *Nat. Commun.* 8:14716.

15

Olfactory Ensheathing Cells Grafted into the Retina of RCS Rats Suppress Inflammation by Down-Regulating the JAK/STAT Pathway

Jing Xie[1,2], Yijian Li[1,2], Jiaman Dai[1,2], Yan He[1,2], Dayu Sun[1,2], Chao Dai[1,2], Haiwei Xu[1,2] and Zheng Qin Yin[1,2]**

[1] *Southwest Eye Hospital, Southwest Hospital, Third Military Medical University (Army Medical University), Chongqing, China,* [2] *Key Laboratory of Visual Damage, Regeneration and Restoration of Chongqing, Chongqing, China*

***Correspondence:**
Haiwei Xu
haiweixu2001@163.com
Zheng Qin Yin
qinzyin@aliyun.com

The inflammatory microenvironment in the retina plays a vital role in the pathogenesis and progression of retinitis pigmentosa (RP). Microglial inflammatory cytokines production leads to gliosis and apoptosis of retinal neurons, and ultimately, visual loss. Cell-based therapies using grafted olfactory ensheathing cells (OECs) have demonstrated modulation of degenerative microenvironments in the central nervous system (CNS), in a number of animal models. However, mechanisms by which grafted OECs can reduce degeneration in the retina are not well understood. In the present study, we set up an *in vitro* OEC/BV2 microglia co-culture system, and an *in vivo* royal college of surgeons (RCS) rat model, used cell transplantation, immunohistochemistry, RT-PCR, western blot to explore the mechanisms by which OECs affect expression of pro- or anti-inflammatory cytokines and polarization of M(IL-6) and M(Arg1) type microglial activation in the retina. We found that compared with the LPS (Lipopolysaccharide) and olfactory nerve fibroblast (ONF), the OEC and BV2 co-culture group modulate microglial cytokines releasing toward the anti-inflammation, and away from the pro-inflammation, which was followed by higher IL-4 and IL-10 and lower TNF-a and IL-6 in their expression levels. *In vivo*, the transplantation group significantly reduced activated resident microglia/infiltrated macrophage, and expression of pro-inflammatory cytokines in RCS rats retina, increased anti-inflammatory cytokines in transplantation area. Additionally, we found that OECs expressed SOCS3 and down-regulated the JAK2/STAT3 (Janus Kinase 2/Signal Transducer and Activator of Transcription 3) pathway. Thirdly, OEC transplantation reduced Caspase-3 expression, protected inner retinal neurons and photoreceptors and therefore, delayed the visual function degeneration. In conclusion, our data suggest that OECs delay retinal degeneration in RP, at least in part through immunomodulation of microglia via the JAK/STAT pathway.

Keywords: retinitis pigmentosa, microglia, infiltrated macrophage neuroinflammation, JAK/STAT pathway, olfactory ensheathing cell

INTRODUCTION

Retinitis pigmentosa (RP) is a heterogeneous group of inherited retinal degenerative diseases, which lead to photoreceptor cell apoptosis and severe vision loss. Photoreceptor degeneration starts from microglial activation, macrophage infiltration, and accumulation of immunoglobulins and complement, which result in sustained inflammation, macroglia proliferation and progressive apoptosis of retinal neurons (Athanasiou et al., 2018; Ben et al., 2019). Therefore, modulation of microglia activation and inflammatory reaction might be a potential intervention for RP.

In royal college of surgeons (RCS) rats with *Mertk* gene mutation of retinal pigmented epithelium (RPE), microglia become activated, and infiltrate into the outer nuclear layer (ONL), to assist with phagocytosis of photoreceptor debris (Zou et al., 2019). As retinal degeneration continues, blood-retinal barrier (BRB) disruption results in the recruitment of blood-borne macrophages. This is an important step in activating the immune cells and releasing pro-inflammatory cytokines, which amplify the disease process, and leading to photoreceptor apoptosis (Kyger et al., 2013). Studies have already shown that blood-derived immune cells are an important component of the disease-associated microenvironment, and they are considered to be critical mediators of neurodegenerative disease progression, not only in CNS but also in retina (Xu et al., 2009; Sevenich, 2018).

Microglia, resident immune population of the retina, are react to injury as specialized scavengers by promote, and resolve inflammation (Ramirez et al., 2017). In the central nervous system (CNS), resident (microglia) and invading innate immune cells (macrophage) coordinated complex responses to injury, and aiming to restore tissue integrity but can also promote destructive neuroinflammation (Wohleb, 2016; Subramaniam and Federoff, 2017). There are two main microglial/macrophage phenotypes (characterized by morphology, cytokine/chemokine expression, and function). The classically activated phenotype, promotes inflammation by releasing numerous pro-inflammatory cytokines (e.g., IL-6, TNF-α, and MCP), as means of M(IL-6). The alternatively activated phenotype promotes tissue repair and regeneration, by releasing protective/trophic factors (e.g., Argl, IL-4, and IL-10), and clearing cellular waste debris through phagocytosis, as means of M(Argl) (Ransohoff, 2016; Edholm et al., 2017). This M(IL-6)/M(Argl) paradigm has been used to describe the *in vitro* perturbation of macrophages, yet there is evidence that microglia can adopt similar phenotypes, and functions *in vivo* (Hu et al., 2015). Due to these opposing effects of different immune cell phenotypes, recent treatment for neuroinflammation are shifting from complete immune cell suppression to find a balance between M(IL-6)/M(Argl) phenotypes and searching regulatory molecules that control the two phenotypes polarization switching (Neves et al., 2016).

Several studies in animal models of neurodegenerative disease, including Alzheimer's disease (Biscaro et al., 2012), Parkinson's disease (Garrido-Mesa et al., 2013) and RP (Peng et al., 2014), have shown that anti-inflammatory therapies, such as minocycline, have neuroprotective properties. However, the duration of action of minocycline is only 5–10 days, and there is therefore considerable interest in developing a cell-based therapy, which can provide sustained modulation of the inflammatory microenvironment in the degenerative retina (Neves et al., 2016; Zhu et al., 2017).

Olfactory ensheathing cells (OECs) are a unique type of glial cell, which share some features and functions with Schwann cells and astrocytes (Li et al., 1997). OECs have been demonstrated to facilitate glial scar rearrangement, blood vessel formation, axon remyelination, and phagocytosis of cellular debris and pathogens (Chuah et al., 2011; Huo et al., 2012). Microarray analysis of the OEC transcriptome indicates that they express higher levels of a number of innate immune factors, compared to astrocytes and Schwann cells, suggesting an enhanced role in modulating immune cells and neuroinflammation (Vincent et al., 2005). Khankan et al. (2016) suggest that grafted OECs reduce macrophage infiltration, maintain serotonergic (5-HT) axons, and reduce inhibitory chondroitin sulfate proteoglycans (CSPGs) in injured rodent spinal cord, leading to the restoration of motor function. Zhang et al. (2017) have shown that OEC transplantation reduces inflammatory cell infiltration in spinal cord injury, and promotes a shift in the macrophage phenotype from M(INF-γ) to M(IL-4). Although microglial phenotype switching is not fully understood, there is thought to be an important role for the JAK/STAT (janus kinase/signal transducer and activator of transcription) pathway and some binding proteins, which may accelerate the pace of immune cell suppression, and toward rebalance M(IL-6)/M(Argl) activity (Hu et al., 2015). In retina, some studies have reported a dynamic shift in microglia changing profile of recognized M(IL-6)- to M(Argl)- following acute light damage (Jiao et al., 2015). In rd1 mice, the microglia orchestrate a continuous spectrum which is activated and polarized to a M(IL-6) phenotype during acute retinal degeneration (Zhou et al., 2017). Recent work has shown that the JAK/STAT pathway is central to the determination of M(IL-6) vs. M(Argl) microglial subtypes (Sica and Bronte, 2007). Upon binding to JAK, members of the IL-6 family of cytokines activate the JAK/STAT signaling pathway (Tam and Ma, 2014; Hu et al., 2015; Qin et al., 2016). It has also been demonstrated that transplanted OECs promote neurological functional recovery in traumatic brain-injured rats via the JAK/STAT3 pathway (Fu et al., 2015).

Our previous research has found that there is microglial activation in the degenerative period of RCS rat retina (Liu et al., 2013; Li et al., 2016). However, as BRB breakdown and macrophages infiltrate into the retina during retinal degeneration (Shen et al., 2010; de Hoz et al., 2016), neither the polarization of resident microglia, and its effect in RCS rat, nor the ways to regulate this polarization, has ever been investigated.

In this study, we investigated the effect of OECs on activated microglia in an *in vitro* co-culture system. Co-culture of OECs with BV2 cells (an immortalized microglial cell line) reduced lipopolysaccharide (LPS)-induced microglial activation and produced microglial polarization toward the M(Argl) phenotype. Immunofluorescent staining showed OECs were SOCS3 positive. These microglial modulation effects of OECs may mediated by downregulation of the JAK2/STAT3 signaling pathway. We also used a rodent model of inherited retinal inflammation and degeneration (RCS rat) to study the *in vivo* effect of OEC implantation into the retina, during the chronic stages of the disease process. Immunohistochemical data, and gene and protein quantification, demonstrated that OEC transplantation delayed the degeneration of retinal neurons and photoreceptors via an anti-apoptotic mechanism, inhibited microglial/macrophage activation, and reduced pro-inflammation cytokines, by down-regulating the JAK2/STAT3 pathway. These results further confirm that microglial activation and a pro-inflammatory environment play a pivotal role in the retinal degeneration of RCS rats, and that OEC transplantation can preserve visual function partly through regulating the microglia-mediated inflammatory environment.

MATERIALS AND METHODS

Animal Models and Ethical Approval

We used black-eyed RCS rats as a model of retinal dystrophy, and "rdy" rats as non-retinal dystrophic controls; both supplied by the Animal Centre of the Third Military Medical University (TMMU). Rats were maintained in the animal facility of the Southwest Eye Hospital, the TMMU. Housing rooms had regular day and night light-cycles (12:12 h). All rats were sacrificed using a carbon dioxide inhalation chamber. All surgical procedures and post-operative care were conducted in accordance with protocols approved by the TMMU Institutional Animal Care and Use Committee.

Chemicals and Reagents

Fetal bovine serum (FBS), Dulbecco's modified Eagle's medium/Ham's Nutrient Mixture (DMEM/F12 plus GlutaMAX), Hank's balanced salt solution (HBSS), phosphate buffer (PBS), and other cell culture reagents were obtained from Thermo-Fisher Corporation (Beijing, China). Trypsin and penicillin/streptomycin solutions were obtained from Hyclone (Bejing, China). Lipopolysaccharide (LPS), 2-cyano-3 (3,4-dihydroxyphenyl)N-(benzyl)2-propenamide (Tyrphostin AG490), poly-L-lysine (PLL), and pentobarbital sodium were purchased from Sigma-Aldrich (Shanghai, China). Glyceraldehyde-3-phosphate dehydrogenase (GAPDH) and primers were purchased from Thermo Fisher Scientific. The primary and secondary antibodies used for immunohistochemistry and WB are listed in **Table 1**.

Primary OEC and ONF Culture and Purification

The olfactory bulbs of adult control rats were used to harvest OECs and olfactory nerve fibroblasts (ONFs, for use as a negative control), and we used differential cell adhesiveness to purify OECs and ONFs (Huo et al., 2011; Dai et al., 2012; Xie et al., 2017). Briefly, olfactory bulbs were removed and put into a 10 mm dish. After carefully isolated of the pia mater and vascular membrane, the olfactory glomerular layers and nerve layer were separated. The nerve layer was cut into 0.5 mm^3 pieces, placed in 0.125% trypsin to digest for 15 min at 37°C. The cells were cultured with DMEM/F-12 containing 10% FBS and 1% penicillin/streptomycin and inoculated into a six-well plate coated with PLL. Culture medium were first changed for the 5th day, and then changed every 3 days until 2 weeks. After several cycles of differential cell adhesiveness, OECs, and ONFs were easily separated. OECs or ONFs were dissociated into suspension and then labeled with lentiviruses carrying the enhanced green fluorescent protein (LV-EGFP) prior to subretinal transplantation. The production, purification and infection of LV-EGFP was according to previous reports (Xie et al., 2017).

Culture of BV2 Microglia

BV2 murine microglial cell line was given by Dr. Guo from the Neurological Surgery Department of Southwest Hospital. Cells were seeded into six-well plates at a concentration of 10^5 cells/well in DMEM containing 10% FBS and 1% penicillin/streptomycin, as previously described (Li et al., 2016).

Co-culture of OEC or ONF Cells With BV2 Microglia Cells

The co-culture experimental design is shown in **Figure 1**. We maintained the BV2 mouse microglia cell line in six-well plates (10,000 cells each). To induce BV2 cell reactivity we used 1 μg/ml of LPS diluted in basal medium for 4 h. OECs or ONFs were seeded into Transwell plate inserts (Millipore) (10,000 cells each). BV2 cultures were co-cultured with the OECs for 24 h (or ONFs as a negative control) to explore their role in immune modulation. To study the role of the JAK/STAT signaling pathway in OECs, the JAK2 antagonist Tyrphostin AG490 (50 μm) (Ito et al., 2010; Hou et al., 2017) was added to the OEC culture 1 h before co-culture with BV2 cells, in relevant experiments.

TABLE 1 | Primary antibodies used.

Antibody	Manufacturer's catalog or lot number	Dilution
Mouse anti-NGFRp75	Santa Cruz, sc-271708	1: 50
Rabbit anti-Iba1	Wako, 019-19741(IFC) 016-20001(WB)	1: 500 1:1000
Rabbit anti-TMEM119	Abcam, ab185333 (IHC) Santa Cruz, sc-244341 (WB)	1: 50 1: 500
Mouse anti-PKCα	Santa Cruz, Santa Cruz, sc-8393	1: 500
Mouse anti-Rhodopsin	Abcam, ab5417	1: 1000
Rabbit anti-S-100β	Abcam, ab868	1: 200
Rabbit anti-Caspase-3	Abcam, ab13847	1: 200
Mouse anti-β-actin	Cell signal technology, 3700	1:2000
Rabbit anti-JAK2	Cell signal technology, 3230	1: 1000
Rabbit anti-STAT3	Cell signal technology, 12640	1: 1000
Rabbit anti-SOCS3	Abcam, ab16030	1:100
Rabbit anti-pJAK2	Cell signal technology, 3776	1: 1000
Rabbit anti-pSTAT3	Cell signal technology, 9145	1: 1000

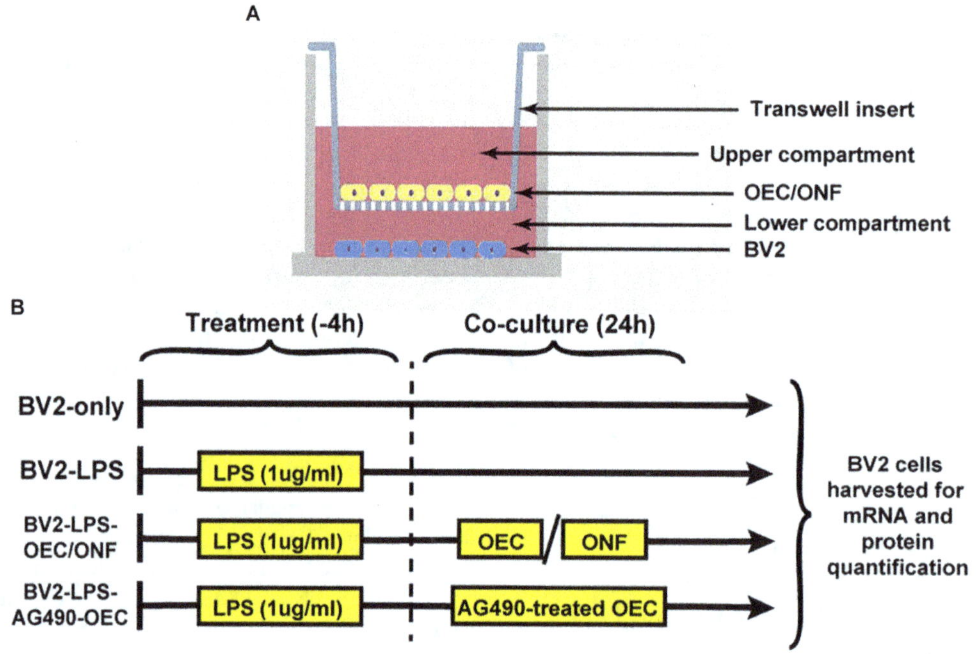

FIGURE 1 | Design of the *in vitro* co-culture experiment. **(A)** Schematic illustrating the co-culture system, with a Transwell insert hanging in each well of a six-well plate. **(B)** Experimental schedule. AG490-OECs are OECs that have been pre-treated for 1 h with AG490 (a JAK2 antagonist) prior to co-culture.

Subretinal Injection

After purification, OECs or ONFs were separately detached and implanted into the subretinal space of P30d RCS rats, as previously described (Xie et al., 2017). Briefly, we exposed the temporal part of the sclera, and a 30G needle was used to reach the subretinal space by incision of the sclera, choroid, and retinal pigment epithelium. A Hamilton syringe with a 33G needle was attached to the choroid layer and slowly injected 10^5 cells (in 3 μl PBS) into the subretinal space of the right eye. The contralateral eye was injected 3 μl PBS with the same surgical procedures. When the fundus was observed, we could see a white flat detachment of the retina at the transplantation site.

Electroretinogram (ERG)

Electroretinogram recordings were performed as previously described (Dai et al., 2017; Xie et al., 2017). We dark-adapted rats for 12 h, and prepared under dim red light. Compound tropicamide eye drops were used to dilate pupils. Before recording, artificial tears was applied to allow the platinum recording electrode contact with the cornea and also prevent dehydration. The needle electrode was fixed under the skin of both sides of the rat's nasal side as a reference electrode, while the ground electrode was placed under the tail. The signal amplifier bandwidth was 0.1–300 Hz, without notch filtering. "RETI-port" software was used to acquire and control stimulus delivery data, running on Roland Electrophysiological Systems hardware (Brandenburg, Germany). The amplitude of the a-wave of ERG was calculated from the baseline to the first trough, and the amplitude of the b-wave was from the trough of the a-wave to the first peak. To improve the signal-to-noise ratio, the inter-stimuli-intervals were longer than 30 s.

Tissue Sample Preparation

For immunofluorescence staining, we firstly intramuscular injection of pentobarbital sodium (10 mg/kg), and then used normal saline and 4% paraformaldehyde (PFA) to transcardially perfused rats. After taking out the eyeball, the cornea, iris, lens were removed, and the eyecups were put in 4% PFA at 4°C for 0.5 h, then dehydrated in a 30% sucrose solution at 4°C overnight. Sections (10 μm thick) were cut using a freezing microtome (Leica, Germany) at −20°C, then air-dried overnight at 25°C, and stored at −20°C for further immunofluorescence staining. For WB and PCR analysis, rats were transcardially perfused with normal saline and retinas were rapidly removed, immediately frozen in liquid nitrogen, and stored at −80°C for further study.

Immunofluorescence Staining

Olfactory ensheathing cells and BV2 cells were plated on PLL-treated cover slips. Following each treatment, cells were fixed in 4% PFA for 10 min and rinsed with PBS for three times. Cover slips and sections were permeabilized with 0.03% Triton X-100 for 10 min, and then blocked with 1% goat serum for 0.5 h at 37°C. Cover slips and sections were incubated in primary antibodies at 4°C overnight (**Table 1** showed primary antibody dilutions). After washing off the primary antibody with PBS, secondary antibodies (1:1000; Invitrogen, United States) were applied at 37°C in the dark for 0.5 h. 4′,6-diamidino-2-phenylindole (DAPI) (Beyotime, China) was used to counterstain nuclei for 10 min in the dark at room temperature. Sections were washed in PBS for three times, and mounted using anti-fade mounting medium (Beyotime, China). Leica SP5 microscope was used to take confocal micrograps at the Central Laboratory in TMMU (Leica Microsystems, Wetzlar, Germany).

RNA Extraction and Real-Time PCR

Real-time polymerase chain reaction (RT-PCR) was implemented as previously described (Fu et al., 2017; He et al., 2017). Briefly, Trizol reagent (Sigma-Aldrich) was used to extract total RNA according to the manufacturer's instructions, and then reverse transcribed with an oligo (dT) primer. Sybr Primix EX TaqTM II and a Takara Thermal Cycler Dice^{TM} Real Time System (Takara Bio Inc. Kusatsu, Shiga, Japan) was used to perform Real-time PCR amplification, according to the manufacturer's protocols. The selected primers are listed in **Tables 2, 3**. All the data were normalized to GAPDH expression. The experimental groups were then normalized to control groups expression.

Western Blot

Samples (either BV2 cells or rat retinas) were collected to a glass homogenizer and grinded with radioimmunoprecipitation assay (RIPA) buffer containing protease inhibitor (Beyotime), and then incubated on ice for 15 min. The supernatant was isolated by supercentrifuge at 15,000 g/min for 5 min at 4°C. The total protein concentration in each sample was quantified using the bicinchoninic acid (BCA) Protein Quantitation Kit (Beyotime). The protein samples were added with 5× SDS (Sodium Dodecyl Sulfate) loading buffer (Beyotime) in a 4:1 ratio and then subjected to SDS-PAGE (Polyacrylamide Gel Electrophoresis; Beyotime). After electroblotting transferred onto polyvinylidene fluoride (PVDF) membranes (Bio-Rad), the 5% non-fat milk in tris-buffered saline Tween (TBST) was used to block and different primary antibodies were incubated with at 4°C overnight (**Table 1**). Washed the membranes with TBST three times and incubated with horseradish peroxidase-conjugated sheep anti-mouse (1:3000; Santa Cruz Biotechnology) or goat anti-rabbit immunoglobulin-G (1:3000; Santa Cruz Biotechnology) as secondary antibodies for 2 h at 25°C. After washing three times with TBST, chemiluminescence detection reagents were used to visualize the bands on the membranes. β-actin was used as a internal control.

TABLE 2 | PCR primer sequences (rat) used to detect pro- and anti-inflammatory cytokines in RCS rats' retina.

Genes (rat)	Forward primer	Reverse primer
GAPDH	AAGGTCGGTGTGAACGGATT	TGAACTTGCCGTGGGTAGAG
TNF-α	CTCAAGCCCTGGTATGAGCC	GGCTGGGTAGAGAACGGATG
IL-6	TCCTACCCCAACTTCCAATGC	TAGCACACTAGGTTTGCCGAG
MCP-1	GCTGTAGTATTTGTCACCAA GCTCAA	GTACTTCTGGACCCATTCCTT ATTG
ICAM-1	AGTGCTGTACCATGATCAGA ATACCT	TAAATGGACGCCACGATCAC
Arg1	CCTGAAGGAACTGAAAGG AAAGTT	GCAAGCCGAT GTACACGATGT
IL-4	ACCCTGTTCTGCTTTCTC	GTTCTCCGTGGTGTTCCT
IL-13	AATCCCTGACCAACATCT	ATAAACTGGGTACTTCG
Iba1	CGAATGCTGGAGAAACTTGG	GTTGGCTTCTGGTGTTCTTTG
TMEM119	GCTACGCTTTCTTCACGTTGC	AACCAATCAGGAAGTGGGGT
PKC-α	TTTCTTCCCCCACCCAATCC	AGGGTCCAAGTCTCTTTGTTTCC
Rhodopsin	AACCTTGAGGGCTTCTTTGCCA	AAGTTGCTCATGGGCTTGGAGA
SOCS3	TTCTTTACCACCGACGGAAC	CACGTTGGAGGAGAGAGGTC

TABLE 3 | PCR primer sequences (mice) used to detect pro- and anti-inflammatory cytokines in BV2 cells.

Genes (mice)	Forward primer	Reverse primer
GAPDH	CAGCAACTCCCACTCTTCCAC	TGGTCCAGGGTTTCTTACTC
TNF-α	TGTGCTCAGAGCTTTCAACAA	CTTGATGGTGGTGCATGAGA
IL-6	TAGTCCTTCCTACCCCAATTTCC	TTGGTCCTTAGCCACTCCTTC
Arg1	ACAAGACAGGGCTCCTTTCAG	GGCTTATGGTTACCCTCCCG
IL-4	ATCCATTTGCATGATGCTCT	GAGCTGCAGAGACTCTTTCG
Iba1	GGATTTGCAGGGAGGAAAAG	TGGGATCATCGAGGAATTG
TMEM119	GTGTCTAACAGGCCCCAGAA	AGCCACGTGGTATCAAGGAG

Statistical Analysis

All data are expressed as mean ± standard deviation (SD). Data were analyzed using SPSS software (v21, IBM, Armonk, NY, United States). The data were evaluated using unpaired two-sample t-tests, to compare the means at the same time points between the experimental and control groups; one-sample t-tests where test distributions were compared to a specific value, or two-way ANOVA followed by Fisher's protected least-significant difference *post hoc* tests. Values of $p < 0.05$ were considered statistically significant.

RESULTS

Resident Microglia Activation and Macrophages Infiltration During Retinal Degeneration

In order to investigate the dynamic change of resident microglia activation and infiltrated macrophages during retinal degeneration, we performed immunohistochemistry on retinal sections from different age of RCS and normal control rats. We stained ionized calcium-binding adaptor molecule-1 (Iba1), a marker for both activated microglia and macrophages (Li et al., 2016), and used TMEM119 to specifically label activated resident microglia (Bennett et al., 2016). In retinas of normal control rats, we rarely saw staining with either Iba1 or TMEM119 (**Figures 2A1,A3,A5,C1,C3,C5**). In contrast, in RCS rats we saw Iba1-positive and TMEM119-positive cells, whose numbers varied during the course of retinal degeneration, over the first 90 post-natal days.

During early-stage degeneration (P30), Iba1-positive cells in RCS rat retinas were localized in the ganglion cell layer (GCL) and ONL (**Figure 2A2**). There was large number of TMEM119-positive cells localized in the ONL only (**Figure 2C2**). At mid-stage (P60), there was an increased number of Iba1-positive cells, and these were found across

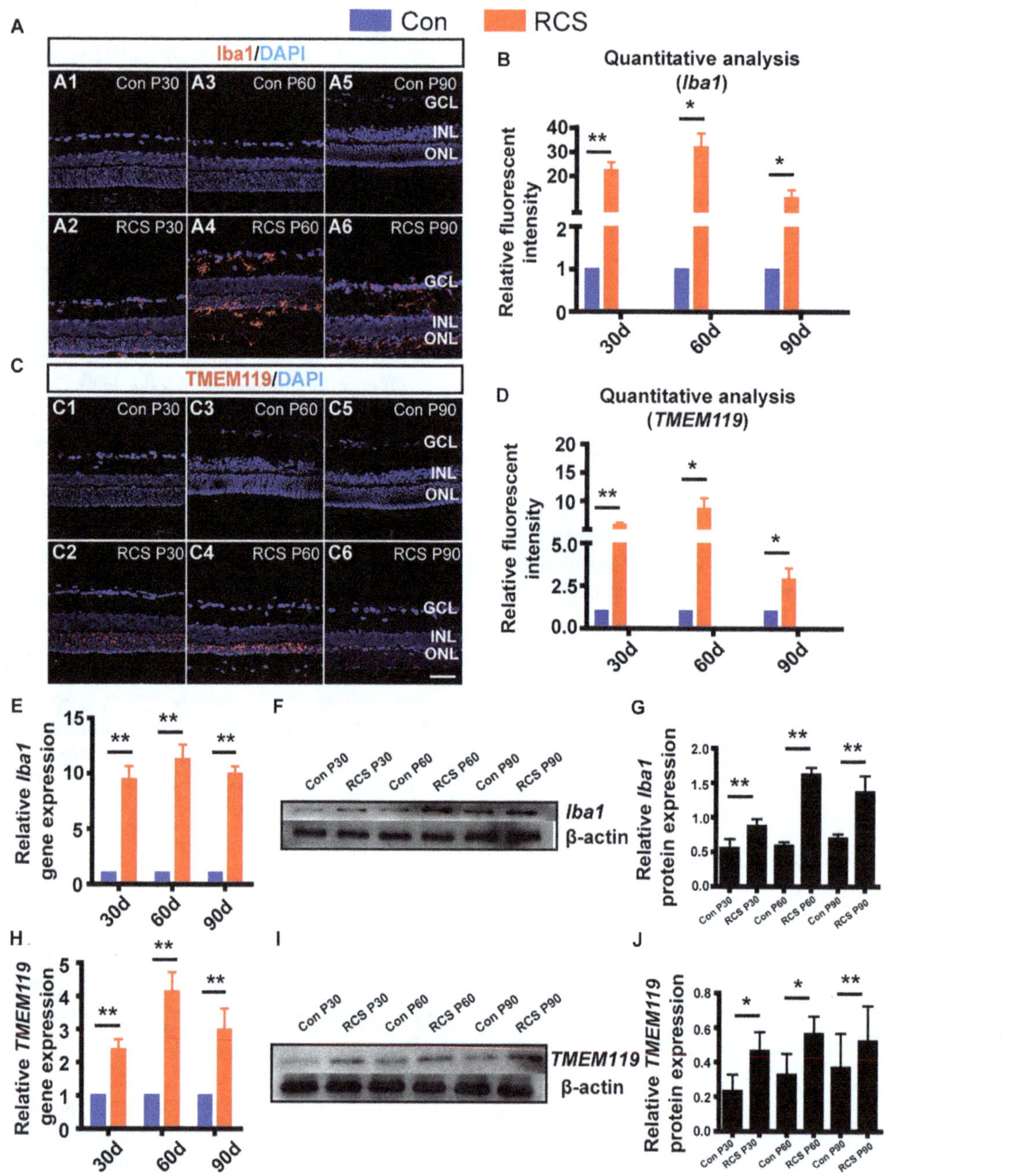

FIGURE 2 | Distribution and activation marker change of resident microglia and infiltrated macrophage in RCS rats. **(A)** Immunohistological images, labeled with Iba1 (red), and DAPI (blue) in retinal slices from control (rdy) rats **(A1,A3,A5)**, and RCS rats of different post-natal ages **(A2,A4,A6)**. **(B)** Quantitative group analysis of relative Iba1 fluorescence intensity at different ages, normalized to each control ($n = 3$ per bar). **(C,D)** Same as **(A,B)**, but for TMEM119 (red) **(E)** mRNA expression level of Iba1 in RCS retinas, relative to expression in control retinas of the same age ($n = 3$ per bar). **(F)** Example western blot of Iba1 protein levels. **(G)** Quantitative group data of Iba1 protein levels in RCS retinas, normalized to β-actin expression levels, and compared to each control retinas ($n = 3$ per bar). **(H–J)** Same as **(E–G)** but for TMEM119. *$p < 0.05$, **$p < 0.01$; scale bars: 50 μm.

the entire retina. These cells had a ramified appearance, suggesting migrating, activated microglia/macrophages (**Figure 2A4**). TMEM119-positive cells remained localized to the ONL but their immunofluorescence became more intense (**Figure 2C4**). At late-stage (P90), there was obvious apoptosis of retinal cells, and the number of both Iba1-positive and TMEM119-positive cells decreased, compared to P60 (**Figures 2A6,C6**).

Secondly, we performed quantitative analysis of relative fluorescence intensity, compared to control retinas. This showed that both Iba1 and TMEM119 intensity was significantly elevated at P30 (Iba1: 22.14 ± 3.3 fold; TMEM119: P30: 5.71 ± 0.42 fold;

$n = 3$ per group; $p < 0.01$ for each vs. controls). Peak expression was at P60 (Iba1: 31.79 ± 5.78 fold; TMEM119: 8.59 ± 1.86 fold; $n = 3$ per group; $p < 0.01$ for each vs. control). Expression then fell at P90, but was still significantly elevated (Iba1: 11.06 ± 3.26-fold; TMEM119: 2.88 ± 0.65-fold $n = 3$ per bar; $p < 0.01$ for each vs. controls) (**Figures 2B,D**).

Thirdly, to investigate changes in Iba1 and TMEM119 at the mRNA and protein level, we also performed RT-PCR and WB. Compared to control rats of the same age, the mRNA expression of Iba1 was increased in RCS retinas by more than ninefold at all ages ($p < 0.01$ for comparisons to control at all ages, $n = 3$ per group; **Figure 2E**). The expression of TMEM119 was increased by more than twofold at all ages ($p < 0.01$ for comparisons to control at all ages, $n = 3$ per group) (**Figure 2H**). In WB analysis (e.g., **Figures 2F,I**), the densities of Iba1 and TMEM119 bands were significantly higher in the RCS rats compared with the same age of control rats (Iba1: Con vs. RCS: P30: 0.87 ± 0.1 vs. 0.56 ± 0.13; P60: 0.59 ± 0.05 vs. 1.63 ± 0.09; P90: 0.7 ± 0.06 vs. 1.38 ± 0.23 $n = 3$ per group; $p < 0.05$; **Figure 2G**) (TMEM119: Con vs. RCS: P30: 0.23 ± 0.09 vs. 0.46 ± 0.11; P60: 0.33 ± 0.12 vs. 0.57 ± 0.1; P90: 0.37 ± 0.2 vs. 0.52 ± 0.2 $n = 3$ per group; $p < 0.05$; **Figure 2J**).

Retinal Degeneration Induces JAK2/STAT3 Pathway Activation and Downstream Cytokines Expression

We next examined the expression of pJAK2, pSTAT3, JAK2 and STAT3 because the JAK/STAT pathway is a well-known modulator of microglia activation and pro-inflammatory cytokines expression. From WB result of retinal tissues from different age of RCS and normal control rats, we can find the expressions of pJAK2, pSTAT3, JAK2, and STAT3 were all up-regulated during retinal degeneration in RCS rats ($n = 3$ per group, **Supplementary Figure S1**) (JAK2: P30: 1.11 ± 0.09 fold; P60: 2.04 ± 0.13 fold; P90: 2.26 ± 0.12 fold; $n = 3$ per group; $p < 0.01$ for P60 and P90 RCS vs. control; STAT3: P30: 1.13 ± 0.12-fold; P60: 1.53 ± 0.05-fold; P90: 1.72 ± 0.11-fold; $n = 3$ per group; $p < 0.01$ for P60 and P90 RCS vs. control) (**Figures 3A–C**). We next examined whether retinal degeneration induced expression of classically identified STAT inducible genes. The results shown in **Figure 3D** reveal that retinal degeneration induced the high expression of pro-inflammatory factors such as TNF-α, IL-6, ICAM-1, and MCP-1 in a time-dependent manner, which is indicative that microglia/macrophages were activated with "classically" phenotype (Qin et al., 2016). In contrast, expression of anti-inflammatory factors such as Arg1, IL-4, and IL-10 tended to be more highly expressed at P30 and were significantly reduced at P60 and P90 ($n = 3$ at each time-point; $p < 0.01$ for each vs. controls, **Figure 3E**). These results collectively demonstrate that retinal degeneration induces activation of JAK2 and STAT3 and downstream gene expression in microglia/macrophages indicative of the pro-inflammatory phenotype. Alternatively activation, responsible for anti-inflammatory effects, are increased in early-stage degeneration (but at much lower multiples of the control level than pro-inflammatory cytokines). Anti-inflammatory cytokine gene expression is then reduced from P60, and this reduction in anti-inflammatory mediators may also have a pro-inflammatory outcome.

OECs Inhibit LPS Induced BV2 Microglia Activation, and Changed the Expression Level of Inflammatory Factors and JAK2/STAT3 Pathway *in vitro*

To examine the immunomodulating potential of OECs against inflammation *in vitro*, LPS-induced microglia activation were established in BV2 cell line (experimental design in **Figure 1**). BV2 microglial cells were treated with LPS (1 μg/mL) for 4 h to release the pro-inflammatory factors (Dai et al., 2011; Li et al., 2016). We then co-cultured the BV2 cells with either OECs, or ONFs (control), for another 24 h, before collecting the BV2 for further use.

Following LPS treatment, we saw increased cell density consistent with microgliosis (**Figures 4A1** vs. **A2**). BV2 cells in the untreated (BV2-only) group showed round/oval or ramified morphology (**Figure 4A1'**). After stimulation with LPS, the morphology of BV2 cells changed into an amoeboid shape (**Figure 4A2'**). BV2 cells in the group co-cultured with OECs (BV2-LPS-OEC) showed similar morphology to the BV2-only group (**Figures 4A3,A3'**), whereas BV2 cells co-cultured with ONFs (control) showed similar morphology to the BV2-LPS group (**Figures 4A4,A4'**). Immunocytochemistry using the microglial marker Iba1 (**Figures 4B1–B4'**) and TMEM119 (**Figures 4C1–C4'**) demonstrated increased expression in the LPS-stimulated BV2 group and BV2-LPS-ONF group, compared to the BV2-only, and BV2-LPS-OEC group. BV2 microglial cells were treated with LPS for 4 h to induce the pro-inflammation phenotype. The mRNA expression of microglial markers (Iba1, TMEM119, **Figures 4D,E**) and pro-inflammation cytokines (TNF-α, IL-6, **Figures 4F,G**) in BV2 cells was enhanced significantly, while the anti-inflammation cytokines (Arg1, IL-4, **Figures 4H,I**) decreased greatly. These trends were markedly reversed after co-culture with OECs. Thirdly, the results from WB analysis revealed that the JAK2 and STAT3 expressions in LPS-stimulated BV2 cells co-cultured with OECs were significantly lower than those in the LPS-stimulated BV2 group (JAK2: 0.62 ± 0.15-fold in BV2-LPS-OEC group; STAT3: 0.81 ± 0.07-fold in BV2-LPS-OEC group; compared with the BV2-LPS group, $p < 0.05$) (**Figures 4J–L**). Moreover, we found purified OECs were immunopositive for the specific NGFRp75 (**Figure 5A1**) and SOCS3 (**Figure 5A2**) using double immunofluorescence staining (**Figure 5A3**). The mRNA expression of SOCS3 in OECs was enhanced significantly from 12 to 48 h after co-cultured with LPS induced BV2 cells, ONFs were used as negative control [SOCS3: 12 h: 3.5 ± 0.74 fold; 24 h: 7.45 ± 2.12 fold; 48 h: 13.24 ± 0.51 fold; $n = 3$ per group; $p < 0.05$ for 12 and 48 h OECs vs. ONFs; $p < 0.01$ for 24 h OECs vs. ONFs) (**Figure 5B**)]. These results suggested that OECs could inhibit microglia activation, reduce expression of pro-inflammatory cytokines (TNF-α, and IL-6), increase the expression of anti-inflammatory cytokines (Arg1 and

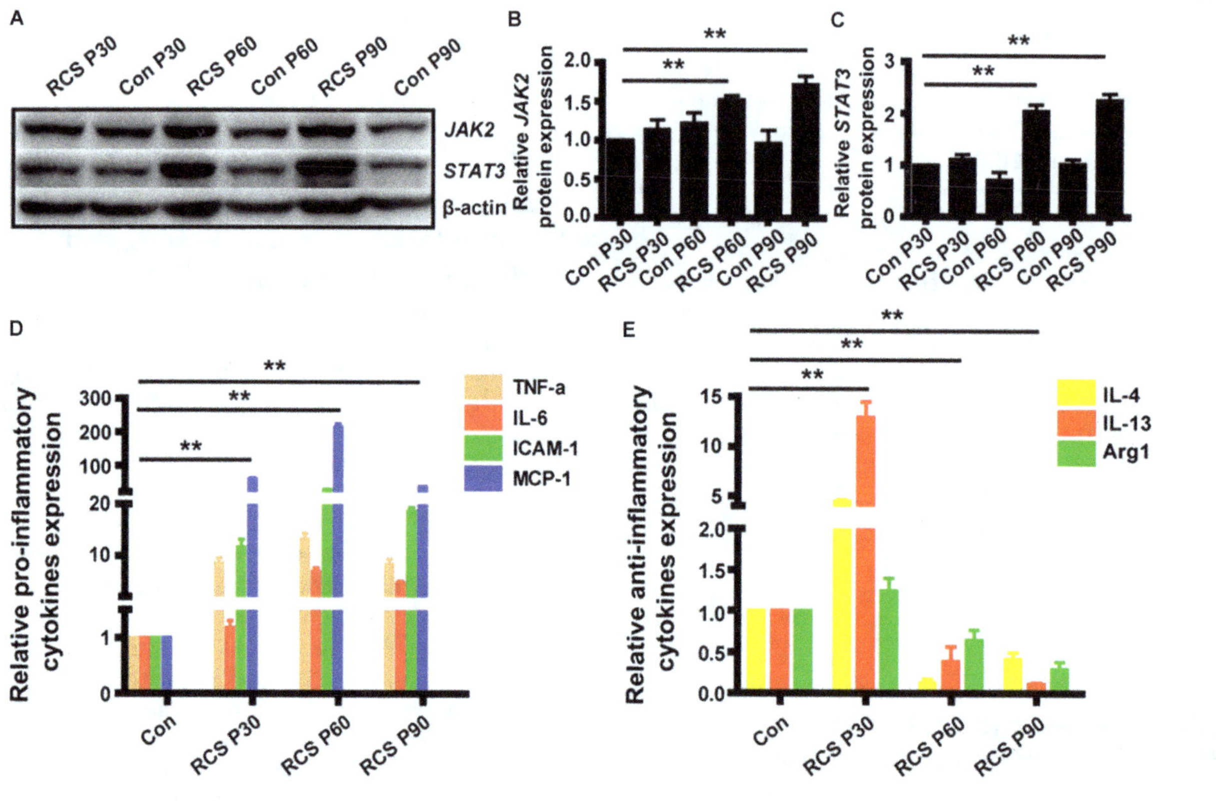

FIGURE 3 | JAK2/STAT3 activation and the change of downstream cytokines expression in RCS rat retinas. **(A)** WB results for protein levels of JAK2 and STAT3 in retinas from RCS rats and control rats of different ages. **(B)** Group data showing JAK2 expression by western blot in RCS rats. **(C)** The same as **(B)**, but for STAT3. **(D)** mRNA expression levels of pro-inflammatory cytokines: TNF-α, IL-6, ICAM-1, and MCP-1 (n = 3 per bar). **(E)** mRNA expression levels of anti-inflammatory cytokines: IL-4, IL-13, and Arg1 (n = 3 per bar). $^*p < 0.05$, $^{**}p < 0.01$.

IL-4), down-regulate JAK2/STAT3 pathway by over-expressed SOCS3 *in vitro*.

OECs Regulate the Expression Level of Pro- and Anti-inflammatory Factors of Microglia in a JAK2/STAT3-Dependent Manner

To determine if JAK2/STAT3 pathway activation was causally involved in the OEC-induced change of pro- and anti-inflammatory factors in microglia, we incubated OECs with AG490, a specific chemical antagonist of JAK2 (at 50 μM, for 1 h) followed by co-culture with LPS-stimulated BV2 for 24 h. We found that the morphology of BV2 co-cultured with AG490-pretreated OECs was amoeboid (**Figure 6A3**), similar to the LPS-induced BV2 group (**Figure 6A**), unlike the morphology of BV2 cells co-cultured with OECs (**Figures 6A2,A2'**). Additionally, immunocytochemistry showed increased expression of Ibal in the BV2-LPS-OEC-AG490 group (**Figure 6B**).

Secondly, we measured the mRNA expression levels of microglial activation markers (Ibal, TMEM119), pro-inflammatory factors (TNF-α, IL-6), and anti-inflammatory factors (Argl and IL-4) of BV2 in each group, using RT-PCR. We found that AG490-pretreatment reversed the changes in mRNA expression by BV2 cells associated with OEC co-culture, such that the BV2-LPS-AG490-OEC group was statistically indistinguishable from the BV2-LPS group, and significantly different from the group treated with normal OECs (**Figures 6C–H**).

Finally, we studied the expression of JAK2/STAT3 proteins in BV2 cells co-cultured with AG490-pretreated OECs, using WB. We found that expression of pJAK2, pSTAT3, JAK2, and STAT3 was very similar between the BV2-LPS, BV2-LPS-ONF groups and the BV2-LPS-OEC-AG490 groups ($n = 3$ per group, **Supplementary Figure S2**) (JAK2: 1.12 ± 0.59-fold in BV2-LPS-OEC-AG490 group; STAT3: 0.94 ± 0.03-fold in BV2-LPS-OEC-AG490 group; compared with the BV2-LPS group, $p > 0.05$) (**Figures 6I–K**). These results collectively demonstrated that pretreatment of OECs with AG490 blocked their effects on pro/anti-inflammatory cytokines expression, suggesting that this effect of OECs is dependent on the JAK2/STAT3 pathway in activated BV2 microglia cells.

OECs Reduce the Classical Immune Cell Activation and the Expression Levels of Pro-inflammatory Factors *in vivo*

In order to understand whether OECs can show similar effect on activated microglia/macrophages *in vivo*, and whether they can modulate the pro-inflammatory microenvironment, we

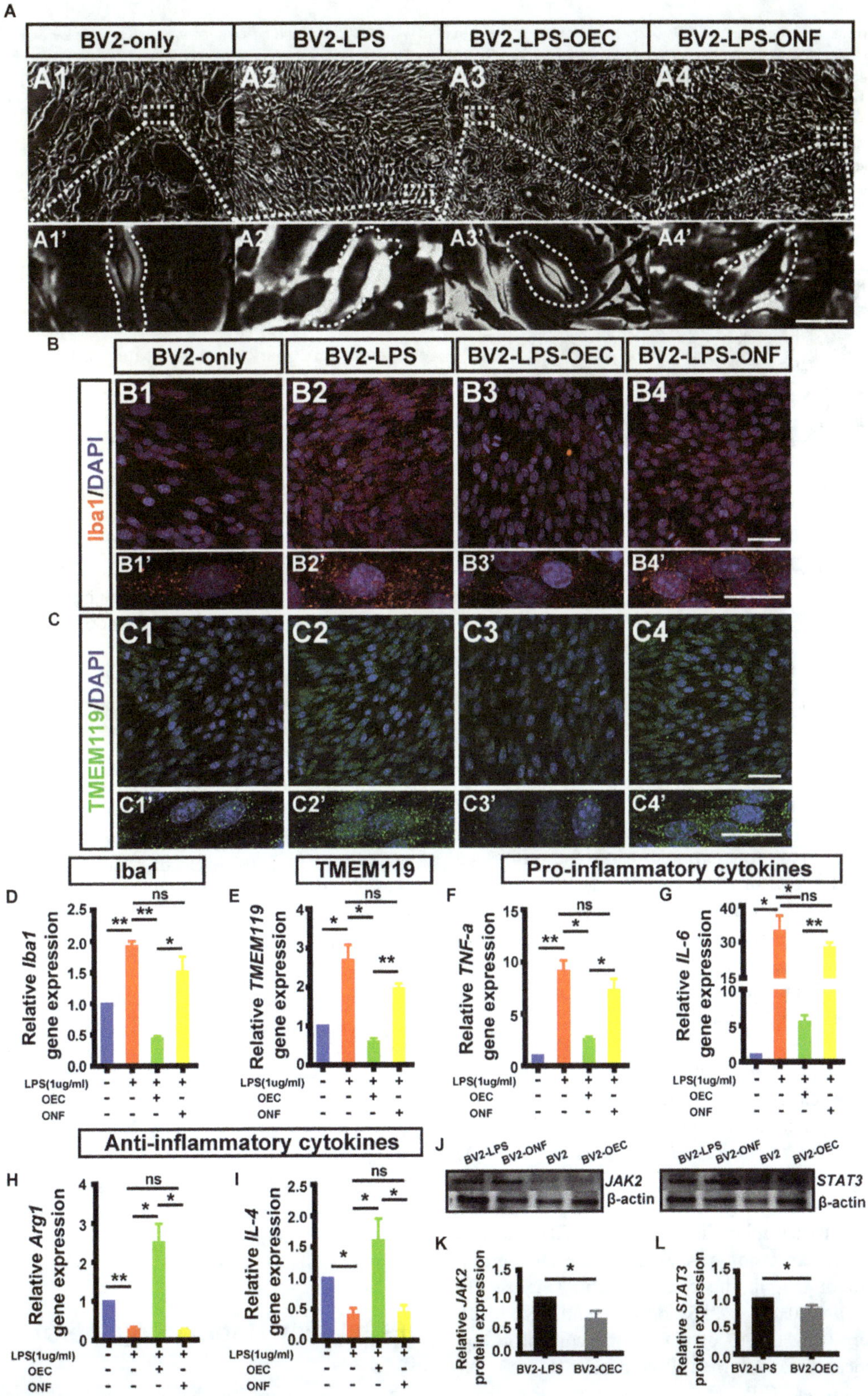

FIGURE 4 | OECs inhibited microglial activation *in vitro* and changed the expression of pro- and anti-inflammatory factors. **(A1)** Optical microscopy image showing density of normal cultured BV2 cells (BV2-only). Lower panel **(A1')** shows an enlargement of the area marked, which shows morphology of microglia. **(A2)** The same *(Continued)*

FIGURE 4 | Continued
for BV2 cells treated with LPS for 4 h (BV2-LPS). **(A3)** The same for BV2 cells treated with LPS for 4 h and co-cultured with OECs (BV2-LPS-OEC). **(A4)** The same for BV2 cells treated with LPS for 4 h and co-cultured with ONFs (BV2-LPS-ONF). Scale bars: 100 μm. in upper panels, 20 μm in lower panels **(B1–B4)** Immunohistochemistry images showing staining with Iba1 (red) and DAPI (blue) for the same groups as in **A1–A4**. Scale bars: 50 μm. **(B1'–B4')** Lower panel shows an enlargement of the cells in **B1–B4**. Scale bars: 20 μm. **(C1–C4)** The same for staining with TMEM119. Scale bars: 50 μm. **(C1'–C4')** Lower panel shows an enlargement of the cells in **C1–C4**. Scale bars: 20 μm. **(D)** mRNA expression level of Iba1 after 24 h co-culture. **(E)** TMEM119 mRNA expression. **(F,G)** Expression of pro-inflammatory cytokines: TNF-α and IL-6, respectively. **(H,I)** Expression of anti-inflammatory cytokines: Arg1 and IL-4, respectively. **(J)** Example WB of JAK2 and STAT3, with β-actin as loading control. **(K)** Quantified JAK2 protein expression by WB (n = 3 per group). **(L)** Same for STAT3 (n = 3 per group). *p < 0.05, **p < 0.01.

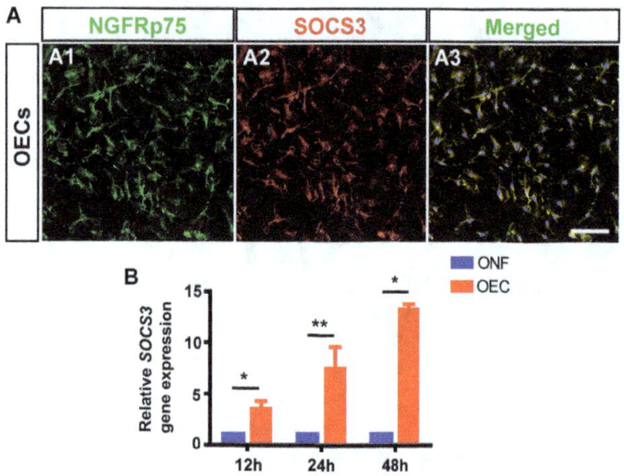

FIGURE 5 | OECs express SOCS3 *in vitro*. **(A)** Immunohistology images, labeled for NGFRp75 (*green*), SOCS3 (*red*) and DAPI (*blue*) in the purified OECs. Scale bar: 50 μm. **(B)** mRNA expression level of SOCS3 after 12, 24, and 48 h co-culture with LPS-BV2 cells compared with ONFs (n = 3 per group). *p < 0.05, **p < 0.01.

assessed the properties of resident microglia, and infiltrated macrophages, and the level of inflammatory markers, 4 weeks after subretinal OEC transplantation in live RCS rats. To do this, we successfully produced EGFP-labeled OECs, using transfection via a lentiviral vector, and these labeled cells were injected subretinally (**Supplementary Figure S3**). Rats were then sacrificed at 4 weeks post-transplantation, and their retinas isolated for further study.

Firstly, and in agreement with the previous *in vitro* results, immunohistochemistry of retinas from the *in vivo* model demonstrated that the number of TMEM119-positive cells (activated resident microglia; *white arrows*, **Figure 7A**) was reduced in RCS rats treated with OECs compared with the control (PBS-injected) group. Secondly, we collected homogenized retinal tissue and used RT-PCR and WB confirmed that, TMEM119 gene expression at 4 weeks in the OEC-group was 0.56 ± 0.146 of that in the control group (**Figure 7B**; n = 3 per group; p < 0.05 vs. PBS control), and TMEM119 protein levels were 0.56 ± 0.13 of that of the control group, both were significantly lower than PBS group (n = 3 per group; p < 0.05; **Figures 7C,D**). However, the number of Iba1-positive cells in the retina did not seem decreased after OEC transplantation (**Figure 7E**). Even Iba1 mRNA levels (**Figure 7F**) and protein levels (**Figures 7G,H**) were comparable between OEC and control groups (p > 0.05 for each).

As we know, microglia are the primary immune cells in the retina, which share many phenotypic and functional properties with macrophages. We have already confirmed OECs have effect on activated microglia *in vitro* and *in vivo*, why the common marker of activated microglia and infiltrated macrophage Iba1 keep stable? We additionally used double-staining of NGFRp75 and Iba1 in purified OECs and in RCS rat after transplantation (**Supplementary Figure S4**). The result showed OECs not only express Iba1 *in vitro*, which merged with it's special marker NGFRp75 (**Supplementary Figure S2A**), but also the Iba1 positive cells *in vivo* stain positive with GFP-OEC after transplantation (**Supplementary Figure S4B**). This circumstantial result indicated that transplanted OECs may also affect Iba1 expression *in vivo*, but due to OEC express Iba1 itself, the Iba1 expression amount keep stable in whole retinal detection.

Having shown *in vitro* that OECs changed the expression level of pro- and anti-inflammatory factors of microglial through the JAK2/STAT3 pathway, we wanted to investigate it *in vivo*. We performed WB analysis of protein lysates from retinas from the OEC group at 4 weeks post-transplantation. This showed that JAK2 expression was reduced to 0.49 ± 0.05 of control group levels (n = 3 per group; p < 0.01; **Figures 8A,B**) and STAT3 was reduced to 0.70 ± 0.06-fold of control group levels (n = 3 per group; p < 0.05; **Figures 8C,D**). These results indicate that OEC treatment reduced the activation of the JAK2/STAT3 pathway in the degenerative RCS retina. RT-PCR analysis showed in the OEC group, there was a significant reduction in mRNA expression of the pro-inflammatory cytokines compared to control (n = 3 per bar; p < 0.01 for each vs. control) (**Figure 8E**). In contrast, we found that anti-inflammatory factors were not affected by OEC transplantation *in vivo* (**Figure 8F**; n = 3 per bar, p > 0.05 for each vs. control). However, it did appear from immunostaining that there were a larger number of Arg1-positive cells around the transplantation site in the OEC group, compared to the PBS group (**Figure 8G**). Maybe the differences between RT-PCR and immunostaining indicated the expression level of anti-inflammatory factors were so low and just in the transplantation area that could not be detected in whole retina tissue.

OECs Delay Retinal Degeneration in RCS Rats

In previous studies, we have shown that transplanted OECs protect photoreceptors by releasing multiple neurotrophic factors, phagocytosing out segment, inhibiting Müller cell gliosis, and suppressing retinal oxidative stress reactions in rat models of retinal degeneration (Huo et al., 2011; Xie et al., 2017;

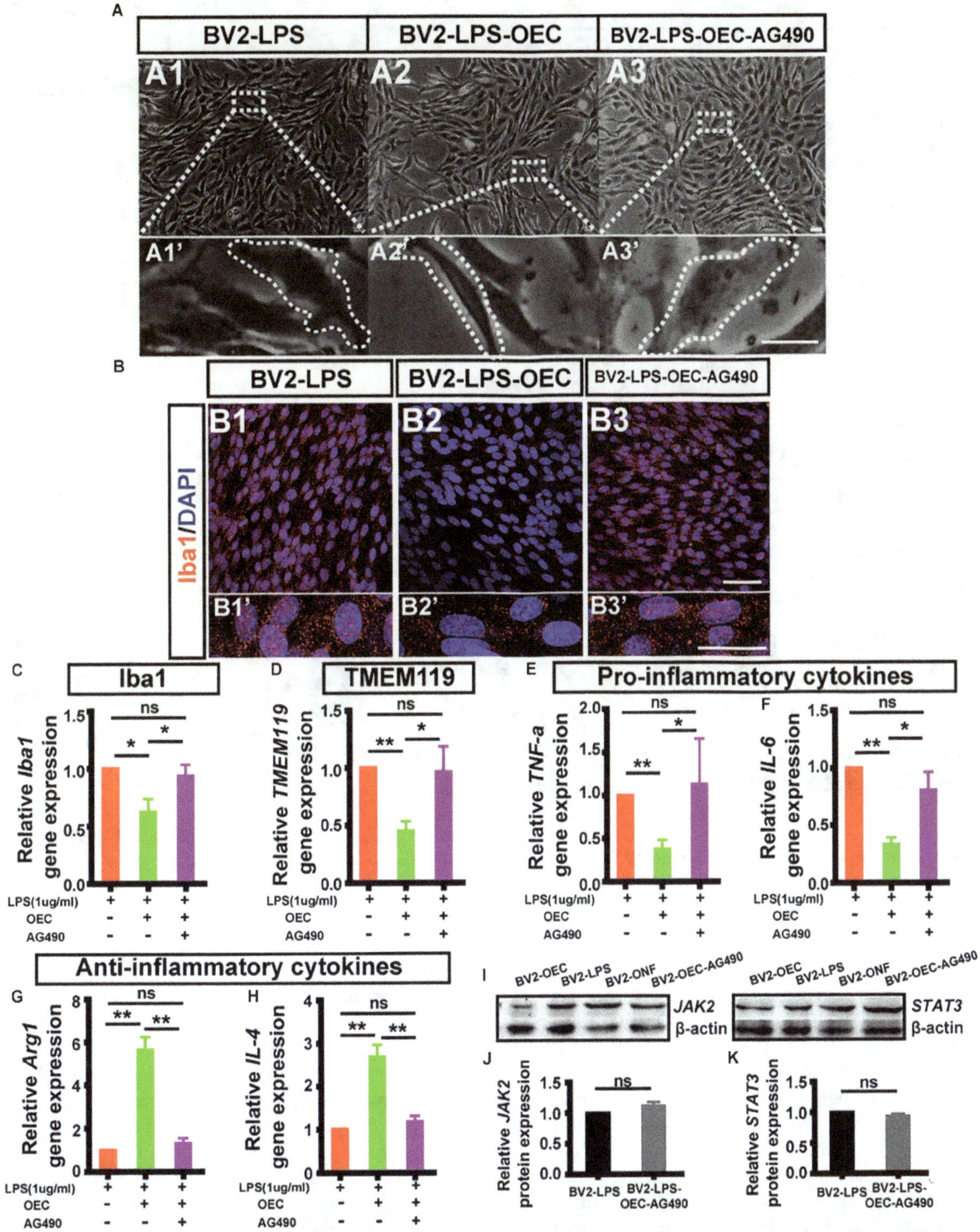

FIGURE 6 | The expression level of pro- and anti-inflammatory factors shift in activated microglia induced by OECs. **(A1)** Optical microscopy image showing density of BV2 cells treated with LPS for 4 h (BV2-LPS). Lower panel **(A1')** shows an enlargement of the area marked which shows the morphology of microglia. **(A2)** The same for BV2 cells treated with LPS for 4 h and co-cultured with OECs (BV2-LPS-OEC). **(A3)** The same for BV2 cells treated with LPS for 4 h and co-cultured with OECs pre-treated with AG490 (BV2-LPS-OEC-AG490). Scale bars: 100 μm in upper panels, 20 μm in lower panels **(B1–B3)** Immunohistochemistry images showing staining with Iba1 (red) and DAPI (blue) for the same groups as in **A1–A3**. Scale bars: 50 μm. **(B1'–B3')** Lower panel shows an enlargement of the cells in **B1–B3**. Scale bars: 20 μm. **(C)** mRNA expression level of Iba1 after 24 h co-culture. **(D)** TMEM119 mRNA expression. **(E,F)** Expression of pro-inflammatory factors: TNF-α and IL-6, respectively. **(G,H)** Expression of anti-inflammatory factors: Arg1 and IL-4, respectively. **(I)** Example WB of JAK2 and STAT3, with β-actin as loading control. **(J)** Quantified JAK2 protein expression by WB (n = 3 per group). **(K)** Same for STAT3 (n = 3 per group). $^{*}p < 0.05$, $^{**}p < 0.01$.

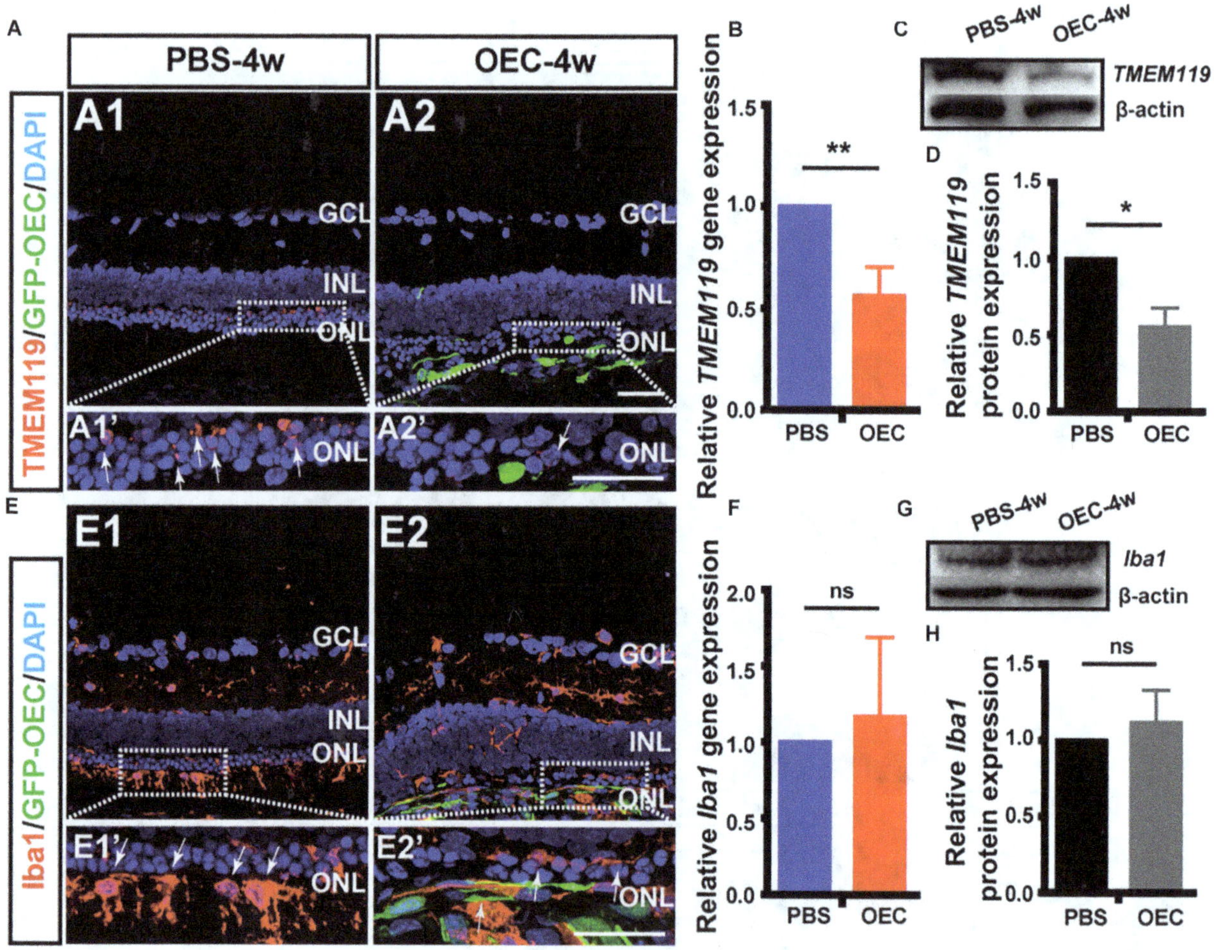

FIGURE 7 | OECs reduce the classical microglial activation *in vivo*. **(A1)** Immunohistology image of a retina from a RCS rat, 4 weeks following subretinal injection of PBS, labeled with Iba1 (red), and DAPI (blue) **(A2)** Same as **A1**, but for a retinal from the OEC-injected group, showing grafted OECs (green, EGFP), and a reduced number of TMEM119-positive cells (white arrows). **(B)** mRNA expression level of TMEM119 in the OEC-group, relative to expression in PBS group ($n = 3$ per bar). **(C)** Example western blot of TMEM119 protein levels. **(D)** Quantitative group data of TMEM119 protein levels in OEC-group, normalized to β-actin expression levels, and compared to PBS-group retinas ($n = 3$ per bar). **(E–H)** Same as **A–D**, but for Iba1. *$p < 0.05$, **$p < 0.01$; Scale bars: 50 μm.

Xue et al., 2017). Regardless of the model of retinal degeneration, we saw peak visual function improvement at 4 weeks post OEC transplantation. In the current study in RCS rats, we have shown a reduction in classically microglial activation and pro-inflammatory cytokine release at the same time point post OEC transplantation, and we therefore also tested visual function and observed retinal anatomy this key time point.

To determine visual function, we measured the amplitudes of the a-wave and b-wave from the ERG (**Figure 9A**). These are electrical features that are associated with the function of the outer and inner retina, respectively. B-wave amplitude was significantly larger in OEC-treated eyes than in PBS-treated controls ($p < 0.05$; **Figure 9C**). This suggested an improvement in inner-retinal function following OEC transplantation. However, there was no significant difference in a-wave amplitude between the transplant group and control group ($p > 0.05$; **Figure 9B**). As an additional negative control, we transplanted ONFs using the same procedure as OECs, and found no improvement in b-wave amplitude at 2, 3, or 4 weeks post-transplantation (**Supplementary Figure S5**).

We next studied changes in the anatomical structure of the retina. During retinal degeneration, the number of photoreceptors in the ONL is reduced and the expression of Caspase-3 is increased (Gu et al., 2017). Using histological methods, we found that OEC reduced the expression of Caspase-3 in transplantation area, and there was noticeably increased preservation of photoreceptor numbers in the ONL (white arrows, **Figures 9D,E**). This suggested that the apoptosis of photoreceptors in RCS rats was reduced by OEC transplantation.

Immunohistochemistry showed that the expression of both PKC-α (**Figure 9F**), a marker of ON-bipolar cells (which contribute to the b-wave), and Rhodopsin (**Figure 9G**), a marker of photoreceptors (which contribute to the a-wave), was increased at 4 weeks post-OEC transplantation, and compared to PBS controls.

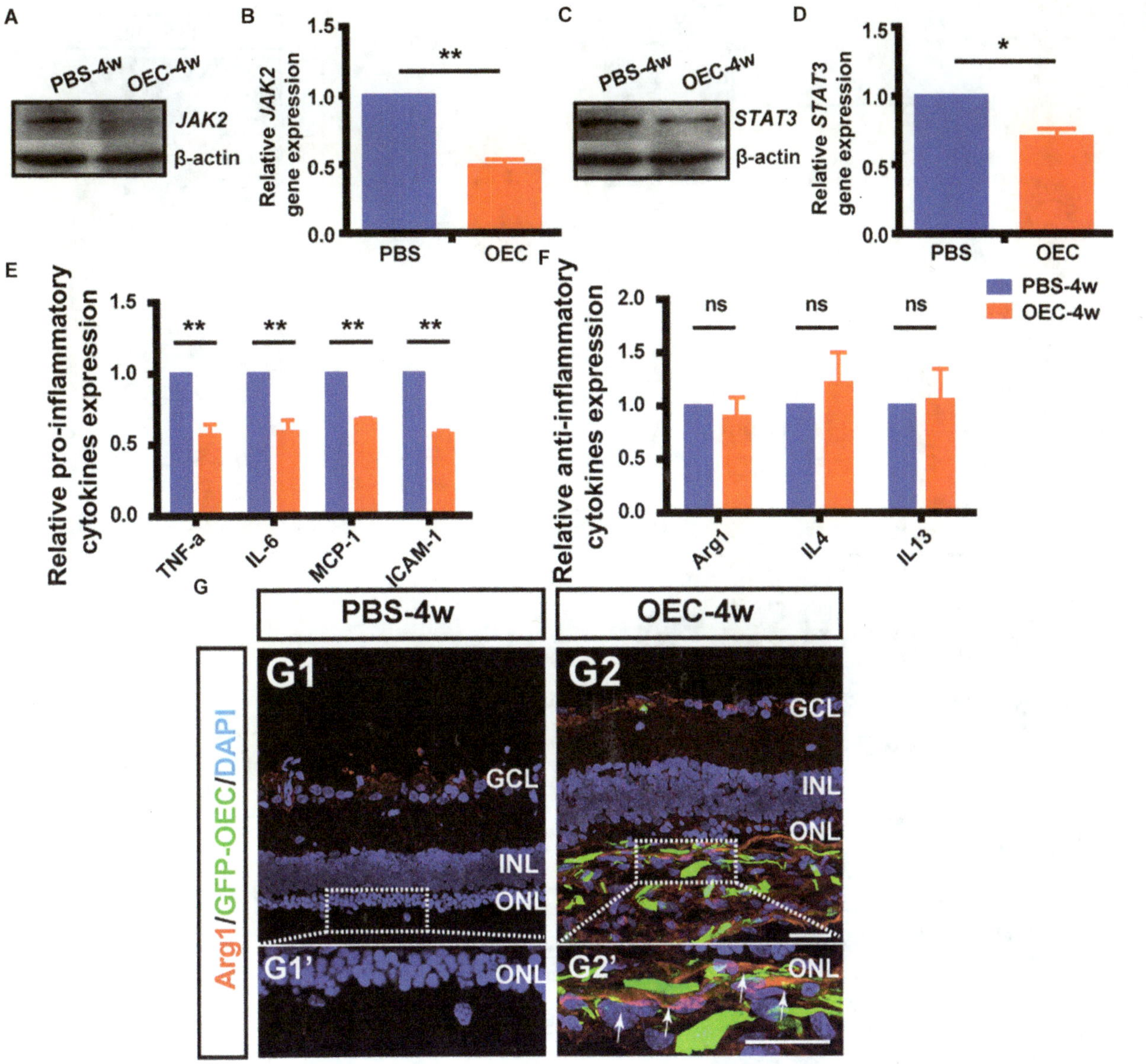

FIGURE 8 | OECs inhibit the JAK2/STAT3 pathway and change the expression level of pro-inflammatory factors *in vivo*. **(A,B)** Example WB and group data showing effect on JAK2 expression of OEC transplantation, 4 weeks post-injection. **(C,D)** As per A, B, but for STAT3 expression. **(E)** mRNA expression levels of pro-inflammatory factors: TNF-α, IL-6, ICAM-1 and MCP-1 ($n = 3$ per bar, $p < 0.05$ for both vs. PBS-group control). **(F)** mRNA expression levels of anti-inflammatory factors: IL-4, IL-13, and Arg1 ($n = 3$ per bar, $p > 0.05$ for both vs. PBS-group control). **(G)** Immunohistology image, labeled with Arg1 (*red*) and DAPI (*blue*), of a retina from the PBS-group **(G1)** and OEC-group **(G2)**, indicating increased Arg1-positive cell numbers in the transplantation area (*white arrows*). $^{*}p < 0.05$, $^{**}p < 0.01$; Scale bars: 50 μm.

Finally, RT-PCR and WB were used to quantify gene and protein expression. The gene expression ratio of both PKC-α and Rhodopsin was significantly increased in the OEC-group compared with the PBS-group (PKC-α: 1.84 ± 0.04 fold, **Figure 9H**, $n = 3$ per group; Rhodopsin: 1.56 ± 0.09 fold, **Figure 9I**, $n = 3$ per group; $p < 0.01$ for each vs. PBS control). WB (**Figures 9J,K**) also showed up-regulation of PKC-α and Rhodopsin protein following OEC transplantation (PKC-α: 1.82 ± 0.12 fold, **Figure 9L**, $n = 3$ per group; Rhodopsin: 1.81 ± 0.15 fold, **Figure 9M**, $n = 3$ per group; $p < 0.05$ for both vs. PBS control). These results collectively demonstrate that OEC transplantation can protect photoreceptors and ON-bipolar cells, and preserve the ERG b-wave in the RCS retina.

DISCUSSION

In the present study, we have shown how OECs can affect LPS-induced microglial activation profiles *in vitro*, how OEC transplantation *in vivo* can moderate the inflammatory microenvironment, protect photoreceptor and improve visual function, in chronic retinal degeneration animal model. Our

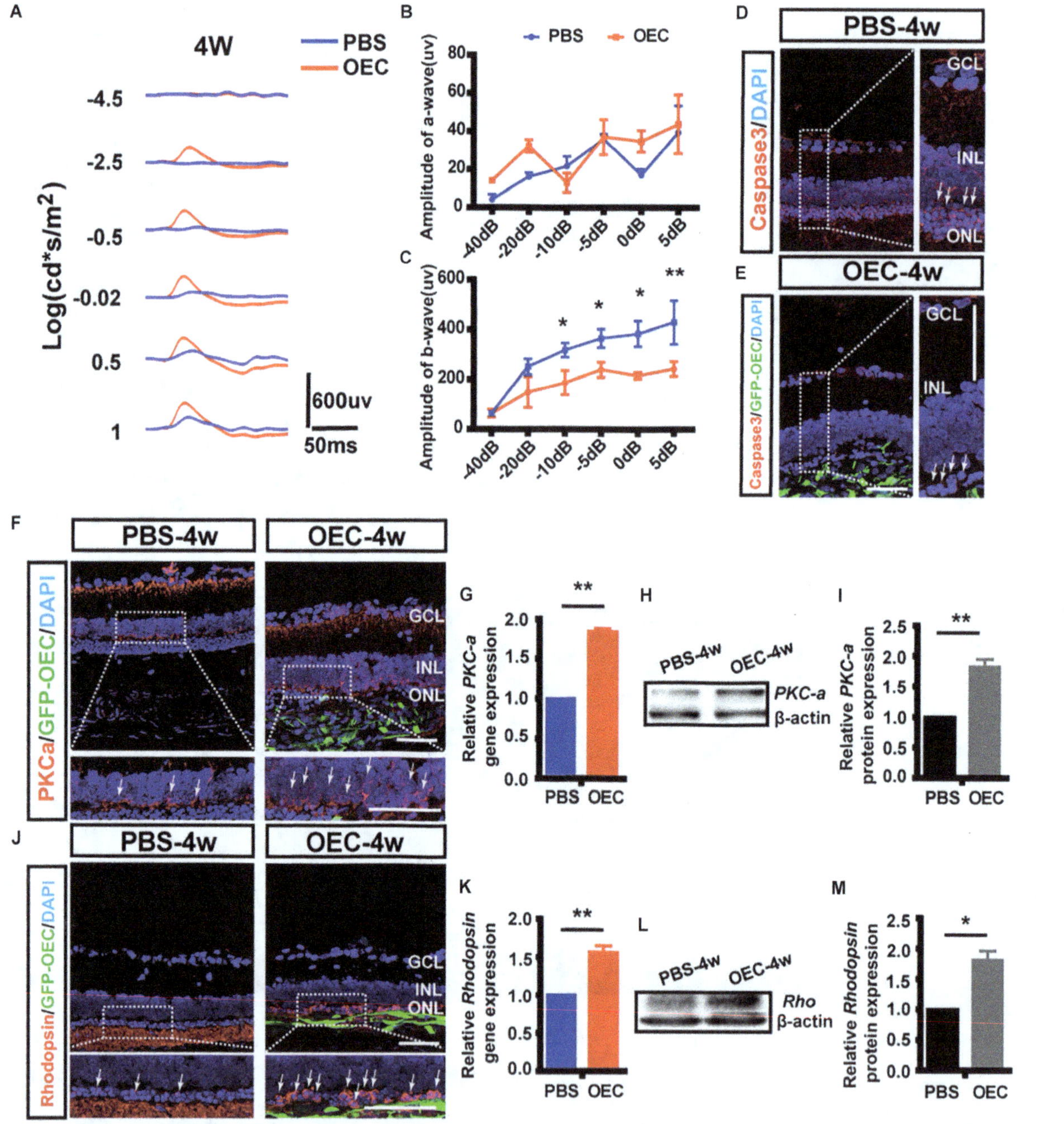

FIGURE 9 | Subretinal OEC transplantation delays functional and anatomical retinal degeneration in RCS rats. **(A)** Representative ERG waveforms of a rat at 4 weeks post-OEC transplantation (red trace), or PBS control injection (blue trace). **(B)** ERG a-wave and **(C)** ERG b-wave amplitude at 4-weeks post OEC transplantation (vs. PBS control injection, $n = 6$). **(D,E)** Immunohistology images, labeled for caspase-3 (red), GFP (green, OECs) and DAPI (blue) in the retinas of the PBS (top), and OEC-treated (bottom) groups. White arrows indicate caspase-3 positive cells. **(F)** Immunohistology images, labeled for PKC-α (red) GFP (green, OECs) and DAPI (blue) in the retinas of the PBS (left) and OEC-treated (right) groups. White arrows indicate PKC-α positive cells. **(G)** Immunohistology images, labeled for Rhodopsin (red) GFP (green, OECs) and DAPI (blue) in the retinas of the PBS (left) and OEC-treated (right) groups. White arrows indicate Rhodopsin positive cells. **(H,I)** The mRNA level of PKC-α **(H)** and Rhodopsin **(I)** in the OEC-group at 4 weeks post-transplantation, expressed relative to the PBS control group. **(J,K)** Example western blot of PKC-α **(J)** and Rhodopsin **(K)** at 4 weeks post-treatment. **(L,M)** Quantification of protein expression level of PKC-α protein **(L)** and Rhodopsin **(M)** at 4 weeks post-treatment, measured relative to β-actin, and expressed relative to the PBS control group. $^*p < 0.05$, $^{**}p < 0.01$. ONL, outer nuclear layer; INL, inner nuclear layer; GCL, ganglion cell layer. Scale bars: 100 μm in **A**; 50 μm in **B,H,I,J,N**; 20 μm in each enlargement.

results demonstrate that during RCS rats' retinal degeneration, classically activated immune cells up-regulated JAK2/STAT3 pathway and produce persistent pro-inflammatory factors, but only transient anti-inflammatory factors in early stage. *In vitro* experiments showed that OECs exert immunomodulatory effects through a change of expression level of inflammatory factors in LPS-induced microglia activation: away from M(IL-6)-type (TNF-α, IL-6, MCP-1 and ICAM-1), toward M(Argl)-type (Argl, IL-4, IL-13), via a mechanism that is dependent on the JAK2/STAT3 pathway. OEC transplantation was able to improve the inflammatory microenvironment of RCS rats *in vivo*: indicated by a reduction in microglial/macrophage activation, down-regulated JAK2/STAT3 pathway and pro-inflammatory factors expression level. Finally, OEC transplantation helps preserve visual function and delay retinal neuron degeneration, partly via these effects on the microglia/macrophage-mediated inflammatory microenvironment.

Resident microglia have been reported to be rapidly activated and migrate toward the inflamed lesion under pathological conditions in the retina (Akhtar-Schäfer et al., 2018; Szepesi et al., 2018). In RCS rats with *Mertk* gene mutation, BRB disruption results in the recruitment of blood-borne macrophages to help phagocytosis the apoptotic photoreceptors (Fernández-Sánchez et al., 2018). In our study, TMEM119 was applied to distinguish resident microglia and infiltrating macrophages. Consistent with previous studies (Greenhalgh et al., 2018; Li et al., 2018), in the normal retina, we observed resident microglia with a quiescent, ramified morphology, with low Ibal, and TMEM119 staining. But in RCS rats, a large number of TMEM119-positive cells were only located in the ONL, whilst Ibal-positive cells located from the GCL to the subretinal space. Cells positive for both types of staining increased from the early stages (P30), and reached peak expression at around P60. As we know, Ibal is the most common marker for activated microglia and macrophages. TMEM119 is a newly discovered and stable marker with unknown function for most or all mouse and human microglia, and it has been demonstrated that bone marrow-derived macrophages in the adult CNS do not express TMEM119 (Bennett et al., 2016). However, no one reported the expression manner of TMEM119 in rat's retina. Therefore, the present study revealed that $TMEM119^{+}$ cells located in ONL during retinal degeneration may stand for a subpopulation of resident microglia which associated with photoreceptor apoptosis, along with the $Iba1^{+}$ microglia/infiltrating macrophages, play a crucial role in releasing of pro-inflammatory cytokines, and amplify the neurodegenerative disease process.

Emerging evidence suggests that there is a dynamic change of microglia/macrophage during the process of retinal degeneration. Jiao et al. (2015) reported the cytokines released by retinal microglia/macrophage changed from pro- to anti-inflammation-type following acute light injury from 24 h to 7 days. In rd1 mice (a rapid-onset model of RP) with rapid rods degenerative process, the activated microglia adopted M(IL-6)-dominent phenotype and lack of M(Argl)-dominent phenotype (Zhou et al., 2017). In the P23H mice (a chronic RP model), retinal neuroinflammation persists throughout the mice life span, even after photoreceptor depletion (Noailles et al., 2016). In the present study, we found evidence of a prominent and persistent M(IL-6) phenotype and a transient M(Argl) phenotype in a RCS rat model of RP. Pro-inflammatory cytokines expression was increased in early life time, and reached a peak (hundreds of times higher than normal rat retinas) around P60. Anti-inflammatory cytokines expression was also increased many-fold at P30, but by P60, fell below control levels. These results suggest that some anti-inflammatory cytokines are activated in the degenerative retina to enhance debris phagocytosis, release protective/trophic factors, and support regeneration in the early stages. However, as degeneration continues, pro-inflammatory cytokines are robustly releasing by M(IL-6)-type microglia/macrophages, causing expanding tissue damage, and worsen disease. Interestingly, the microglia/macrophage activation pattern found in this study, with an early peak in anti-inflammatory factors expression, is in contrast to the more commonly observed pro- to anti-inflammatory cytokines shift described in the injury in CNS, including in spinal cord injury (SCI) (Kigerl et al., 2009), stroke (Hu et al., 2012), and traumatic brain injury (Wang et al., 2013).

Two recent studies have demonstrated that OECs can change the microglia/macrophage polarization toward M(Argl) in a SCI model, and this is accompanied by an inhibition of local inflammatory responses (Khankan et al., 2016; Zhang et al., 2017), but they didn't mention the possible mechanism. To investigate the specific immunomodulatory effect of OECs on microglia in retina, we used an *in vitro* Transwell co-culture system and an *in vivo* OEC transplantation model. *In vitro*, we found that OECs not only reduced Ibal and TMEM119 expression induced by LPS which stand for microglia cell line activation, but also changed the cytokines expression level of BV2 cells from the destructive M(IL-6) phenotype toward the neuroprotective and tissue-reparative M(Argl) phenotype. Because of the complex microenvironment, the result of *in vivo* studies were not all the same as *in vitro* ones. In RCS rats, we confirmed a reduction of $TMEM119^{+}$ resident microglial activation and a stable expression of Iba1. As we reported before, OECs, like microglia, and have the function of phagocytosis (Li et al., 2017). Some anatomy studies also suggested OECs are the primary immune cells in olfactory nervous system (Gladwin and Choi, 2015; Barton et al., 2017), but there is no reported correlation between Iba1 and OEC in past studies. From our study, immunohistochemistry demonstrates OECs co-expressed Ibal and its specific marker NGFRp75. Presumably although the transplanted OECs suppressed the Ibal level expressed by both resident microglia and infiltrated macrophages in transplanted area, due to the OEC itself also expressed Ibal, the whole detection level in post-transplanted retina did not vary significantly. RT-PCR result showed a decrease in pro-inflammatory cytokines and a stable in anti-inflammatory cytokines expression from whole retinal detection. Although the mRNA level of anti-inflammatory cytokines did not show significant change, the immunohistochemistry showed that Argl-positive cells were notably increased surrounding transplantation area. We suspected OEC may increase anti-inflammatory cytokines in the transplantation area, but as the amount is limited, it couldn't be detected in whole retinal tissue.

Mechanistically, the most critical question to address is how OECs regulate microglial activation and the M(IL-6)/M(Argl) phenotype switch. Among numerous pathways, the JAK/STAT pathway is thought to be key regulator. It is reported that IFN-γ induced STAT1 activation leads to an increase in production of pro-inflammatory cytokines, and programs microglia/macrophages to the M(IL-6) phenotype (Qin et al., 2012). In contrast, STAT6 activation contributes to anti-inflammatory M(Argl) phenotype and the release of neurotrophic factors, Argl, IL-4, and IL-13. STAT3 involve both Argl-stimulated M2 polarization and IL-6-stimulated M1 polarization (Martinez and Gordon, 2014). In our research, *in vivo* experiments demonstrated that the JAK2/STAT3 pathway became activated during degenerative period which was the same as research on retinal degeneration has shown that JAK2 and STAT3 proteins are involved in photoreceptor apoptosis, especially in rd1 mice model (Samardzija et al., 2006; Lange et al., 2010). After OEC co-cultured or transplanted, both JAK2 and STAT3 protein expression was strongly reduced. By using a specific chemical antagonist of JAK2, AG490 to pretreat OEC, we were able to demonstrate that the mechanism by which OECs appear to reduce retinal microglial activation and inflammation is JAK2-dependent. Recent studies about macrophage polarization have established strong potential in suppressing of cytokine signaling (SOCS) proteins. SOCS proteins, especially SOCS3, not only suppresses JAK tyrosine kinase activity and negatively regulate JAK/STAT pathway, but also inhibits gp130-related cytokine receptors and abrogates IL-6-induced pro-inflammatory effects (Carow and Rottenberg, 2014). In macrophages, SOCS3 inhibits STAT3 activity preserving the natural cytotoxicity of M1 and develop characteristics of M2a activated cells (Liu et al., 2008). Qin et al. reported astrocytes expressed SOCS1 and SOCS3 (Qin et al., 2008) while Girolami et al. (2010) found SOCS3 expression is restricted mainly to Schwann cells in peripheral nerve injury. Consistent with both astrocytes and Schwann cells, SOCS3 immunoreactivity was observed in OECs. The mRNA expression of SOCS3 in OECs was enhanced significantly from 12 to 48 h after co-cultured with LPS induced BV2 cells. As during retinal degeneration, the apoptosis of photoreceptors and glia cell produce a large number of IL-6, which band to gp130, and form a complex to activate the kinase function of JAK2. JAK2 then phosphorylate STAT3, two phosphorylated STAT3 translocate into the cell nucleus, and promote pro-inflammatory factors gene transcription (Yin et al., 2015). We speculated as the OECs high expressing SOCS3, after co-cultured or transplanted, the IL-6 was regulated, and the cascade JAK/STAT pathway was inhibited. Suppression of the JAK/STAT pathway result in a switch of the proinflammatory phenotype to the anti-inflammatory M2 phenotype, as M2-related genes, such as IL-4 and Argl, were altered significantly *in vitro*.

Finally, we found substantial improvement in visual function and anatomical retinal structure following OEC transplantation in RCS rats. OECs led to nearly a twofold increase in the amplitude of the ERG b-wave, a significant inhibition of Caspase-3 expression, a up-regulated gene and protein expression of PKC-α and Rhodopsin protein which stand for ON-bipolar cells and photoreceptors. Similarly, OECs have been found to modulate the host immune response, and promotes preservation of neurons and axons in a rat model of SCI (Khankan et al., 2016). In addition, transplanted OEC promote the macrophage shift from M(INF-γ) to M(IL-4) and significantly improve motor function after SCI (Zhang et al., 2017). Therefore, combined with the previous research, we conducted the OEC injection modulate microenvironment through secrete neurotrophic factors, inhibit Müller cell gliosis, anti-oxidant, and immunomodulatory mechanisms in retinal degenerative models (Huo et al., 2011, 2012; Xie et al., 2017; Xue et al., 2017) to protect photoreceptors and inner neurons. Finally, in this study, we demonstrated that OECs have effect on activated microglia polarization from M(IL-6) to M(Argl) through JAK2/STAT3 pathway. In addition, alleviation of activated resident microglia and reduction of pro-inflammatory microenvironment by OEC transplantation contributed to substantial visual function and structure improvement.

AUTHOR CONTRIBUTIONS

JX conceived and designed the study, collected and assembled the data, carried out the data analysis, interpreted the data, and wrote the manuscript. YL, JD, YH, DS, and CD collected and assembled the data. HX conceived and designed the study, carried out the data analysis, interpreted the data, and wrote and approved the final version of the manuscript. ZY conceived and designed the study, was responsible for the financial support, carried out the data analysis, interpreted the data, and approved the final version of the manuscript.

ACKNOWLEDGMENTS

The authors would like to thank all the laboratory staff who were involved in this work.

REFERENCES

Akhtar-Schäfer, I., Wang, L., Krohne, T. U., Xu, H., and Langmann, T. (2018). Modulation of three key innate immune pathways for the most common retinal degenerative diseases. *EMBO Mol. Med.* 10:e8259. doi: 10.15252/emmm.201708259

Athanasiou, D., Aguila, M., Bellingham, J., Li, W., McCulley, C., Reeves, P. J., et al. (2018). The molecular and cellular basis of rhodopsin retinitis pigmentosa reveals potential strategies for therapy. *Prog. Retin Eye Res.* 62, 1–23. doi: 10.1016/j.preteyeres.2017.10.002

Barton, M. J., John, J. S., Clarke, M., Wright, A., and Ekberg, J. (2017). The glia response after peripheral nerve injury: a comparison between schwann cells and olfactory ensheathing cells and their uses for neural regenerative therapies. *Int. J. Mol. Sci.* 18:E287. doi: 10.3390/ijms18020287

Ben, M., Barek, K., and Monville, C. (2019). Cell therapy for retinal dystrophies: from cell suspension formulation to complex retinal tissue bioengineering. *Stem. Cells Int.* 2019:4568979. doi: 10.1155/2019/4568979

Bennett, M. L., Bennett, F. C., Liddelow, S. A., Ajami, B., Zamanian, J. L., Fernhoff, N. B., et al. (2016). New tools for studying microglia in the mouse and human CNS. *Proc. Natl. Acad. Sci. U. S. A.* 113, E1738–E1746. doi: 10.1073/pnas.1525528113

Biscaro, B., Lindvall, O., Tesco, G., Ekdahl, C. T., and Nitsch, R. M. (2012). Inhibition of microglial activation protects hippocampal neurogenesis and improves cognitive deficits in a transgenic mouse model for Alzheimer's disease. *Neurodegener. Dis.* 9, 187–198. doi: 10.1159/000330363

Carow, B., and Rottenberg, M. E. (2014). SOCS3, a major regulator of infection and inflammation. *Front. Immunol.* 5:58. doi: 10.3389/fimmu.2014.00058

Chuah, M. I., Hale, D. M., and West, A. K. (2011). Interaction of olfactory ensheathing cells with other cell types in vitro and after transplantation: glial scars and inflammation. *Exp. Neurol.* 229, 46–53. doi: 10.1016/j.expneurol.2010.08.012

Dai, C., Khaw, P. T., Yin, Z. Q., Li, D., Raisman, G., and Li, Y. (2012). Olfactory ensheathing cells rescue optic nerve fibers in a rat glaucoma model. *Transl. Vis. Sci. Technol.* 1:3. doi: 10.1167/tvst.1.2.3

Dai, J., Fu, Y., Zeng, Y., Li, S., and Qin Yin, Z. (2017). Improved retinal function in RCS rats after suppressing the over-activation of mGluR5. *Sci. Rep.* 7:3546. doi: 10.1038/s41598-017-03702-z

Dai, J. N., Zong, Y., Zhong, L. M., Li, Y. M., Zhang, W., Bian, L. G., et al. (2011). Gastrodin inhibits expression of inducible NO synthase, cyclooxygenase-2 and proinflammatory cytokines in cultured LPS-stimulated microglia via MAPK pathways. *PLoS One* 6:e21891. doi: 10.1371/journal.pone.0021891

de Hoz, R., Rojas, B., Ramírez, A. I., Salazar, J. J., Gallego, B. I., Trivio, A., et al. (2016). Retinal macroglial responses in health and disease. *Biomed. Res. Int.* 2016:2954721. doi: 10.1155/2016/2954721

Edholm, E. S., Rhoo, K. H., and Robert, J. (2017). Evolutionary aspects of macrophages polarization. *Results Probl. Cell Differ.* 62, 3–22. doi: 10.1007/978-3-319-54090-0_1

Fernández-Sánchez, L., Esquiva, G., Pinilla, I., Lax, P., and Cuenca, N. (2018). Retinal vascular degeneration in the transgenic p23h rat model of retinitis pigmentosa. *Front. Neuroanat.* 12:55. doi: 10.3389/fnana.2018.00055

Fu, X. M., Liu, S. J., Dan, Q. Q., Wang, Y. P., Lin, N., Lv, L. Y., et al. (2015). Combined bone mesenchymal stem cell and olfactory ensheathing cell transplantation promotes neural repair associated with cntf expression in traumatic brain-injured rats. *Cell Transplant.* 24, 1533–1544. doi: 10.3727/096368914X679345

Fu, Y., Hou, B., Weng, C., Liu, W., Dai, J., Zhao, C., et al. (2017). Functional ectopic neuritogenesis by retinal rod bipolar cells is regulated by miR-125b-5p during retinal remodeling in RCS rats. *Sci. Rep.* 7:1011. doi: 10.1038/s41598-017-01261-x

Garrido-Mesa, N., Zarzuelo, A., and Gálvez, J. (2013). Minocycline: far beyond an antibiotic. *Br. J. Pharmacol.* 169, 337–352. doi: 10.1111/bph.12139

Girolami, E. I., Bouhy, D., Haber, M., Johnson, H., and David, S. (2010). Differential expression and potential role of SOCS1 and SOCS3 in wallerian degeneration in injured peripheral nerve. *Exp. Neurol.* 223, 173–182. doi: 10.1016/j.expneurol.2009.06.018

Gladwin, K., and Choi, D. (2015). Olfactory ensheathing cells: part I—current concepts and experimental laboratory models. *World Neurosurg.* 83, 114–119. doi: 10.1016/j.wneu.2013.03.010

Greenhalgh, A. D., Zarruk, J. G., Healy, L. M., Baskar Jesudasan, S. J., Jhelum, P., Salmon, C. K., et al. (2018). Peripherally derived macrophages modulate microglial function to reduce inflammation after CNS injury. *PLoS Biol.* 16:e2005264. doi: 10.1371/journal.pbio.2005264

Gu, R., Tang, W., Lei, B., Ding, X., Jiang, C., and Xu, G. (2017). Glucocorticoid-induced leucine zipper protects the retina from light-induced retinal degeneration by inducing Bcl-xL in rats. *Invest. Ophthalmol. Vis. Sci.* 58, 3656–3668. doi: 10.1167/iovs.17-22116

He, X., Sun, D., Chen, S., and Xu, H. (2017). Activation of liver X receptor delayed the retinal degeneration of rd1 mice through modulation of the immunological function of glia. *Oncotarget* 8, 32068–32082. doi: 10.18632/oncotarget.16643

Hou, L., Zhou, X., Zhang, C., Wang, K., Liu, X., Che, Y., et al. (2017). NADPH oxidase-derived H2O2 mediates the regulatory effects of microglia on astrogliosis in experimental models of Parkinson's disease. *Redox Biol.* 12, 162–170. doi: 10.1016/j.redox.2017.02.016

Hu, X., Leak, R. K., Shi, Y., Suenaga, J., Gao, Y., Zheng, P., et al. (2015). Microglial and macrophage polarization—new prospects for brain repair. *Nat. Rev. Neurol.* 11, 56–64. doi: 10.1038/nrneurol.2014.207

Hu, X., Li, P., Guo, Y., Wang, H., Leak, R. K., Chen, S., et al. (2012). Microglia/macrophage polarization dynamics reveal novel mechanism of injury expansion after focal cerebral ischemia. *Stroke* 43, 3063–3070. doi: 10.1161/STROKEAHA.112.659656

Huo, S. J., Li, Y., Raisman, G., and Yin, Z. Q. (2011). Transplanted olfactory ensheathing cells reduce the gliotic injury response of müller cells in a rat model of retinitis pigmentosa. *Brain Res.* 1382, 238–244. doi: 10.1016/j.brainres.2010.12.079

Huo, S. J., Li, Y. C., Xie, J., Li, Y., Raisman, G., Zeng, Y. X., et al. (2012). Transplanted olfactory ensheathing cells reduce retinal degeneration in royal college of surgeons rats. *Curr. Eye Res.* 37, 749–758. doi: 10.3109/02713683.2012.697972

Ito, T., Ikeda, K., Tomita, K., and Yokoyama, S. (2010). Interleukin-6 upregulates the expression of PMP22 in cultured rat Schwann cells via a JAK2-dependent pathway. *Neurosci. Lett.* 472, 104–108. doi: 10.1016/j.neulet.2010.01.061

Jiao, H., Natoli, R., Valter, K., Provis, J. M., and Rutar, M. (2015). Spatiotemporal cadence of macrophage polarisation in a model of light-induced retinal degeneration. *PLoS One* 10:e0143952. doi: 10.1371/journal.pone.0143952

Khankan, R. R., Griffis, K. G., Haggerty-Skeans, J. R., Zhong, H., Roy, R. R., Edgerton, V. R., et al. (2016). Olfactory ensheathing cell transplantation after a complete spinal cord transection mediates neuroprotective and immunomodulatory mechanisms to facilitate regeneration. *J. Neurosci.* 36, 6269–6286. doi: 10.1523/JNEUROSCI.0085-16.2016

Kigerl, K. A., Gensel, J. C., Ankeny, D. P., Alexander, J. K., Donnelly, D. J., and Popovich, P. G. (2009). Identification of two distinct macrophage subsets with divergent effects causing either neurotoxicity or regeneration in the injured mouse spinal cord. *J. Neurosci.* 29, 13435–13444. doi: 10.1523/JNEUROSCI.3257-09.2009

Kyger, M., Worley, A., and Adamus, G. (2013). Autoimmune responses against photoreceptor antigens during retinal degeneration and their role in macrophage recruitment into retinas of RCS rats. *JNeuroimmunol.* 254, 91–100. doi: 10.1016/j.jneuroim.2012.10.007

Lange, C., Thiersch, M., Samardzija, M., and Grimm, C. (2010). The differential role of Jak/STAT signaling in retinal degeneration. *Adv. Exp. Med. Biol.* 664, 601–607. doi: 10.1007/978-1-4419-1399-9_69

Li, Q., Lan, X., Han, X., and Wang, J. (2018). Expression of Tmem119/Sall1 and Ccr2/CD69 in FACS-Sorted microglia- and monocyte/macrophage-enriched cell populations after intracerebral hemorrhage. *Front. Cell Neurosci.* 12:520. doi: 10.3389/fncel.2018.00520

Li, Y., Field, P. M., and Raisman, G. (1997). Repair of adult rat corticospinal tract by transplants of olfactory ensheathing cells. *Science* 277, 2000–2002. doi: 10.1126/science.277.5334.2000

Li, Y., Zou, T., Xue, L., Yin, Z. Q., Huo, S., and Xu, H. (2017). TGF-β1 enhances phagocytic removal of neuron debris and neuronal survival by olfactory ensheathing cells via integrin/MFG-E8 signaling pathway. *Mol. Cell. Neurosci.* 85, 45–56. doi: 10.1016/j.mcn.2017.08.006

Li, Z., Zeng, Y., Chen, X., Li, Q., Wu, W., Xue, L., et al. (2016). Neural stem cells transplanted to the subretinal space of rd1 mice delay retinal degeneration by suppressing microglia activation. *Cytotherapy* 18, 771–784. doi: 10.1016/j.jcyt.2016.03.001

Liu, Y., Stewart, K. N., Bishop, E., Marek, C. J., Kluth, D. C., Rees, A. J., et al. (2008). Unique expression of suppressor of cytokine signaling 3 is essential for classical macrophage activation in rodents in vitro and in vivo. *J. Immunol.* 180, 6270–6278. doi: 10.4049/jimmunol.180.9.6270

Liu, Y., Yang, X., Utheim, T. P., Guo, C., Xiao, M., Liu, Y., et al. (2013). Correlation of cytokine levels and microglial cell infiltration during retinal degeneration in RCS rats. *PLoS One* 8:e82061. doi: 10.1371/journal.pone.0082061

Martinez, F. O., and Gordon, S. (2014). The M1 and M2 paradigm of macrophage activation: time for reassessment. F1000Prime Rep. 6:13. doi: 10.12703/P6-13

Neves, J., Zhu, J., Sousa-Victor, P., Konjikusic, M., Riley, R., Chew, S., et al. (2016). Immune modulation by MANF promotes tissue repair and regenerative success in the retina. *Science* 353:aaf3646. doi: 10.1126/science.aaf3646

Noailles, A., Maneu, V., Campello, L., Gómez-Vicente, V., Lax, P., and Cuenca, N. (2016). Persistent inflammatory state after photoreceptor loss in an animal model of retinal degeneration. *Sci. Rep.* 6:33356. doi: 10.1038/srep33356

Peng, B., Xiao, J., Wang, K., So, K. F., Tipoe, G. L., and Lin, B. (2014). Suppression of microglial activation is neuroprotective in a mouse model of human retinitis pigmentosa. *J. Neurosci.* 34, 8139–8150. doi: 10.1523/JNEUROSCI.5200-13.2014

Qin, H., Buckley, J. A., Li, X., Liu, Y., Fox, T. H. III, Meares, G. P., et al. (2016). Inhibition of the JAK/STAT Pathway protects against α-Synuclein-Induced neuroinflammation and dopaminergic neurodegeneration. *J. Neurosci.* 36, 5144–5159. doi: 10.1523/JNEUROSCI.4658-15.2016

Qin, H., Holdbrooks, A. T., Liu, Y., Reynolds, S. L., Yanagisawa, L. L., and Benveniste, E. N. (2012). SOCS3 deficiency promotes M1 macrophage polarization and inflammation. *J. Immunol.* 189, 3439–3448. doi: 10.4049/jimmunol.1201168

Qin, H., Niyongere, S. A., Lee, S. J., Baker, B. J., and Benveniste, E. N. (2008). Expression and functional significance of SOCS-1 and SOCS-3 in astrocytes. *J. Immunol.* 181, 3167–3176. doi: 10.4049/jimmunol.181.5.3167

Ramirez, A. I., de Hoz, R., Salobrar-Garcia, E., Salazar, J. J., Rojas, B., Ajoy, D., et al. (2017). The Role of microglia in retinal neurodegeneration: alzheimer's disease, parkinson, and glaucoma. *Front. Aging Neurosci.* 9:214. doi: 10.3389/fnagi.2017.00214

Ransohoff, R. M. (2016). A polarizing question: do M1 and M2 microglia exist. *Nat. Neurosci.* 19, 987–991. doi: 10.1038/nn.4338

Samardzija, M., Wenzel, A., Aufenberg, S., Thiersch, M., Remé, C., and Grimm, C. (2006). Differential role of Jak-STAT signaling in retinal degenerations. *FASEB J.* 20, 2411–2413. doi: 10.1096/fj.06-5895fje

Sevenich, L. (2018). Brain-resident microglia and blood-borne macrophages orchestrate central nervous system inflammation in neurodegenerative disorders and brain cancer. *Front. Immunol.* 9:697. doi: 10.3389/fimmu.2018.00697

Shen, W., Li, S., Chung, S. H., and Gillies, M. C. (2010). Retinal vascular changes after glial disruption in rats. *J. Neurosci. Res.* 88, 1485–1499. doi: 10.1002/jnr.22317

Sica, A., and Bronte, V. (2007). Altered macrophage differentiation and immune dysfunction in tumor development. *J. Clin. Invest.* 117, 1155–1166. doi: 10.1172/jci31422

Subramaniam, S. R., and Federoff, H. J. (2017). Targeting microglial activation states as a therapeutic avenue in parkinson's disease. *Front. Aging Neurosci.* 9:176. doi: 10.3389/fnagi.2017.00176

Szepesi, Z., Manouchehrian, O., Bachiller, S., and Deierborg, T. (2018). Bidirectional microglia-neuron communication in health and disease. *Front. Cell Neurosci.* 12:323. doi: 10.3389/fncel.2018.00323

Tam, W. Y., and Ma, C. H. (2014). Bipolar/rod-shaped microglia are proliferating microglia with distinct M1/M2 phenotypes. *Sci. Rep.* 4:7279. doi: 10.1038/srep07279

Vincent, A. J., Taylor, J. M., Choi-Lundberg, D. L., West, A. K., and Chuah, M. I. (2005). Genetic expression profile of olfactory ensheathing cells is distinct from that of Schwann cells and astrocytes. *Glia* 51, 132–147. doi: 10.1002/glia.20195

Wang, G., Zhang, J., Hu, X., Zhang, L., Mao, L., Jiang, X., et al. (2013). Microglia/macrophage polarization dynamics in white matter after traumatic brain injury. *J. Cereb. Blood Flow Metab.* 33, 1864–1874. doi: 10.1038/jcbfm.2013.146

Wohleb, E. S. (2016). Neuron-microglia interactions in mental health disorders: "for better, and for worse". *Front. Immunol.* 7:544. doi: 10.3389/fimmu.2016.00544

Xie, J., Huo, S., Li, Y., Dai, J., Xu, H., and Yin, Z. Q. (2017). Olfactory ensheathing cells inhibit gliosis in retinal degeneration by downregulation of the müller cell notch signaling pathway. *Cell Transplant.* 26, 967–982. doi: 10.3727/096368917x694994

Xu, H., Chen, M., and Forrester, J. V. (2009). Para-inflammation in the aging retina. *Prog. Retin Eye Res.* 28, 348–368. doi: 10.1016/j.preteyeres.2009.06.001

Xue, L., Zeng, Y., Li, Q., Li, Y., Li, Z., Xu, H., et al. (2017). Transplanted olfactory ensheathing cells restore retinal function in a rat model of light-induced retinal damage by inhibiting oxidative stress. *Oncotarget* 8, 93087–93102. doi: 10.18632/oncotarget.21857

Yin, Y., Liu, W., and Dai, Y. (2015). SOCS3 and its role in associated diseases. *Hum. Immunol.* 76, 775–780. doi: 10.1016/j.humimm.2015.09.037

Zhang, J., Chen, H., Duan, Z., Chen, K., Liu, Z., Zhang, L., et al. (2017). The Effects of Co-transplantation of olfactory ensheathing cells and schwann cells on local inflammation environment in the contused spinal cord of rats. *Mol. Neurobiol.* 54, 943–953. doi: 10.1007/s12035-016-9709-5

Zhou, T., Huang, Z., Sun, X., Zhu, X., Zhou, L., Li, M., et al. (2017). Microglia polarization with m1/m2 phenotype changes in rd1 mouse model of retinal degeneration. *Front. Neuroanat.* 11:77. doi: 10.3389/fnana.2017.00077

Zhu, J., Cifuentes, H., Reynolds, J., and Lamba, D. A. (2017). Immunosuppression via loss of IL2rγ enhances long-term functional integration of hESC-derived photoreceptors in the mouse retina. *Cell Stem Cell.* 20:374-384.e5. doi: 10.1016/j.stem.2016.11.019

Zou, T., Gao, L., Zeng, Y., Li, Q., Li, Y., Chen, S., et al. (2019). Organoid-derived C-Kit+/SSEA4- human retinal progenitor cells promote a protective retinal microenvironment during transplantation in rodents. *Nat. Commun.* 10:1205. doi: 10.1038/s41467-019-08961-0

16

Effects of Exosomes on Neurological Function Recovery for Ischemic Stroke in Pre-Clinical Studies

*Mudan Huang†, Zhongqiu Hong†, Chongjun Xiao, Lili Li, Lilin Chen, Shimei Cheng, Tingting Lei and Haiqing Zheng**

Department of Rehabilitation Medicine, The Third Affiliated Hospital, Sun Yat-sen University, Guangzhou, China

Correspondence:
Haiqing Zheng
zhenghaiqing0909@aliyun.com

†These authors have contributed equally to this work

Background: Exosomes, especially stem cell-derived exosomes, have been widely studied in pre-clinical research of ischemic stroke. However, their pooled effects remain inconclusive.

Methods: Relevant literature concerning the effects of exosomes on neurological performance in a rodent model of ischemic stroke was identified via searching electronic databases, including PubMed, Embase, and Web of Science. The primary outcomes included neurological function scores (NFS) and infarct volume (IV), and the secondary outcomes were several pro-inflammatory factors and terminal deoxynucleotidyl transferase deoxyuridine triphosphate nick end labeling-positive cells. Subgroup analyses regarding several factors potentially influencing the effects of exosomes on NFS and IV were also conducted.

Results: We identified 21 experiments from 18 studies in the meta-analysis. Pooled analyses showed the positive and significant effects of exosomes on NFS (standardized mean difference −2.79; 95% confidence interval −3.81 to −1.76) and IV (standardized mean difference −3.16; 95% confidence interval −4.18 to −2.15). Our data revealed that the effects of exosomes on neurological outcomes in rodent stroke models might be related to routes of administration and exosomes sources. In addition, there was significant attenuation in pro-inflammatory factors, including interleukin-6, tumor necrosis factor-α and interleukin-1β, and terminal deoxynucleotidyl transferase deoxyuridine triphosphate nick end labeling-positive cells when undergoing exosomes treatment.

Conclusion: Cell-derived exosomes treatment demonstrated statistically significant improvements in structural and neurological function recovery in animal models of ischemic stroke. Our results also provide relatively robust evidence supporting cell-derived exosomes as a promising therapy to promote neurological recovery in stroke individuals.

Keywords: exosomes, animal models, ischemic stroke, cell-derived exosomes, meta-analysis

INTRODUCTION

Ischemic stroke continues to be a leading cause of adult death and permanent disability throughout the world (Chen L. et al., 2016). In most instances, ischemic stroke is caused by blood-vessel occlusion, most commonly due to an embolus or local thrombosis (Dirnagl et al., 1999). Given the limitation of available therapeutic options, patients still retain long-term dysfunction after stroke, although they have received recombinant tissue plasminogen activator or endovascular intervention (Schwamm et al., 2013; Nogueira et al., 2018).

To replace the loss of the functional neurons, stem cell transplantation represents a promising treatment option (Boltze et al., 2014). Several studies have identified that transplantation of mesenchymal stem cells (MSCs) not only facilitated functional recovery in rodents after stroke but also improved neurological outcomes for post-stroke survivors (Liu et al., 2011; Levy et al., 2019). However, only a small percentage of MSCs engrafted into the ischemic hemisphere exerted neuroprotective effects (Acosta et al., 2015; Gervois et al., 2016). In addition, cell-based therapy was found with several potential risks and limitations, such as tumor formation, occlusion in the microvasculature, and weak capacity to cross the blood–brain barrier (Lukomska et al., 2019).

Emerging studies demonstrated that exosomes released from MSCs are the major biological principle underlying the therapeutic benefits of MSC transplantation (Xin et al., 2013b; Doeppner et al., 2015). Exosomes are nanosized vesicles with a size of 40 to 150 nm and secreted by different cell types. They play a pivotal role in mediating intercellular communication by delivering a variety of functional biomolecules, including messenger RNAs, microRNAs (miRNAs), long non-coding RNA, proteins, and lipids, to recipient cells (Raposo and Stoorvogel, 2013; Tkach and Théry, 2016). Until now, exosomal miRNAs have been investigated more fully than other cargo of exosomes in the light of their functional significance. Indeed, increasing data showed that exosomal miRNAs improved neurological function and promoted neurovascular remodeling in the ischemic brain (Xin et al., 2013b, 2017a). Furthermore, nanometer-sized exosomes are admitted to cross the blood-brain barrier and prevent tumor formation. Therefore, cell-derived exosomes are expected to become an efficient and safe treatment for ischemic stroke.

A considerable number of animal studies have been performed to explore the efficacy of exosomes on an ischemic stroke model, but they have various cell origins, administration routes, injection doses, and timepoints of intervention and evaluation (Song et al., 2019; Zhao et al., 2020). However, there are a few clinical trials concerning cell-derived exosomes in the treatment of ischemic stroke, and most of them are in the initial recruitment stage of the trials. To characterize and quantify the pre-clinical evidence and provide a vital theoretical basis for clinical application, we therefore aimed to conduct this meta-analysis of animal studies.

MATERIALS AND METHODS

The protocol was established according to the Preferred Reporting Items for Systematic Review and Meta-Analysis guidelines (Moher et al., 2009).

Search Strategy

We conducted a literature search in the electronic database, including PubMed, Embase, and Web of Science. The search strategy was as follows: ("exosomes" or "exosome" or "exosomal" or "secretome") and ("brain ischemia" or "cerebral ischemia" or "ischemia stroke" or "brain infarct" or "cerebral infarct" or "middle cerebral artery occlusion" or "MCAO"). All of the search strategies were performed from the initiation to June 2020, and the literature was published in English. The references of selected articles were further searched by hand to obtain additional citations.

Study Selection

Two authors (MH and ZH) independently abstracted all data from any eligible publication, according to a standard protocol. Discrepancies were resolved by discussion with the third reviewer (HZ). The inclusion criteria for this research were (1) population-rodent models with ischemic stroke (transient or permanent); (2) intervention unmodified cell-derived exosomes; (3) comparison-vehicle, phosphate-buffered saline, or positive control drug or no treatment; and (4) outcome measure—the primary outcomes were neurological function score (NFS) and infarct volume (IV). The secondary outcomes were the percentage of apoptotic cells, levels of interleukin (IL)-1β, IL-6, and tumor necrosis factor (TNF)-α. The exclusion criteria were (1) ischemic stroke was not conducted on rodent models; (2) artificially modified cell-derived exosomes; (3) repeated data or insufficient information; (4) the intervention was a combination of exosomes and another drug; and (5) clinical articles, case reports, commentary, reviews, conference abstracts, correspondence, expert opinion, and *in vitro* studies.

Data Extraction

The following data extraction from the included studies was conducted independently by two reviewers (MH and ZH). The following details were collected: (1) name of the first author and the publication year; (2) country or region; (3) animal sex and species; (4) number of animals in the study, control group, and exosomes treatment group; (5) comorbid status of animals; (6) method of an ischemic stroke model; (7) occlusion time of animal with ischemic stroke; (8) treatment time; (9) origins of exosome; (10) administration route of exosome; (11) administration dose of exosome; (12) measurement time; and (13) outcome measures. For included studies without available numerical values in the text, we contacted with authors twice by email and performed the WebPlotDigitizer software (https://automeris.io/WebPlotDigitizer/) to extract data from the figures or graphs (Burda et al., 2017).

Quality Assessment

The methodological quality of each eligible studies was assessed independently by two investigators (MH and ZH) via using the 10-item Collaborative Approach to Meta-Analysis and Review of Animal Data from Experimental Studies quality checklist and detailed items as follows: *A*, publication in a peer-reviewed journal; *B*, statement of temperature control; *C*, randomization to treatment or control group; *D*, blinding of the ischemic model establishment; *E*, blinding of outcome assessment; *F*, anesthetic without obvious intrinsic vascular protection activity; *G*, an appropriate animal model such as advanced age, diabetic, or hypertensive; *H*, estimation of the sample size; *I*, compliance with animal welfare regulations; and *J*, declaration of potential conflicts of interest (Zhao et al., 2020).

Effect Sizes Estimation

The effect size was calculated using standardized mean difference (SMD) with 95% confidence intervals (95% CIs) for continuous outcomes because a single outcome measure was assessed and reported across trials using different measurement tools. For quantitative synthesis, the pooled effect estimation was calculated by comparing the change from baseline with the endpoint of the study between the intervention group and the control group.

The primary outcome was served to assess functional and structural recovery, which included NFSs and IV. If more than one measure for NFS were used in an individual trial, a modified neurological severity score was considered as a priority outcome measure because it is more appropriate to reflect neurological impairment; otherwise, neurological severity score and various behavioral tests (adhesive removal test, foot-fault test). We choose the 2,3,5-triphenyl tetrazolium chloride method as IV assessment after considering MRI and cresyl violet staining. Several pro-inflammatory factors, such as IL-6, TNF-α, and IL-1β, and the percentage of terminal deoxynucleotidyl transferase deoxyuridine triphosphate nick end labeling-positive cells, were detected as the secondary outcomes. Besides, several subgroup analyses were conducted to investigate whether study characteristics (administration routes, type of ischemia, exosomes sources, intervention time, and different species) had the effects on exosomes in improving neurological performance via using random-effects models. The authors were contacted twice by email for original data if the published study data were insufficient for data analyses.

Statistical Analysis

All statistical analyses were performed by Review Manager 5.3 (The Cochrane Collaboration, Oxford, UK) and Stata version 14.0 (Stata Corp, College Station, TX, USA). We used the SMD with the 95% CI to record the continuous outcomes. Statistical heterogeneity was analyzed by using the I-square test. A random-effects model test was performed when heterogeneity was significant ($I^2 > 50\%$). Subgroup and sensitivity were conducted to investigate the origin of heterogeneity. A value of $p < 0.05$ was considered statistically significant.

RESULTS

Search Results and Study Characteristics

The process of study selection is outlined in **Figure 1**. A total of 18 studies with 21 comparisons satisfied the inclusion criteria (Xin et al., 2013a, 2017a,b; Chen K. H. et al., 2016; Jiang et al., 2018; Deng et al., 2019; Li et al., 2019; Nalamolu et al., 2019a,b; Pei et al., 2019; Song et al., 2019; Sun et al., 2019; Venkat et al., 2019; Zheng et al., 2019; Li G. et al., 2020; Ling et al., 2020; Safakheil and Safakheil, 2020; Zhao et al., 2020). The characteristics of the included studies are presented in **Table 1**. Moreover, the Preferred Reporting Items for Systematic Review and Meta-Analysis flow diagram was showed in **Supplementary Table 1**. All these studies were published between 2013 and 2020, and 367 animals were included in this meta-analysis. Among the included studies, six of them used bone marrow mesenchymal stromal cells—exosome (Xin et al., 2013a, 2017a,b; Deng et al., 2019; Safakheil and Safakheil, 2020; Zhao et al., 2020), one adopted adipose-derived mesenchymal stem cells—exosome (Chen K. H. et al., 2016), one applied human umbilical cord mesenchymal stem cells—exosome (Li G. et al., 2020), one used human umbilical cord blood mesenchymal stromal cells—exosome (Nalamolu et al., 2019b), one used adipose-derived stem cells—exosome(Jiang et al., 2018), one used neural stem cells—exosome (Sun et al., 2019), one used human urine-derived stem cells—exosome (Ling et al., 2020), one used mouse brain endothelial cells—exosome (Venkat et al., 2019), one used macrophages—exosome (Zheng et al., 2019), one used microglia—exosome, and one used astrocytes—exosome (Pei et al., 2019; Song et al., 2019), and the two remaining used another origin of exosomes (Li et al., 2019; Nalamolu et al., 2019a).

The ischemic stroke model was established with a unifilar nylon suture (Xin et al., 2013a, 2017a,b; Chen K. H. et al., 2016; Jiang et al., 2018; Deng et al., 2019; Nalamolu et al., 2019a,b; Pei et al., 2019; Song et al., 2019; Sun et al., 2019; Zheng et al., 2019; Li G. et al., 2020; Ling et al., 2020; Safakheil and Safakheil, 2020; Zhao et al., 2020), bipolar electrocoagulation forceps (Li et al., 2019), or thrombus (Venkat et al., 2019). A total of 13 studies used the transient ischemic stroke model (Xin et al., 2013a, 2017a,b; Chen K. H. et al., 2016; Deng et al., 2019; Nalamolu et al., 2019a,b; Song et al., 2019; Sun et al., 2019; Zheng et al., 2019; Ling et al., 2020; Safakheil and Safakheil, 2020; Zhao et al., 2020), and four studies used the permanent model (Jiang et al., 2018; Li et al., 2019; Pei et al., 2019; Venkat et al., 2019). Most studies chose to perform exosomes injection immediately after the injury model was established (Jiang et al., 2018; Li et al., 2019; Nalamolu et al., 2019a,b; Song et al., 2019; Zheng et al., 2019; Li G. et al., 2020), and several studies reported the same dose of exosomes (100 μg) injected intravenously (Xin et al., 2013a, 2017a,b; Chen K. H. et al., 2016; Safakheil and Safakheil, 2020). A variety of miRNA types have been reported in exosomes, such as miR-17-92 (Xin et al., 2017a), miR-126 (Venkat et al., 2019), miR-124 (Song et al., 2019), miR-26a (Ling et al., 2020), miR-26b-5p (Li G. et al., 2020), miR-30d-5p (Jiang et al., 2018), and miR-138-5p (Deng et al., 2019).

Quality Assessment

The median quality score across the 18 studies was 5.5, and the range was from 4 to 7. The most widely met criteria included peer-reviewed publications (100% of studies), appropriate animal model (100% of studies), statement of compliance with animal welfare regulations (100% of studies), and statement of potential conflict of interests (100% of studies). Moreover, 72.22% of the included studies were randomly divided into the treatment and

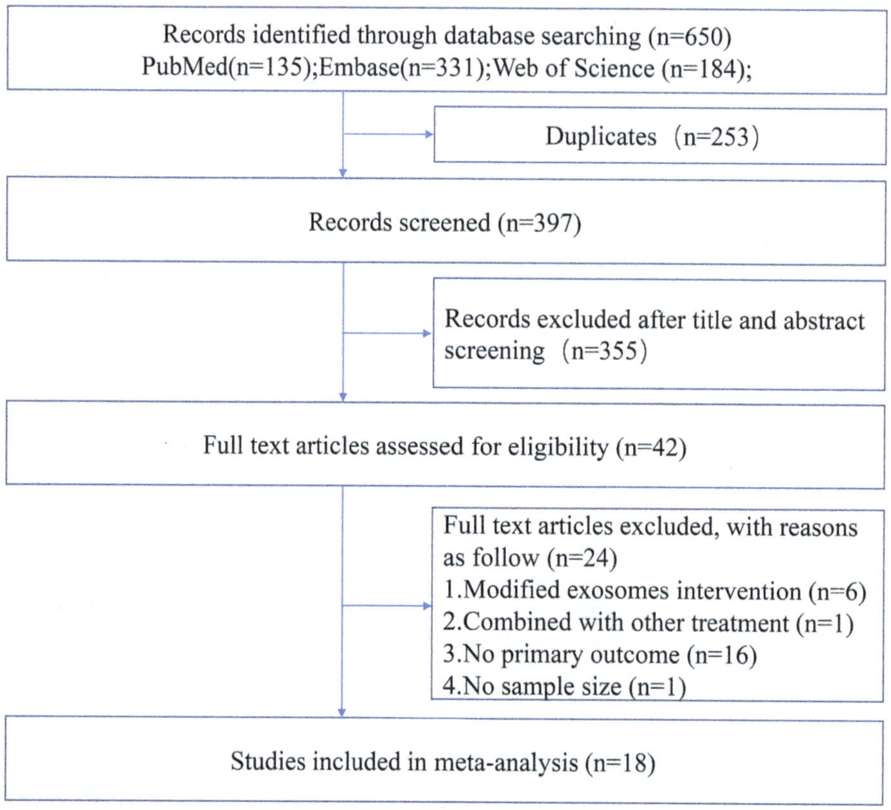

FIGURE 1 | Flow chart of the study selection.

control groups, whereas the randomization was not mentioned in the remaining studies (27.78%). However, the calculation of sample size, the blinded induction of stroke model, and the blinded assessment of the experimental outcome had not been reported in all included studies. Further details of the study quality score are presented in **Table 2**.

Primary Outcomes

Neurological Function Scores

Fourteen studies with 17 comparisons reported the NFSs. The pooled analysis showed that exosomes can significantly improve the neurological function compared with the control (SMD −2.79; 95% CI −3.81 to −1.76; $p < 0.001$; $I^2 = 91.3\%$; **Figure 2A**). The subgroup analysis showed that all cell-derived exosomes are effective in improving neurological function (**Figure 2B**). Moreover, the subgroup analysis showed that all administrational routes help improve the neurological function, except for internal jugular vein administration (**Figure 2C**). In addition, the subgroup analysis presented that immediate or delayed treatment with cell-derived exosomes could improve the neural recovery after transient or permanent stroke (**Supplementary Figures 1A,B**; **Supplementary Table 2**). The sensitivity analysis showed that none of the single studies significantly influenced the result (**Supplementary Figure 1C**).

Infarct Volume

Meta-analysis of 15 studies with 18 comparisons showed significant effects of exosomes for reducing IV compared with control groups (SMD −3.16; 95% CI −4.18 to −2.15; $p < 0.001$; $I^2 = 90.8\%$; **Figure 3A**). The subgroup analysis presented that all cell-derived exosomes benefit from reducing IV except for mouse brain endothelial cells (**Figure 3B**). Moreover, the subgroup analysis showed that all administrational routes effectively reduce IV (**Figure 3C**). In addition, the subgroup analysis indicated that immediate or delayed treatment with cell-derived exosomes reduced IV after transient or permanent stroke (**Supplementary Figures 2A,B**; **Supplementary Table 2**). None of the single studies significantly influenced the result via sensitivity analysis (**Supplementary Figure 2C**).

Secondary Outcomes

Five studies evaluated the anti-inflammatory effect of exosomes by measuring IL-6 levels compared with the control. Meta-analysis presented a significant reduction in the exosome group (SMD −2.9; 95% CI −4.58 to −1.22; $p < 0.001$; $I^2 = 85.8\%$; **Figure 4A**). A meta-analysis of seven studies with eight comparisons indicated that the level of TNF-α was significantly decreased in the exosome group compared with the control group (SMD −3.10; 95% CI −3.94 to −2.26; $p < 0.05$; $I^2 = 52.9\%$; **Figure 4B**). Moreover, four studies with

TABLE 1 | Characteristics of included studies.

References	Country or region	Gender and species	Animal number	Comorbid status	Occlusion method	Occlusion time	Therapy time	Exosomes source	Administration method	Therapy dose	Measure time	Outcome measure
Xin et al. (2017a)	USA	Male Wistar rats	T:24 C:8 E:8	Healthy	Filament insertion	2 h	24 h after operation	BMSCs	Intravenous administration	100 μg	1, 3, 7, 14, 21, and 28 days after stroke (C:8 and E:8 /timepoint)	1. Behavioral tests (mNSS, Foot-fault test); 2. Axonal density, phosphorylated NF-H, synaptophysin; 3. Neurite branching, spine density; 4. NeuN/Brdu, NG2/Brdu, MBP/Brdu
Zhao et al. (2020)	China	Male SD rats	T:24 C:8 E:8	Healthy	Filament insertion	90 min	2 h after operation	BMSCs	Tail vein	120 μg	1, 3, and 7 days after stroke (C:8 and E:8 /timepoint)	1. Behavioral tests (NSS, shuttle box test); 2. CD206 and CD86; 3. NO, IL-1β, TNF-α, IL-12, IL-10, TGF-β, BDNF, GDNF (ELISA, brain tissue).
Song et al. (2019)	China	Male ICR mice	T:48 C:8 E:8	Healthy	Filament insertion	1 h	Immediately after operation	M2 microglia	Tail vein	100 μg/day, 3 days	3 days after stroke (C:8 and E:8)	1. Infarct volumes (Cresyl Violet Staining); 2. Behavioral tests (mNSS); 3. Apoptosis (TUNEL)
Pei et al. (2019)	China	Male C57BL/6 mice	T:60 C:10 E:10	Healthy	Filament insertion	Permanent	60 min after operation	Astrocytes	Tail vein	80 μg	3 days after stroke (C:10 and E:10)	1. Infarct volumes (TTC); 2. Apoptosis (TUNEL); 3. GFAP, Iba-1; 4. Beclin-1, LC3-I/II, P62; 5. TNF-α, IL-6. IL-1β (ELISA, brain tissue).
Safakheil and Safakheil (2020)	Iran	Male Wistar rats	T:60 C:12 E:12	Healthy	Filament insertion	60 min	24 h after operation	BMSCs	Stereotaxic (brain cortex)	100 μg	24 h,7 days after stroke (C:12 and E:12 /timepoint)	1. Behavioral tests (EBST, Garcia); 2. Infarct volume (TTC); 3. Dead cells (Cresyl Violet Staining); 4. NLRP1, NLRP3; 5. GFAP-positive cells; 6.MDA, SOD
Deng et al. (2019)	China	NR C57BL/6 mice	T:40 C:10 E:10	Healthy	Filament insertion	30 min	NR	BMSCs	NR	NR	4 weeks after treatment (C:10 and E:10)	1. Infarct volumes (TTC); 2. Apoptosis (TUNEL); 3. LDH, cleaved caspase-3, Bax, Bcl-2; 4. IL-6, IL-1β, TNF-α (WB, brain tissue).
Jiang et al. (2018)	China	Male SD rats	T:24 C:6 E:6	Healthy	Filament insertion	Permanent	Immediately after operation	ADSCs	Tail vein	80 μg	3 h, 72 h after stroke (C:6 and E:6 /timepoint)	1. Infarct volumes (TTC); 2. Apoptosis (TUNEL); 3. TNF-α, IL-6, IL-4, IL-10 (ELISA, serum). 4. CD206, iNOS, Beclin-1, Atg5, LC3

(Continued)

TABLE 1 | Continued

References	Country or region	Gender and species	Animal number	Comorbid status	Occlusion method	Occlusion time	Therapy time	Exosomes source	Administration method	Therapy dose	Measure time	Outcome measure
Zheng et al. (2019)	China	Male SD rats	T:18 C:6 E:6	Healthy	Filament insertion	2 h	Immediately after operation	Macrophages	Tail vein	2 mg	6 h (C:6 and E:6), 24 h after stroke (C:3 andE:3)	1. Infarct volumes (TTC); 2. Behavioral tests (Zea-Longa, Ludmila Belayev); 3. CD 80, CD206, NeuN-positive cells; 4. IL-6, TNF-α, NF-κB p65 (WB, brain tissue).
Li et al. (2019)	China	Male C57BL/6 mice	T:42-48 C:14-16 E:14-16	Healthy	Bipolar electrocoat-gulation	Permanent	Immediately after operation	RIPC mice plasma	Tail vein	10 μg/day, 14 days	24 h (C:14-16 and E:14-16) and 3, 7, 14, 21, 28 days after stroke (C:8-10 and E:8-10 /timepoint)	1. Infarct volumes (TTC); 2. Behavioral tests (rotarod test, adhesive removal test); 3. HIF-1α
Li et al. (2019)	China	Male C57BL/6 mice	T:42-48 C:14-16 E:14-16	Healthy	Bipolar electrocoat-gulation	Permanent	Immediately after operation	Non-RIPC mice plasma	Tail vein	10 μg/day, 14 days	24 h (C:14-16 and E:14-16) and 3, 7, 14, 21, and 28 days after stroke (C:8-10 and E:8-10 /timepoint)	1. Infarct volumes (TTC); 2. Behavioral tests (rotarod test, adhesive removal test); 3. HIF-1α
Sun et al. (2019)	United States	Male CB57/B6 mice	T:30 C:15 E:15	Healthy	Filament insertion	1 h	2 h after operation	NSCs	Internal jugular vein	10 μg	24 h, 4 days after stroke (C:15 and E:15 /timepoint)	1. Infarct volumes (TTC); 2. Behavioral tests (neurologic deficit score)
Nalamolu et al. (2019b)	USA	Male SD rats	T:30 C:15 E:15	Healthy	Filament insertion	2 h	Immediately after operation	HUCB-MSCs	Tail vein	150 μg	1 (C:15 and E:15), 3, 5, and 7 days after stroke (C:9 and E:9 /timepoint)	1. Infarct volumes (TTC); 2. Brain swelling; 3. Behavioral tests (mNSS, adhesive removal test, Beam-walking, Rotarod); 4. Body weight changes and mortality
Nalamolu et al. (2019a)	USA	Male SD rats	T:36 C:12 E:12	Healthy	Filament insertion	2 h	Immediately after operation	Cocultures of normal and OGD-induced HUCB-MSCs	Tail vein	150 μg	1 (C:12 and E:12), 3 (C:10 and E:10), 5, and 7 days after stroke (C:8 and E:8 /timepoint)	1. Infarct volumes (TTC); 2. Brain swelling; 3. Behavioral tests (mNSS, adhesive removal test, Beam-walking); 4. Body weight changes and mortality
Nalamolu et al. (2019a)	USA	Male SD rats	T:36 C:12 E:12	Healthy	Filament insertion	2 h	Immediately after operation	OGD-induced HUCB-MSCs	Tail vein	150 μg	1 (C:12 and E:12), 3 (C:10 and E:10), 5, and 7 days after stroke (C:8 and E:8 /timepoint)	1. Behavioral tests (mNSS, adhesive removal test, Beam-walking); 2. Body weight changes and mortality
Li G. et al. (2020)	China	Wild Balc/C mice	T:124 C:20 E:20	Healthy	Filament insertion	NR	Immediately after operation	hUCMSCs	Tail vein	5 μg	24 h after stroke (C:20 and E:20)	1. Infarct volumes (TTC); 2. Brain edema; 3. Behavioral tests (Longa)
Li S. et al. (2020)	China	Wild Balc/C mice	T:124 C:20 E:20	Healthy	Filament insertion	NR	Immediately after operation	hUCMSCs	Tail vein	50 μg	24 h after stroke (C:20 and E:20)	1. Infarct volumes (TTC); 2. Brain edema; 3. Behavioral tests (Longa); 4. TNF-?, IL-6, and CCL-2 (ELISA, brain tissue). 5. iNOS, Arg1, CD38^{+}, CD206^{-}

(Continued)

TABLE 1 | Continued

References	Country or region	Gender and species	Animal number	Comorbid status	Occlusion method	Occlusion time	Therapy time	Exosomes source	Administration method	Therapy dose	Measure time	Outcome measure
Chen K. H. et al. (2016)	Taiwan	Male SD rats	T:60 C:12 E:12	Healthy	Filament insertion	50 min	3 h after operation	Xenogenic ADMSCs	Intravenous administration	100 μg	1, 3, 14, 28, and 60 days after stroke (C:12 and E:12/timepoint)	1. Infarct volumes (MRI); 2. Behavioral tests (Corner test); 3. MMP-9, IL-1β, TNF-α, RANTES, PAI-1, NF-κB and iNOS (WB, brain tissue). 4. NOX-1, NOX-2 and oxidized protein; 5. Cleaved caspase 3 and cleaved PARP, Smad3, TGF-β, Smad1/5 and BMP-2; 6. γ-H2AX, cytosolic cytochrome C, CD31, eNOS, VEGF, CXCR4; 7. CD11, CD68, XRCC1/CD90, p53BP1/CD90, GFAP, AQP4
Ling et al. (2020)	China	Male SD rats	T:40 C:10 E:10	Healthy	Filament insertion	2 h	4 h after operation	hUSCs	Intravenous administration	1×10^{11} particle	1, 2, 3, 7, 14, 21, and 28 days after stroke (C:10 and E:10/timepoint)	1. Infarct volumes (MRI/cresyl violet staining); 2. Behavioral tests (mNSS, foot-fault test); 3. $EdU^+/Nestin^+$, $EdU^+/Sox2^+$.
Xin et al. (2013a)	USA	Male Wistar rats	T:12 C:6 E:6	Healthy	Filament insertion	2 h	24 h after operation	BMSCs	Tail vein	100 μg	1, 3, 7, 14, 21, and 28 days after stroke (C:6 and E:6/timepoint)	1. Infarct volumes (HE); 2. Behavioral tests (mNSS, Foot-fault test); 3. Axonal density, synaptophysin; 4. DCX/BrdU and vWF/Brdu
Venkat et al. (2019)	USA	male BKS.Cg-m+/+$Lepr^{db}$ /J mice	T:25 C:7 E:6	T2DM	Photothro-mbotic	Permanent	3 days after operation	Mouse brain endothelial cells	Intravenous administration	3×10^{10}	1, 7, 14, 21, 25, 26, 27, and 28 days after stroke (C:7 and E:6/timepoint)	1. Infarct volumes (HE); 2. Behavioral tests (adhesive removal test, novel odor recognition test); 3. Axon and Myelin Density, Vascular Density, Arterial Diameter, Vessel Patency; 4. ED1 and CD163
Xin et al. (2017b)	USA	Male Wistar rats	T:24 C:6 E:6	Healthy	Filament insertion	2 h	24 h after operation	BMSCs	Intra-arterial administration	100 μg	1, 3, 7, 14, 21, and 28 days after stroke (C:6 and E:6/timepoint)	1. Behavioral tests (mNSS, Foot-fault test); 2. Axonal density, synaptophysin, SMI-31

TTC, 2,3,5-triphenyl tetrazolium chloride; BMSCs, bone marrow mesenchymal stem cells; EBST, elevated body swing test; LDH, lactate dehydrogenase; NR, not reported; SD, Sprague-Dawley; C, control group; T, treatment group; ADSCs, adipose-derived stem cells; ADMSCs, adipose-derived mesenchymal stem cells; RIPC, remote ischemic pre-conditioning; NSCs, neural stem cells, HUCB-MSCs human umbilical cord blood-mesenchymal stem cells; mNSS, modified neurological severity scores; OGD, oxygen–glucose-deprived, hUCMSCs, human umbilical cord mesenchymal stem cells; hUSCs, human urine-derived stem cells; HE, hematoxylin and eosin; vWF, von Willebrand factor; T2DM, type 2 diabetes mellitus; BDNF, brain-derived neurotrophic factor; GDNF, glial cell-derived neurotrophic factor; TGF, transforming growth factor; ELISA, enzyme-linked immunosorbent assay; WB Western blot analysis; T, total animal in this study; C, control group; E, exosomes treatment group; TUNEL, terminal deoxynucleotidyl transferase deoxyuridine triphosphate nick end labeling.

TABLE 2 | Quality assessment of eligible studies.

References	A	B	C	D	E	F	G	H	I	J	Total
Xin et al. (2017a)	√		√			√	√		√	√	6
Zhao et al. (2020)	√	√				√	√		√	√	6
Song et al. (2019)	√		√			√	√		√	√	6
Pei et al. (2019)	√	√	√			√	√		√	√	7
Safakheil and Safakheil (2020)	√	√				√	√		√	√	6
Deng et al. (2019)	√		√			√	√		√	√	6
Jiang et al. (2018)	√	√				√	√		√	√	6
Zheng et al. (2019)	√		√			√	√		√	√	6
Li et al. (2019)	√		√				√		√	√	5
Sun et al. (2019)	√		√			√	√		√	√	6
Nalamolu et al. (2019b)	√	√	√			√	√		√	√	7
Nalamolu et al. (2019a)	√		√				√		√	√	5
Li G. et al. (2020)	√	√	√				√		√	√	6
Chen K. H. et al. (2016)	√	√				√	√		√	√	6
Ling et al. (2020)	√		√				√		√	√	5
Xin et al. (2013a)	√						√		√	√	4
Venkat et al. (2019)	√		√			√	√		√	√	6
Xin et al. (2017b)	√		√				√		√	√	5

A peer-reviewed journal; B temperature control; C animals were randomly allocated; D blind established model; E blinded outcome assessment; F use of anesthetic without significant intrinsic vascular protection activity; G appropriate animal model (diabetic, advanced age, or hypertensive); H calculation of sample size; I statement of compliance with animal welfare regulations; J statement of potential conflict of interests.

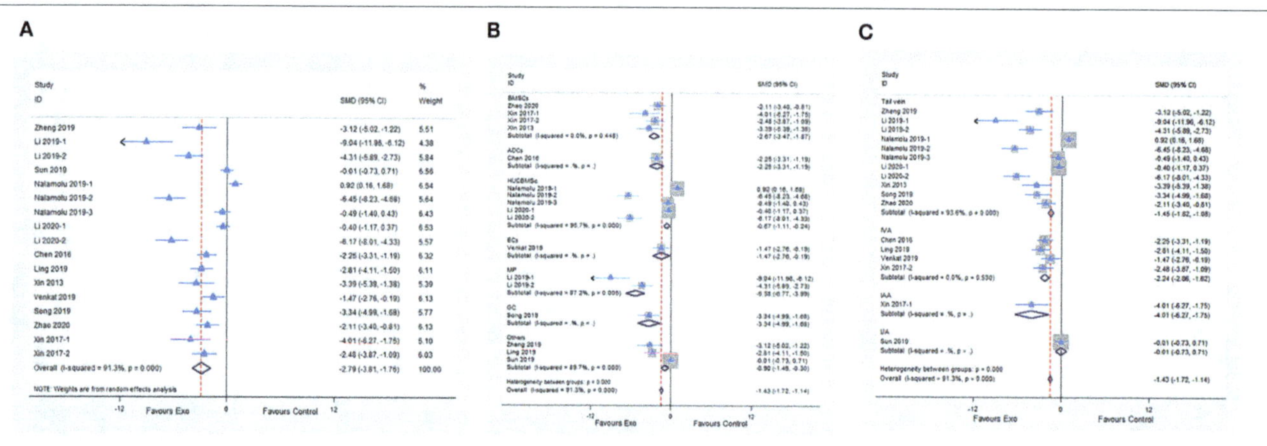

FIGURE 2 | (A) Forest plot shows the efficacy of exosomes in improving the neurological function in the ischemic stroke model. **(B)** Forest plot shows the efficacy of exosomes derived from various cells in improving the neurological function in the ischemic stroke model. **(C)** Forest plot shows the efficacy of exosomes via different administrational routes in improving the neurological function in the ischemic stroke model.

five comparisons assessed the IL-1β level. Meta-analysis showed a significant decrease in the exosome treatment group (SMD −2.29; 95% CI −3.97 to −0.61; $p < 0.001$; $I^2 = 85.1\%$; **Figure 4C**). A meta-analysis of four studies showed a significant reduction of terminal deoxynucleotidyl transferase deoxyuridine triphosphate nick end labeling-positive cell percentage between the exosome group and the control group (SMD −3.20; 95% CI −4.29 to −2.1; $I^2 = 47.8\%$; $p = 0.125$; **Figure 4D**; **Supplementary Table 2**).

Publication Bias

The funnel plot was conducted to examine publication bias for the primary outcomes. Potential publication bias of primary outcomes was tested using Egger's test ($p < 0.0001$; **Figures 5A,B**). The publication bias was more likely to generate from different types of exosome applications or inconsistent measurements. Hence, we performed several subgroup analyses to figure out the potential sources of publication bias, which are described above in primary outcomes.

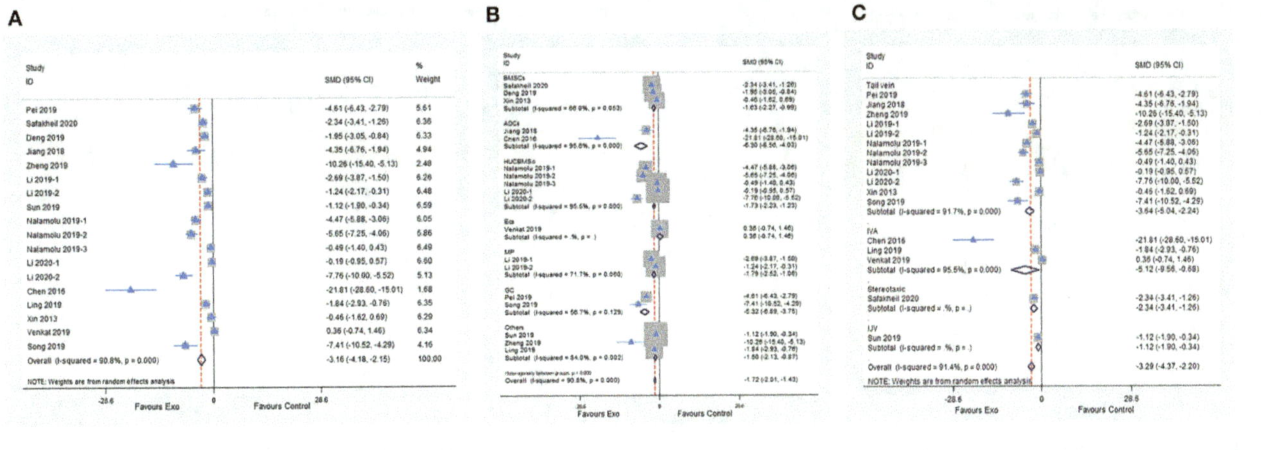

FIGURE 3 | (A) Forest plot shows the efficacy of exosomes in reducing infarct volume in the ischemic stroke model. **(B)** Forest plot shows the efficacy of exosomes derived from various cells in reducing infarct volume in the ischemic stroke model. **(C)** Forest plot shows the efficacy of exosomes via different administrational routes in reducing infarct volume in the ischemic stroke model.

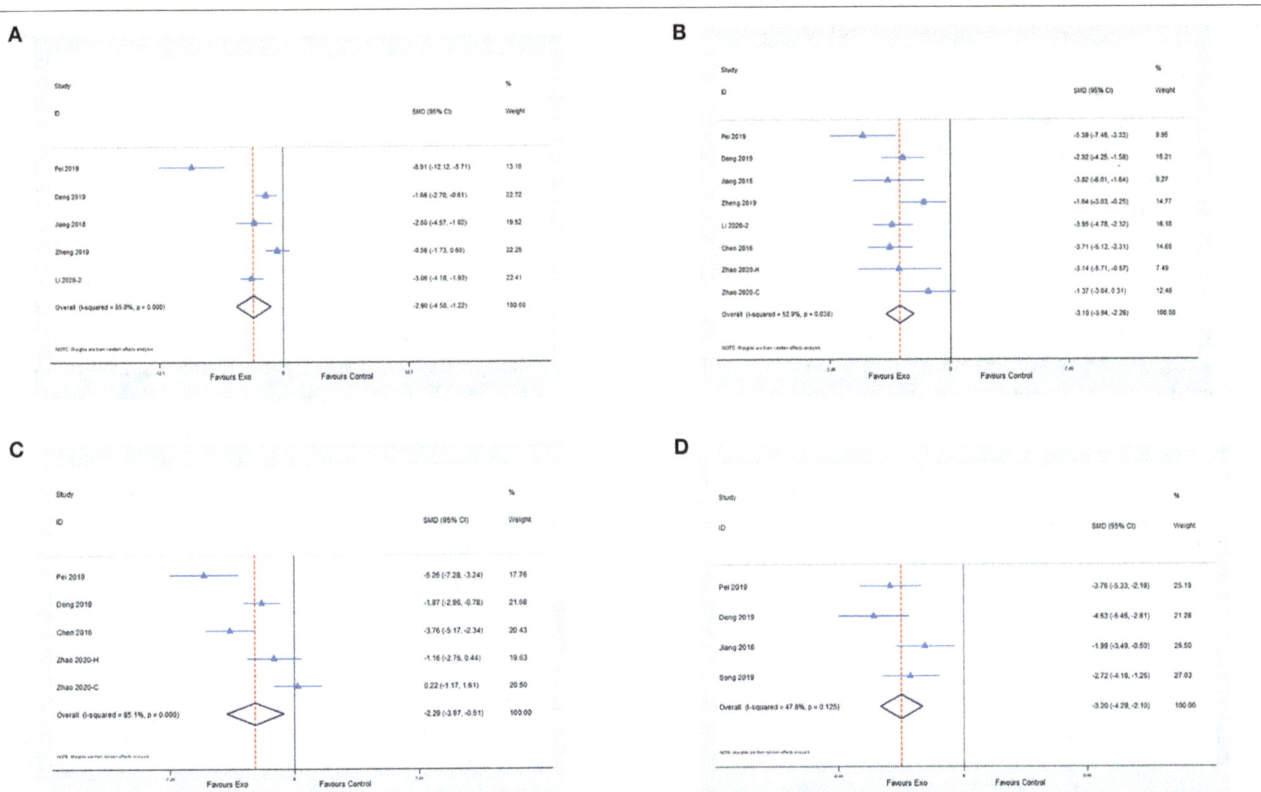

FIGURE 4 | (A) Pooled estimate of exosomes on anti-inflammation, according to IL-6. **(B)** Pooled estimate of exosomes on anti-inflammation, according to TNF-α. **(C)** Pooled estimate of exosomes on anti-inflammation according to IL-1β. **(D)** Pooled estimate of exosomes on anti-apoptosis according to terminal deoxynucleotidyl transferase deoxyuridine triphosphate nick end labeling-positive cells.

DISCUSSION

To the best of our knowledge, this is the first meta-analysis to systematically describe ischemic stroke with unmodified cell-derived exosomes treatment. A meta-analysis of 18 studies with 21 comparisons presented a comprehensive summary of the pooled effect of cell-derived exosomes on rodents with ischemic stroke. The available evidence from our included studies indicated that stem cell-derived exosomes substantially improved neurological function and reduced IV in a rodent with cerebral

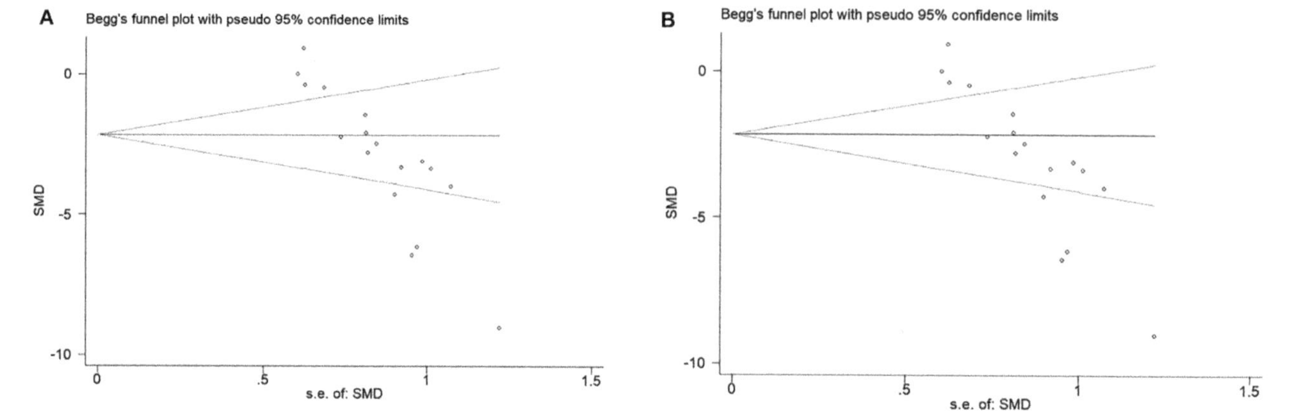

FIGURE 5 | (A) Beggs funnel plot with neural function score. **(B)** Beggs funnel plot with infarct volume.

ischemia. Moreover, pooled analyses demonstrated that exosome therapy not only ameliorates the inflammatory response but also reduces cell apoptosis after ischemic stroke. Therefore, our meta-analysis provided crucial clues for human clinical trials on exosomes and ischemic stroke.

A previous meta-analysis evaluated the efficacy of extracellular vesicles therapy for stroke (Thomas et al., 2020). Both modified and unmodified cell-derived extracellular vesicles were included to explore their benefits in ischemic and hemorrhagic stroke in this publication, whereas it was tough to identify the real therapeutic effects induced by extracellular vesicles or signaling molecule added artificially. In contrast, we specifically investigated the effects of unmodified cell-derived exosomes on an ischemic stroke to reduce the sources of heterogeneity and improve the overall quality of the included studies. Moreover, our meta-analysis further evaluated the effect of exosomes on pro-inflammatory cytokines and cell apoptosis, providing useful information for further laboratory studies and clinical trials.

In the current study, the subgroup analysis showed that all administrational routes helped improve the neurological function, except for internal jugular vein administration. The result that internal jugular vein administration is not effective might be relevant to the short interval between exosome intervention (2 h) and functional assessment (24 h) and/or the relative insensitivity of a four-point test using neurologic deficit score. Besides, the organ sequestration of circulating exosomes injected intravenously or inadequate dosing may be contributed to this negative result (Sun et al., 2019). In this meta-analysis, 12 included studies applied to the rats, whereas nine studies used mice. The subgroup analysis showed that either mice or rats have significant and positive effects on IVs and neurological scores (**Supplementary Figure 3**), but the effects of exosomes on IV between-group difference did reach statistical significance (QB = 7.38, $P = 0.007$), which demonstrated in terms of reducing IVs, a larger treatment effect of exosomes on rats than mice.

Remarkably, all studies were performed on male rodents. Indeed, hormone and sex hormone-independent mechanisms played a central role in regulating sex differences in stroke (Wilson, 2013). Immune response, a part of sex hormone-independent mechanisms, is a vital factor in determining the difference for males and females to respond to experimental stroke (Dotson and Offner, 2017). A study reported by Dotson et al. revealed that splenectomy before middle cerebral artery occlusion significantly decreased the peripheral macrophages/monocytes and activated T cells in male C57BL/6J mice but not female, which resulted in attenuation of neurological impairment in male mice but not female (Dotson et al., 2015).

Exosomes, endosome-derived nanosized membrane vesicles, are released by most cell types. Moreover, exosomes are vital carriers for regulating angiogenesis, neurogenesis, inflammation, and cell apoptosis (Deng et al., 2019; Tian et al., 2019; Ling et al., 2020; Zhao et al., 2020). Emerging studies demonstrated that stem cell transplantation improved neurological recovery after ischemic stroke, and a previous meta-analysis showed that stem cell therapy was safe and effective in animals and humans with cerebral ischemia (Xin et al., 2013a). More importantly, recent studies reported no significant difference in efficacy between stem cell treatment and stem cell-derived exosomes in a rodent model with ischemic stroke. Stem cell-derived exosomes take part in cell-to-cell communication and are considered as the paracrine effectors of cell-based therapy by transferring a great deal of signaling factors including proteins, lipids, mRNA, and non-coding RNAs (such as miRNAs) (Liang et al., 2014; Xin et al., 2014). Compared with cell-based treatment, cell-derived exosomes have lower immunogenicity. In addition, exosome administration could reduce or avoid some of the risks related to cell translation, such as tumorigenesis and cytokine release syndrome (Phinney and Pittenger, 2017). Thus, exosomes seem to be a promising therapy for the recovery of ischemic stroke.

Increasing studies have demonstrated that RNAs derived from exosomes are the critical component for their therapeutic effect. As the content of exosomes, miRNAs are ~22 nucleotides in length and function as non-coding RNAs regulating the level of protein expression by modulating mRNA translation (Ha and Kim, 2014). The miRNAs participate in post-transcriptional regulation of gene expressions typically by binding to the 3′-untranslated region of mRNA sequences (Rupaimoole and Slack, 2017). Given the dramatic effects of miRNA, exosomal miRNAs have gained more attention than proteins and lipids. More and more studies demonstrated that the exosomal miRNAs were involved in the processes of neurological recovery and neural remodeling after stroke (Xin et al., 2017a; Zhang et al., 2019).

In our included studies, there are numerous miRNAs correlated with the neuroprotective effects of exosomes, such as miR-17-92 (Xin et al., 2017a), miR-126 (Venkat et al., 2019), miR-124 (Song et al., 2019), miR-26a (Ling et al., 2020), miR-26b-5p (Li G. et al., 2020), miR-30d-5p (Jiang et al., 2018), and miR-138-5p (Deng et al., 2019). For example, miR-17-92 contained in exosomes enhanced neural plasticity and improved neurological function after stroke, possibly via downregulating PTEN expression and subsequent activating its downstream proteins, the PI3K/Akt/mTOR/GSK-3β signaling pathway (Xin et al., 2017a). Song et al. (2019) demonstrated that M2 microglia-derived exosomes alleviated ischemia-reperfusion injury and protected neuronal survival, and the underlying mechanism might be relevant to exosomal miR-124 and its downstream target ubiquitin-specific protease 14. Moreover, a study reported by Venkat et al. (2019) demonstrated that endothelial cells-Exo promoted capillary tube formation and improved axonal remodeling. However, the downregulation of miR-126 in endothelial cells-Exo reduced Exo-afforded effects of capillary tube formation and axonal regeneration. Thus, it means that exosomal miRNAs play a key role in the neuroprotective function and neural recovery after stroke.

Inflammation is widely involved in the pathogenesis of numerous cerebrovascular disorders, such as ischemic stroke and intracerebral hemorrhage (Kleinig and Vink, 2009), leading to secondary injury to the brain. Increasing evidence demonstrated that microglia-modulated inflammation actively played a part in cerebral infarction (Cai et al., 2017). When ischemic stroke occurred, microglia were polarized toward the M1 phenotype, which produces multiple pro-inflammatory cytokines, including IL-6, TNF-α, and IL-1β, thus aggravating neurofunctional impairment (Tobin et al., 2014). In contrast, the microglia M2 phenotype promotes the secretion of anti-inflammatory cytokines, such as IL-4 and IL-10, thus alleviating neuroinflammation and improving neuronal survival (Schmieder et al., 2012). Li et al. reported that the pro-inflammatory cytokines TNF-α, IL-1β, and IL-6 in the brain tissue were significantly elevated in the rats with ischemic stroke (Li S. et al., 2020).

In this study, six included studies demonstrated that exosome treatment significantly decreased the production of pro-inflammatory factors in brain tissue after cerebral ischemia (Chen K. H. et al., 2016; Deng et al., 2019; Pei et al., 2019; Zheng et al., 2019; Li G. et al., 2020; Zhao et al., 2020). On the other hand, Zhao et al. found that exosome treatment promoted the secretion of anti-inflammatory cytokines (TGF-β and IL-10) in the ischemic hemisphere and repressed microglia M1 polarization to inhibit microglial inflammation via regulating the CysLT 2R-ERK1/2 pathway (Zhao et al., 2020). Moreover, a study reported by Jiang et al. demonstrated that exosomes secreted from adipose-derived stem cells upregulated the expression of anti-inflammatory cytokines (IL-4 and IL-10) and enhanced M2 microglia/macrophage polarization (Jiang et al., 2018). Therefore, exosomes exert a potential neuroprotective effect partially through attenuating brain inflammation.

The ischemic core and penumbra are two separate vital areas of the brain during ischemic stroke. The ischemic core experiences an extreme and rapid decrease of blood flow with irreversible cell death a few minutes after cerebral ischemia. However, apoptosis within the penumbra may appear after several hours or days with reversible cell death (Bandera et al., 2006). Thus, apoptosis is another essential part of the pathogenesis of ischemic stroke, and inhibition of apoptosis could alleviate cerebral injury in stroke models (Uzdensky, 2019). Our meta-analysis demonstrated that cell-derived exosomes contributed to the reduction of cell apoptosis. Song et al. indicated that treatment with M2 microglia-derived exosomes greatly attenuated apoptotic neurons and protected from ischemia-reperfusion injury after stroke compared with phosphate-buffered saline treatment (Song et al., 2019). Moreover, Pei et al. (2019) demonstrated that astrocyte-derived exosomes significantly decreased middle cerebral artery occlusion that induced neuron apoptosis via inhibiting autophagy in experimental ischemic stroke. Additional research reported by Xin et al. (2017a) demonstrated that exosomes not only promoted neurogenesis and oligodendrogenesis but also improved neuronal dendrite plasticity in the ischemic boundary zone, which is possibly related to the PI3K/Akt/mTOR/GSK-3β signaling pathway. Besides, endothelial cell-derived exosomes significantly increased artery diameter and vessel density in the ischemic boundary zone compared with stroke mice without any treatment (Venkat et al., 2019).

Exosomes are emerging as essential intercellular carriers in exerting neurorestorative effects after stroke (Tian et al., 2018; Song et al., 2019). The present meta-analysis result showed that both immediate and delayed treatments with cell-derived exosomes provide therapeutic benefits. Moreover, almost all cell-derived exosomes are effective in improving neurological recovery. However, there are multiple translational challenges needed to be solved before the clinical application of cell-derived exosomes. Firstly, the approach of isolation and storage may have an impact on the characteristics and quality of exosomes (Maroto et al., 2017). Appropriate methods of isolation and storage need to be standardized. Secondly, it is vital to develop the optimal methods to produce a sufficient quantity of exosomes to meet the clinical requirement. Thirdly, in most studies, the follow-up time was <1 month. Thus, the long-term effects of exosomes are a pivotal challenge that needs to be further explored. Fourthly, dose-response and therapeutic window researches are required before the clinical application.

Our study has several potential limitations. First, the earliest start time of intervention available in large human clinical trials was a median of 45 min after ischemic stroke onset (Saver et al., 2015). However, there were ten animal experiments adopting exosome therapy immediately after brain ischemia, which may lead to the overestimation of exosome intervention effects achievable. Second, the ischemic stroke model was performed on young, healthy male rodents in most of the included studies. Nevertheless, ischemic stroke usually occurs in elderly male or female individuals with several risk factors of cerebrovascular diseases (such as hypertension, hyperlipemia, and diabetes), indicating that the heterogenicity of the post-stroke patient requires more complicated interventions. Third, it is important to realize that scaling the production of exosomes for human clinical trials would be required. The general consensus on exosome isolation, purification, and normalization procedures would overcome the crucial bottlenecks in translational studies. Finally, several limitations for this meta-analysis itself should be considered. On the one hand, the heterogeneity between the included studies cannot be evidently decreased, even if we performed subgroup and sensitivity analyses. On the other hand, we performed the data extraction from graphics using WebPlotDigitizer software, which may alter the original data and affect the results.

CONCLUSION

Pooled data from the present meta-analysis demonstrated that cell-derived exosomes improve neurological function and reduce IV in rodent models of ischemic stroke. Furthermore, we found that cell-derived exosomes have potential neuroprotective effects, which were mainly mediated through anti-inflammation and anti-apoptosis. Despite the fact that some factors, such as publication bias and study quality, may affect the validity of positive results, this meta-analysis provides important clues to translate new therapeutic options for cerebral ischemia.

AUTHOR CONTRIBUTIONS

MH and ZH carried out the design, collected the data, performed the statistical analyses, and drafted the manuscript. CX and LL assisted with statistical analyses and drafted the manuscript. LC, SC, and TL participated in the collection of data. HZ conceived the design of the study and contributed to draft the manuscript. All authors contributed to the article and approved the submitted version.

SUPPLEMENTARY MATERIAL

Supplementary Figure 1 | (A) The forest plot shows the efficacy of exosomes via different administrational time in improving the neurological function in the ischemic stroke model. **(B)** The forest plot shows the efficacy of exosomes to different types of stroke in improving the neurological function in the ischemic stroke model. **(C)** The sensitivity analysis of included studies in neurological function scores.

Supplementary Figure 2 | (A) The forest plot shows the efficacy of exosomes via different administrational time in reducing infarct volume in the ischemic stroke model. **(B)** The forest plot shows the efficacy of exosomes to different types of stroke in reducing infarct volume in the ischemic stroke model. **(C)** The sensitivity analysis of included studies in infarct volume.

Supplementary Figure 3 | (A) The forest plot shows the efficacy of exosomes in improving the neurological function in rats and mice with ischemic stroke. **(B)** The forest plot shows the efficacy of exosomes in reducing infarct volume in rats and mice with ischemic stroke.

Supplementary Table 1 | PRISMA 2009 checklist.

Supplementary Table 2 | Meta-analytic results from included studies.

REFERENCES

Acosta, S. A., Tajiri, N., Hoover, J., Kaneko, Y., and Borlongan, C. V. (2015). Intravenous bone marrow stem cell grafts preferentially migrate to spleen and abrogate chronic inflammation in stroke. *Stroke* 46, 2616–2627. doi: 10.1161/STROKEAHA.115.009854

Bandera, E., Botteri, M., Minelli, C., Sutton, A., Abrams, K. R., and Latronico, N. (2006). Cerebral blood flow threshold of ischemic penumbra and infarct core in acute ischemic stroke: a systematic review. *Stroke* 37, 1334–1339. doi: 10.1161/01.STR.0000217418.29609.22

Boltze, J., Lukomska, B., and Jolkkonen, J. (2014). Mesenchymal stromal cells in stroke: improvement of motor recovery or functional compensation? *J. Cereb. Blood Flow Metab.* 34, 1420–1421. doi: 10.1038/jcbfm.2014.94

Burda, B. U., O'Connor, E. A., Webber, E. M., Redmond, N., and Perdue, L. A. (2017). Estimating data from figures with a web-based program: considerations for a systematic review. *Res. Synth. Methods* 8, 258–262. doi: 10.1002/jrsm.1232

Cai, J., Xu, D., Bai, X., Pan, R., Wang, B., Sun, S., et al. (2017). Curcumin mitigates cerebral vasospasm and early brain injury following subarachnoid hemorrhage via inhibiting cerebral inflammation. *Brain Behav.* 7:e00790. doi: 10.1002/brb3.790

Chen, K. H., Chen, C. H., Wallace, C. G., Yuen, C. M., Kao, G. S., Chen, Y. L., et al. (2016). Intravenous administration of xenogenic adipose-derived mesenchymal stem cells (ADMSC) and ADMSC-derived exosomes markedly reduced brain infarct volume and preserved neurological function in rat after acute ischemic stroke. *Oncotarget* 7, 74537–74556. doi: 10.18632/oncotarget.12902

Chen, L., Zhang, G., Khan, A. A., Guo, X., and Gu, Y. (2016). Clinical efficacy and meta-analysis of stem cell therapies for patients with brain ischemia. *Stem Cells Int.* 2016:6129579. doi: 10.1155/2016/6129579

Deng, Y., Chen, D., Gao, F., Lv, H., Zhang, G., Sun, X., et al. (2019). Exosomes derived from microRNA-138-5p-overexpressing bone marrow-derived mesenchymal stem cells confer neuroprotection to astrocytes following ischemic stroke via inhibition of LCN2. *J. Biol. Eng.* 13:71. doi: 10.1186/s13036-019-0193-0

Dirnagl, U., Iadecola, C., and Moskowitz, M. A. (1999). Pathobiology of ischaemic stroke: an integrated view. *Trends Neurosci.* 22, 391–397. doi: 10.1016/S0166-2236(99)01401-0

Doeppner, T. R., Herz, J., Görgens, A., Schlechter, J., Ludwig, A. K., Radtke, S., et al. (2015). Extracellular vesicles improve post-stroke neuroregeneration and prevent postischemic immunosuppression. *Stem Cells Transl. Med.* 4, 1131–1143. doi: 10.5966/sctm.2015-0078

Dotson, A. L., and Offner, H. (2017). Sex differences in the immune response to experimental stroke: implications for translational research. *J. Neurosci. Res.* 95, 437–446. doi: 10.1002/jnr.23784

Dotson, A. L., Wang, J., Saugstad, J., Murphy, S. J., and Offner, H. (2015). Splenectomy reduces infarct volume and neuroinflammation in male but not female mice in experimental stroke. *J. Neuroimmunol.* 278, 289–298. doi: 10.1016/j.jneuroim.2014.11.020

Gervois, P., Wolfs, E., Ratajczak, J., Dillen, Y., Vangansewinkel, T., Hilkens, P., et al. (2016). Stem Cell-Based Therapies for ischemic stroke: preclinical results and the potential of imaging-assisted evaluation of donor cell fate and mechanisms of brain regeneration. *Med. Res. Rev* 36, 1080–1126. doi: 10.1002/med.21400

Ha, M., and Kim, V. N. (2014). Regulation of microRNA biogenesis. *Nat. Rev. Mol. Cell Biol.* 15, 509–524. doi: 10.1038/nrm3838

Jiang, M., Wang, H., Jin, M., Yang, X., Ji, H., Jiang, Y., et al. (2018). Exosomes from MiR-30d-5p-ADSCs reverse acute ischemic stroke-induced, autophagy-mediated brain injury by promoting m2 microglial/macrophage polarization. *Cell. Physiol. Biochem.* 47, 864–878. doi: 10.1159/000490078

Kleinig, T. J., and Vink, R. (2009). Suppression of inflammation in ischemic and hemorrhagic stroke: therapeutic options. *Curr. Opin. Neurol.* 22, 294–301. doi: 10.1097/WCO.0b013e32832b4db3

Levy, M. L., Crawford, J. R., Dib, N., Verkh, L., Tankovich, N., and Cramer, S. C. (2019). Phase I/II study of safety and preliminary efficacy of intravenous allogeneic mesenchymal stem cells in chronic stroke. *Stroke* 50, 2835–2841. doi: 10.1161/STROKEAHA.119.026318

Li, G., Xiao, L., Qin, H., Zhuang, Q., Zhang, W., Liu, L., et al. (2020). Exosomes-carried microRNA-26b-5p regulates microglia M1 polarization after cerebral ischemia/reperfusion. *Cell Cycle* 19, 1022–1035. doi: 10.1080/15384101.2020.1743912

Li, S., Lu, Y., Ding, D., Ma, Z., Xing, X., Hua, X., et al. (2020). Fibroblast growth factor 2 contributes to the effect of salidroside on dendritic and synaptic plasticity after cerebral ischemia/reperfusion injury. *Aging* 12, 10951–10968. doi: 10.18632/aging.103308

Li, Y., Ren, C., Li, H., Jiang, F., Wang, L., Xia, C., et al. (2019). Role of exosomes induced by remote ischemic preconditioning in neuroprotection against cerebral ischemia. *Neuroreport* 30, 834–841. doi: 10.1097/WNR.0000000000001280

Liang, X., Ding, Y., Zhang, Y., Tse, H. F., and Lian, Q. (2014). Paracrine mechanisms of mesenchymal stem cell-based therapy: current status and perspectives. *Cell Transplant* 23, 1045–1059. doi: 10.3727/096368913X667709

Ling, X., Zhang, G., Xia, Y., Zhu, Q., Zhang, J., Li, Q., et al. (2020). Exosomes from human urine-derived stem cells enhanced neurogenesis via miR-26a/HDAC6 axis after ischaemic stroke. *J. Cell. Mol. Med.* 24, 640–654. doi: 10.1111/jcmm.14774

Liu, Z., Li, Y., Zhang, R. L., Cui, Y., and Chopp, M. (2011). Bone marrow stromal cells promote skilled motor recovery and enhance contralesional axonal connections after ischemic stroke in adult mice. *Stroke* 42, 740–744. doi: 10.1161/STROKEAHA.110.607226

Lukomska, B., Stanaszek, L., Zuba-Surma, E., Legosz, P., Sarzynska, S., and Drela, K. (2019). Challenges and controversies in human mesenchymal stem cell therapy. *Stem Cells Int.* 2019:9628536. doi: 10.1155/2019/9628536

Maroto, R., Zhao, Y., Jamaluddin, M., Popov, V. L., Wang, H., Kalubowilage, M., et al. (2017). Effects of storage temperature on airway exosome integrity for diagnostic and functional analyses. *J. Extracell. Vesicles* 6:1359478. doi: 10.1080/20013078.2017.1359478

Moher, D., Liberati, A., Tetzlaff, J., and Altman, D. G. (2009). Preferred reporting items for systematic reviews and meta-analyses: the PRISMA statement. *PLoS Med.* 6:e1000097. doi: 10.1371/journal.pmed.1000097

Nalamolu, K. R., Venkatesh, I., Mohandass, A., Klopfenstein, J. D., Pinson, D. M., Wang, D. Z., et al. (2019a). Exosomes secreted by the cocultures of normal and oxygen-glucose-deprived stem cells improve post-stroke outcome. *Neuromolecular Med.* 21, 529–539. doi: 10.1007/s12017-019-08540-y

Nalamolu, K. R., Venkatesh, I., Mohandass, A., Klopfenstein, J. D., Pinson, D. M., Wang, D. Z., et al. (2019b). Exosomes treatment mitigates ischemic brain damage but does not improve post-stroke neurological outcome. *Cell. Physiol. Biochem.* 52, 1280–1291. doi: 10.33594/000000090

Nogueira, R. G., Jadhav, A. P., Haussen, D. C., Bonafe, A., Budzik, R. F., Bhuva, P., et al. (2018). Thrombectomy 6 to 24 hours after stroke with a mismatch between deficit and infarct. *N. Engl. J. Med.* 378, 11–21. doi: 10.1056/NEJMoa1706442

Pei, X., Li, Y., Zhu, L., and Zhou, Z. (2019). Astrocyte-derived exosomes suppress autophagy and ameliorate neuronal damage in experimental ischemic stroke. *Exp. Cell Res.* 382:111474. doi: 10.1016/j.yexcr.2019.06.019

Phinney, D. G., and Pittenger, M. F. (2017). Concise review: MSC-derived exosomes for cell-free therapy. *Stem Cells* 35, 851–858. doi: 10.1002/stem.2575

Raposo, G., and Stoorvogel, W. (2013). Extracellular vesicles: exosomes, microvesicles, and friends. *J. Cell Biol* 200, 373–383. doi: 10.1083/jcb.201211138

Rupaimoole, R., and Slack, F. J. (2017). MicroRNA therapeutics: towards a new era for the management of cancer and other diseases. *Nat. Rev. Drug Discov.* 16, 203–222. doi: 10.1038/nrd.2016.246

Safakheil, M., and Safakheil, H. (2020). The effect of exosomes derived from bone marrow stem cells in combination with rosuvastatin on functional recovery and neuroprotection in rats after ischemic stroke. *J. Mol. Neurosci.* 70, 724–737. doi: 10.1007/s12031-020-01483-1

Saver, J. L., Starkman, S., Eckstein, M., Stratton, S. J., Pratt, F. D., Hamilton, S., et al. (2015). Prehospital use of magnesium sulfate as neuroprotection in acute stroke. *N. Engl. J. Med.* 372, 528–536. doi: 10.1056/NEJMoa1408827

Schmieder, A., Michel, J., Schönhaar, K., Goerdt, S., and Schledzewski, K. (2012). Differentiation and gene expression profile of tumor-associated macrophages. *Semin. Cancer Biol.* 22, 289–297. doi: 10.1016/j.semcancer.2012.02.002

Schwamm, L. H., Ali, S. F., Reeves, M. J., Smith, E. E., Saver, J. L., Messe, S., et al. (2013). Temporal trends in patient characteristics and treatment with intravenous thrombolysis among acute ischemic stroke patients at Get With The Guidelines-Stroke hospitals. *Circ. Cardiovasc. Qual. Outcomes* 6, 543–549. doi: 10.1161/CIRCOUTCOMES.111.000095

Song, Y., Li, Z., He, T., Qu, M., Jiang, L., Li, W., et al. (2019). M2 microglia-derived exosomes protect the mouse brain from ischemia-reperfusion injury via exosomal miR-124. *Theranostics* 9, 2910–2923. doi: 10.7150/thno.30879

Sun, X., Jung, J. H., Arvola, O., Santoso, M. R., Giffard, R. G., Yang, P. C., et al. (2019). Stem cell-derived exosomes protect astrocyte cultures from *in vitro* ischemia and decrease injury as post-stroke intravenous therapy. *Front. Cell. Neurosci.* 13:394. doi: 10.3389/fncel.2019.00394

Thomas, J. M., Cunningham, C. J., Lawrence, C. B., Pinteaux, E., and Allan, S. M. (2020). Therapeutic potential of extracellular vesicles in preclinical stroke models: a systematic review and meta-analysis. *BMJ Open Sci.* 4:e100047. doi: 10.1136/bmjos-2019-100047

Tian, T., Zhang, H. X., He, C. P., Fan, S., Zhu, Y. L., Qi, C., et al. (2018). Surface functionalized exosomes as targeted drug delivery vehicles for cerebral ischemia therapy. *Biomaterials* 150, 137–149. doi: 10.1016/j.biomaterials.2017.10.012

Tian, Y., Zhu, P., Liu, S., Jin, Z., Li, D., Zhao, H., et al. (2019). IL-4-polarized BV2 microglia cells promote angiogenesis by secreting exosomes. *Adv. Clin. Exp. Med.* 28, 421–430. doi: 10.17219/acem/91826

Tkach, M., and Théry, C. (2016). Communication by extracellular vesicles: where we are and where we need to go. *Cell* 164, 1226–1232. doi: 10.1016/j.cell.2016.01.043

Tobin, M. K., Bonds, J. A., Minshall, R. D., Pelligrino, D. A., Testai, F. D., and Lazarov, O. (2014). Neurogenesis and inflammation after ischemic stroke: what is known and where we go from here. *J. Cereb. Blood Flow Metab.* 34, 1573–1584. doi: 10.1038/jcbfm.2014.130

Uzdensky, A. B. (2019). Apoptosis regulation in the penumbra after ischemic stroke: expression of pro- and antiapoptotic proteins. *Apoptosis* 24, 687–702. doi: 10.1007/s10495-019-01556-6

Venkat, P., Cui, C., Chopp, M., Zacharek, A., Wang, F., Landschoot-Ward, J., et al. (2019). MiR-126 mediates brain endothelial cell exosome treatment-induced neurorestorative effects after stroke in type 2 diabetes mellitus mice. *Stroke* 50, 2865–2874. doi: 10.1161/STROKEAHA.119.025371

Wilson, M. E. (2013). Stroke: understanding the differences between males and females. *Pflugers Arch.* 465, 595–600. doi: 10.1007/s00424-013-1260-x

Xin, H., Katakowski, M., Wang, F., Qian, J. Y., Liu, X. S., Ali, M. M., et al. (2017a). MicroRNA cluster miR-17-92 cluster in exosomes enhance neuroplasticity and functional recovery after stroke in rats. *Stroke* 48, 747–753. doi: 10.1161/STROKEAHA.116.015204

Xin, H., Li, Y., and Chopp, M. (2014). Exosomes/miRNAs as mediating cell-based therapy of stroke. *Front. Cell. Neurosci.* 8:377. doi: 10.3389/fncel.2014.00377

Xin, H., Li, Y., Cui, Y., Yang, J. J., Zhang, Z. G., and Chopp, M. (2013a). Systemic administration of exosomes released from mesenchymal stromal cells promote functional recovery and neurovascular plasticity after stroke in rats. *J. Cereb. Blood Flow Metab.* 33, 1711–1715. doi: 10.1038/jcbfm.2013.152

Xin, H., Li, Y., Liu, Z., Wang, X., Shang, X., Cui, Y., et al. (2013b). MiR-133b promotes neural plasticity and functional recovery after treatment of stroke with multipotent mesenchymal stromal cells in rats via transfer of exosome-enriched extracellular particles. *Stem Cells* 31, 2737–2746. doi: 10.1002/stem.1409

Xin, H., Wang, F., Li, Y., Lu, Q. E., Cheung, W. L., Zhang, Y., et al. (2017b). Secondary release of exosomes from astrocytes contributes to the increase in neural plasticity and improvement of functional recovery after stroke in rats treated with exosomes harvested from microRNA 133b-overexpressing multipotent mesenchymal stromal cells. *Cell Transplant.* 26, 243–257. doi: 10.3727/096368916X693031

Zhang, H., Wu, J., Wu, J., Fan, Q., Zhou, J., Wu, J., et al. (2019). Exosome-mediated targeted delivery of miR-210 for angiogenic therapy after cerebral ischemia in mice. *J. Nanobiotechnology* 17:29. doi: 10.1186/s12951-019-0461-7

Zhao, Y., Gan, Y., Xu, G., Yin, G., and Liu, D. (2020). MSCs-derived exosomes attenuate acute brain injury and inhibit microglial inflammation by reversing cysLT2R-ERK1/2 mediated microglia M1 polarization. *Neurochem. Res.* 45, 1180–1190. doi: 10.1007/s11064-020-02998-0

17

Small-Medium Extracellular Vesicles and their miRNA Cargo in Retinal Health and Degeneration: Mediators of Homeostasis and Vehicles for Targeted Gene Therapy

Yvette Wooff[1,2†], Adrian V. Cioanca[1†], Joshua A. Chu-Tan[1,2], Riemke Aggio-Bruce[1,2], Ulrike Schumann[1] and Riccardo Natoli[1,2]*

[1] The John Curtin School of Medical Research, The Australian National University, Canberra, ACT, Australia, [2] The ANU Medical School, The Australian National University, Canberra, ACT, Australia

***Correspondence:**
Riccardo Natoli
riccardo.natoli@anu.edu.au

† These authors have contributed equally to this work

Photoreceptor cell death and inflammation are known to occur progressively in retinal degenerative diseases such as age-related macular degeneration (AMD). However, the molecular mechanisms underlying these biological processes are largely unknown. Extracellular vesicles (EV) are essential mediators of cell-to-cell communication with emerging roles in the modulation of immune responses. EVs, including exosomes, encapsulate and transfer microRNA (miRNA) to recipient cells and in this way can modulate the environment of recipient cells. Dysregulation of EVs however is correlated to a loss of cellular homeostasis and increased inflammation. In this work we investigated the role of isolated retinal small-medium sized EV (s-mEV) which includes exosomes in both the healthy and degenerating retina. Isolated s-mEV from normal retinas were characterized using dynamic light scattering, transmission electron microscopy and western blotting, and quantified across 5 days of photo-oxidative damage-induced degeneration using nanotracking analysis. Small RNAseq was used to characterize the miRNA cargo of retinal s-mEV isolated from healthy and damaged retinas. Finally, the effect of exosome inhibition on cell-to-cell miRNA transfer and immune modulation was conducted using systemic daily administration of exosome inhibitor GW4869 and *in situ* hybridization of s-mEV-abundant miRNA, miR-124-3p. Electroretinography and immunohistochemistry was performed to assess functional and morphological changes to the retina as a result of GW4869-induced exosome depletion. Results demonstrated an inverse correlation between s-mEV concentration and photoreceptor survivability, with a decrease in s-mEV numbers following degeneration. Small RNAseq revealed that s-mEVs contained uniquely enriched miRNAs in comparison to in whole retinal tissue, however, there was no differential change in the s-mEV miRNAnome following photo-oxidative damage. Exosome inhibition via the use of GW4869 was also found to exacerbate retinal degeneration, with reduced retinal function and increased levels of inflammation and cell death demonstrated following photo-oxidative damage in exosome-inhibited mice. Further, GW4869-treated mice displayed impaired translocation of photoreceptor-derived miR-124-3p to the inner retina during damage. Taken together, we propose that retinal s-mEV and their miRNA cargo play an essential role in maintaining retinal homeostasis through immune-modulation, and have the potential to be used in targeted gene therapy for retinal degenerative diseases.

Keywords: retina, neurodegeneration, extracellular vesicle, microRNA, immuno-modulation, gene therapy

INTRODUCTION

Retinal degenerative diseases comprise a heterogeneous group of visual disorders associated with neuroinflammation and the progressive death of retinal neurons, often resulting in irreversible blindness (Ratnapriya and Swaroop, 2013). Despite advances in understanding the pathogenesis and progression of retinal degenerative diseases (Donoso et al., 2006; Janik-Papis et al., 2009; Knickelbein et al., 2015), the precise molecular mechanisms that propagate retinal inflammation and subsequent cell death remain unknown. The study of extracellular vesicles (EV), including exosomes, might provide greater clarity in unpicking these mechanisms given their role as endogenous modulators of biological processes, including in inflammation (Chan et al., 2019), and intracellular communication pathways (Yuana et al., 2013; Isola and Chen, 2017; Mathieu et al., 2019).

Extracellular vesicle are small membrane-enclosed delivery vehicles which have been widely investigated for their vital role in mediating cell-to-cell communication in both healthy as well as diseased states (Yuana et al., 2013; Isola and Chen, 2017; Mathieu et al., 2019). Exosomes, the smallest EV fraction (40–200 nm in diameter) (Théry et al., 2002; Vella et al., 2017; Hessvik and Llorente, 2018; Słomka et al., 2018), in particular, have recently been implicated in the pathogenesis of retinal degenerative diseases. These include Age-Related Macular Degeneration (AMD) (Wang et al., 2009; Klingeborn et al., 2017; Hsu et al., 2018; Li et al., 2019), Diabetic Retinopathy (Huang et al., 2018), and Retinitis Pigmentosa (Vidal-Gil et al., 2019).

Exosomes are released by nearly all cell types and are formed via the endocytic pathway, with the invagination of the endosomal membrane allowing intraluminal vesicle formation in multivesicular bodies (MVB) (Van Niel et al., 2006). Biogenesis occurs either in an endosomal sorting complex required for transport (ESCRT)-dependent or ESCRT-independent manner (Kowal et al., 2014; Li et al., 2019), the latter of which can be blocked using GW4869, a non-competitive inhibitor of neutral sphingomyelinase 2 (nSMase2) - the key enzyme for generating exosomes via the ESCRT-independent pathway (Luberto et al., 2002). Following biogenesis, the MVB fuses with the host cell plasma membrane and releases the internalized exosomes into the extracellular environment (Van Niel et al., 2006; Abels and Breakefield, 2016; Hessvik and Llorente, 2018). From here, exosomes can travel to nearby or distant target cells, often through biological fluids such as blood (Charlotte et al., 2016), to exert their biological effects (Van der Pol et al., 2012; Yáñez-Mó et al., 2015; Abels and Breakefield, 2016). As exosomes selectively incorporate proteins, mRNA and non-coding RNA such as microRNA (miRNA) from their host cell (reviewed in Yáñez-Mó et al., 2015; Abels and Breakefield, 2016), the transfer of exosomal contents can alter the environment of target recipient cells. In a healthy state, the transfer of exosomal nucleic acid contents, including miRNA, is required for homeostatic maintenance (Desdín-Micó and Mittelbrunn, 2017; Fleshner and Crane, 2017). However, in disease, aberrations to this process or the selective encapsulation of toxic proteins and/or dysregulated miRNA, can cause progressive inflammation (Gupta and Pulliam, 2014; Alexander et al., 2015).

miRNA have been labeled as 'master regulators' of gene expression, due to their ability to target and repress multiple genes within and across different biological pathways (Christopher et al., 2016). This regulatory power makes them ideal therapeutic and diagnostic molecules (Rupaimoole and Slack, 2017), in particular to combat dysregulated immune responses, such as those occurring in retinal degenerations (Bartel, 2009; Christopher et al., 2016). miRNA have been reported to be enriched in exosomes over their host cells, suggesting that they are selectively incorporated to serve a dynamic, fast-response biological need (Squadrito et al., 2014; Bhome et al., 2018).

The biological importance of exosomes and exosomal-miRNA (exoMiR) is reflected by their association with a range of inflammatory (Van Hezel et al., 2017; Cypryk et al., 2018), autoimmune (Long et al., 2018; Anel et al., 2019) and neurodegenerative diseases (Jan et al., 2017; Soria et al., 2017; Jain et al., 2019). These include diabetes (Saeedi et al., 2018), rheumatoid arthritis (Skriner et al., 2006; Zhang et al., 2006; Wan et al., 2019), and Alzheimer's and Parkinson's disease (Gupta and Pulliam, 2014; Li et al., 2018; Jain et al., 2019; Sackmann et al., 2019). exoMiR have been reported to play pathogenic roles in these diseases by promoting angiogenesis (Kosaka et al., 2013; Atienzar–Aroca et al., 2016), and modulating immune responses, including in the recruitment of immune cells (Théry et al., 2002; Abusamra et al., 2005; Robbins and Morelli, 2014; Ye et al., 2014; Wong et al., 2016; Van Hezel et al., 2017); features that contribute to cell death. To date however, the identification and role of exosomes and their miRNA cargo in the retina in both healthy and diseased states, is largely unexplored (Klingeborn et al., 2017; Li et al., 2019). While small EV fractions isolated following high speed $> 100{,}000 \times g$ ultracentrifugation and expressing tetraspanin markers CD9, CD63, and CD81 (such as those isolated in this work) are commonly referred to as exosomes, without evidence of endosomal origin and in complying with

MISEV 2018 guidelines (Théry et al., 2018), are herein referred to as small-to-medium EV, or s-mEV. In reference to other works, EV terminology will be referred to as published in the original papers.

Characterizing the role of s-mEV and their miRNA cargo in both the normal and degenerating retina will aid in elucidating novel cell-to-cell communication pathways that could play a role in propagating inflammation during retinal degenerative diseases. Furthermore, uncovering the miRNA signature within retinal s-mEV as well as their potential binding partners may reveal novel regulatory mechanisms underpinning retinal degenerations, ultimately leading to the discovery of therapeutic targets. This study characterizes for the first time, retinal-derived s-mEV from both the healthy and degenerating mouse retina using a previously established model of photo-oxidative damage-induced retinal degeneration (Natoli et al., 2016). Photo-oxidative damage models such as the one employed in this study accurately replicate key pathological changes seen in AMD, including the upregulation of oxidative stress and inflammatory pathways, progressive centralized focal photoreceptor cell loss and microglial/macrophage recruitment and activation (Marc et al., 2008; Tanito et al., 2008; Natoli et al., 2016; Abokyi et al., 2020).

We show that s-mEV secretion is inversely correlated to photoreceptor survivability, with the severity of retinal degeneration directly correlating to decreased retinal s-mEV numbers. We used small RNAseq to characterize the miRNA cargo of retinal s-mEV (exoMiR). Although we demonstrated that there was no change in s-mEV miRNA-cargo in response to retinal degeneration, we found that s-mEVs contain a set of uniquely enriched miRNAs. Further, we show that miRNA contained in retinal s-mEV were associated with the regulation of inflammatory, cell survival and motility pathways. Upon systemic exosome inhibition using GW4869, retinal function in healthy and photo-oxidative damaged mice was significantly reduced compared to controls. In addition, photoreceptor cell death and inflammation were significantly increased in GW4869-injected photo-oxidative damaged mice, compared to controls. Using *in situ* hybridization, we further demonstrated that the expression of miR-124-3p in the inner retina was reduced in GW4869-injected photo-oxidative damaged mice, suggesting that miR-124-3p movement could be mediated through s-mEV-dependent transport.

Taken together, these results suggest a novel role for s-mEV and s-mEV-miRNA-mediated cell-to-cell communication in the retina. We demonstrate that maintaining and transporting necessary levels of s-mEV cargo is required for normal retinal homeostasis and immunomodulation, with insufficient bioavailability of s-mEV potentially leading to inflammatory cell death (see **Figure 10**). In addition, this work elucidates downstream biological targets of s-mEV-miRNA that are required for retinal homeostatic maintenance, and identifies potential miRNA and mRNA therapeutic targets for further investigations.

MATERIALS AND METHODS

Animal Handling and Photo-Oxidative Damage

All experiments were conducted in accordance with the ARVO Statement for the Use of Animals in Ophthalmic and Vision Research and with approval from the Australian National University's (ANU) Animal Experimentation Ethics Committee (AEEC) (Ethics ID: A2017/41; Rodent models and treatments for retinal degenerations). Adult male and female C57BL/6J wild-type (WT) mice (aged between 50 and 90 postnatal days) were bred and reared under 12 h light/dark cycle conditions (5 lux) with free access to food and water. The C57BL/6J colony was genotyped for the presence of both the $Rpe65^{450Met}$ polymorphism or the deleterious $Crb1^{rd8}$ mutation using previously published primer sets (Kim et al., 2004; Mattapallil et al., 2012). Sequencing for these was conducted at the ACRF Biomolecular Resource Facility, ANU. All animals used possessed the $Rpe65^{450Met}$ polymorphism but were free of the $Crb1^{rd8}$ mutation. Littermate age-matched WT mice were randomly assigned to photo-oxidative damage (PD) and dim-reared control (DR) groups. Animals in the photo-oxidative damage group were continuously exposed to 100k lux white LED light for a period of 1, 3, or 5 days as described previously (Natoli et al., 2016), with the majority of experiments conducted for 5 days. Dim-reared control mice were maintained in 12 h light (5 lux)/dark cycle conditions.

Retinal s-mEV Isolation

Mice were euthanized with CO_2 following experimental runs. Either two (from one mouse) or four (from two mice - used for high-throughput sequencing) retinas were pooled and collected in Hanks Buffered Saline Solution (HBSS, Gibco; Thermo Fisher Scientific, MA, United States). Retinas were transferred to 500 μL digestion solution [(HBSS containing 2.5 mg/mL papain (Worthington Biochemical, NJ, United States), 200U DNase I (Roche Diagnostics, NSW, Australia), 5 μg/mL catalase (Sigma-Aldrich, MO, United States), 10 μg/mL gentamycin (Sigma-Aldrich, MO, United States) and 5 μg/mL superoxide dismutase (Worthington Biochemical, NJ, United States)] and finely chopped using scissors. Retinas were incubated at 37°C for 8 min, followed by 20 min at 8°C, to allow for the breakdown of the extracellular matrix and s-mEV release. Following digestion, tissue suspensions were neutralized by diluting in 11.5 mL of HBSS and centrifuged at 1000 × *g* for 10 minutes at 4°C to remove cells and cell debris. The supernatant was transferred to 14 × 89 mm Beckman Ultra-Clear ultracentrifuge tubes (Beckman Coulter, CA United States) and centrifuged at 10,000 × *g* for 30 min at 4°C in a Beckman Coulter Optima XE-100 [fitted with a SW41Ti Rotor (Beckman Coulter, CA, United States)], to collect large EVs and remaining cell debris. The s-mEV-containing supernatant was transferred to new ultracentrifuge tubes and centrifuged for

1.5 h at 150,000 × g at 4°C. The supernatant was carefully decanted, and the s-mEV pellet resuspended via tituration for 1 min in 500 μL Ultrapure Endotoxin-free 0.1M PBS (Thermo Fisher Scientific, MA, United States) and used immediately for quantification.

For RNA isolation, the s-mEV pellet was resuspended immediately in 100 μl RNAse A (10 μg/ml in Ultrapure Endotoxin-free 0.1M PBS) and incubated for 30 minutes at 37°C to digest any RNA contamination. Following RNase treatment, s-mEV RNA was extracted using the mirVana miRNA Isolation Kit (Thermo Fisher Scientific, MA, United States) as per section "RNA Extraction."

S-mEV Characterization

NanoSight

The size and concentration of s-mEV were measured using nanoparticle tracking analysis on a NanoSight NS300 (Malvern Instruments, Malvern, United Kingdom). s-mEV samples were diluted 1:20 (retinal s-mEV) or 1:40 (cell culture s-mEV) in 1 ml Ultrapure Endotoxin-free 0.1M PBS to achieve a particle per frame value between 20 and 100. Samples were analyzed under constant flow provided by a syringe pump set at a speed of 35 (equating to ~3.1 μl/min; Gerritzen et al., 2017). A total of nine 30 s long videos were captured (camera setting: 14) for each sample. The detection threshold was set between 4 and 5 and was not altered between measurements of the same experiment. The concentration values, modal and mean sizes were exported to Prism V7 (GraphPad Software, CA, United States) for statistical analysis and plotting.

Zetasizer

Dynamic light scattering measurements were performed using a Zetasizer Nano ZS 90 (Malvern Instruments, Malvern, United Kingdom). A 500 μL undiluted retinal s-mEV suspension in Ultrapure Endotoxin-free 0.1M PBS was prepared, loaded in a low-volume disposable sizing cuvette (ZEN0112, Malvern Instruments, Malvern, United Kingdom) and agitated before measurements. Measurement parameters were set as follows: Material Refractive Index – 1.46, Dispersant Refractive Index – 1.330, Viscosity (cP) – 0.888, Temperature (°C) – 25, and Measurement Duration (s) – 60. The acquired intensity data was transformed using the General-Purpose Model within the Zetasizer analysis software to generate the size distribution of the s-mEV.

Transmission Electron Microscopy (TEM)

A 30 μl retinal s-mEV suspension was placed on a 200-mesh carbon-coated copper grid (Sigma-Aldrich, MO, United States) pre-treated with glow discharge using an Emtech K100X system (Quorum Technologies, Sussex, United Kingdom). After 20 min, s-mEV were contrasted with 2% uranyl acetate solution for 1 minute, followed by three washes in 0.22 μm filtered PBS (Thermo Fisher Scientific, MA, United States). Excess PBS was removed by placing a piece of absorbent paper at the edge of the grid. The grids were imaged on a Hitachi 7100 FA transmission electron microscope (Hitachi, Tokyo, Japan) at 100 kV. The images were captured with a side mounted Gatan Orius CCD camera (Gatan, CA, United States) at 4008 × 2672 pixels resolution using a 2 s exposure operated through Gatan Microscopy Suite (Gatan, CA, United States). A total of 20 images were captured at 100,000x magnification from four different grids, each containing s-mEV isolated from a different retinal s-mEV preparation (2 mouse retinas/preparation). The images were imported into ImageJ V2.0 software (National Institutes of Health, Bethesda, MD, United States), scale-calibrated and the diameter of approximately 230 s-mEV was measured. The size distribution was plotted in a histogram with 20 nm wide bins using Prism V7.0 (GraphPad Software, CA, United States).

Western Blot

s-mEV pellets (see section "Retinal s-mEV Isolation") were immediately lysed in 50 μL CellLyticTM Cell Lysis Buffer (supplemented with 1:100 protease inhibitor cocktail; Sigma-Aldrich, MO, United States). The blot was performed as previously described (Jiao et al., 2018). Briefly, 10 μg of denatured protein was loaded onto Novex 4-20% Tris-Glycine Mini Gels (Thermo Fisher Scientific, MA, United States) and subjected to electrophoresis (45 min, 150 V). The protein bands were transferred (45 min, 20 V) to a nitrocellulose membrane (Bio-Rad, CA, United States) using the Power Blotter semi-dry system (Thermo Fisher Scientific, MA, United States). Membranes were then washed in PBS-Tween (0.025%; PBS-T), blocked in 3% BSA for 1 h and then incubated overnight at 4°C with primary s-mEV marker antibodies CD63 (1:1000, Ts63, Thermo Fisher Scientific, MA, United States), CD81 (1:2000, ab109201, Abcam, Cambridge, United Kingdom) or CD9 (1:2000, ab92726, Abcam, Cambridge, United Kingdom). Following three washes in PBS-T, blots were incubated in appropriate secondary antibodies, HRP-conjugated Goat Anti-Rabbit IgG (H + L) (1:1000, 170-6515, Bio-Rad, CA, United States) or Goat-anti-Mouse IgG (1:1000, 170-6516, Bio-Rad, CA, United States) for 2 h at room temperature. Membranes were washed in PBS-T and developed for 2 min with Clarity™ Western ECL Substrate (Bio-Rad, CA, United States). Imaging was performed using a ChemiDoc™ MP Imaging System with Image Lab™ software (Bio-Rad, CA, United States).

Exosome Inhibition

Exosome inhibition was performed using GW4869 (Sigma-Aldrich, MO, United States), a known inhibitor of exosome biogenesis and release (Catalano and O'Driscoll, 2020). GW4869 was reconstituted in dimethyl sulfoxide (DMSO; Sigma-Aldrich, MO, United States) to a concentration of 5 mM and used as a stock solution for further dilution in Ultrapure Endotoxin-free 0.1M PBS. Mice were injected with 1.25 mg/kg GW4869 via intraperitoneal (I.P.) injection daily for 5 days. 10.3% DMSO in Ultrapure Endotoxin-free 0.1M PBS (corresponding to the final volume of DMSO in GW4869 preparations) was used as a negative control. All mice were monitored daily for signs of distress or sickness.

Retinal Assessment

Retinal Function via Electroretinography (ERG)

To assess retinal function full-field scotopic ERG was performed as previously described (Natoli et al., 2017). Briefly, mice were dark-adapted overnight before being anesthetized with an intraperitoneal injection of Ketamine (100 mg/kg; Troy Laboratories, NSW, Australia) and Xylazil (10 mg/kg; Troy Laboratories, NSW, Australia). Both pupils were dilated with one drop each of 2.5% w/v Phenylephrine hydrochloride and 1% w/v Tropicamide (Bausch and Lomb, NY, United States).

Anesthetized and pupil dilated mice were placed on the thermally regulated stage of the Celeris ERG system (Diagnosys LLC, MA, United States). The Celeris ERG system has combined Ag/AgCl electrode-stimulator eye probes which measure the response from both eyes simultaneously, and uses 32-bit ultra-low noise amplifiers fitted with impedance testing. Eye probes were cleaned with 70% ethanol and then a 0.3% Hypromellose eye drop solution (GenTeal; Novartis, NSW, Australia) was applied to both probes. The probes were then placed covering and just touching the surface of each eye. A single- or twin-flash paradigm was used to elicit a mixed response from rods and cones. Flash stimuli for mixed responses were provided using 6500K white flash luminance range over stimulus intensities from -0.01 – 40 log.cd.s.m^{-2}. Responses were recorded and analyzed using Espion V6 Software (Diagnosys LLC, MA, United States). Statistics were performed in Prism V7.0 using a two-way analysis of variance (ANOVA) to test for differences in a-wave and b-wave responses. Data was expressed as the mean wave amplitude $\pm$ SEM (μV).

Optical Coherence Tomography (OCT)

Cross-sectional images of live mouse retinas were taken at 1 mm increments from the optic nerve using a Spectralis HRA + OCT device (Heidelberg Engineering, Heidelberg, Germany) as previously described (Natoli et al., 2016). Eye gel (GenTeal; Novartis, NSW, Australia) was administered to both eyes for recovery.

Using OCT cross-sectional retinal images, and ImageJ V2.0 software (National Institutes of Health, Bethesda, MD, United States), the thickness of the outer nuclear layer (ONL), was calculated as the ratio to the whole retinal thickness (outer limiting membrane to the inner limiting membrane).

Retinal Tissue Collection and Preparation

Animals were euthanized with CO_2 following functional ERG analysis. The superior surface of the left eye was marked and enucleated, then immersed in 4% paraformaldehyde for 3 h. Eyes were then cryopreserved in 15% sucrose solution overnight, embedded in OCT medium (Tissue Tek, Sakura, Japan) and cryosectioned at 12 μm in a parasagittal plane (superior to inferior) using a CM 1850 Cryostat (Leica Biosystems, Germany). To ensure accurate comparisons were made for histological analysis, only sections containing the optic nerve head were used for analysis. The retina from the right eye was excised through a corneal incision and placed into RNAlater solution (Thermo Fisher Scientific, MA, United States) at 4°C overnight and then stored at -80°C until further use.

Immunolabeling

Immunohistochemical analysis of retinal cryosections was performed as previously described (Rutar et al., 2015). Fluorescence was visualized and images taken using a laser-scanning A1^{+} confocal microscope at 20x magnification (Nikon, Tokyo, Japan). Images panels were analyzed using ImageJ V2.0 software and assembled using Photoshop CS6 software (Adobe Systems, CA, United States).

IBA-1 Immunohistochemistry

Immunolabeling for IBA-1 (1:500, 019-19741, Wako, Osaka, Japan) and quantification was performed as previously described (Rutar et al., 2015). The number of IBA-1^{+} cells (a marker of retinal microglia and macrophages) was counted across the superior and inferior retina using two retinal sections per mouse and then averaged. Retinal cryosections were stained with the DNA-specific dye bisbenzimide (1:10000, Sigma-Aldrich, MO, United States) to visualize the cellular layers.

TUNEL Assay

Terminal deoxynucleotidyl transferase (Tdt) dUTP nick end labeling (TUNEL), was used as a measure of photoreceptor cell death. TUNEL *in situ* labeling was performed on retinal cryosections using a Tdt enzyme (Cat# 3333566001, Sigma-Aldrich, MO, United States) and biotinylated deoxyuridine triphosphate (dUTP) (Cat# 11093070910, Sigma-Aldrich, MO, United States) as previously described (Natoli et al., 2010). Images of TUNEL staining were captured with the A1^{+} Nikon confocal microscope at 20x magnification. The total number of TUNEL^{+} cells were counted including both the superior and inferior retina using two retinal sections per animal, and is represented as the average number of TUNEL^{+} cells per retinal section.

To further quantify photoreceptor survival, the thickness of the ONL on retinal cryosections was determined by counting the number of nuclei rows (photoreceptor cell bodies) in the area of retinal lesion development (1 mm superior to the optic nerve head). Photoreceptor cell row quantification was performed five times per retina using two retinal cryosections at comparable locations per mouse. The thickness the ONL, inner nuclear layer (INL), and the combined ganglion cell layer (GCL)-outer plexiform layer (OPL) thickness were also measured at the lesion site on the superior retina, and expressed as a ratio to whole retinal thickness.

In situ Hybridization

Localization of miR-124-3p within the retina was determined by *in situ* hybridization. A double DIG-labeled miR-124-3p miRCURY LNA miRNA Detection Probe (Exiqon, Vedbaek, Denmark) was used on retinal cryosections, which were hybridized for 1 h at 53°C as previously described (Chu-Tan et al., 2018). The bound probe was visualized using 5-bromo-4-chloro-3 indoyl phosphate (NBT/BCIP; Sigma-Aldrich Corp., St. Louis, MO, United States). Bright field images were captured on the A1^{+} Nikon confocal microscope fitted with a DS-Ri1-U3 color camera at 20x magnification and 4076 × 3116 pixel resolution. All images were centered at the site of lesion located approximately 1mm superiorly to the optic nerve head. The images were imported into ImageJ V2.0 software, converted to 8-bit format and then the densitometry was calculated. Mean gray values were measured at five different locations along the INL, ONL and the outer limiting membrane/photoreceptor inner segment region with background levels subtracted prior.

TABLE 1 | TaqMan hydrolysis probes (Thermo Fisher Scientific, MA, United States) used for qPCR.

Gene symbols	Gene name	Catalog number
miR-124a	Mmu_miR-124-3p	001182 (assay ID)
U6 snRNA	Small nuclear RNA U6	001973 (assay ID)
Gapdh	Glyceraldehyde-3-phosphatase dehydrogenase	Mm01536933_m1
Pdcd6ip	Programmed cell death 6 interacting domain	Mm00478032_m1
Smpd3	Sphingomyelin phosphodiesterase 3	Mm00491359_m1

RNA Extraction

RNA (enriched for miRNA) extraction and purification from retinas or RNase A treated s-mEV pellets was performed using an acid-phenol:chloroform extraction method with the mirVana miRNA Isolation Kit (Thermo Fisher Scientific, MA, United States) according to the manufacturer's instructions. The concentration and purity of each RNA sample was assessed using the ND-1000 spectrophotometer (Nanodrop Technologies, DE, United States). The size distribution and concentration of s-mEV miRNA was further assessed using a 2100 Agilent Bioanalyzer with an Agilent Small RNA Kit (Agilent Technologies, CA, United States), according to the manufacturers' instruction.

cDNA Synthesis From mRNA and miRNA Templates

Following purification of RNA, cDNA was synthesized from 1 μg RNA using either the Tetro cDNA Synthesis Kit (Bioline Reagents, London, United Kingdom) from an mRNA template, or using the TaqMan MicroRNA RT kit (Thermo Fisher Scientific) from a miRNA template, according to manufacturers' instructions.

Quantitative Real-Time Polymerase Chain Reaction

The expression of ESCRT-independent exosome biogenesis pathways genes was measured by qRT-PCR. We targeted *Pdcd6ip* (also known as *Alix*), which encodes an accessory protein in the ESCRT-dependent pathway, and *Smpd3*, which encodes nSMase2 in the ESCRT-independent pathway (Hessvik and Llorente, 2018). The expression of miR-124-3p was also investigated in retinal lysates from exosome-inhibited mice, and controls. The expression of these genes and miRNA was measured using mouse specific TaqMan hydrolysis probes (**Table 1**) and TaqMan Gene Expression Master Mix (Thermo Fisher Scientific, MA, United States). Reactions were performed in technical duplicates in a 384-well format using a QuantStudio 12 K Flex RT-PCR machine (Thermo Fisher Scientific, MA, United States). Data was analyzed using the comparative C_t method ($\Delta\Delta C_t$) and results are presented as percent change relative to control. Expression was normalized to reference gene glyceraldehyde-3-phosphate dehydrogenase (*Gapdh*) for mRNA, and small nuclear RNA U6 for miRNA.

Small RNA High-Throughput Sequencing

Library Preparation and Sequencing

cDNA libraries from miRNA-enriched s-mEV RNA samples were prepared by the John Curtin School of Medical Research Biomolecular Research Facility (JCSMR, BRF, ACT, Australia), using the Capture and Amplification by Tailing and Switching method (CATS RNA-seq Kit v2 × 24, Diagenode Cat# C05010041, Leige, Belgium). 10 ng RNA was used as input and the dephosphorylation step omitted to select for 3′-OH RNA species (miRNA). The library was amplified with 9 PCR cycles and cleaned with 0.9x AMPure® XP beads (A63881, Beckman Coulter, CA, United States) to enrich for DNA fragments shorter than 50 nt. Libraries were multiplexed and sequenced on a single lane using the Illumina NextSeq500 (Illumina, CA, United States) acquiring 50 base-pairs single-end reads. The sequencing depth was between 8 and 15 million reads/sample with an average phred read quality of 33 (**Supplementary Figure S3A**). Sequencing libraries prepared from whole retinal tissue were retrieved from BioProject database (Accession ID: PRJNA606092)[1]. These libraries were previously prepared by our group using the same library construction method, and the same bioinformatic analysis pipeline was applied as stated below, see Section "Bioinformatics."

Bioinformatics

Sequencing reads were initially checked for quality scores, adapter/index content and K-mer content using FastQC v0.11.8 (Babraham Bioinformatics, Cambridge, United Kingdom), then imported into Partek® Flow® (Partek Inc, MO, United States) for all subsequent analyses. Base composition analysis indicated the presence of an enriched poly-A tail and the CATS template-switching nucleotide within the first three base-pairs of the reads (**Supplementary Figure S3B**), indicating a successful library preparation. A trimming pipeline was created according to the following cutadapt code "cutadapt –trim-n -a GATCGGAAGAGCACACGTCTG -a AGAGCACACGTCTG < input.file > | cutadapt -u 3 -a AAAAAAAAAACAAAAAAAAAA -e 0.16666666666666666 - | cutadapt -g GTTCAGAGTTCTACAGTCCGACGATC -m 18 -o < output.file > -", to remove the template switching nucleotide and the 3′ and 5′ adapters associated with the CATS library preparation. After trimming, the base composition was centered around 25% for each nucleotide (**Supplementary Figure S3C**), and FastQC analysis confirmed the effective removal of the 3′ and 5′ adapters and indices.

Trimmed reads were aligned against the mature mouse miRNA downloaded from miRbase v.22 using the Burrows–Wheeler Aligner (BWA) in backtrack mode with parameters -n 1(base), -o 1, -e -1, -d 10, -i 5, -k 2, -E 4, -R 30, -t 24 (as recommended by Ziemann et al., 2016). miRNAs with less than 10 alignments across all samples were discarded from the subsequent analyses. Reads aligned to miRbase v.22 (mature

[1]https://www.ncbi.nlm.nih.gov/bioproject/?term=PRJNA606092

miRNAs) with BWA had a length distribution between 18 and 24 nucleotides consistent with the expected length of mature miRNA (**Supplementary Figure S3D**). Aligned reads were normalized using the Trimmed means of M (TMM) and Upper Quartile (UQ) methods as recommended (Tam et al., 2015), with the latter being chosen as the preferred method as it produced less variable means and distributions (**Supplementary Figures S3E–G**). After normalization, a two-dimensional principal component analysis (PCA) was performed in Partek® Flow® (Partek Inc, MO, United States) to assess the clustering of the samples and identify outliers. Fold changes and statistical significance were computed using the Gene Specific Analysis (GSA) tool within Partek® Flow® (Partek Inc, MO, United States). This function uses the corrected Akaike Information Criterion to select the best statistical model for each gene from the available Normal, Negative Binomial, Lognormal or Lognormal with shrinkage options. Lognormal with shrinkage (Wu et al., 2013), produced the best fit and thus was selected to perform differential analysis. A dataset was created containing the expression values for each miRNA in all samples and imported into R (R Core Team, 2019). The counts were $\log_2$ transformed to display the distribution of miRNAs in dim-reared and photo-oxidative samples using the packages ggbeeswarm (Erik, 2017) and ggplot2 (Wickham, 2016). Hierarchical clustering analysis (HCA) was performed in Partek® Flow® (Partek Inc, MO, United States) using the Euclidian distance as a point metric distance and the average linkage as the agglomerative method.

Sequencing data can be accessed from BioProject (Accession ID: PRJNA615966)[2].

Network and Pathway Enrichment Analysis

The miRNet platform (Fan et al., 2016) was used to elucidate potential interactions between s-mEV miRNA and retinal mRNA. The mRNA dataset is available from BioProject (Accession ID: PRJNA606092) (see text footnote 1) and comprises of all retinal genes with a $\log_2$(count per million) value > -2.4 (**Supplementary Table S2**). The retinal targetome of the top 10 s-mEV miRNAs as well as the targetome of the s-mEVs enriched miRNA were separately imported into Enrichr (Chen et al., 2013) and analyzed for over-expressed pathways annotated in Wikipathways (Slenter et al., 2018) (mouse annotation), and gene-disease associations listed in DisGeNet database (Piñero et al., 2016).

In vitro Experiments

661W Cell Culture

Murine photoreceptor-derived 661W cells (kindly gifted by Dr. Muayyad R. Al-Ubaidi, Department of Biomedical Engineering, University of Houston, Houston, TX, United States) (Al-Ubaidi et al., 1992), were used for *in vitro* experiments at passage 1–5. The authenticity of the of the cells was validated by short tandem repeat analysis (CellBank, Sydney, Australia). Cells were cultured in growth media [Dulbecco's Modified Eagle Medium (DMEM; Sigma-Aldrich, MO, United States) supplemented with 10% fetal bovine serum (FBS; Sigma-Aldrich, MO, United States), 6 mM L-glutamine (Thermo Fisher Scientific, MA, United States) and antibiotic-antimycotic (100 U/ml penicillin, 100 μg/ml streptomycin; Thermo Fisher Scientific, MA, United States)], as previously published (Lu et al., 2018). Cells were maintained and all incubation steps were performed in dark conditions in a humidified atmosphere of 5% CO_2 at 37°C, unless otherwise stated. Cells were passaged by trypsinization every 3 – 4 days.

To deplete FBS of s-mEV, the serum was centrifuged (200,000 × *g*, at 4°C for 18 h) using a Beckman Coulter Optima XE-100 Ultracentrifuge (Beckman Coulter, CA, United States), with a SW41Ti rotor (Beckman Coulter, CA United States) and the supernatant used as FBS supplement in all GW4869 experiments (adjusted growth media).

In vitro Photo-Oxidative Damage

661W cells were seeded in 96 well plates (Nunc, Thermo Fisher Scientific, MA, United States) at 2×10^4 cells/well 24 h prior to photo-oxidative damage experiments. Cells were exposed for 4 h to 15,000 lux light (2.2 mW/cm^2; irradiance measured with PM100D optical power meter, Thorlabs, NJ, United States) from two white fluorescent lamps (2 × 10W T4 tri-phosphor 6500K daylight fluorescent tubes; Crompton, NSW, Australia) as published previously (Lu et al., 2017; Fernando et al., 2018). Control cells were completely wrapped in aluminum foil with six small incisions to allow air/gas exchange.

Cell Viability by MTT Assay

Cell viability was tested by 3-(4,5-dimethylthiazol-2-yl)-2,5-diphenyltetrazolium bromide (MTT) assay (Roche, Mannheim, Germany), according to manufacturer's instructions. Briefly, conditioned media from 661W cells undergoing photo-oxidative damage or grown under dim conditions and treated with GW4869 at 5, 10, 20, and 40 μM concentrations or equivalent concentrations of DMSO was discarded and 100 μl growth media containing 10% MTT was added to all wells. The cells were incubated for 4 h to allow for the formation of formazan crystals then 100 μl of solubilization buffer was added to each well and cells incubated for 18 h. The absorbance at 570 nm of each well was measured with a 670 nm reference wavelength using an Infinite® 200 Pro plate reader (Tecan, Mannedorf, Switzerland). Background absorbance (wells containing medium with MTT reagent only) was subtracted from all sample wells and the viability calculated by dividing the absorbance of treated wells (DMSO or GW4869) by controls (cell treated with medium only).

Exosome Inhibition in 661W Cells

To test the effect of GW4869 on exosome inhibition, 661W cells were seeded at a density of 1×10^5 cells/well in 6 well plates (Nunc, Thermo Fisher Scientific, MA, United States) and grown to 80% confluence in growth medium. Media was removed and cells treated with 20 μM GW4869, or equivalent DMSO, in adjusted growth media for 4 h. The conditioned medium from four GW4869-treated or control wells (661W cells treated with equivalent DMSO concentration) was pooled and processed for s-mEV isolation and characterization as described in Sections "Retinal s-mEV isolation" and "NanoSight."

[2]https://www.ncbi.nlm.nih.gov/bioproject/?term=PRJNA615966

Statistical Analyses

All graphing and statistical analyses were performed using Prism V7.0, unless otherwise specified. An unpaired Student's *t*-test, one-way analysis of variance (ANOVA), or two-way ANOVA with Tukey's multiple comparison *post hoc* test was utilized as appropriate to determine the statistical outcome. Non-adjusted *p*-values ($P < 0.05$) or false-discovery adjusted *p*-values ($FDR < 0.1$) were deemed statistically significant. All data was expressed as the mean ± SEM.

RESULTS

Retinal s-mEV Isolation

In this work we demonstrate a novel protocol for the isolation of s-mEV from mouse retinas (**Figure 1A**). EVs isolated from dim-reared retinas displayed properties of s-mEV, including the distinctive round, cup-shaped morphology of exosomes (Jung and Mun, 2018) as seen in representative negative-stained TEM images (**Figure 1B** and **Supplementary Figure S1A**). Isolated vesicles were within the expected size-range of s-mEV including exosomes (Théry et al., 2002; Hessvik and Llorente, 2018; Słomka et al., 2018; Théry et al., 2018), as shown by TEM size distribution histogram, nanotracking analysis using NanoSight NS300 and dynamic light scatter using a Zetasizer Nano instrument (**Figures 1C–F**). Further, isolated vesicles displayed known s-mEV markers CD63, CD81, and CD9, as determined by western blotting (**Figure 1G**).

Retinal s-mEV Decrease in Concentration but Not Size, During Photo-Oxidative Damage-Induced Degeneration

Following the characterization of retinal s-mEV isolated from dim-reared retinas, the effect of photo-oxidative damage on the secretion of s-mEV was investigated (**Figure 2A**). The relative retinal s-mEV concentration was found to decrease in a damage-dependent manner, reducing significantly from dim-reared controls by 17 ± 2.7% after up to 3 days of photo-oxidative damage and 38 ± 4.0% after 5 days of photo-oxidative damage (**Figure 2B**). S-mEV concentration after 5 days of photo-oxidative damage was decreased from $(4.27 \pm 0.2) \times 10^{10}$ vesicles/ml to $(2.82 \pm -0.3) \times 10^{10}$ vesicles/ml (**Figure 2C**). The analysis of the size distribution of s-mEV at these two timepoints showed that vesicles with diameter of up to 200 nm had a more pronounced decrease in numbers (**Figure 2D**). The mean and modal size of s-mEV remained unchanged after 5 days of photo-oxidative damage (**Figure 2E**). Taken together these results suggest that while there is no change in the size of isolated s-mEV, there was a progressive reduction in the concentration of exosomes during photo-oxidative damage-induced degeneration.

GW4869 Inhibits s-mEV Bioavailability *in vitro*

Given the correlation between increased photoreceptor cell death and reduced retinal s-mEV concentration during photo-oxidative damage, the effect of s-mEV, specifically exosome inhibition was investigated using GW4869. To determine efficiency, GW4869 was added in culture to 661W photoreceptor-like cells (**Figure 3A**). s-mEV isolated from 661W were characterized using TEM (**Figure 3B** and **Supplementary Figure S1B**). Compared to DMSO controls, 661W s-mEV concentration (**Figures 3C,D**) was significantly reduced in GW4869-treated cells, demonstrating a high level of efficacy of this inhibitor. Furthermore, increased GW4869 concentrations did not affect cell viability compared to DMSO controls under dim conditions. However, following photo-oxidative damage (4 h of 15,000k lux white light), an inverse dose-dependent correlation was observed between GW4869 concentration and cell viability (**Figures 3E,F**). Overall GW4869 was shown to reduce the bioavailability of isolated s-mEV *in vitro*, with the reduction likely attributed to a decrease in exosome production.

Endocytic Pathway Inhibition Results in Reduced s-mEV Bioavailability *in vivo*

The contribution of exosomes to retinal s-mEV population was investigated *in vivo* (**Figures 4A,B**). s-mEV isolates from retinas of GW4869-treated mice (daily IP injections for 5 days) were reduced in concentration compared to those from DMSO-injected controls. This was seen in both dim-reared and photo-oxidative damaged retinas (**Figures 4C,D**). Additionally, the expression of genes associated with ESCRT-dependent and ESCRT-independent exosome biogenesis pathways were measured by qRT-PCR. We targeted *Pdcd6ip* (also known as *Alix*), which encodes an accessory protein in the ESCRT-dependent pathway, and *Smpd3*, which encodes nSMase2 in the ESCRT-independent pathway (Hessvik and Llorente, 2018). Retinal lysates from dim-reared and 5-day photo-oxidative damaged retinas as well as lysates from GW4869 and DMSO-injected photo-oxidative damaged retinas were used. The expression of *Pdcd6ip* and *Smpd3* was significantly increased in 5-day photo-oxidative damaged retinas compared to dim-reared controls (**Figure 4E**). Conversely, *Pdcd6ip* and *Smpd3* were found to be significantly reduced in GW486-injected retinas compared to DMSO-injected controls (**Figure 4F**). These results demonstrate that GW4869 can be used as an inhibitor of exosome production, and that exosomes likely contribute to the population as well as effects of retinal s-mEV.

Exosome Inhibition Reduces Retinal Function in Dim-Reared Mice

The effect of GW4869-mediated exosome inhibition on retinal health was investigated in dim-reared mice (**Figure 5A**). Retinal function was assessed using electroretinography (ERG). We show that GW4869-treated mice had reduced retinal function for both a-wave and b-wave measures (**Figures 5B–D**), compared to DMSO injected controls. To investigate this further, we measured photoreceptor cell death by counting TUNEL$^+$ cells in the outer nuclear layer (ONL). The inflammatory response was also measured by counting IBA-1$^+$ cells in the outer retina. No significant difference in either measure was observed between GW4869-injected and DMSO treated retinas

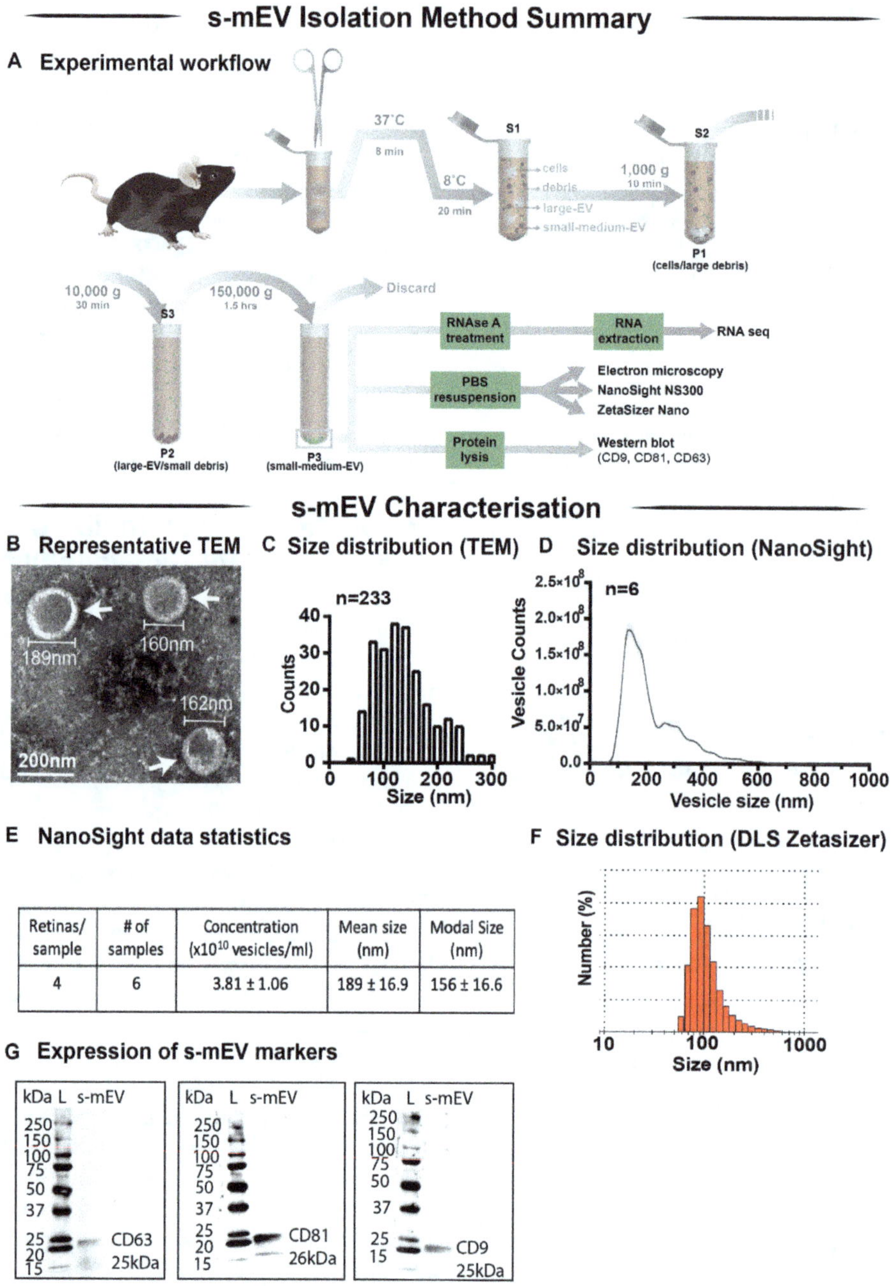

Retinas/ sample	# of samples	Concentration ($x10^{10}$ vesicles/ml)	Mean size (nm)	Modal Size (nm)
4	6	3.81 ± 1.06	189 ± 16.9	156 ± 16.6

FIGURE 1 | Retinal s-mEV isolation and characterization. **(A)** Workflow for the isolation and characterization of retinal s-mEV. s-mEV were isolated from papain-digested retinas using differential ultracentrifugation. s-mEV pellets (P3) were processed for RNA extraction and RNA sequencing (RNA seq) and size and morphology analysis by electron microscopy, NanoSight NS300 and Zetasizer Nano. s-mEV pellets were also digested in protein lysis buffer and analyzed for the expression of s-mEV markers CD9, CD81, and CD63 by western blot. S, supernatant; P, pellet. **(B)** Representative Transmission Electron Microscopy (TEM) micrograph showing isolated retinal vesicles (white arrows) with consistent round, cup-shaped morphology and size of s-mEV. **(C)** Diameter distribution of retinal s-mEV measured from TEM micrographs (n = 233). **(D)** Size distribution of retinal s-mEV (n = 6 samples, 4 retinas/sample) using nanotracking analysis performed on a NanoSight NS300. **(E)** Average concentration, mean and modal size of s-mEV isolated from mouse retinas measured by nanotracking analysis (n = 6 samples, 4 retinas/sample). **(F)** Representative plot of the size distribution of s-mEV measured by dynamic light scatter (DLS) using a Zetasizer Nano instrument (n = 5). **(G)** Full-length western blots showing the presence of s-mEV markers CD63 (25 kDa), CD9 (26 kDa), and CD81 (25 kDa) in retinal s-mEV protein lysates. Molecular weight standard (L). Scale bar = 200 nm.

(**Figures 5E–G**). Additionally, photoreceptor cell death was also assessed by measuring the ratio of ONL thickness to total retinal thickness from OCT images. No significant difference was detected between the two groups (**Figures 5H,I**). Overall, while GW4869 impaired retinal function in dim-reared mice, no difference was observed in histological measures between

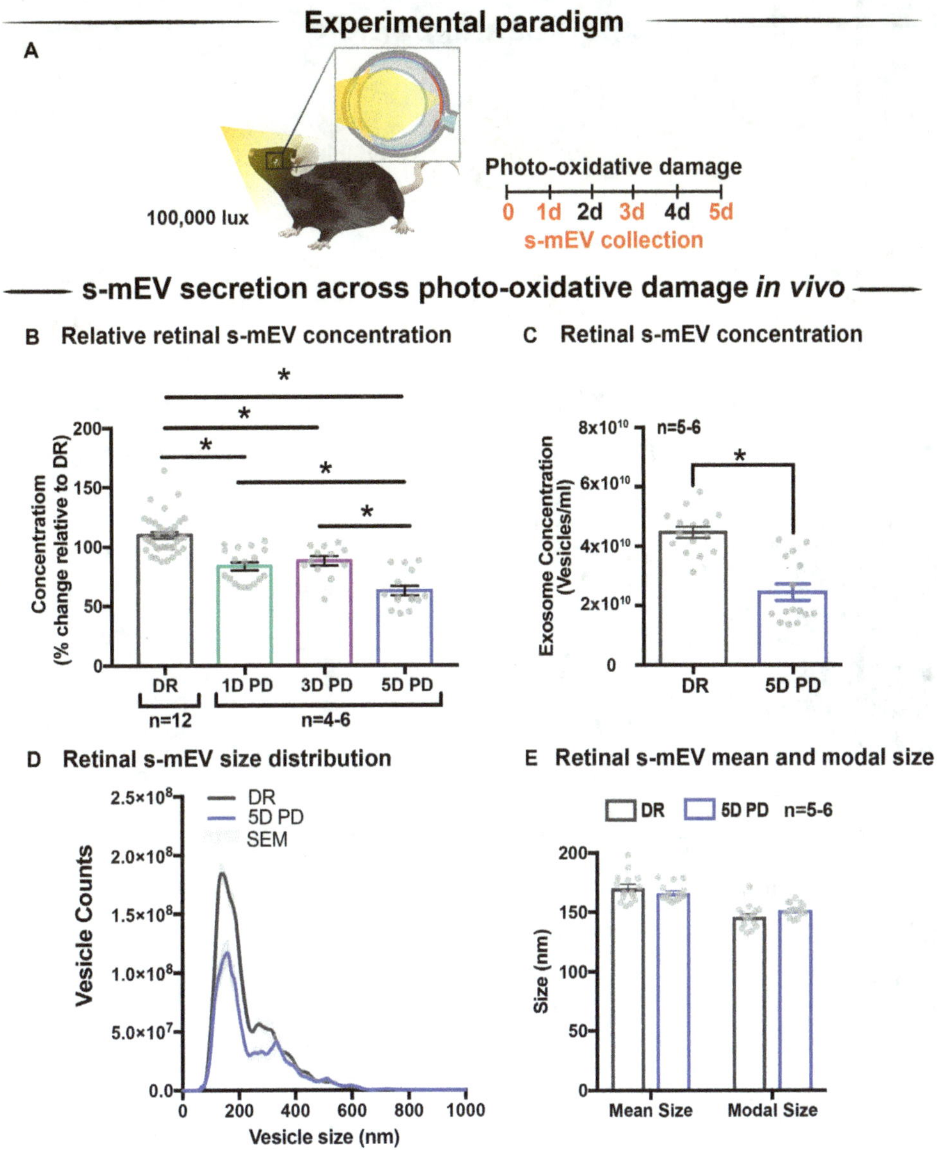

FIGURE 2 | Retinal s-mEV decrease in concentration but not size, over photo-oxidative damage-induced degeneration. **(A)** Experimental paradigm, with red retinal labeling indicating lesion site. **(B)** Concentration of retinal s-mEV decreased progressively across time in PD, with significantly reduced s-mEV at 1D, 3D, and 5D PD compared to DR controls ($P < 0.05$, $n = 3–12$). Concentration values are presented as percentage of DR. **(C)** Significant reduction in the concentration of s-mEV from 5D PD retinas compared to DR controls ($P < 0.05$, $n = 5–6$). **(D)** No significant change in the size distribution of s-mEV isolated from 5D PD retinas compared to DR controls, however vesicles with size <200 nm show a marked decrease in concentration following 5D PD ($P > 0.05$, $n = 5–6$). **(E)** No difference in either the mean or modal size of retinal s-mEV between DR and 5D PD groups ($P > 0.05$, $n = 5–6$). All results obtained using nanotracking analysis (NanoSight NS300).
* denotes = $P < 0.05$.

GW4869-injected mice and DMSO controls over the 5-day period investigated.

Exosome Inhibition Reduces Retinal Function and Increases Cell Death and Immune Cell Recruitment in 5-Day Photo-Oxidative Damaged Mice

Given the reduction in ERG responses in GW4869-treated dim-reared mice, the effect of exosome inhibition was further investigated in mice subjected to 5 days of photo-oxidative damage (**Figure 6A**). GW4869-injected 5-day photo-oxidative damaged mice showed significantly reduced retinal function for both a-wave and b-wave measures (**Figures 6B–D**), compared to DMSO controls. Interestingly, and contrasting to GW4869-treated dim-reared mice, we observed significantly higher levels of photoreceptor cell death (TUNEL$^+$) and immune cell recruitment (IBA-1$^+$) in GW4869-injected 5-day photo-oxidative damaged mice compared to DMSO controls (**Figure 6E**). The increased number of IBA-1$^+$ cells in the

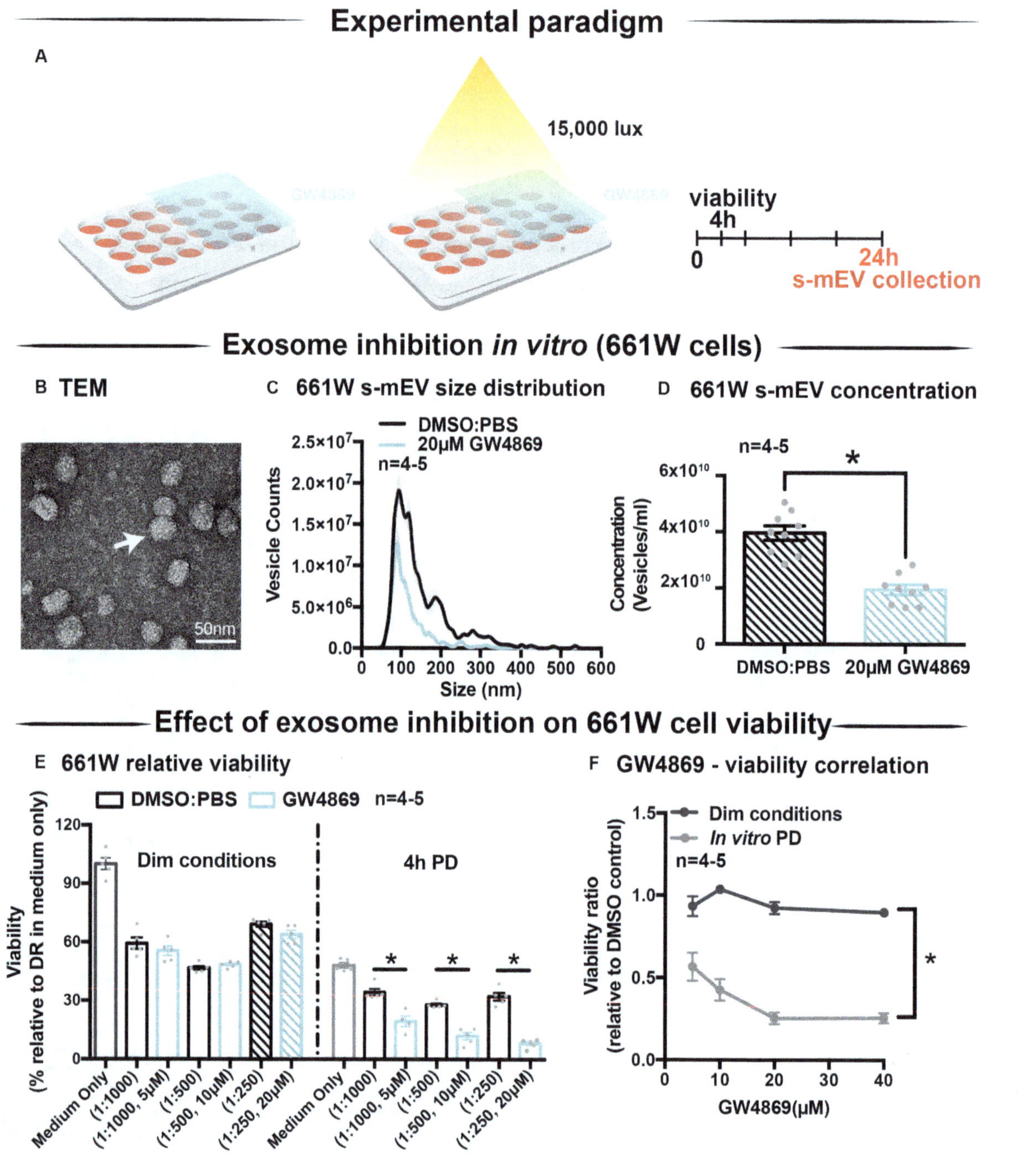

FIGURE 3 | GW4869 inhibits bioavailability of s-mEV *in vitro*. **(A)** Experimental paradigm. **(B)** Representative TEM micrograph showing isolated 661W vesicles. **(C,D)** s-mEV isolated from 661W photoreceptor-like cells in the presence of exosome inhibitor GW4869 (20 μM) or DMSO control for 24 h showed a significant reduction in concentration, as measured by NanoSight NS300. ($P < 0.05$, n = 4–5). **(E)** MTT assay to determine viability of 661W cells treated with increasing concentrations of GW4869 or DMSO. Results demonstrated no dose response in viability levels for either treatment under dim conditions, however, significant dose-dependent reduction in cell viability when treated with GW4869 under PD conditions ($P < 0.05$). **(F)** The viability ratio (viability$_{GW4869}$/viability$_{DMSO}$) of 661W cells treated with increasing concentrations of GW4869 in dim conditions or under 4 h PD showed no change while in dim conditions, however, is reduced between doses of 0–20 μM GW4869 before reaching a plateau by 40 μM while in PD conditions. Scale bar = 50 nm. *Patterned bars in **(E)** represent equivalent dosages/concentrations and conditions of GW4869/DMSO used in **(D)**. * denotes = $P < 0.05$.

outer retina included a higher ratio of amoeboid to ramified microglia and an increased presence of subretinal macrophages (**Figure 6E**). Further, there was a significant thinning of the ONL, indicating higher levels of photoreceptor cell death in GW4869-treated mice (**Figure 6F**). However, no difference in the thickness of the INL or GCL was observed (**Figure 6F**). Increased

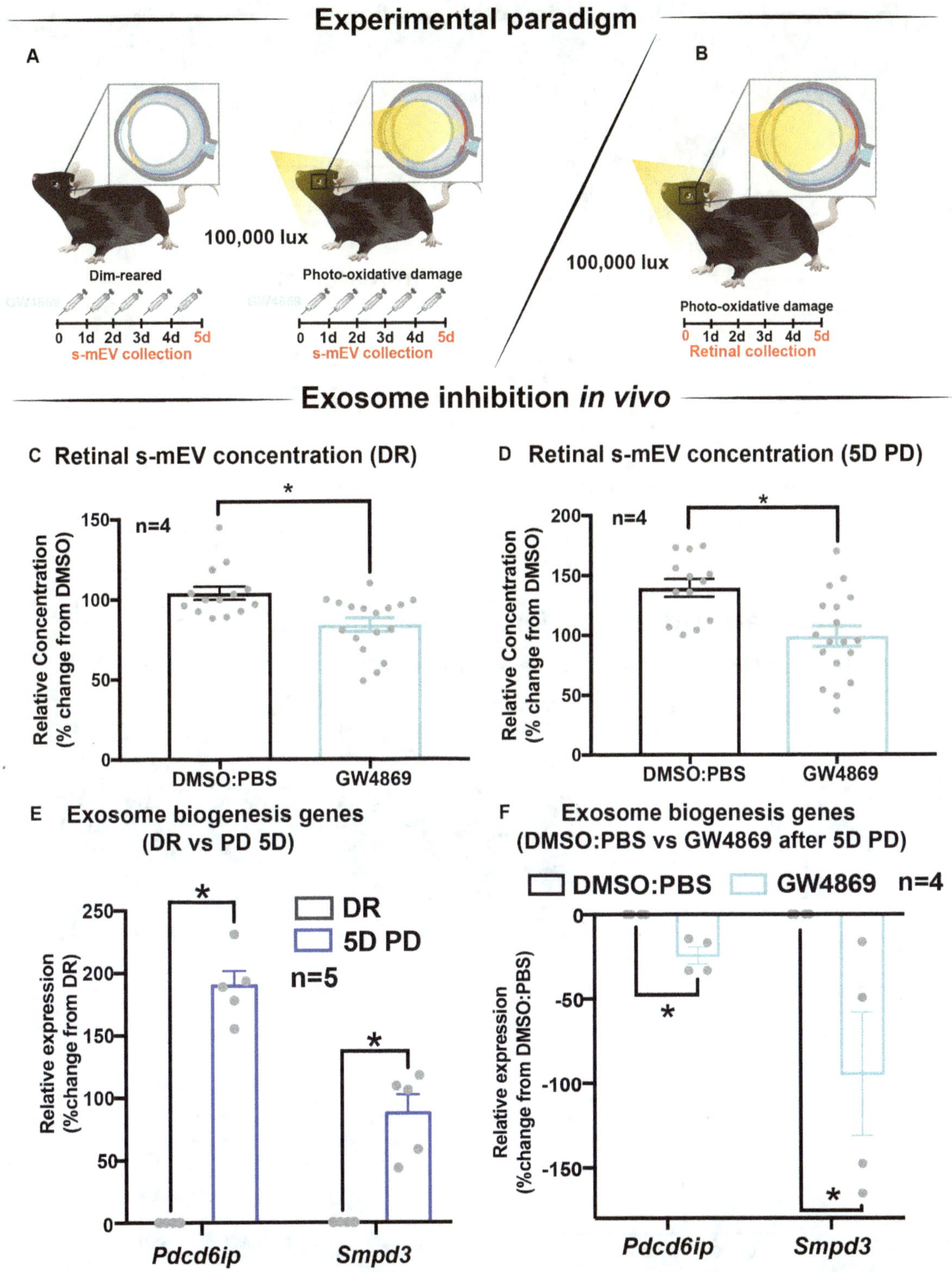

FIGURE 4 | GW4869 inhibits bioavailability of retinal s-mEV. **(A,B)** Experimental paradigm with red retinal labeling indicating lesion site. **(C,D)** Nanotracking analysis (NanoSight NS300) quantification revealed a significant reduction in the concentration of s-mEV isolated from both DR and 5D PD retinas following 5 daily intraperitoneal injections of GW4869 compared to respective DMSO controls ($P < 0.05$, $n = 4$). The expression of exosome biogenesis genes *Pdcd6ip* and *Smpd3* as measured by qRT-PCR was **(E)** significantly increased following 5D PD relative to DR controls ($P < 0.05$, $n = 5$) and **(F)** significantly reduced following daily injections of GW4869 while undergoing 5D PD, relative to DMSO injected controls ($P < 0.05$, $n = 4$). * denotes = $P < 0.05$.

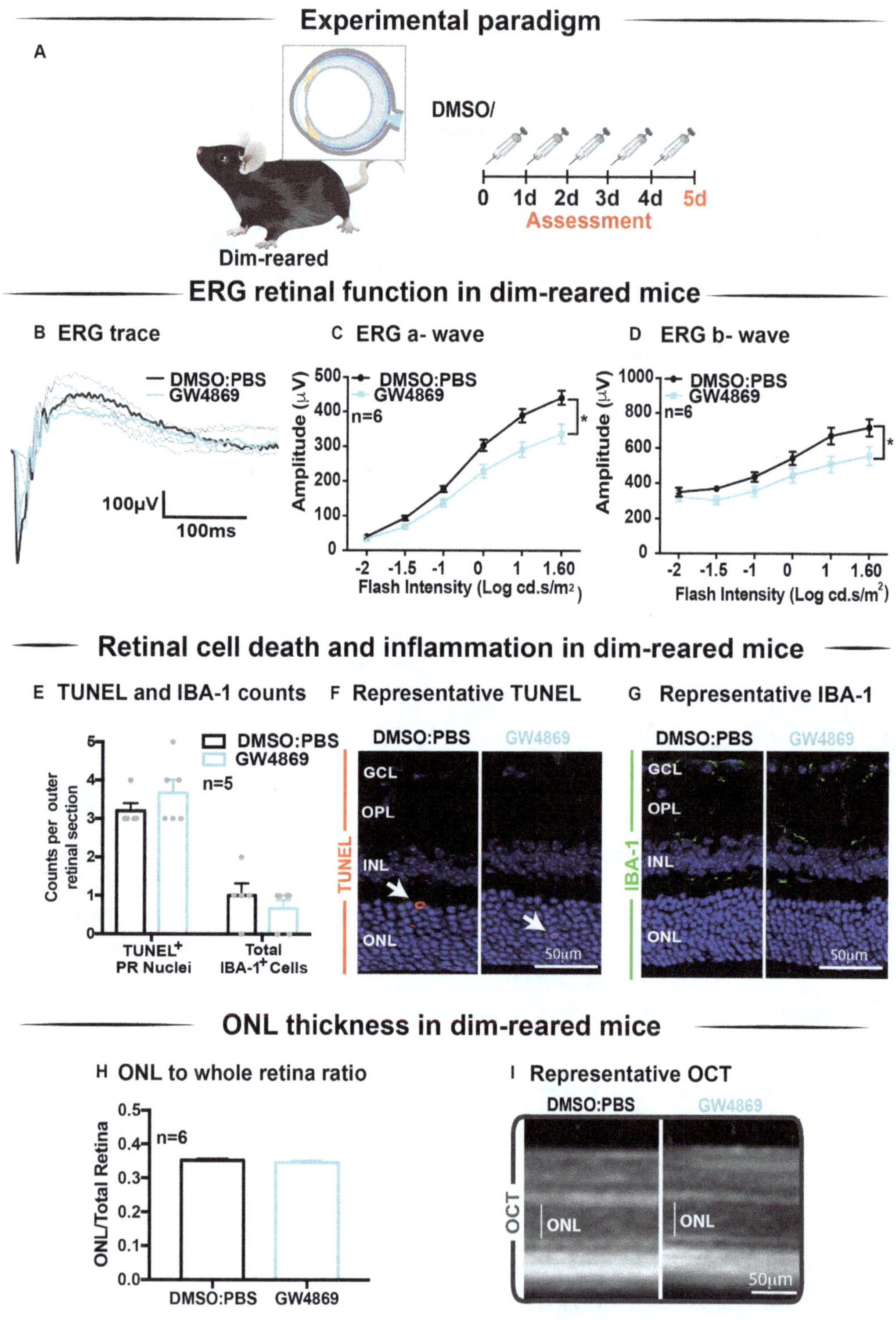

FIGURE 5 | Exosome inhibition via the use of GW4869 in dim-reared mice reduces retinal function. **(A)** Experimental paradigm. Following 5 daily intraperitoneal injections of GW4869 (1.25 mg/kg), retinal ERG function, **(B)** shown in representative ERG traces, was significantly reduced for both **(C)** a-wave and **(D)** b-wave responses, compared to DMSO controls in DR mice ($P < 0.05$, $n = 6$). **(E)** No significant changes were seen in either the number of TUNEL$^+$ cells in the outer nuclear layer, or in the number of IBA-1$^+$ cells in the outer retina following 5 daily intraperitoneal injections of GW4869 compared to DMSO controls ($P > 0.05$, $n = 5$). Representative confocal images show **(F)** TUNEL$^+$ cells (white arrows) and **(G)** IBA-1$^+$ cells in the outer retina, in both GW4869 and DMSO injected DR mice. Representative confocal images are taken in the superior outer retina, 1 mm from optic nerve head. **(H)** There was no change in ONL thickness (represented as ratio to total retinal thickness), as measured by ocular coherence tomography (OCT) between GW4869 or DMSO injected DR mice ($P > 0.05$, $n = 6$). **(I)** Representative images of OCT scans show ONL thickness in superior retina, (white line) in GW4869 and DMSO injected DR mice. Scale bars = 50 µM. * denotes = $P < 0.05$.

levels of photoreceptor cell death in GW4869-injected photo-oxidative damage mice were further demonstrated by a reduction of photoreceptor nuclei in the ONL (**Figure 6G**). Representative confocal images demonstrate an increase in photoreceptor cell death with significantly more TUNEL$^+$ cells and thinner ONL (**Figure 6H**), as well as increased IBA-1$^+$ cells (**Figure 6I**), in GW4869-treated mice compared to controls. Taken together these results show that exosome inhibition in retinas subjected to photo-oxidative damage causes reduced retinal function, concomitant with increased photoreceptor cell death and increased recruitment and activation of microglia/macrophages.

Retinal s-mEVs Possess a Distinct miRNA Profile Compared to the Retina Which Remains Unchanged Over 5-Days of Photo-Oxidative Damage

To obtain pure s-mEV-derived RNA, s-mEV isolated from dim-reared and 5 days photo-oxidative damaged retinas were treated with RNaseA prior to RNA isolation to remove potential contaminating RNA species attached to the outside. Bioanalyzer electropherograms show that RNaseA treatment successfully removed contaminating RNA, whereas s-mEV RNA was protected. Further, retinal s-mEV RNA was enriched for RNA species between 10 to 40 nucleotides (**Supplementary Figure S2**).

Databases of retinal and s-mEV miRNA were obtained using small RNAseq (**Figure 7A**), and were compared to determine if there was any selective enrichment of miRNA in isolated s-mEV. After generating normalized miRNA counts, a two-dimensional principal component analysis (PCA) was employed to visualize global differences between the expression of miRNA from dim-reared retinal s-mEVs and whole retinas. This analysis showed a clear segregation between the miRNA profiles of these groups (**Figure 7B**). Next, we examined the correlation between the expression levels of miRNAs from dim-reared retinal s-mEVs and whole retinas, finding a significant correlation (r = 0.71, $p < 0.0001$) between these groups. In particular, miR-183-5p, let-7b/c-5p, miR-124-3p, miR-125a/b-5p, and miR-204-5p were highly expressed in both groups (**Figure 7C**). This correlation analysis also indicated that some miRNA species were enriched in s-mEVs over the retina, with differential expression analysis uncovering that 28 miRNAs are over-represented in s-mEVs (FC $> +2$, FDR < 0.05), including miR-3770b, miR-706, miR-669c-3p, miR-7020-3p, and miR-3770 which were all over 100-fold more abundant (**Figure 7D**).

Following, we examined the differential expression in miRNA between dim-reared and photo-oxidative damaged-isolated s-mEVs. PCA showed no clustering of the samples suggesting that the miRNA contents of retinal s-mEV were not substantially altered following photo-oxidative damage (**Figure 7E**). Hierarchical clustering analysis (HCA) further illustrated this, with no significant clustering between dim-reared and 5 days photo-oxidative damage groups (**Figure 7F**). In line with the PCA and HCA, differential expression analysis revealed that only miR-1249-3p was differentially expressed (p-value = 0.046, fold change = -2.08) (**Figure 7G**). However, when adjusting for multiple comparisons (Benjamini-Hochberg), this miRNA did not pass the significance threshold (**Supplementary Table S1**).

A total of 154 miRNAs were detected in retinal s-mEV (**Supplementary Table S1**) showing a similar mean count distributions in both groups (**Figure 7H**). The top 10 most highly expressed miRNAs were consistent in both dim-reared and 5 days photo-oxidative damaged retinal s-mEV, and accounted for approximately 67% of the total miRNA counts (**Figure 7I**). miR-124-3p was the most abundant s-mEV-miRNA representing 17.5 and 16.3% in dim-reared and 5 days photo-oxidative damaged retinal s-mEV, respectively (**Figure 7I**). Four members of the let-7 family (let-7c-5p, let-7a-5p, let-7b-5p, and let-7f-5p) as well as miR-183-5p, miR-125a-5p, miR-125b-5p, miR-706, and miR-204-5p comprised the rest of the top 10 most highly abundant s-mEV-miRNA (**Figure 7I**). These results demonstrate the selective enrichment of miRNA in s-mEV from the retina, but that the miRnome of s-mEV does not change in response to photo-oxidative damage.

s-mEV miRnome Is Associated With Inflammatory, Cell Death and Motility Pathways

Considering that s-mEV miRNAs were not altered following 5 days of photo-oxidative damage and that the top 10 most abundant miRNAs accounted for 67% of the total s-mEV miRNAome, we focused on these top 10 s-mEV-miRNAs. A network analysis was performed to understand the interactions between the top 10 s-mEV-miRNAs and the retinal transcriptome (**Figure 8A** and **Supplementary Table S2**). This was also performed for the enriched s-mEV miRNA. Network analyses of the top 10 most abundant s-mEV miRNA revealed that eight miRNAs form a regulatory network containing 1326 targets (**Figure 8Bi**), of which, miR-124-3p, mir-706 and let-7b-5p, have 472, 235 and 353 predicted interactions with retinal transcripts, respectively (**Supplementary Table S3**). Similarly, the retinal targetome of the miRNA showing preferential enrichment in s-mEVs was explored with miRNet showing that miR-466i-3p, miR-446i-5p, miR-let-7b-5p, miR-1195, and miR-706 all had over 100 predicted targets in the retina (**Figure 8Bii** and **Supplementary Table S4**). The predicted targets of both sets of miRNA were separately used for enrichment analyses against WikiPathways (mouse pathway annotation) and DisGeNET (database containing gene-disease associations) on the Enrichr platform. A total of 50 pathways were significantly over-represented in WikiPathways (**Supplementary Tables S5, S6**). Notably, pathways pertaining to inflammatory processes including IL-1 to IL-6 signaling, Toll-like receptor signaling and chemokine signaling showed a significant enrichment. Pathways related to cell survival and motility were also significantly enriched (**Figures 8C,D**), with apoptosis significantly associated with miRNA targets from s-mEV-enriched miRNA (**Figure 8C**). Finally, DisGeNET database analysis revealed a significant association between s-mEV-miRNA targets and 6 retinal diseases (**Figure 8E** and **Supplementary Table S7**), with the majority associated with an

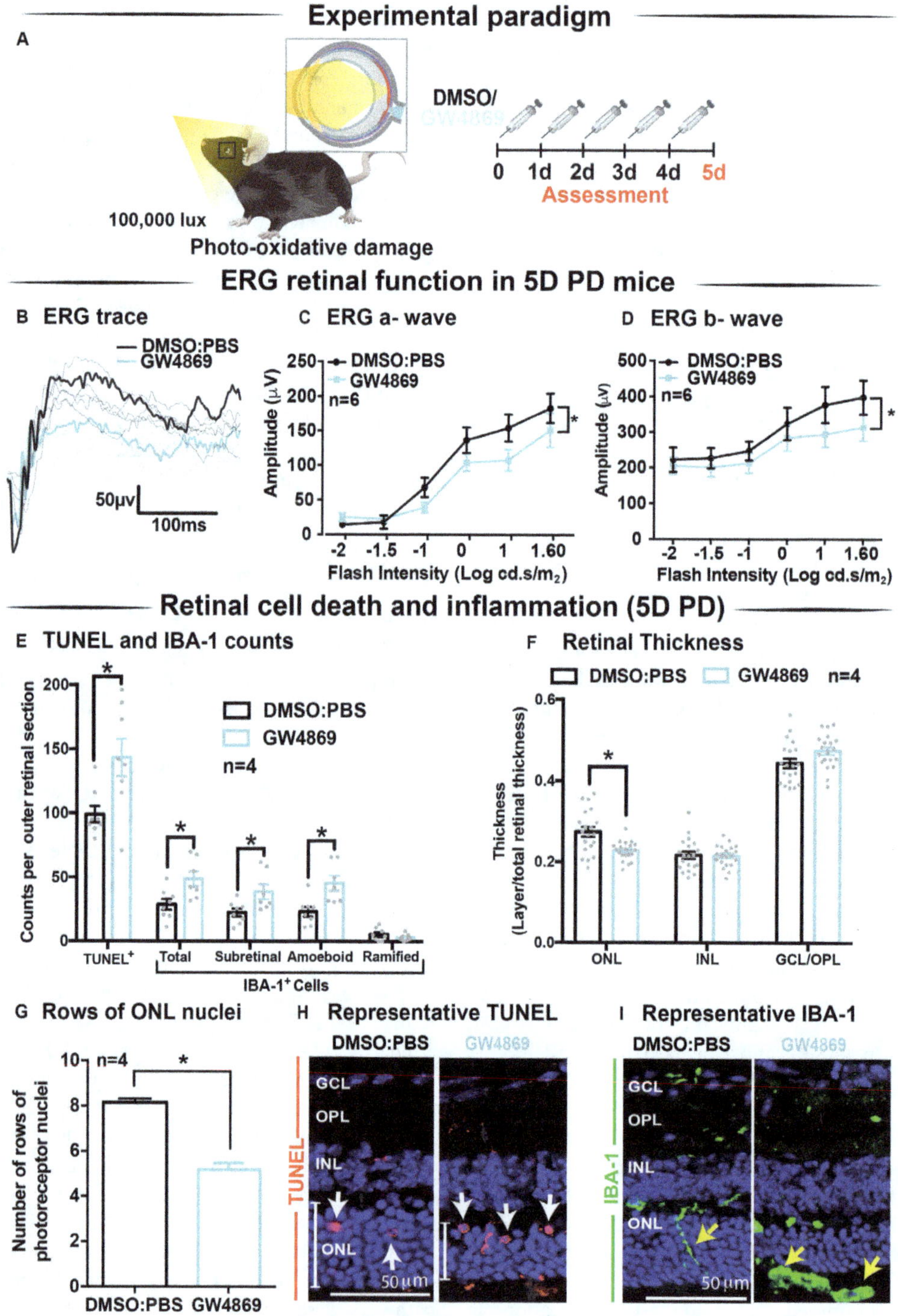

FIGURE 6 | Exosome inhibition via the use of GW4869 in mice subjected to 5 days photo-oxidative damage reduces retinal function and increases cell death and immune cell recruitment. **(A)** Experimental paradigm with red retinal labeling indicating lesion site. Following 5 daily intraperitoneal injections of GW4869 (1.25 mg/kg), retinal ERG function, **(B)** shown in representative ERG traces, was significantly reduced for both **(C)** a-wave and **(D)** b-wave responses, compared to DMSO controls in 5D PD mice ($P < 0.05$, $n = 6$). **(E)** There was a significant increase in the number of TUNEL$^+$ cells in the ONL, as well as a significant increase in the total number, number of subretinal, and number of amoeboid but not ramified IBA-1$^+$ cells in the outer retina in GW4869-injected mice compared to DMSO controls following 5D PD ($P < 0.05$, $n = 4$). **(F)** There was a significant reduction in the thickness of the ONL of GW4869-injected 5D PD mice compared to DMSO controls ($P < 0.05$, $n = 4$), however, no change was observed in the thickness of the INL or GCL between these groups ($P > 0.05$, $n = 4$). **(G)** The number of photoreceptor row nuclei in the ONL was also significantly reduced in GW4869-injected 5D PD mice compared to DMSO controls ($P < 0.05$, $n = 4$). Representative confocal images show **(H)** increased levels of TUNEL$^+$ cells (while arrow) and reduced ONL thickness (white line) as well as **(I)** increased number of IBA-1$^+$ cells in the outer retina, in GW4869-injected mice following 5 days of PD. Representative confocal images are taken in the superior outer retina, 1 mm from optic nerve head. Scale bars = 50 µM. * denotes = $P < 0.05$.

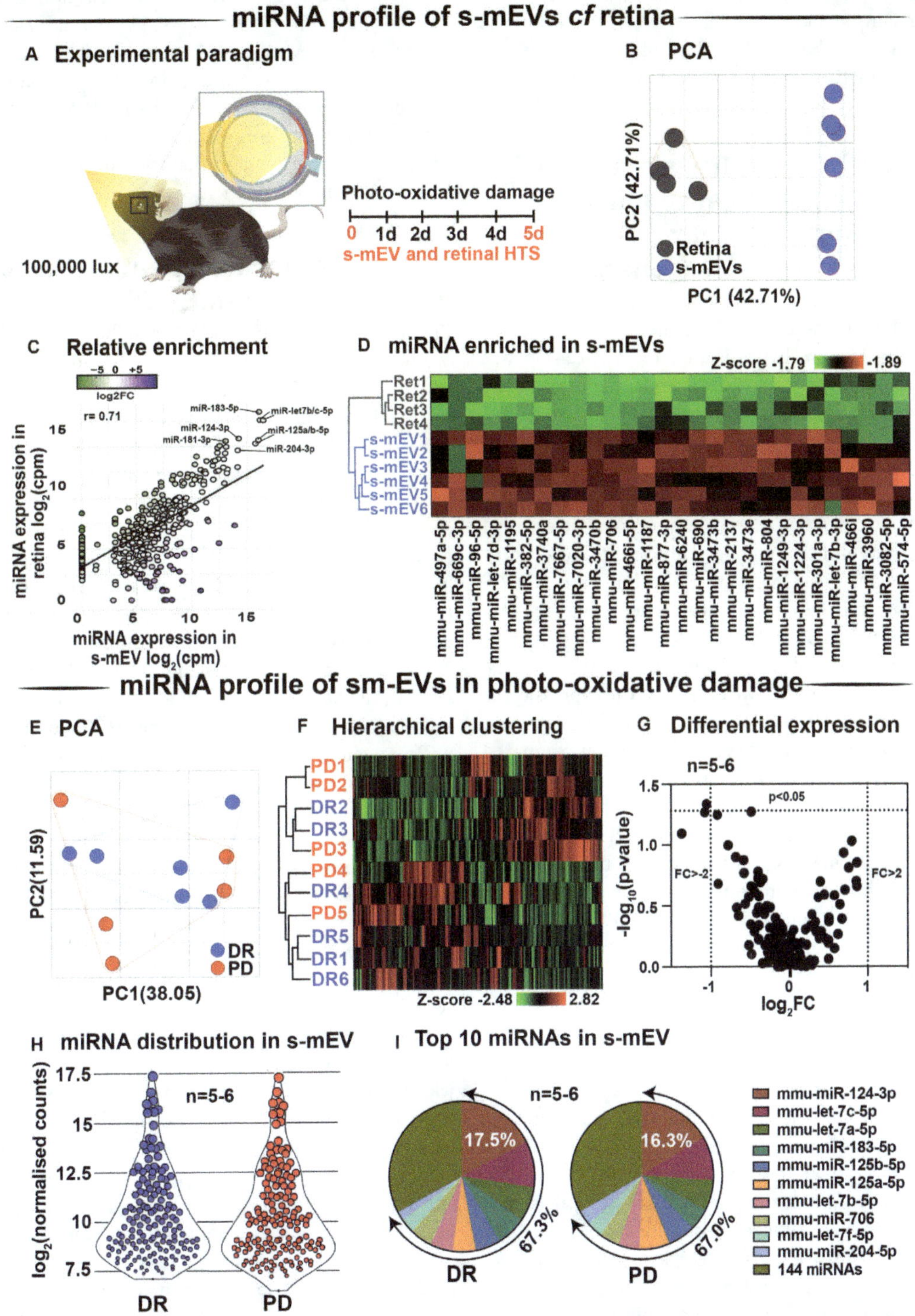

FIGURE 7 | s-mEV miRNA profiling and enrichment analysis. **(A)** Experimental paradigm with red retinal labeling indicating lesion site. **(B)** Principal component analysis (PCA) of retinal miRNA and s-mEVs miRNA showed clear clustering suggesting that each group possesses uniquely enriched miRNA. **(C)** A positive correlation was observed ($r = 0.71$, $p < 0.0001$) between miRNA counts in the dim-reared retina and s-mEVs with miR-124-3p, miR-181-3p, miR-204-3p, miR-125a/b-5p, and miR-let7a/b-5p being abundant in both. **(D)** Hierarchical clustering of miRNA showed significant enrichment of miRNA in s-mEVs compared to the dim-reared retina (FC > 2, FDR < 0.05). **(E)** PCA and **(F)** hierarchical clustering analysis show no grouping or miRNA between DR and PD samples and identified no outliers. **(G)** Volcano plot indicating that no differential analysis was observed in s-mEV from DR and 5D PD retinas ($P > 0.05$, n = 5-6). **(H)** Violin plots showing the distribution of miRNA transcripts in s-mEV from DR and 5D PD retinas (n = 5–6). Each data point represents the $\log_2$ transformed average number of counts for each miRNA retained after removing miRNAs with low expression and the size of each point is proportional to the average expression value of each miRNA. s-mEV from both DR and 5 days PD retinas contain similar amounts of miRNA transcripts with low, medium and high expression. **(I)** The top 10 most abundant miRNA were conserved in retinal s-mEV from DR and 5 days PD retinas and accounted for approximately 67% of the total counts.

inflammatory and/or proliferative profile. In summary, s-mEV miRNA cargo remains stable in the degenerating retina but the most abundant and enriched miRNA in s-mEV are predicted to modulate inflammation, cell death responses and cellular motility pathways.

s-mEV May Mediate the Translocation of miR-124-3p From the Outer to Inner Retina During Photo-Oxidative Damage-Induced Retinal Degeneration

We hypothesized that s-mEV may mediate the transportation of miRNA, including miR-124-3p, through the retina during damage. To test this, *in situ* hybridization of miR-124-3p was performed on retinal cryosections from mice injected daily with GW4869 and subjected to 5 days of photo-oxidative damage (**Figure 9A**). As previously reported (Chu-Tan et al., 2018), miR-124-3p expression was significantly increased in the INL after 5 days photo-oxidative damage compared to dim-reared controls (**Figures 9B,C**). Interestingly, GW4869 treatment significantly reduced photo-oxidative damaged-induced miR-124-3p expression in the ONL and INL but not ILM/IS (**Figures 9D,E**). This indicates that the within-retinal expression, and potentially movement of miR-124-3p in response to photo-oxidative damage may be mediated by s-mEV such as exosomes. To further explore our hypothesis, we examined the expression of miR-124-3p in photo-oxidative damaged retinas after treatment with GW4869 and observed no differential expression compared to DMSO-injected controls (**Figure 9F**). This indicates that the reduced expression of miR-124-3p in the ONL and INL is likely due to lack of transport into the INL rather than a decrease in total retinal expression. Given the increased levels of cell death and inflammation in mice treated with GW4869, we speculate that s-mEV shuttling of miRNA including miR-124-3p is required for normal retinal homeostasis and immune modulation.

DISCUSSION

This study describes for the first time, the isolation and characterization of mouse retinal s-mEV and the important role they play in retinal health and degeneration. We demonstrate four key findings from this study: Firstly, we demonstrate that s-mEV isolated from mouse retinas decrease in concentration progressively as a consequence of retinal degeneration. Secondly, we show that partial s-mEV depletion, via systemic administration of the exosome inhibitor GW4869, resulted in reduced retinal function in the normal retina and further exacerbated functional losses in mice subjected to photo-oxidative damage. Notably, significant photoreceptor cell death and inflammation were observed in GW4869-injected mice undergoing photo-oxidative damage, which was mirrored *in vitro* by 661W photoreceptor-like cells displaying increased susceptibility to photo-oxidative damage following exosome inhibition. Thirdly, we used small RNA sequencing and bioinformatic analyses to report on the potential regulatory roles of s-mEV miRNAs in retinal degeneration. Lastly, using the s-mEV-abundant miRNA miR-124-3p as a measure, we provide evidence that retinal s-mEV could be involved in the dynamic cell-to-cell transfer of miRNA in the degenerating retina. Taken together, we propose that retinal s-mEV and their miRNA cargo play an essential role in maintaining retinal homeostasis through immune-modulation, and that this effect is potentially mediated by s-mEV populations including exosomes.

Photoreceptor Cell Death Is Associated With Reduced s-mEV Bioavailability

Previous studies have demonstrated that miRNA-laden exosomes are found in abundance in the CNS and are central to cell-to-cell communication (Fauré et al., 2006; Lachenal et al., 2011). Although it is unclear which retinal cell type(s) s-mEV isolated in this work are secreted by, our evidence supports a photoreceptor-derived contribution. We demonstrate a progressive decrease in s-mEV numbers during photo-oxidative damage, which correlates with increasing levels of photoreceptor cell death as previously described in this model (Natoli et al., 2016). Therefore, we attribute the decrease in isolated retinal s-mEV to a loss of photoreceptors. In addition, the high abundance of photoreceptor prominent miRNAs miR-124-3p (Chu-Tan et al., 2018) and miR-183-5p (Xiang et al., 2017) in isolated retinal s-mEV, further attests to a potential photoreceptor origin, with miR-124-3p comprising 17.5 and 16.3% of the total s-mEV-miRnome in normal and damaged retinas, respectively.

Previous studies by Vidal-Gil et al. (2019) report that the exosome protein marker CD9 was localized to the photoreceptor inner segments and ONL, further supporting a photoreceptor-origin hypothesis. However, it should be noted that CD9 was also found in the INL, GCL and choroid. Using a genetic model (*rd10* mice) of retinitis pigmentosa, which results in progressive rod photoreceptor-specific cell death (Gargini et al., 2007; Vidal-Gil et al., 2019) showed that EVs isolated from retinal explants contain rhodopsin protein. This exosomal rhodopsin expression increased following photoreceptor rescue. Therefore, the authors describe EV secretion as a function of photoreceptor cell death, with significantly lower levels of CD9 expressing vesicles isolated from *rd10* mouse retinas at postnatal day 18 (P18) compared to at P16 (Vidal-Gil et al., 2019). Taken together these results demonstrate a clear relationship between retinal s-mEV secretion and photoreceptor cell survival, suggesting a significant s-mEV population may derive from photoreceptors in the retina.

Contrary to our findings, it is well established that a pro-oxidative stress or a proinflammatory environment often results in the increased release of exosomes (Lehmann et al., 2008; Essandoh et al., 2015; Benedikter et al., 2018; Hessvik and Llorente, 2018). We do observe increased expression of exosome biogenesis genes *Pdcd6ip* and *Smpd3* in retinal lysates following 5 days of photo-oxidative damage. We therefore suggest that all retinal cells including photoreceptors, as demonstrated by Vidal-Gil et al. (2019), may be potentially increasingly releasing s-mEV including exosomes as a consequence of the heightened presence of inflammatory and oxidative stress stimuli created in response to acute retinal degenerations. Further, we hypothesize

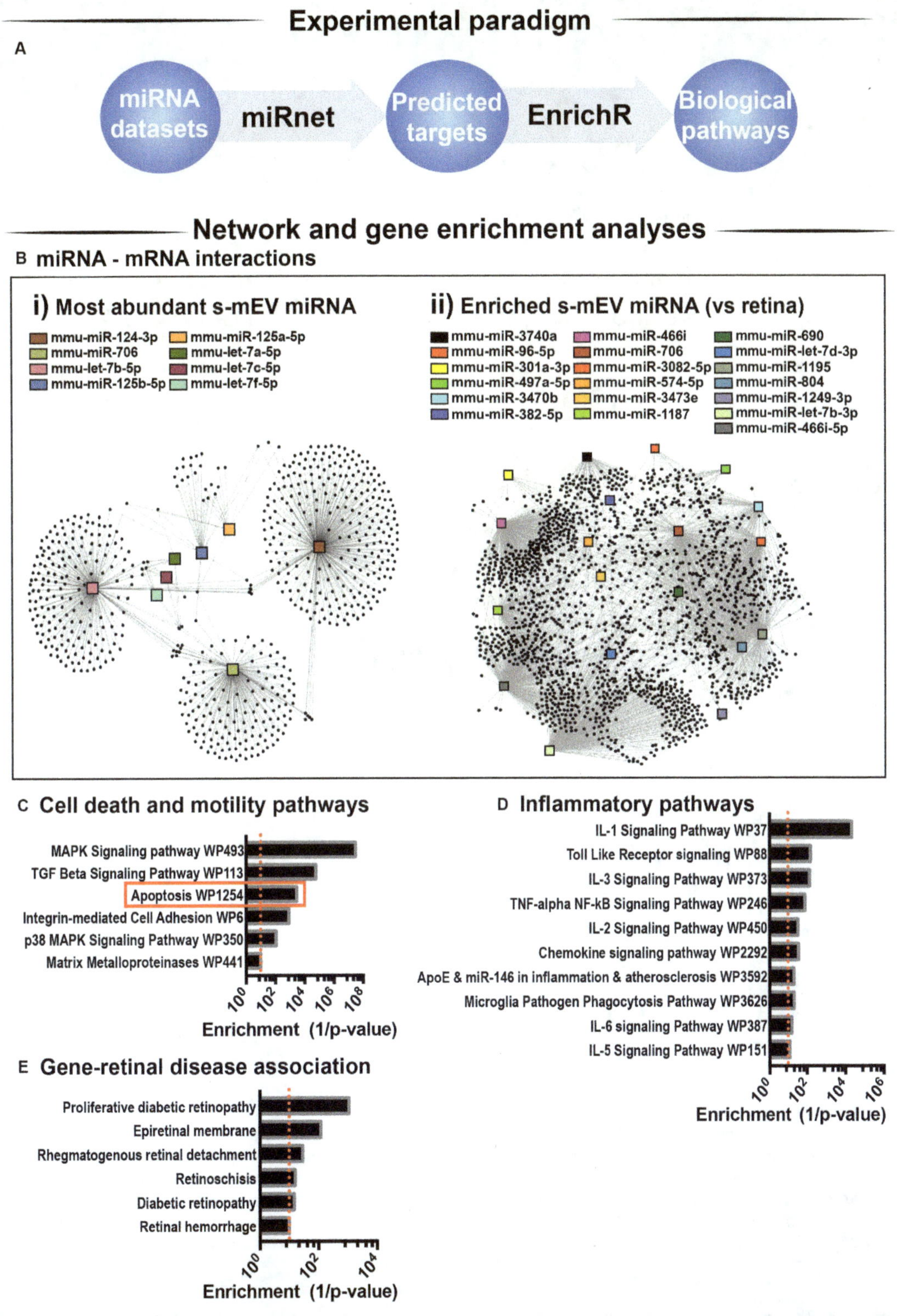

FIGURE 8 | s-mEV miRNA target and pathway analysis. **(A)** Analysis pipeline. **(B)** miRNet network analysis using the **(i)** top 10 most highly expressed s-mEV-miRNAs and **(ii)** miRNAs enriched in s-mEVs showed their predicted interactions with target mRNAs. From the top 10 most abundant s-mEV miRNAs, eight formed an interconnected network, with miR-124-3p, miR-706 and let-7b-5p having the largest number of predicted targets. From the most enriched s-mEVs miRNAs, miR-466i-3p, miR-466i-5p and miR-1195 had the largest targetomes. Pathway analysis was performed using the predicted targets of the most abundant **(Bi)** and most enriched **(Bii)** miRNA, collectively showing a significant enrichment of pathways pertaining to **(C)** cell survival and motility, **(D)** inflammatory processes and oxidative stress ($P < 0.05$). **(E)** Enrichment analysis was also performed against DisGeNet (database containing gene-disease associations) revealing that a number of retinal diseases were significantly associated with the targets of s-mEV-miRNAs ($P < 0.05$).

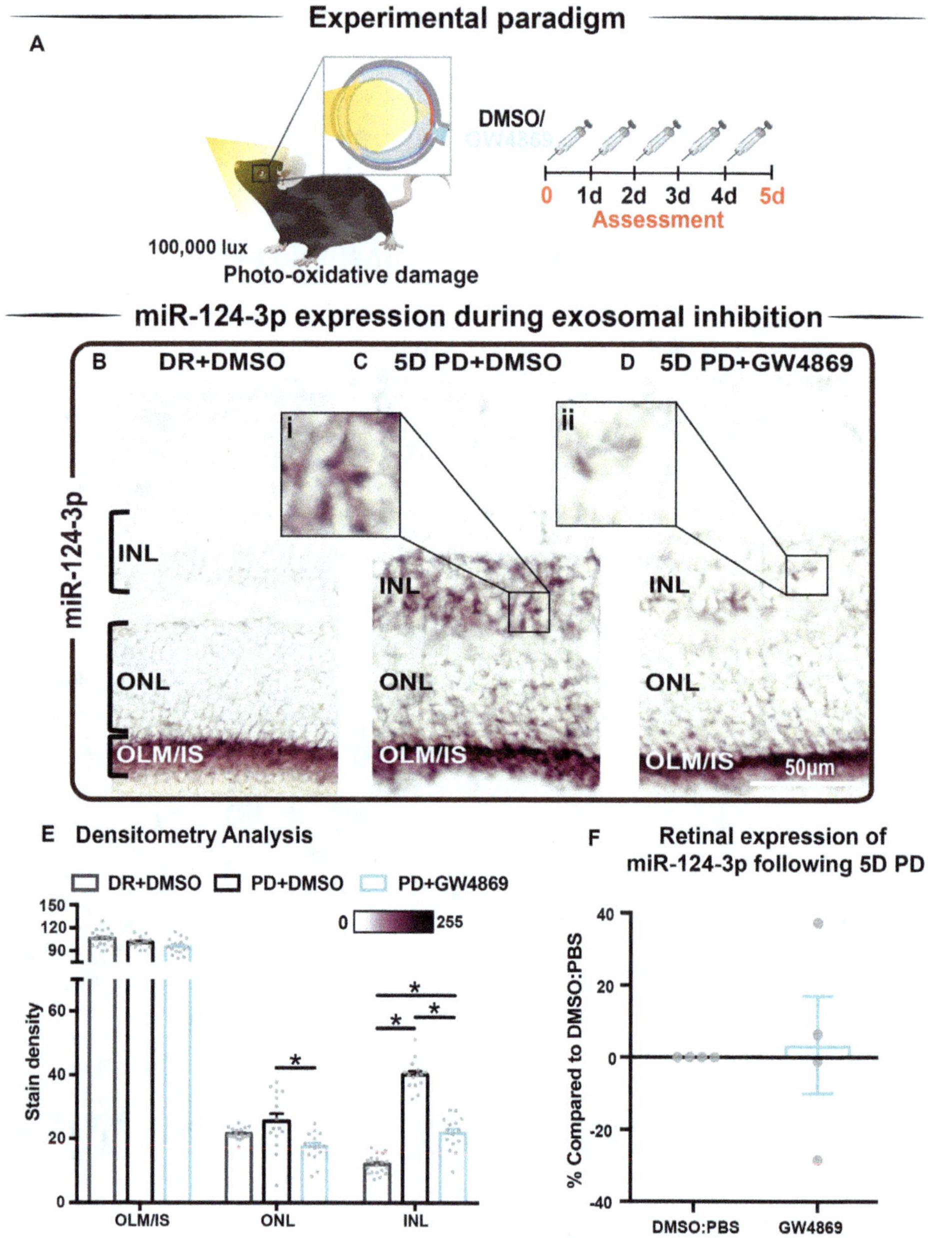

FIGURE 9 | s-mEV may mediate the translocation of miR-124-3p from the outer to inner retina during photo-oxidative damage-induced retinal degeneration. **(A)** Experimental paradigm with red retinal labeling indicating lesion site. **(B)** Using *in situ* hybridization, miR-124-3p expression in DR retinal cryosections was localized to the outer limiting membrane/photoreceptor inner segments (OLM, IS), however **(C)** following 5D PD, miR-124-3p was seen in the INL (inlet, **i**). **(D)** Following intraperitoneal administration of GW4869 exosome inhibitor, miR-124-3p expression is significantly reduced in the ONL and INL (inlet, **ii**). Representative images are taken in the superior outer retina, 1 mm from optic nerve head. **(E)** Densitometry analysis of miR-124-3p in retinal layers confirmed a decrease in miR-124-3p expression in the ONL and INL of mice treated with GW4869. **(F)** Retinal expression of miR-124-3p was conducted using qRT-PCR in photo-oxidative damaged mice treated with GW4869 or DMSO control, showing no change in total retinal expression ($P > 0.05$). Scale bar = 50 μM. * denotes = $P < 0.05$.

that photoreceptors, as the most abundant retinal cell type, are the largest contributor to the pool of retinal s-mEV and thus their loss leads to a net decrease in the total bioavailability of retinal s-mEV. Independent isolation and quantification of s-mEV and in particular exosomes from each major retinal cell type may shed some light in this regard and will be investigated in the future. We also note that it is also possible that the increased expression of *Smpd3* may be occurring as a result of degeneration, with nSMase2/ceramide-induced apoptosis reported in models of retinal degenerations (German et al., 2006; Chen et al., 2012; Simón et al., 2019). While we do not see a protection from cell death following treatment with GW4869, we suspect that reduced exosomal communication may supersede possible protective effects from nSMase2/ceramide inhibition. Investigations into the

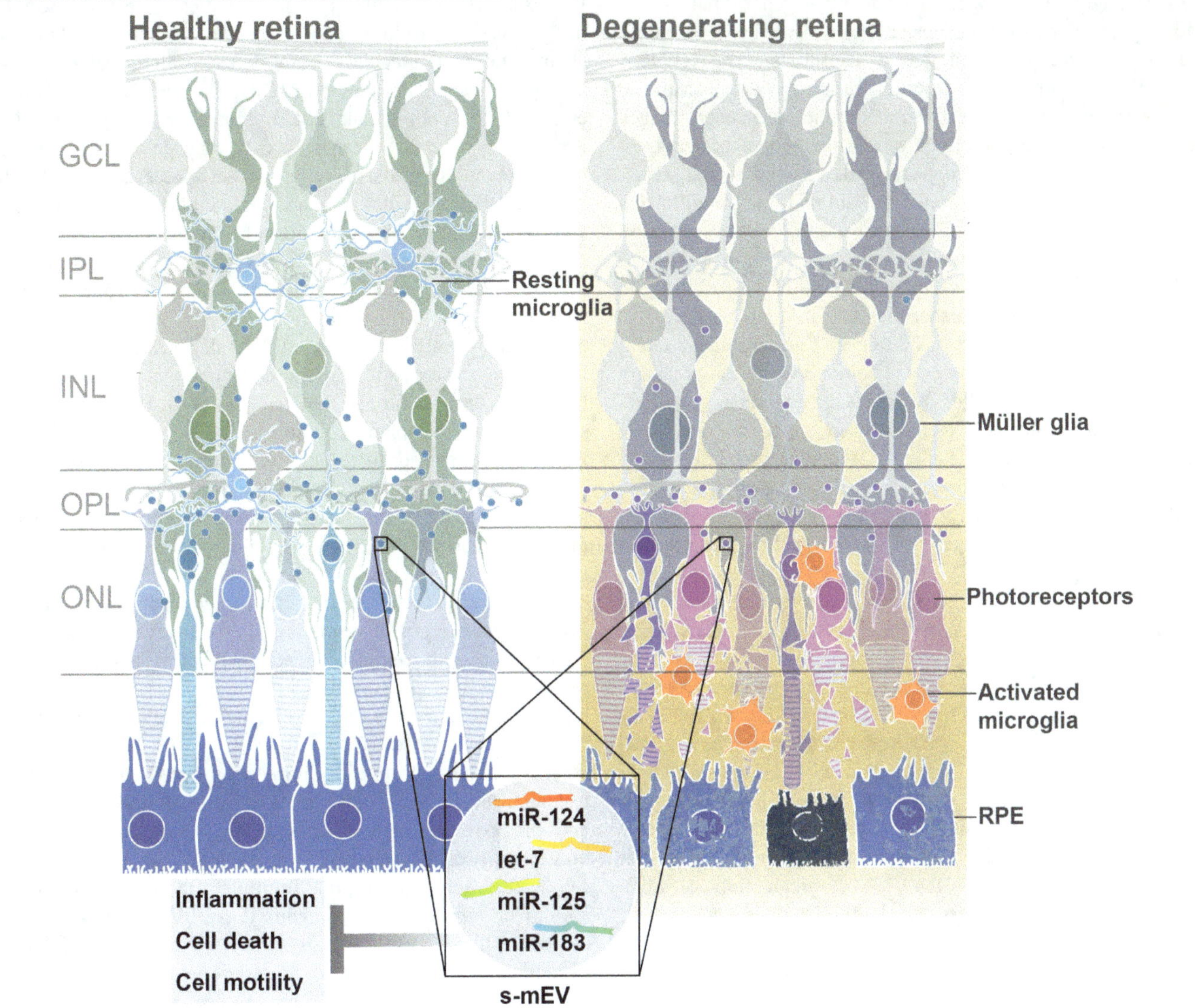

FIGURE 10 | Proposed functional role of s-mEV. Under normal homeostatic conditions, s-mEV laden with miRNA, including miR-124, let-7, miR-125, and miR-183, potentially originating from photoreceptors, are continuously trafficked between retinal neurons and glial cells. However, in retinal degenerations, a loss of photoreceptors leads to a decrease in s-mEV bioavailability, and consequently an insufficient transfer of s-mEV miRNA cargo to their cell targets. As the retinal targetome of s-mEV-miRNAs is associated with inflammation, survival and cell motility pathways, inadequate pathway regulation following s-mEV depletion leads to exacerbated retinal damage.

interplay between exosome release and retinal ceramide levels will be explored in future works.

An additional hypothesis is that in response to excessive light, increased numbers of retinal s-mEV are mobilized as early responders to stress. However, even by 1 day of photo-oxidative damage, have already been depleted past a homeostatic level, ensuing an inflammatory response. As we previously demonstrated the potential translocation of miR-124-3p to the INL within 24 h of damage (Chu-Tan et al., 2018), it seems possible that s-mEV transport of this miRNA could have occurred before this early time point as a rapid response to stress. In fact, Krol et al. (2010) demonstrated that photoreceptor-specific miRNAs such as miR-183/182/96 as well as miR-204 and miR-211 are modulated in response to light exposure as short as 3 h. We also identified mir-183/182/96, miR-204 and miR-211 in retinal s-mEV however, while Krol et al. (2010) do not propose a mechanism responsible for the miRNA turnover observed in their work, our work presented here suggests that s-mEV may be involved. Quantifying s-mEV secretion during early stress responses is required to determine if s-mEV depletion is a cause or consequence of retinal degeneration, and represents an important piece of the therapeutic puzzle.

Retinal s-mEV as Mediators of Immuno-Modulation

Regardless of origin, to date the role that s-mEV play in retinal health and disease is still largely unclear. Results from this work strongly support a mechanism by which retinal s-mEV and in particular exosomes mediate homeostasis and immunomodulation, with the inhibition of exosomes using GW4869 both *in vitro*, and *in vivo*, resulting in increased cell death, as well as recruitment and activation of microglia/macrophages. Importantly these observations were only evident under stress conditions, with both control 661W cells and dim-reared retinas displaying no major signs of cell death or inflammation following exosome-inhibition in the absence of photo-oxidative damage. We suggest that unlike in the degenerating retina, that exosome inhibition had no major effects on cell health. This is likely to be a consequence of the experimental paradigm used in this study and the short period of inhibition, or alternatively could indicate that under stress exosomal communication is necessary for cell survival. A smaller average size was seen in 661W-isolated s-mEV compared to those from the retina, and while we attribute this to the heterogenous nature of whole tissue, it could suggest that photoreceptors primarily secrete a smaller s-mEV fraction such as exosomes, and it is this population that may mediate retinal damage. This hypothesis however requires further investigation.

We hypothesize that as a consequence of longer-term exosome inhibition, inadequate translocation of miRNA cargo via s-mEVs results in the dysregulation of immune pathways. We have previously reported that in response to photo-oxidative damage, miR-124-3p upregulation in the INL may occur via outer-to-inner retinal translocation, with miR-124-3p acting as an anti-inflammatory regulator of C-C Motif Chemokine Ligand 2 (Ccl2) to prevent the recruitment of microglia/macrophages (Chu-Tan et al., 2018). In this present study, we provide further evidence of s-mEV-mediated miRNA translocation, demonstrating that in mice treated with GW4869, INL upregulation of miR-124-3p, the most highly expressed miRNA in isolated retinal s-mEV, was reduced. Although correlative, we suggest that insufficient gene regulation due to reduced s-mEV/miRNA bioavailability could contribute to the increased presence of immune cells and inflammation as seen in retinal degenerations. While we do not exclude the possibility that miR-124-3p could be upregulated in the INL in response to photo-oxidative damage, and downregulated following treatment with GW4869, we believe that this is unlikely, given the lack of differential change in the expression of miR-124-3p in photo-oxidative damaged mice injected with GW4869 compared to controls. The s-mEV-dependent transport of miRNA in mediating immune regulation however requires further exploration particularly in regard to the hypothesized movement of s-mEVs in retinal damage. Future experiments should utilize EV reporter strains such as those generated by Men et al. (2019) as well as intra-ocular administration of fluorescently tagged s-mEVs to obtain direct evidence of s-mEV movement and uptake in the degenerating retina, and will be investigated in further detail in the future.

To further support the hypothesis that retinal s-mEV mediate homeostasis and immune responses, network and pathway analysis of retinal s-mEV-miRNAs revealed that the top 10 miRNA and miRNA enriched in s-mEVs were associated with the regulation of inflammatory and cell survival pathways. For example, our network analysis indicates that the top 10 miRNA are controlling genes associated with Interleukin and chemokine signaling, with both of these biological pathways involved in a plethora of retinal degenerative diseases (Rutar et al., 2015; Wooff et al., 2019). Moreover, MAPK signaling and TGF-β have also been shown to play pivotal roles in the development of retinal degenerations (Kyosseva, 2016; Wang et al., 2018), with both pathways showing a strong association with the targetome of the top 10 most highly expressed s-mEV-miRNAs. As the top 10 miRNAs in both dim-reared and photo-oxidative damage retinal s-mEV make up approximately 70% of the total s-mEV-MiRnome, it is not surprising that a reduction in s-mEV numbers during degeneration could lead to immune dysfunction due to inadequate gene regulation. In fact, as recently demonstrated by Bian et al. (2020) exosomes secreted from transplanted neural stem/progenitor cells (NPCs) in the subretinal space resulted in delayed photoreceptor cell death, preserved retinal function and the suppressed activation of retinal microglia. Further, this group demonstrated that the miRNA cargo of NPC exosomes mediated this protection, with many NPC-derived miRNA found to be similar to s-mEV-derived miRNA reported in this study (Bian et al., 2020). Enrichment analysis of retinal s-mEV-MiR revealed significant associations with retinal degenerative diseases such diabetic retinopathy, retinal detachment and retinal hemorrhage further highlighting their potential involvement in modulating retinal inflammatory diseases.

From these collective findings, we propose that in normal retinal health, s-mEV are secreted from photoreceptor cells, and are released to the surrounding retina to maintain a homeostatic environment. However, following photoreceptor cell death, we hypothesize that immune responses are no longer able to be regulated due to reduced s-mEV numbers and the bioavailability of miRNA cargo; leading to the upregulation of inflammatory pathways, infiltration and activation of microglia/macrophages and progressive retinal cell death (**Figure 10**); characteristic features of retinal degenerative diseases (Kauppinen et al., 2016; Rivera et al., 2017; Rübsam et al., 2018; Wooff et al., 2019).

s-mEV as Mediators of Phototransduction

In addition to a potential role as regulators of retinal homeostasis through immune modulation, we highlight a possible involvement of retinal s-mEV, and in particular exosomes, in phototransduction. In both dim-reared and photo-oxidative damaged mice, exosome inhibition resulted in reduced retinal function as measured by ERG. While we could attribute the loss of retinal function in photo-oxidative damage mice to increased levels of photoreceptor cell death and immune cell recruitment, dim-reared mice showed no detectable signs of either cell death or inflammation at the time-point investigated, yet still demonstrated significantly lower functional responses.

While the role of s-mEV and exosomes in phototransduction in the retina is currently unknown, in the CNS the communication of exosomal cargo from oligodendrocytes and astrocytes has been reported to affect neurotransmission pathways (Antonucci et al., 2012; Fröhlich et al., 2014; Pascua-Maestro et al., 2018). In particular, Frühbeis et al. (2013) suggested that the exosome communication between oligodendrocytes and neuronal axons is triggered by glutamate release (Frühbeis et al., 2013). While oligodendrocytes are not present in the retina, Müller glia perform a similar role, responsible for the reuptake of glutamate released from photoreceptors terminals and the synaptic clefts of second order neurons as part of normal phototransduction (Bringmann et al., 2013). We speculate that, perturbations of exosomal communication systems could lead to aberrant glutamate reuptake by Muller cells and impaired signal transduction, explaining the observed reduction in retinal function of exosome-inhibited mice. However, this area requires further in-depth exploration.

The Role of Exosomes in Retinal Degenerative Diseases and Their Potential as Therapeutic Gene Therapy Vehicles

To date, little progress has been made in uncovering the role that s-mEV may play in the progression of retinal degenerations, and further how we can utilize s-mEV, their cargo, and/or biological targets in diagnostic panels and in therapeutic development. It has been recently shown in cell culture models (Kang et al., 2014; Atienzar–Aroca et al., 2016; Zhang et al., 2019), that following high levels of oxidative stress (pathologically relevant to the onset of many retinal degenerative diseases), increased exosome numbers containing high expression of vascular endothelial growth factor receptor (VEGFR) are secreted from a retinal pigmented epithelium (RPE) cell culture line (aRPE19) to promote neovascularization, a hallmark feature of several retinal pathologies (Atienzar–Aroca et al., 2016). It has also been reported that exosomes containing retinal proteins were found in abundance in the vitreous humor of the mouse and human, demonstrating a level of communication that exists between these ocular tissues (Zhao et al., 2018). While it has yet to be discovered if there is any differential change in exosome composition or concentration in the vitreous during retinal degenerative diseases, exosomes identified in the aqueous humor of AMD patients were found to contain a cross-over set of proteins also found in the culture medium of aRPE19 cells (Kang et al., 2014). This finding, while largely correlative, suggests that exosomes derived from the RPE may play a role in disease pathogenesis. Furthermore, these results demonstrate the possibility that access to exosome populations in biological fluids of the eye could provide a representation of what may be occurring in the retina and therefore serves a diagnostic potential.

Exosome based gene therapies are at the forefront of therapeutic development, with multiple clinical trials underway (Lener et al., 2015; Sarko and McKinney, 2017), including for the treatment of neurodegenerative diseases (Sarko and McKinney, 2017; Mead and Tomarev, 2020). However, their use for the treatment of retinal degenerations is largely in its infancy (reviewed in Mead and Tomarev, 2020). Hajrasouliha et al. (2013) provides evidence for the immunomodulatory properties of exosomes derived from cultured retinal astrocytes (Hajrasouliha et al., 2013). Using a laser induced model of choroidal neovascularization Hajrasouliha et al. (2013), demonstrated that the periocular injection of astrocytic exosomes reduced the CCL2-dependent migration of macrophages to the lesion site and attenuated angiogenesis (Hajrasouliha et al., 2013). As CCL2 has been implicated as a key chemokine in the pathogenesis of multiple retinal degenerative diseases (Newman et al., 2012; Rutar et al., 2012; Zuzic et al., 2019), and is a known target of miR-124-3p (Zuzic et al., 2019); exosome-based therapies that replenish retinal levels of this miRNA may prove efficacious as a possible therapeutic, as evidenced in other works (Chu-Tan et al., 2018). In addition, exosomes derived from microglial cells and injected into the vitreous of mice subjected to oxygen-induced retinopathy showed protective effects, reducing avascular regions in the retina, VEGF expression, and photoreceptor apoptosis, compared to controls (Xu et al., 2019). It was hypothesized by these authors that exosomal-miR-24-3p mediated this protection against hypoxia-induced cell death (Xu et al., 2019).

The combined findings of our work uncover a novel role for retinal s-mEV in both health and degeneration, unveiling a panel of s-mEV-miRNA required for retinal homeostasis, and target networks of these gene regulators comprising inflammatory, oxidative stress and cell survival pathways. As we elude to retinal health requiring optimal levels of s-mEV, and their cargo; replenishing s-mEV loads in the retina itself may prove as efficacious therapy, and will be the focus of future works. Further, both the unique s-mEV-miRNA signature and downstream target pathways open additional avenues for therapeutic development.

CONCLUSION

Results from this work suggest that s-mEV are released from photoreceptor cells to maintain retinal homeostasis. However, as a consequence of photoreceptor cell death, s-mEV secretion and/or bioavailability becomes reduced. Consequently, retinal s-mEV cargo, which contains regulatory miRNA and other molecules, are unable to regulate immune responses, subsequently contributing to progressive retinal cell death. We hypothesis that this mechanism is likely to be involved in many retinal degenerative and inflammatory diseases.

AUTHOR CONTRIBUTIONS

YW, AC, and RN: conceptualization, investigation, and writing – original draft. YW, AC, JC-T, and RN: methodology and data analysis. YW, AC, JC-T, RA-B, US, and RN: writing – review and editing. RN: supervision and funding acquisition.

ACKNOWLEDGMENTS

The authors would like to acknowledge the facilities and the scientific and technical assistance of Microscopy Australia at the Advanced Imaging Precinct, Australian National University. This microscopy facility is supported by the University and State and Federal Governments. In addition, the facilities and staff of the John Curtin School of Medical Research Biomolecular Research Facility (JCSMR, BRF, ACT, AUS).

REFERENCES

Abels, E. R., and Breakefield, X. O. (2016). *Introduction to Extracellular Vesicles: Biogenesis, RNA Cargo Selection, Content, Release, and Uptake*. Berlin: Springer.

Abokyi, S., To, C.-H., Lam, T. T., and Tse, D. Y. (2020). Central role of oxidative stress in age-related macular degeneration: evidence from a review of the molecular mechanisms and animal models. *Oxidat. Med. Cell. Long.* 2020:7901270.

Abusamra, A. J., Zhong, Z., Zheng, X., Li, M., Ichim, T. E., Chin, J. L., et al. (2005). Tumor exosomes expressing Fas ligand mediate CD8+ T-cell apoptosis. *Blood Cell. Mol. Dis.* 35, 169–173. doi: 10.1016/j.bcmd.2005.07.001

Alexander, M., Hu, R., Runtsch, M. C., Kagele, D. A., Mosbruger, T. L., Tolmachova, T., et al. (2015). Exosome-delivered microRNAs modulate the inflammatory response to endotoxin. *Nat. Commun.* 6:7321.

Al-Ubaidi, M. R., Hollyfield, J. G., Overbeek, P. A., and Baehr, W. (1992). Photoreceptor degeneration induced by the expression of simian virus 40 large tumor antigen in the retina of transgenic mice. *Proc. Natl. Acad. Sci. U.S.A.* 89, 1194–1198. doi: 10.1073/pnas.89.4.1194

Anel, A., Gallego-Lleyda, A., de Miguel, D., Naval, J., and Martínez-Lostao, L. (2019). Role of exosomes in the regulation of T-cell mediated immune responses and in autoimmune disease. *Cells* 8:154. doi: 10.3390/cells8020154

Antonucci, F., Turola, E., Riganti, L., Caleo, M., Gabrielli, M., Perrotta, C., et al. (2012). Microvesicles released from microglia stimulate synaptic activity via enhanced sphingolipid metabolism. *EMBO J.* 31, 1231–1240. doi: 10.1038/emboj.2011.489

Atienzar–Aroca, S., Flores–Bellver, M., Serrano–Heras, G., Martinez–Gil, N., Barcia, J. M., Aparicio, S., et al. (2016). Oxidative stress in retinal pigment epithelium cells increases exosome secretion and promotes angiogenesis in endothelial cells. *J. Cell. Mol. Med.* 20, 1457–1466. doi: 10.1111/jcmm.12834

Bartel, D. P. (2009). MicroRNAs: target recognition and regulatory functions. *Cell* 136, 215–233. doi: 10.1016/j.cell.2009.01.002

Benedikter, B. J., Weseler, A. R., Wouters, E. F. M., Savelkoul, P. H. M., Rohde, G. G. U., and Stassen, F. R. M. (2018). Redox-dependent thiol modifications: implications for the release of extracellular vesicles. *Cell. Mol. Life Sci.* 75, 2321–2337. doi: 10.1007/s00018-018-2806-z

Bhome, R., Del Vecchio, F., Lee, G.-H., Bullock, M. D., Primrose, J. N., Sayan, A. E., et al. (2018). Exosomal microRNAs (exomiRs): Small molecules with a big role in cancer. *Cancer Lett.* 420, 228–235. doi: 10.1016/j.canlet.2018.02.002

Bian, B., Zhao, C., He, X., Gong, Y., Ren, C., Ge, L., et al. (2020). Exosomes derived from neural progenitor cells preserve photoreceptors during retinal degeneration by inactivating microglia. *J. Extracell. Vesic.* 9:1748931. doi: 10.1080/20013078.2020.1748931

Bringmann, A., Grosche, A., Pannicke, T., and Reichenbach, A. (2013). GABA and Glutamate Uptake and Metabolism in Retinal Glial (Müller) Cells. *Front. Endocrinol.* 4:48–48.

Catalano, M., and O'Driscoll, L. (2020). Inhibiting extracellular vesicles formation and release: a review of EV inhibitors. *J. Extracell. Vesic.* 9:1703244. doi: 10.1080/20013078.2019.1703244

Chan, B. D., Wong, W.-Y., Lee, M. M.-L., Cho, W. C.-S., Yee, B. K., Kwan, Y. W., et al. (2019). Exosomes in inflammation and inflammatory disease. *Proteomics* 19:1800149. doi: 10.1002/pmic.201800149

Charlotte, L., Jose, M. V., Derek, M. Y., and Sean, M. D. (2016). Microvesicles and exosomes: new players in metabolic and cardiovascular disease. *J. Endocrinol.* 228, R57–R71.

Chen, E. Y., Tan, C. M., Kou, Y., Duan, Q., Wang, Z., Meirelles, G. V., et al. (2013). Enrichr: interactive and collaborative HTML5 gene list enrichment analysis tool. *BMC Bioinform.* 14:128–128.

Chen, H., Tran, J.-T. A., Brush, R. S., Saadi, A., Rahman, A. K., Yu, M., et al. (2012). *Ceramide Signaling in Retinal Degeneration. Retinal Degenerative Diseases.* Berlin: Springer, 553–558.

Christopher, A. F., Kaur, R. P., Kaur, G., Kaur, A., Gupta, V., and Bansal, P. (2016). MicroRNA therapeutics: discovering novel targets and developing specific therapy. *Perspect. Clin. Res.* 7, 68–74.

Chu-Tan, J. A., Rutar, M., Saxena, K., Aggio-Bruce, R., Essex, R. W., Valter, K., et al. (2018). MicroRNA-124 dysregulation is associated with retinal inflammation and photoreceptor death in the degenerating RetinaMicroRNA-124 dysregulation in the degenerating retina. *Investigat. Ophthalmol. Sci.* 59, 4094–4105.

Cypryk, W., Nyman, T. A., and Matikainen, S. (2018). From inflammasome to exosome—does extracellular vesicle secretion constitute an inflammasome-dependent immune response? *Front. Immunol.* 9:2188.

Desdín-Micó, G., and Mittelbrunn, M. (2017). Role of exosomes in the protection of cellular homeostasis. *Cell Adhes. Migr.* 11, 127–134. doi: 10.1080/19336918.2016.1251000

Donoso, L. A., Kim, D., Frost, A., Callahan, A., and Hageman, G. (2006). The role of inflammation in the pathogenesis of age-related macular degeneration. *Surv. Ophthalmol.* 51, 137–152. doi: 10.1016/j.survophthal.2005.12.001

Erik, C. S. S.-M. (2017). *ggbeeswarm: Categorical Scatter (Violin Point) Plots.* Available online at: https://github.com/eclarke/ggbeeswarm (Augest 7, 2017).

Essandoh, K., Yang, L., Wang, X., Huang, W., Qin, D., Hao, J., et al. (2015). Blockade of exosome generation with GW4869 dampens the sepsis-induced inflammation and cardiac dysfunction. *Biochim. Biophys. Acta* 1852, 2362–2371. doi: 10.1016/j.bbadis.2015.08.010

Fan, Y., Siklenka, K., Arora, S. K., Ribeiro, P., Kimmins, S., and Xia, J. (2016). miRNet - dissecting miRNA-target interactions and functional associations through network-based visual analysis. *Nucleic Acids Res.* 44, W135–W141.

Fauré, J., Lachenal, G., Court, M., Hirrlinger, J., Chatellard-Causse, C., Blot, B., et al. (2006). Exosomes are released by cultured cortical neurones. *Mol. Cell. Neurosci.* 31, 642–648. doi: 10.1016/j.mcn.2005.12.003

Fernando, N., Wooff, Y., Aggio-Bruce, R., Chu-Tan, J. A., Jiao, H., Dietrich, C., et al. (2018). Photoreceptor survival is regulated by GSTO1-1 in the degenerating retina. *Investigat. Ophthalmol. Vis. Sci.* 59, 4362–4374.

Fleshner, M., and Crane, C. R. (2017). Exosomes, DAMPs and miRNA: features of stress physiology and immune homeostasis. *Trends Immunol.* 38, 768–776. doi: 10.1016/j.it.2017.08.002

Fröhlich, D., Kuo, W. P., Frühbeis, C., Sun, J.-J., Zehendner, C. M., Luhmann, H. J., et al. (2014). Multifaceted effects of oligodendroglial exosomes on neurons: impact on neuronal firing rate, signal transduction and gene regulation. *Philos. Transact. R. Soc. Lond. Ser. Biol. Sci.* 369:20130510. doi: 10.1098/rstb.2013.0510

Frühbeis, C., Fröhlich, D., Kuo, W. P., Amphornrat, J., Thilemann, S., Saab, A. S., et al. (2013). Neurotransmitter-triggered transfer of exosomes mediates oligodendrocyte–neuron communication. *PLoS Biol.* 11:e1001604. doi: 10.1371/journal.pbio.1001604

Gargini, C., Terzibasi, E., Mazzoni, F., and Strettoi, E. (2007). Retinal organization in the retinal degeneration 10 (rd10) mutant mouse: a morphological and ERG study. *J. Compar. Neurol.* 500, 222–238. doi: 10.1002/cne.21144

German, O. L., Miranda, G. E., Abrahan, C. E., and Rotstein, N. P. (2006). Ceramide is a mediator of apoptosis in retina photoreceptors. *Investigat. Ophthalmol. Vis. Sci.* 47, 1658–1668.

Gerritzen, M. J. H., Martens, D. E., Wijffels, R. H., and Stork, M. (2017). High throughput nanoparticle tracking analysis for monitoring outer membrane vesicle production. *J. Extracell. Ves.* 6:1333883. doi: 10.1080/20013078.2017.1333883

Gupta, A., and Pulliam, L. (2014). Exosomes as mediators of neuroinflammation. *J. Neuroinflam.* 11, 68–68.

Hajrasouliha, A. R., Jiang, G., Lu, Q., Lu, H., Kaplan, H. J., Zhang, H.-G., et al. (2013). Exosomes from retinal astrocytes contain antiangiogenic components that inhibit laser-induced choroidal neovascularization. *J. Biol. Chem.* 288, 28058–28067. doi: 10.1074/jbc.m113.470765

Hessvik, N. P., and Llorente, A. (2018). Current knowledge on exosome biogenesis and release. *Cell. Mol. Life Sci.* 75, 193–208. doi: 10.1007/s00018-017-2595-9

Hsu, M.-Y., Chiu, C.-C., Wang, J.-Y., Huang, C.-T., Huang, Y.-F., Liou, J.-C., et al. (2018). Paper-based microfluidic platforms for understanding the role of exosomes in the pathogenesis of major blindness-threatening diseases. *Nanomaterials (Basel Switzerland)* 8:310. doi: 10.3390/nano8050310

Huang, C., Fisher, K. P., Hammer, S. S., Navitskaya, S., Blanchard, G. J., and Busik, J. V. (2018). Plasma exosomes contribute to microvascular damage in diabetic retinopathy by activating the classical complement pathway. *Diabetes* 67, 1639–1649. doi: 10.2337/db17-1587

Isola, L., and Chen, S. (2017). Exosomes: the messengers of health and disease. *Curr. Neuropharmacol.* 15, 157–165. doi: 10.2174/1570159x14666160825160421

Jain, G., Stuendl, A., Rao, P., Berulava, T., Pena Centeno, T., Kaurani, L., et al. (2019). A combined miRNA–piRNA signature to detect Alzheimer's disease. *Transl. Psychiatry* 9:250.

Jan, A. T., Malik, M. A., Rahman, S., Yeo, H. R., Lee, E. J., Abdullah, T. S., et al. (2017). Perspective insights of exosomes in neurodegenerative diseases: a critical appraisal. *Front. Aging Neurosci.* 9:317–317.

Janik-Papis, K., Ulińska, M., Krzyzanowska, A., Stoczyńska, E., Borucka, A., Woźniak, K., et al. (2009). Role of oxidative mechanisms in the pathogenesis of age-related macular degeneration. *Klinika Oczna* 111, 168–173.

Jiao, H., Rutar, M., Fernando, N., Yednock, T., Sankaranarayanan, S., Aggio-Bruce, R., et al. (2018). Subretinal macrophages produce classical complement activator C1q leading to the progression of focal retinal degeneration. *Mol. Neurodegener.* 13:45.

Jung, M. K., and Mun, J. Y. (2018). Sample preparation and imaging of exosomes by transmission electron microscopy. *J. Vis. Exp. JoVE* 131:56482.

Kang, G.-Y., Bang, J. Y., Choi, A. J., Yoon, J., Lee, W.-C., Choi, S., et al. (2014). Exosomal proteins in the aqueous humor as novel biomarkers in patients with neovascular age-related macular degeneration. *J. Prot. Res.* 13, 581–595. doi: 10.1021/pr400751k

Kauppinen, A., Paterno, J. J., Blasiak, J., Salminen, A., and Kaarniranta, K. (2016). Inflammation and its role in age-related macular degeneration. *Cell. Mol. Life Sci. CMLS* 73, 1765–1786.

Kim, S. R., Fishkin, N., Kong, J., Nakanishi, K., Allikmets, R., and Sparrow, J. R. (2004). Rpe65 Leu450Met variant is associated with reduced levels of the retinal pigment epithelium lipofuscin fluorophores A2E and iso-A2E. *Proc. Natl. Acad. Sci. U.S.A.* 101, 11668–11672. doi: 10.1073/pnas.0403499101

Klingeborn, M., Dismuke, W. M., Rickman, C. B., and Stamer, W. D. (2017). Roles of exosomes in the normal and diseased eye. *Prog. Retin. Eye Res.* 59, 158–177. doi: 10.1016/j.preteyeres.2017.04.004

Knickelbein, J. E., Chan, C.-C., Sen, H. N., Ferris, F. L., and Nussenblatt, R. B. (2015). Inflammatory mechanisms of age-related macular degeneration. *Int. Ophthalmol. Clin.* 55:63.

Kosaka, N., Iguchi, H., Hagiwara, K., Yoshioka, Y., Takeshita, F., and Ochiya, T. (2013). Neutral sphingomyelinase 2 (nSMase2)-dependent exosomal transfer of angiogenic microRNAs regulate cancer cell metastasis. *J. Biol. Chem.* 288, 10849–10859. doi: 10.1074/jbc.m112.446831

Kowal, J., Tkach, M., and Théry, C. (2014). Biogenesis and secretion of exosomes. *Curr. Opin. Cell Biol.* 29, 116–125. doi: 10.1016/j.ceb.2014.05.004

Krol, J., Busskamp, V., Markiewicz, I., Stadler, M. B., Ribi, S., Richter, J., et al. (2010). Characterizing light-regulated retinal MicroRNAs reveals rapid turnover as a common property of neuronal MicroRNAs. *Cell* 141, 618–631. doi: 10.1016/j.cell.2010.03.039

Kyosseva, S. V. (2016). Targeting MAPK signaling in age-related macular degeneration. *Ophthalmol. Eye Dis.* 8, 23–30.

Lachenal, G., Pernet-Gallay, K., Chivet, M., Hemming, F. J., Belly, A., Bodon, G., et al. (2011). Release of exosomes from differentiated neurons and its regulation by synaptic glutamatergic activity. *Mol. Cell. Neurosci.* 46, 409–418. doi: 10.1016/j.mcn.2010.11.004

Lehmann, B. D., Paine, M. S., Brooks, A. M., McCubrey, J. A., Renegar, R. H., Wang, R., et al. (2008). Senescence-associated exosome release from human prostate cancer cells. *Cancer Res.* 68:7864. doi: 10.1158/0008-5472.can-07-6538

Lener, T., Gimona, M., Aigner, L., Börger, V., Buzas, E., Camussi, G., et al. (2015). Applying extracellular vesicles based therapeutics in clinical trials–an ISEV position paper. *J. Extracell. Ves.* 4:30087.

Li, J. J., Wang, B., Kodali, M. C., Chen, C., Kim, E., Patters, B. J., et al. (2018). In vivo evidence for the contribution of peripheral circulating inflammatory exosomes to neuroinflammation. *J. Neuroinflam.* 15:8.

Li, N., Zhao, L., Wei, Y., Ea, V. L., Nian, H., and Wei, R. (2019). Recent advances of exosomes in immune-mediated eye diseases. *Stem Cell Res. Ther.* 10:278.

Long, H., Wang, X., Chen, Y., Wang, L., Zhao, M., and Lu, Q. (2018). Dysregulation of microRNAs in autoimmune diseases: pathogenesis, biomarkers and potential therapeutic targets. *Cancer Lett.* 428, 90–103. doi: 10.1016/j.canlet.2018.04.016

Lu, Y.-Z., Fernando, N., Natoli, R., Madigan, M., and Valter, K. (2018). 670nm light treatment following retinal injury modulates Müller cell gliosis: Evidence from in vivo and in vitro stress models. *Exp. Eye Res.* 169, 1–12. doi: 10.1016/j.exer.2018.01.011

Lu, Y.-Z., Natoli, R., Madigan, M., Fernando, N., Saxena, K., Aggio-Bruce, R., et al. (2017). Photobiomodulation with 670 nm light ameliorates Müller cell-mediated activation of microglia and macrophages in retinal degeneration. *Exp. Eye Res.* 165, 78–89. doi: 10.1016/j.exer.2017.09.002

Luberto, C., Hassler, D. F., Signorelli, P., Okamoto, Y., Sawai, H., Boros, E., et al. (2002). Inhibition of tumor necrosis factor-induced cell death in MCF7 by a novel inhibitor of neutral sphingomyelinase. *J. Biol. Chem.* 277, 41128–41139. doi: 10.1074/jbc.m206747200

Marc, R. E., Jones, B., Watt, C., Vazquez-Chona, F., Vaughan, D., and Organisciak, D. (2008). Extreme retinal remodeling triggered by light damage: implications for age related macular degeneration. *Mol. Vis.* 14:782.

Mathieu, M., Martin-Jaular, L., Lavieu, G., and Théry, C. (2019). Specificities of secretion and uptake of exosomes and other extracellular vesicles for cell-to-cell communication. *Nat. Cell Biol.* 21, 9–17. doi: 10.1038/s41556-018-0250-9

Mattapallil, M. J., Wawrousek, E. F., Chan, C.-C., Zhao, H., Roychoudhury, J., Ferguson, T. A., et al. (2012). The Rd8 mutation of the Crb1 gene is present in vendor lines of C57BL/6N mice and embryonic stem cells, and confounds ocular induced mutant phenotypes. *Investigat. Ophthalmol. Vis. Sci.* 53, 2921–2927.

Mead, B., and Tomarev, S. (2020). Extracellular vesicle therapy for retinal diseases. *Prog. Retin. Eye Res.* 10:100849. doi: 10.1016/j.preteyeres.2020.100849

Men, Y., Yelick, J., Jin, S., Tian, Y., Chiang, M. S. R., Higashimori, H., et al. (2019). Exosome reporter mice reveal the involvement of exosomes in mediating neuron to astroglia communication in the CNS. *Nat. Commun.* 10:4136.

Natoli, R., Fernando, N., Madigan, M., Chu-Tan, J. A., Valter, K., Provis, J., et al. (2017). Microglia-derived IL-1β promotes chemokine expression by Müller cells and RPE in focal retinal degeneration. *Mol. Neurodegen.* 12:31.

Natoli, R., Jiao, H., Barnett, N. L., Fernando, N., Valter, K., Provis, J. M., et al. (2016). A model of progressive photo-oxidative degeneration and inflammation in the pigmented C57BL/6J mouse retina. *Exp. Eye Res.* 147, 114–127. doi: 10.1016/j.exer.2016.04.015

Natoli, R., Zhu, Y., Valter, K., Bisti, S., Eells, J., and Stone, J. (2010). Gene and noncoding RNA regulation underlying photoreceptor protection: microarray study of dietary antioxidant saffron and photobiomodulation in rat retina. *Mol. Vis.* 16:1801.

Newman, A. M., Gallo, N. B., Hancox, L. S., Miller, N. J., Radeke, C. M., and Maloney, M. A. (2012). Systems-level analysis of age-related macular degeneration reveals global biomarkers and phenotype-specific functional networks. *Genome Med.* 4:16. doi: 10.1186/gm315

Pascua-Maestro, R., González, E., Lillo, C., Ganfornina, M. D., Falcón-Pérez, J. M., and Sanchez, D. (2018). Extracellular vesicles secreted by astroglial cells transport apolipoprotein D to neurons and mediate neuronal survival upon oxidative stress. *Front. Cell Neurosci.* 12:526.

Piñero, J. À, Bravo, N., Queralt-Rosinach, A., Gutiérrez-Sacristán, J., Deu-Pons, E., Centeno, J., et al. (2016). DisGeNET: a comprehensive platform integrating information on human disease-associated genes and variants. *Nucleic Acids Res.* 45, D833–D839.

R Core Team (2019). *R: A Language and Environment for Statistical Computing.* Vienna: R Core Team.

Ratnapriya, R., and Swaroop, A. (2013). Genetic architecture of retinal and macular degenerative diseases: the promise and challenges of next-generation sequencing. *Genome Med.* 5:84.

Rivera, J. C., Holm, M., Austeng, D., Morken, T. S., Zhou, T. E., Beaudry-Richard, A., et al. (2017). Retinopathy of prematurity: inflammation, choroidal degeneration, and novel promising therapeutic strategies. *J. Neuroinflam.* 14, 165–165.

Robbins, P. D., and Morelli, A. E. (2014). Regulation of immune responses by extracellular vesicles. *Nature Rev. Immunol.* 14, 195–208. doi: 10.1038/nri3622

Rübsam, A., Parikh, S., and Fort, P. E. (2018). Role of inflammation in diabetic retinopathy. *Int. J. Mol. Sci.* 19:942. doi: 10.3390/ijms19040942

Rupaimoole, R., and Slack, F. J. (2017). MicroRNA therapeutics: towards a new era for the management of cancer and other diseases. *Nat. Rev. Drug Discov.* 16:203. doi: 10.1038/nrd.2016.246

Rutar, M., Natoli, R., Chia, R., Valter, K., and Provis, J. M. (2015). Chemokine-mediated inflammation in the degenerating retina is coordinated by Müller cells, activated microglia, and retinal pigment epithelium. *J. Neuroinflam.* 12:8. doi: 10.1186/s12974-014-0224-1

Rutar, M. V., Natoli, R. C., and Provis, J. M. (2012). Small interfering RNA-mediated suppression of Ccl2 in Muller cells attenuates microglial recruitment and photoreceptor death following retinal degeneration. *J. Neuroinflam.* 9, 5754–5760.

Sackmann, V., Sardar Sinha, M., Sackmann, C., Civitelli, L., Bergström, J., Ansell-Schultz, A., et al. (2019). Inhibition of nSMase2 reduces the transfer of oligomeric α-synuclein irrespective of hypoxia. *Front. Mol. Neurosci.* 12:200.

Saeedi, B. M. J. E., Esfandiary, G., Taheripak, P., Codoñer–Franch, E., and Mirzaei, H. (2018). Molecular aspects of diabetes mellitus: Resistin, microRNA, and exosome. *J. Cell. Biochem.* 119, 1257–1272. doi: 10.1002/jcb.26271

Sarko, D. K., and McKinney, C. E. (2017). Exosomes: origins and therapeutic potential for neurodegenerative disease. *Front. Neurosci.* 11:82.

Simón, M. V., Spalm, F. H. P., Vera, M. S., and Rotstein, N. P. (2019). Sphingolipids as emerging mediators in retina degeneration. *Front. Cell. Neurosci.* 13:246. doi: 10.3389/fncel.2019.00246

Skriner, K., Adolph, K., Jungblut, P. R., and Burmester, G. R. (2006). Association of citrullinated proteins with synovial exosomes. *Arthr. Rheum. Off. J. Am. Coll. Rheumatol.* 54, 3809–3814. doi: 10.1002/art.22276

Slenter, D. N., Kutmon, M., Hanspers, K., Riutta, A., Windsor, J., Nunes, N., et al. (2018). WikiPathways: a multifaceted pathway database bridging metabolomics to other omics research. *Nucleic Acids Res.* 46, D661–D667.

Słomka, A., Urban, S. K., Lukacs-Kornek, V., Żekanowska, E., and Kornek, M. (2018). Large extracellular vesicles: have we found the holy grail of inflammation? *Front. Immunol.* 9:2723.

Soria, F. N., Pampliega, O., Bourdenx, M., Meissner, W. G., Bezard, E., and Dehay, B. (2017). Exosomes, an unmasked culprit in neurodegenerative diseases. *Front. Neurosci.* 11:26–26.

Squadrito, M. L., Baer, C., Burdet, F., Maderna, C., Gilfillan, G. D., Lyle, R., et al. (2014). Endogenous RNAs modulate microRNA sorting to exosomes and transfer to acceptor cells. *Cell Rep.* 8, 1432–1446. doi: 10.1016/j.celrep.2014.07.035

Tam, S., Tsao, M.-S., and McPherson, J. D. (2015). Optimization of miRNA-seq data preprocessing. *Brief. Bioinform.* 16, 950–963. doi: 10.1093/bib/bbv019

Tanito, M., Kaidzu, S., Ohira, A., and Anderson, R. E. (2008). Topography of retinal damage in light-exposed albino rats. *Exp. Eye Res.* 87, 292–295. doi: 10.1016/j.exer.2008.06.002

Théry, C., Witwer, K. W., Aikawa, E., Alcaraz, M. J., Anderson, J. D., Andriantsitohaina, R., et al. (2018). Minimal information for studies of extracellular vesicles 2018 (MISEV2018): a position statement of the International Society for Extracellular Vesicles and update of the MISEV2014 guidelines. *J. Extracell. Ves.* 7:1535750.

Théry, C., Zitvogel, L., and Amigorena, S. (2002). Exosomes: composition, biogenesis and function. *Nat. Rev. Immunol.* 2:569. doi: 10.1038/nri855

Van der Pol, E., Böing, A. N., Harrison, P., Sturk, A., and Nieuwland, R. (2012). Classification, functions, and clinical relevance of extracellular vesicles. *Pharmacol. Rev.* 64, 676–705. doi: 10.1124/pr.112.005983

Van Hezel, M. E., Nieuwland, R., Van Bruggen, R., and Juffermans, N. P. (2017). The ability of extracellular vesicles to induce a pro-inflammatory host response. *Int. J. Mol. Sci.* 18:1285. doi: 10.3390/ijms18061285

Van Niel, G., Porto-Carreiro, I., Simoes, S., and Raposo, G. (2006). Exosomes: a common pathway for a specialized function. *J. Biochem.* 140, 13–21. doi: 10.1093/jb/mvj128

Vella, L. J., Scicluna, B. J., Cheng, L., Bawden, E. G., Masters, C. L., Ang, C.-S., et al. (2017). A rigorous method to enrich for exosomes from brain tissue. *J. Extracell. Vesic.* 6:1348885. doi: 10.1080/20013078.2017.1348885

Vidal-Gil, L., Sancho-Pelluz, J., Zrenner, E., Oltra, M., and Sahaboglu, A. (2019). Poly ADP ribosylation and extracellular vesicle activity in rod photoreceptor degeneration. *Sci. Rep.* 9:3758.

Wan, P., Su, W., Zhang, Y., Li, Z., Deng, C., Li, J., et al. (2019). LncRNA H19 initiates microglial pyroptosis and neuronal death in retinal ischemia/reperfusion injury. *Cell Death Differ.* 27:1.

Wang, A. L., Lukas, T. J., Yuan, M., Du, N., Tso, M. O., and Neufeld, A. H. (2009). Autophagy and exosomes in the aged retinal pigment epithelium: possible relevance to drusen formation and age-related macular degeneration. *PLoS One* 4:e4160. doi: 10.1371/journal.pone.0004160

Wang, K., Li, H., Sun, R., Liu, C., Luo, Y., Fu, S., et al. (2018). Emerging roles of transforming growth factor β signaling in wet age-related macular degeneration. *Acta Biochim. Biophys. Sinica* 51, 1–8. doi: 10.1093/abbs/gmy145

Wickham, H. (2016). *ggplot2: Elegant Graphics for Data Analysis.* New York, NY: Springer-Verlag.

Wong, W. Y., Lee, M. M. L., Chan, B. D., Kam, R. K. T., Zhang, G., Lu, A. P., et al. (2016). Proteomic profiling of dextran sulfate sodium induced acute ulcerative colitis mice serum exosomes and their immunomodulatory impact on macrophages. *Proteomics* 16, 1131–1145. doi: 10.1002/pmic.201500174

Wooff, Y., Man, S. M., Natoli, R., Aggio-Bruce, R., and Fernando, N. (2019). IL-1 Family members mediate cell death, inflammation and angiogenesis in retinal degenerative diseases. *Front. Immunol.* 10:1618.

Wu, H., Wang, C., and Wu, Z. (2013). A new shrinkage estimator for dispersion improves differential expression detection in RNA-seq data. *Biostatistics (Oxford England)* 14, 232–243. doi: 10.1093/biostatistics/kxs033

Xiang, L., Chen, X.-J., Wu, K.-C., Zhang, C.-J., Zhou, G.-H., Lv, J.-N., et al. (2017). miR-183/96 plays a pivotal regulatory role in mouse photoreceptor maturation and maintenance. *Proc. Natl. Acad. Sci. U.S.A.* 114:6376. doi: 10.1073/pnas.1618757114

Xu, W., Wu, Y., Hu, Z., Sun, L., Dou, G., Zhang, Z., et al. (2019). Exosomes from microglia attenuate photoreceptor injury and neovascularization in an animal model of retinopathy of prematurity. *Mol. Ther. Nucleic Acids* 16, 778–790. doi: 10.1016/j.omtn.2019.04.029

Yáñez-Mó, M., Siljander, P. R.-M., Andreu, Z., Bedina Zavec, A., Borràs, F. E., Buzas, E. I., et al. (2015). Biological properties of extracellular vesicles and their physiological functions. *J. Extracell. Ves.* 4:27066.

Ye, S.-B., Li, Z.-L., Luo, D.-H., Huang, B.-J., Chen, Y.-S., Zhang, X.-S., et al. (2014). Tumor-derived exosomes promote tumor progression and T-cell dysfunction through the regulation of enriched exosomal microRNAs in human nasopharyngeal carcinoma. *Oncotarget* 5, 5439. doi: 10.18632/oncotarget.2118

Yuana, Y., Sturk, A., and Nieuwland, R. (2013). Extracellular vesicles in physiological and pathological conditions. *Blood Rev.* 27, 31–39. doi: 10.1016/j.blre.2012.12.002

Zhang, H.-G., Liu, C., Su, K., Yu, S., Zhang, L., Zhang, S., et al. (2006). A membrane form of TNF-α presented by exosomes delays T cell activation-induced cell death. *J. Immunol.* 176, 7385–7393. doi: 10.4049/jimmunol.176.12.7385

Zhang, W., Ma, Y., Zhang, Y., Yang, J., He, G., and Chen, S. (2019). Photo-oxidative blue-light stimulation in retinal pigment epithelium cells promotes exosome secretion and increases the activity of the NLRP3 inflammasome. *Curr. Eye Res.* 44, 67–75. doi: 10.1080/02713683.2018.1518458

Zhao, Y., Weber, S. R., Lease, J., Russo, M., Siedlecki, C. A., Xu, L.-C., et al. (2018). Liquid biopsy of vitreous reveals an abundant vesicle population consistent with the size and morphology of exosomes. *Transl. Vis. Sci. Technol.* 7, 6–6.

Ziemann, M., Kaspi, A., and El-Osta, A. (2016). Evaluation of microRNA alignment techniques. *RNA* 22, 1120–1138. doi: 10.1261/rna.055509.115

Zuzic, M., Rojo, A. J. E., Wohl, S. G., and Busskamp, V. (2019). Retinal miRNA functions in health and disease. *Genes* 10:377. doi: 10.3390/genes10050377

Permissions

All chapters in this book were first published by Frontiers; hereby published with permission under the Creative Commons Attribution License or equivalent. Every chapter published in this book has been scrutinized by our experts. Their significance has been extensively debated. The topics covered herein carry significant findings which will fuel the growth of the discipline. They may even be implemented as practical applications or may be referred to as a beginning point for another development.

The contributors of this book come from diverse backgrounds, making this book a truly international effort. This book will bring forth new frontiers with its revolutionizing research information and detailed analysis of the nascent developments around the world.

We would like to thank all the contributing authors for lending their expertise to make the book truly unique. They have played a crucial role in the development of this book. Without their invaluable contributions this book wouldn't have been possible. They have made vital efforts to compile up to date information on the varied aspects of this subject to make this book a valuable addition to the collection of many professionals and students.

This book was conceptualized with the vision of imparting up-to-date information and advanced data in this field. To ensure the same, a matchless editorial board was set up. Every individual on the board went through rigorous rounds of assessment to prove their worth. After which they invested a large part of their time researching and compiling the most relevant data for our readers.

The editorial board has been involved in producing this book since its inception. They have spent rigorous hours researching and exploring the diverse topics which have resulted in the successful publishing of this book. They have passed on their knowledge of decades through this book. To expedite this challenging task, the publisher supported the team at every step. A small team of assistant editors was also appointed to further simplify the editing procedure and attain best results for the readers.

Apart from the editorial board, the designing team has also invested a significant amount of their time in understanding the subject and creating the most relevant covers. They scrutinized every image to scout for the most suitable representation of the subject and create an appropriate cover for the book.

The publishing team has been an ardent support to the editorial, designing and production team. Their endless efforts to recruit the best for this project, has resulted in the accomplishment of this book. They are a veteran in the field of academics and their pool of knowledge is as vast as their experience in printing. Their expertise and guidance has proved useful at every step. Their uncompromising quality standards have made this book an exceptional effort. Their encouragement from time to time has been an inspiration for everyone.

The publisher and the editorial board hope that this book will prove to be a valuable piece of knowledge for researchers, students, practitioners and scholars across the globe.

List of Contributors

Sook Hyun Chung, Weiyong Shen, Belinda Yau and Mark Gillies
Save Sight Institute, Department of Clinical Ophthalmology and Eye Health, The University of Sydney, Sydney, NSW, Australia

Kathryn C. Davidson
Centre for Eye Research Australia, Royal Victorian Eye and Ear Hospital, East Melbourne, VIC, Australia

Alice Pébay
Centre for Eye Research Australia, Royal Victorian Eye and Ear Hospital, East Melbourne, VIC, Australia
Department of Anatomy and Neuroscience, The University of Melbourne, Parkville, VIC, Australia
Department of Surgery, Royal Melbourne Hospital, The University of Melbourne, Parkville, VIC, Australia

Raymond C. B. Wong
Centre for Eye Research Australia, Royal Victorian Eye and Ear Hospital, East Melbourne, VIC, Australia
Department of Surgery, The University of Melbourne, Parkville, VIC, Australia
Shenzhen Eye Hospital, Shenzhen, China

Qing Zeng, Yuqing Zhou, He He, Xiaoli Liu, Rui Zhu, Meimei Zhang and Guozhi Huang
Department of Rehabilitation Medicine, Zhujiang Hospital, Southern Medical University, Guangzhou, China
Rehabilitation Medical School, Southern Medical University, Guangzhou, China

Donghui Liang
Department of Traditional Chinese Medicine, Zhujiang Hospital, Southern Medical University, Guangzhou, China

Xun Luo
Kerry Rehabilitation Medicine Research Institute, Shenzhen, China
Shenzhen Sanming Project Group, Spaulding Rehabilitation Hospital, Harvard Medical School, Charlestown, MA, United States

Yao Wang
Department of Rehabilitation Medicine, Shenzhen Dapeng New District Nan'ao People's Hospital, Shenzhen, China

Inês Barros and Catarina O. Miranda
CNC—Center for Neuroscience and Cell Biology, University of Coimbra, Coimbra, Portugal
CIBB—Center for Innovative Biomedicine and Biotechnology, University of Coimbra, Coimbra, Portugal
III—Institute for Interdisciplinary Research, University of Coimbra, Coimbra, Portugal

Adriana Marcelo, Teresa P. Silva and João Barata
CNC—Center for Neuroscience and Cell Biology, University of Coimbra, Coimbra, Portugal
CIBB—Center for Innovative Biomedicine and Biotechnology, University of Coimbra, Coimbra, Portugal

David Rufino-Ramos
CNC—Center for Neuroscience and Cell Biology, University of Coimbra, Coimbra, Portugal
CIBB—Center for Innovative Biomedicine and Biotechnology, University of Coimbra, Coimbra, Portugal
Faculty of Pharmacy, University of Coimbra, Coimbra, Portugal

Luís Pereira de Almeida
CNC—Center for Neuroscience and Cell Biology, University of Coimbra, Coimbra, Portugal
CIBB—Center for Innovative Biomedicine and Biotechnology, University of Coimbra, Coimbra, Portugal
Faculty of Pharmacy, University of Coimbra, Coimbra, Portugal
Viravector—Viral Vector for Gene Transfer Core Facility, University of Coimbra, Coimbra, Portugal

Yulou Yu, Ran Zhao, Qiong Wang, Zishan Dong, Ping Lv, Hailin Zhang and Chuan Wang
The Key Laboratory of Neural and Vascular Biology, Ministry of Education, Hebei Medical University, Shijiazhuang, China
The Key Laboratory of New Drug Pharmacology and Toxicology, Department of Pharmacology, Hebei Medical University, Shijiazhuang, China

Jing Yang and Wei Wang
Department of Physiology, Hebei Medical University, Shijiazhuang, China
School of Pharmaceutical Sciences, Binzhou Medical University, Yantai, China

Feng Luan
Department of Otolaryngology, The Third Hospital of Hebei Medical University, Shijiazhuang, China

Guoqiang Gu
Department of Cardiology, The Second Hospital of Hebei Medical University, Shijiazhuang, China

Junming Tang
Hubei Key Laboratory of Embryonic Stem Cell Research, Hubei University of Medicine, Shiyan, China

Jinpeng Sun
Key Laboratory Experimental Teratology of the Ministry of Education, Department of Biochemistry and Molecular Biology, School of Basic Medical Sciences, Cheeloo College of Medicine, Shandong University, Jinan, China

Yonglong Guo
Ophthalmology Department, The First Affiliated Hospital of Jinan University, Guangzhou, China
Key Laboratory for Regenerative Medicine of Ministry of Education, Jinan University, Guangzhou, China

Peiyuan Wang, Shiwei Liu and Jixing Cao
Ophthalmology Department, The First Affiliated Hospital of Jinan University, Guangzhou, China

Deliang Zhu
Key Laboratory for Regenerative Medicine of Ministry of Education, Jinan University, Guangzhou, China

Jacey Hongjie Ma
Aier School of Ophthalmology, Central South University, Changsha, China
Shenzhen Aier Eye Hospital, Shenzhen, China

Shibo Tang and Zekai Cui
Aier School of Ophthalmology, Central South University, Changsha, China
Aier Eye Institute, Changsha, China

Quan Yu
Centric Laboratory, Medical College, Jinan University, Guangzhou, China

Yunxia Xue and Zhijie Li
Institute of Ophthalmology, Medical College, Jinan University, Guangzhou, China

Jiansu Chen
Ophthalmology Department, The First Affiliated Hospital of Jinan University, Guangzhou, China
Key Laboratory for Regenerative Medicine of Ministry of Education, Jinan University, Guangzhou, China
Aier School of Ophthalmology, Central South University, Changsha, China
Aier Eye Institute, Changsha, China
Institute of Ophthalmology, Medical College, Jinan University, Guangzhou, China

Yohei Hayashi, Miho Takami and Mami Matsuo-Takasaki
iPS Advanced Characterization and Development Team, RIKEN BioResource Research Center, Tsukuba, Japan

Jianyang Liu, Yan Huang, Jialin He, Lite Ge and Zhiping Hu
Department of Neurology, The Second Xiangya Hospital, Central South University, Changsha, China

Yi Zhuo, Wei Chen and Da Duan
Developmental Biology of Ministry of Education, College of Life Sciences, Hunan Normal University, Changsha, China

Ming Lu
Developmental Biology of Ministry of Education, College of Life Sciences, Hunan Normal University, Changsha, China
Hunan Provincial Key Laboratory of Neurorestoratology, Second Affiliated Hospital of Hunan Normal University, Changsha, China

Nana Yang and Jielun Yu
School of Bioscience and Technology, Weifang Medical University, Weifang, China
Medical Laboratory Animal Center, Weifang Medical University, Weifang, China

Shiyu Sun, Kaixuan Lv and Linlin Zhang
School of Bioscience and Technology, Weifang Medical University, Weifang, China

Guangqing Duan
School of Clinical Medicine, Weifang Medical University, Weifang, China

Chen Liang
School of Chemical Engineering, Qingdao University of Science & Technology, Qingdao, China

Yaohui Tang
Med-X Research Institute and School of Biomedical Engineering, Shanghai Jiao Tong University, Shanghai, China

Guohua Lu
School of Psychology, Weifang Medical University, Weifang, China

Wei Zhai, Linghui Qu, Yijian Li, Yuxiao Zeng, Qiyou Li, Haiwei Xu and Zheng Qin Yin
Southwest Hospital/Southwest Eye Hospital, Third Military Medical University (Army Medical University), Chongqing, China
Key Lab of Visual Damage and Regeneration & Restoration of Chongqing, Chongqing, China

Lixiong Gao
Southwest Hospital/Southwest Eye Hospital, Third Military Medical University (Army Medical University), Chongqing, China
Key Lab of Visual Damage and Regeneration & Restoration of Chongqing, Chongqing, China
Department of Ophthalmology, The 6th Medical Center of PLA General Hospital, Beijing, China

Li Chen
Department of Neurology, The First Affiliated Hospital of Zhengzhou University, Zhengzhou, China

Yali Wang
Department of Neurology, Affiliated Suzhou Hospital of Nanjing Medical University, Suzhou, China

Jie Xie
Help Stem Cell Innovations, Nanjing Life Science and Technology Innovation Park, Nanjing, China

Shaoxue Zeng and Yingying Chen
Save Sight Institute, The University of Sydney, Sydney, NSW, Australia
Department of Ophthalmology, West China Hospital, Sichuan University, Chengdu, China

Ting Zhang, Matthew Pierce, Mark C. Gillies and Ling Zhu
Save Sight Institute, The University of Sydney, Sydney, NSW, Australia

Michele C. Madigan and Lianlin Huang
Save Sight Institute, The University of Sydney, Sydney, NSW, Australia
School of Optometry and Vision Sciences, University of New South Wales, Sydney, NSW, Australia

Nilisha Fernando
The John Curtin School of Medical Research, The Australian National University, Acton, ACT, Australia

Riemke Aggio-Bruce and Riccardo Natoli
The John Curtin School of Medical Research, The Australian National University, Acton, ACT, Australia
The Australian National University Medical School, The Australian National University, Acton, ACT, Australia

Fanfan Zhou
Sydney Pharmacy School, The University of Sydney, Sydney, NSW, Australia

Yurong Zhang, Mengdi Zhang, Jie Yu, Qiaoyun Wang, Xiaohong Pan, Xue Gao and Hongliu Sun
School of Pharmaceutical Sciences, Binzhou Medical University, Yantai, China

Wei Zhu
Shandong Academy of Medical Sciences, Jinan, China

Lucy S. French, Carla B. Mellough and Livia S. Carvalho
Centre for Ophthalmology and Visual Sciences (incorporating Lions Eye Institute), The University of Western Australia, Nedlands, WA, Australia

Fred K. Chen
Centre for Ophthalmology and Visual Sciences (incorporating Lions Eye Institute), The University of Western Australia, Nedlands, WA, Australia
Department of Ophthalmology, Royal Perth Hospital, Perth, WA, Australia
Department of Ophthalmology, Perth Children's Hospital, Nedlands, WA, Australia

Fan Li
Menzies Institute for Medical Research, University of Tasmania, Hobart, TAS, Australia
State Key Laboratory of Ophthalmology, Zhongshan Ophthalmic Centre, Sun Yat-sen University, Guangzhou, China

Kristof Wing, Qi Wang, Qinyi Lu, Minh Thuan Nguyen Tran and Kaylene M. Young
Menzies Institute for Medical Research, University of Tasmania, Hobart, TAS, Australia

Alex W. Hewitt
Menzies Institute for Medical Research, University of Tasmania, Hobart, TAS, Australia
Centre for Eye Research Australia, Royal Victorian Eye and Ear Hospital, East Melbourne, VIC, Australia

Jiang-Hui Wang
Centre for Eye Research Australia, Royal Victorian Eye and Ear Hospital, East Melbourne, VIC, Australia

Guei-Sheung Liu
Menzies Institute for Medical Research, University of Tasmania, Hobart, TAS, Australia
Ophthalmology, Department of Surgery, The University of Melbourne, Parkville, VIC, Australia

Chi D. Luu and Sandy S. C. Hung
Centre for Eye Research Australia, Royal Victorian Eye and Ear Hospital, East Melbourne, VIC, Australia
Ophthalmology, Department of Surgery, The University of Melbourne, Parkville, VIC, Australia

James A. Bender and Anthony L. Cook
Wicking Dementia Research and Education Centre, University of Tasmania, Hobart, TAS, Australia

Jinying Chen
Menzies Institute for Medical Research, University of Tasmania, Hobart, TAS, Australia
Department of Ophthalmology, The First Affiliated Hospital of Jinan University, Guangzhou, China

Jing Xie, Jiaman Dai, Yan He, Dayu Sun and Chao Dai
Southwest Eye Hospital, Southwest Hospital, Third Military Medical University (Army Medical University), Chongqing, China
Key Laboratory of Visual Damage, Regeneration and Restoration of Chongqing, Chongqing, China

Mudan Huang, Zhongqiu Hong, Chongjun Xiao, Lili Li, Lilin Chen, Shimei Cheng, Tingting Lei and Haiqing Zheng
Department of Rehabilitation Medicine, The Third Affiliated Hospital, Sun Yat-sen University, Guangzhou, China

Adrian V. Cioanca and Ulrike Schumann
The John Curtin School of Medical Research, The Australian National University, Canberra, ACT, Australia

Yvette Wooff and Joshua A. Chu-Tan
The John Curtin School of Medical Research, The Australian National University, Canberra, ACT, Australia
The ANU Medical School, The Australian National University, Canberra, ACT, Australia

Index

www.ingramcontent.com/pod-product-compliance
Lightning Source LLC
LaVergne TN
LVHW080954210926
848485LV00004B/831
* 9 7 8 1 6 4 6 4 6 5 7 6 7 *